VOLUME EIGHTY-TWO

# VITAMINS AND HORMONES

## Hormones of the Limbic System

# VITAMINS AND HORMONES

VOLUME EIGHTY-TWO

# VITAMINS AND HORMONES

# Hormones of the Limbic System

*Editor-in-Chief*

GERALD LITWACK
*Chair, Department of Basic Sciences*
*The Commonwealth Medical College*
*Scranton, Pennsylvania*

ELSEVIER

AMSTERDAM • BOSTON • HEIDELBERG • LONDON
NEW YORK • OXFORD • PARIS • SAN DIEGO
SAN FRANCISCO • SINGAPORE • SYDNEY • TOKYO
Academic Press is an imprint of Elsevier

*Cover photo credit:*
Kobayashi, K. Hippocampal mossy fiber synaptic transmission and its modulation. *Vitamins and Hormones* (2010) **82,** pp. 65–86.

Academic Press is an imprint of Elsevier
32 Jamestown Road, London, NWI 7BY, UK
Radarweg 29, PO Box 211, 1000 AE Amsterdam, The Netherlands
Linacre House, Jordan Hill, Oxford OX2 8DP, UK
30 Corporate Drive, Suite 400, Burlington, MA 01803, USA
525 B Street, Suite 1900, San Diego, CA 92101-4495, USA

First edition 2010



ISBN: 978-0-12-381515-6
ISSN: 0083-6729

For information on all Academic Press publications
visit our website at elsevierdirect.com

10 11 12 10 9 8 7 6 5 4 3 2 1

Printed and bound by CPI Group (UK) Ltd, Croydon, CR0 4YY

Transferred to digital print 2013

# Contents

# CONTRIBUTORS

**Garnik K. Akopian**
Neuroscience Program, University of Southern California, Los Angeles, California, USA

**Eliana P. Araújo**
Department of Nursing, FCM UNICAMP, Campinas, Sao Paulo, Brazil

**Carolin Bürck**
Hochschule Furtwangen University, Department Mechanical & Process Engineering, Jakob-Kienzle-Straße 17, Villingen-Schwenningen, Germany

**Ai-Min Bao**
Department of Neurobiology, Institute of Neuroscience, Zhejiang University School of Medicine, Hangzhou, Zhejiang, People's Republic of China

**Michel Baudry**
Neuroscience Program, University of Southern California, Los Angeles, California, USA

**Erik B. Bloss**
Fishberg Department of Neuroscience and Kastor Neurobiology of Aging Laboratory, Mount Sinai School of Medicine, New York, New York, USA

**Shuken Boku**
Department of Psychiatry, Hokkaido University Graduate School of Medicine, Sapporo, Japan

**Monika Brandt**
Hochschule Furtwangen University, Department Mechanical & Process Engineering, Jakob-Kienzle-Straße 17, Villingen-Schwenningen, Germany

**S. Marc Breedlove**
Neuroscience Program, Michigan State University, East Lansing, Michigan, USA, and Psychology Department, Michigan State University, East Lansing, Michigan, USA

**Elizabeth C. Cottrell**
Institute of Metabolic Science, Metabolic Research Laboratories, University of Cambridge, Addenbrooke's Hospital, Cambridge, United Kingdom, and Centre for Cardiovascular Science, Queen's Medical Research Institute, University of Edinburgh, Edinburgh, Scotland, United Kingdom

**Catharine H. Duman**
Department of Psychiatry, Yale University, Connecticut Mental Health Center, New Haven, Connecticut, USA

**Michael R. Foy**
Department of Psychology, Loyola Marymount University, Los Angeles, California, USA

**Anastasia Galeeva**
Pavlov Institute of Physiology, Russian Academy of Science, Makarova 6, Saint-Petersburg, Russia, and Medical School of Tampere University, University of Tampere, Tampere, Finland

**Marcel Gehlhaus**
Pathologisches Institut, Abt. Neuropathologie, Neurozentrum, Universität Freiburg, Breisacherstraße 64, Freiburg, Germany

**Michael J. Glass**
Department of Neurology and Neuroscience, Weill Cornell Medical College, New York, USA

**Richard G. Hunter**
Laboratory of Neuroendocrinology, Laboratory of Neurobiology and Behavior, Rockefeller University, New York, New York, USA

**Ryan T. Johnson**
Neuroscience Program, Michigan State University, East Lansing, Michigan, USA

**Cynthia L. Jordan**
Neuroscience Program, Michigan State University, East Lansing, Michigan, USA, and Psychology Department, Michigan State University, East Lansing, Michigan, USA

**Kazunori Kageyama**
Department of Endocrinology and Metabolism, Hirosaki University Graduate School of Medicine, Hirosaki, Aomori, Japan

**Nina Killer**
Hochschule Furtwangen University, Department Mechanical & Process Engineering, Jakob-Kienzle-Straße 17, Villingen-Schwenningen, Germany

**Rolf Knoth**
Pathologisches Institut, Abt. Neuropathologie, Neurozentrum, Universität Freiburg, Breisacherstraße 64, Freiburg, Germany

**Katsunori Kobayashi**
Department of Pharmacology, Nippon Medical School, Tokyo, Japan, and Japan Science and Technology Agency, CREST, Saitama, Japan

**Tsukasa Koyama**
Department of Psychiatry, Hokkaido University Graduate School of Medicine, Sapporo, Japan

**Julian G. Mercer**
Division of Obesity and Metabolic Health, Rowett Institute of Nutrition and Health, University of Aberdeen, Bucksburn, Aberdeen, Scotland, United Kingdom

**Ralf P. Meyer**
Medizinische Fakultät der Universität Freiburg, Breisacherstraße 64, Freiburg, Germany

**Dai Mitsushima**
Department of Physiology, Yokohama City University Graduate School of Medicine, Kanazawa-ku, Yokohama, Japan

**Jerzy W. Mozrzymas**
Laboratory of Neuroscience, Department of Biophysics, Wroclaw Medical University, Wroclaw, Poland, and Department of Animal Molecular Physiology, Wroclaw University, Wroclaw, Poland

**Shin Nakagawa**
Department of Psychiatry, Hokkaido University Graduate School of Medicine, Sapporo, Japan

**Alexander G. Nikonenko**
Department of Cytology, Bogomoletz Institute of Physiology, Kiev, Ukraine

**Natalia Ordyan**
Pavlov Institute of Physiology, Russian Academy of Science, Makarova 6, Saint-Petersburg, Russia

**Susan E. Ozanne**
Institute of Metabolic Science, Metabolic Research Laboratories, University of Cambridge, Addenbrooke's Hospital, Cambridge, United Kingdom

**Georgios Pantazis**
Institut für Hirnforschung, Universitätsklinikum Tübingen, Calwerstraße 3, Tübingen, Germany

**Marie-Christine Pardon**
School of Biomedical Sciences, Institute of Neuroscience, Queen's Medical Centre, University of Nottingham Medical School, Nottingham, United Kingdom

**Raúl G. Paredes**
Instituto de Neurobiología, Universidad Nacional Autónoma de México, Querétaro, Querétaro, Mexico

**Markku Pelto-Huikko**
Medical School of Tampere University, University of Tampere, Tampere, Finland

**Svetlana Pivina**
Pavlov Institute of Physiology, Russian Academy of Science, Makarova 6, Saint-Petersburg, Russia

**Justyna Rogalska**
Department of Animal Physiology, Institute of General and Molecular Biology, N. Copernicus University, Torun, Poland

**Ricarda Schwab**
Hochschule Furtwangen University, Department Mechanical & Process Engineering, Jakob-Kienzle-Straße 17, Villingen-Schwenningen, Germany

**Galyna G. Skibo**
Department of Cytology, Bogomoletz Institute of Physiology, Kiev, Ukraine

**Toshihiro Suda**
Department of Endocrinology and Metabolism, Hirosaki University Graduate School of Medicine, Hirosaki, Aomori, Japan

**Dick F. Swaab**
Netherlands Institute for Neuroscience, An Institute of the Royal Netherlands Academy of Arts and Sciences, Amsterdam, The Netherlands

**Yasin Temel**
Departments of Neuroscience and Neurosurgery, Maastricht University Medical Center, Maastricht, The Netherlands, and European Graduate School of Neuroscience (EURON), Maastricht, The Netherlands

**Gert J. ter Horst**
Department of Neuroscience, Neuroimaging Center, University Medical Center Groningen, Groningen, The Netherlands

**Richard F. Thompson**
Neuroscience Program, University of Southern California, Los Angeles, California, USA

**Márcio A. Torsoni**
Laboratory of Cell Signaling, FCM UNICAMP, Campinas, Sao Paulo, Brazil

**Lício A. Velloso**
Laboratory of Cell Signaling, FCM UNICAMP, Campinas, Sao Paulo, Brazil

**Tomasz Wójtowicz**
Laboratory of Neuroscience, Department of Biophysics, Wroclaw Medical University, Wroclaw, Poland

# Preface

The limbic system, consisting mainly of the hippocampus, amygdala, and hypothalamus, is the system responsible for many important neurological disorders and functions such as stress, depression, obesity, and memory. Several hormones are involved in this system, notably the sex hormones and the corticosteroid receptors. Many of these aspects are reviewed in this volume.

The following are the opening manuscripts. C. H. Duman reviews "Models of Depression." R. T. Johnson, S. M. Breedlove, and C. L. Jordan discuss "Glia in the Amygdala." "Limbic Effects of High Frequency Stimulation of the Subthalamic Nucleus" is a contribution of Y. Temel. K. Kobayashi continues with "Hippocampal Mossy Fiber Synaptic Transmission and its Modulation."

Some overall effects on the system are discussed by R. P. Meyer, G. Pantazis, N. Killer, C. Burck, R. Schwab, M. Brandt, R. Knoth, and M. Gehlhaus in "Xenobiotics in the Limbic System – Affecting Brain's Network Function." Next, G. G. Skibo and A. G. Nikonenko offer "Brain Plasticity after Ischemic Episode." Then, "Hypothalamic Inflammation and Obesity" is reviewed by E. P. Araujo, M. A. Torsoni, and L. A. Velloso.

Neurohormone and neurotransmitter receptors come into play with several papers. M. J. Glass introduces "The Role of Functional Postsynaptic NMDA Receptors in the Central Nucleus of the Amygdala in Opioid Dependence." E. B. Bloss and R. G. Hunter review "Hippocampal Kainate Receptors." "The Role of Neurotrophic Factors in Behavioral Processes: Implications for the Treatment of Psychiatric and Neurodegenerative Disorders" is described by M. C. Pardon. "Postnatal Development of Hypothalamic Leptin Receptors" is the topic of E. C. Cottrell, J. G. Mercer, and S. E. Orzanne. M. R. Foy, M. Baudry, G. Akopian, and R. F. Thompson contribute "Regulation of Hippocampal Synaptic Plasticity by Estrogen and Progesterone." R. G. Paredes offers "Hormones and Sexual Reward." D. Mitsushima writes on "Sex Steroids and Acetylcholine Release in the Hippocampus" and T. Wojtowicz and J. W. Mozrzymas review "Estradiol and GABAergic Transmission in the Hippocampus." "Transcriptional Regulation of Hypothalamic Corticotropin-Releasing Factor Gene" is described by K. Kageyama and T. Suda. G. J. ter Horst writes on "Estrogen in the limbic system." A. M. Bao and D. F. Swaab discuss "Corticotrophin-Releasing Hormone and Arginine Vasopressin in Depression: Focus on the Human Postmortem Hypothalamus." "Postnatal Ontogeny of the Glucocorticoid Receptor in the Hippocampus" is described by A. Galeeva,

M. Pelto-Huikko, S. Pinina, and N. Ordyan. The final two papers are Mineralocorticoid and Glucocorticoid Receptors in Hippocampus: their "Impact on Neurons Survival and Behavioral Impairment after Neonatal Brain Injury" by J. Rogalska and "Glucocorticoids and Lithium in Adult Hippocampal Neurogenesis" by S. Boku, S. Nakagawa, and T. Koyama.

Collaboration with Narmada Thangavelu and Lisa Tickner at Elsevier made this volume possible.

The figure on the cover of this volume is Figure 2 from Dr. Katsunori Kobayashi's manuscript entitled "Hippocampal Mossy Fiber Synaptic Transmission and its Modulation.

*Gerald Litwack*
Scranton, PA
January 22, 2010

CHAPTER ONE

# Models of Depression

Catharine H. Duman

## Contents



## Abstract

The incidence of depressive illness is high in the United States and worldwide, and the inadequacy of currently available drug treatments contributes to the significant health burden associated with depression. A basic understanding of the underlying disease processes in depression is lacking, and therefore, recreating the disease in animal models is not possible. Currently used models of depression attempt to produce quantifiable correlates of human symptoms in experimental animals. The models differ in the degree to which they produce features that resemble a depressive-like state, and models that include stress exposure are widely used. Paradigms that employ acute or subchronic stress exposure include learned helplessness, forced swim test, and tail suspension test, which employ relatively short-term exposure to inescapable or uncontrollable stress and can reliably detect antidepressant drug response. Longer-term models include chronic mild stress models, early-life stress models, and social conflict models, which may more accurately simulate processes that lead to depression.

Department of Psychiatry, Yale University, Connecticut Mental Health Center, New Haven, Connecticut, USA

*Vitamins and Hormones*, Volume 82
ISSN 0083-6729, DOI: 10.1016/S0083-6729(10)82001-1

These models each have varying degrees of face, construct, and predictive validity for depression and contribute differently to our understanding of antidepressant processes. 

## I. Introduction

Major depressive disorder is a leading cause of disability worldwide, with a lifetime population prevalence as high as 20% (Kessler *et al.*, 2005). Depression is a heterogeneous group of illnesses that vary in symptomology and most likely in etiology. Symptoms of depression include depressed mood, loss of pleasure, inability to concentrate, lack of energy, dysregulated sleep or appetite, feelings of worthlessness or guilt, and thoughts of suicide (DSM-IV). In addition to the personal suffering and loss associated with depression, the high incidence and chronic nature of depressive illness result in a significant public health burden. This is estimated to be tens of billions of dollars each year for the United States, largely due to loss of productivity in the workplace (Wang *et al.*, 2003).

The public health impact of depression is partially due to the fact that available treatments are suboptimal. Antidepressant drugs in use today are based on the strategy of blocking the reuptake or degradation of monoamine neurotransmitters (Morilak and Frazer, 2004). Tricyclic antidepressants and monoamine oxidase inhibitors were used as antidepressants starting several decades ago and led to the development of the serotonin-selective and norepinephrine-selective reuptake inhibitors (SSRIs and NRIs, respectively) that are widely used today. Significant percentages of depressed patients do not respond to any of the available drugs. For patients that do respond, therapeutic effect develops slowly, usually over several weeks of chronic drug treatment and patients are at risk during this time. The SSRIs and NRIs have improved safety and side-effect profiles compared to the older drugs, but because the primary mechanisms are similar, the therapeutic efficacy and the drawbacks (slow therapeutic onset, low remission rates, and treatment resistant patients) are also similar (Nestler *et al.*, 2002; Sonawalla and Fava, 2001). Improvement of therapeutic options (especially designing treatments for patients that do not respond to currently available drugs) depends on the identification of underlying pathological processes in depression and potential mechanisms for their reversal. Both human and animal studies are essential to these goals.

## II. General Considerations in Modeling Depression

A basic understanding of the underlying disease processes in depression is currently lacking, and therefore recreating the disease in animal models is not possible. Currently used models of depression attempt by various means to produce quantifiable correlates of human symptoms in experimental animals. Some of the most prominent symptoms of depression are subjective feelings, which are not readily assessed in animals. This makes it necessary to model symptoms of depression that can easily translate to behaviors that are measurable in animals. Psychomotor, sleep, and appetite changes are not uniformly useful for the investigation of neurobiological mechanisms as these behavioral alterations can occur in either direction in depression. Currently used animal models for depression research vary considerably in the extent to which they produce features that resemble a depressive-like state. Examples of measures that can be assessed in rodent behavioral models include motor responses to stress, reward-related responding and social interaction, with the rationale that they reflect levels of helplessness or despair, anhedonia, and social withdrawal, respectively, all relevant to human depression.

The models are generally evaluated for their reliability or reproducibility, their ability to accurately predict outcome in humans (predictive validity), their ability to reproduce in animals aspects of the illness in humans (face validity), and the extent to which they model the true disease process or its etiology in humans (construct or etiologic validity) (McKinney, 2001; Willner, 1984, 1997). Predictive validity includes the ability of a model to accurately detect treatments that are useful clinically. While the utility of many of the models to be discussed here is based on their predictive validity for pharmacological treatments, an important feature that has been lacking in the more widely used models is an accurate reflection of the temporal characteristics of treatment effectiveness as it occurs in humans. This consideration has become increasingly important as mechanisms of neural plasticity are implicated as central to antidepressant effectiveness (Pittenger and Duman, 2008).

The degree of construct and/or etiologic validity is generally low in most currently used animal models for depression, mainly because the pathophysiological basis for depression is not known. Although the models attempt to produce specific behavioral or physiological features of depression, the features in the animal models likely come about through processes that are very different from those operative in human depression. Therefore, results need to be carefully interpreted for relevance that may be more specific to the model than for human depression. This is an important limitation of currently used models. Because they are used largely based on their abilities to detect mechanisms of current antidepressant drugs, they select for those (known) mechanisms and may lack the ability to detect potentially novel mechanisms. This emphasizes the importance of using animal models with features that result from processes believed to be relevant to human depression.

## III. Stress and Models of Depression

Exposure to stress is a main environmental risk factor associated with the occurrence of depression (Keller *et al.*, 2007; Kendler *et al.*, 1999; Kessler, 1997). Recent work has indicated that stress exposure may interact with genetic risk factors to increase susceptibility to depression (Caspi *et al.*, 2003; Kaufman *et al.*, 2006). For these reasons, many animal models have attempted to reproduce some core components of major depressive disorder through exposure to stress. Experimentally, the outcome of stress exposure is influenced by several variables, including the nature of the stress (physical/systemic vs. cognitive/psychological), the severity of the stress, and exposure parameters. Different neural circuits are activated by different types of stressors. For example, differential involvement of limbic pathways is thought to occur for the processing of stressors that differ in their systemic versus cognitive/psychological nature (Anisman and Matheson, 2005; Herman and Cullinan, 1997). The degree of control an animal has over stress exposure has been demonstrated to be important to the consequences of stress exposure, for example, behavioral impairment and increased brain amine utilization can be seen after exposure to uncontrollable stress but are not apparent when a subject is able to control the stress exposure (Anisman and Matheson, 2005). The degree of predictability is also thought to affect the outcome-repeated stress exposure. Greater unpredictability can reduce the probability of adaptive processes occurring upon repeated stress exposure and promote the appearance of stress effects such as brain monoamine utilization and behavioral changes (Anisman and Matheson, 2005; Willner *et al.*, 1987).

The inclusion of a stress component and an appropriate temporal profile can strengthen the validity of a model. This chapter discusses rodent behavioral models of depression that include a stress component, including learned helplessness (LH), forced swim test (FST), and tail suspension test (TST), in which rodents are exposed to relatively acute or subchronic stress. The FST and TST have been widely used as screening tests for antidepressant activity. Chronic paradigms such as chronic unpredictable mild stress exposure, early-life stress (ELS) paradigms, and social defeat/conflict models are also discussed (Table 1.1). These models are considered more naturalistic in the induction of a depressive-like state and are suggested to have better potential homology to the human situation. Surgical, pharmacological, immune, or genetic models, while useful to the field, have been reviewed elsewhere and are not covered in this chapter (Crowley and Lucki, 2005; Dunn *et al.*, 2005; El Yacoubi and Vaugeois, 2007; Henn and Vollmayr, 2005; Overstreet *et al.*, 2005; Song and Leonard, 2005).

**Table 1.1** Stress-based models of depression and antidepressant response

| Model | Stress | Test duration | Primary endpoint | Advantages |
|---|---|---|---|---|
| LH | Inescapable footshock | Subchronic | Active avoidance | Depression-like syndrome |
| FST | Water immersion | Acute | Immobility | Ease of use<br>Predictive validity |
| TST | Suspension | Acute | Immobility | Ease of use<br>Predictive validity |
| NSF/NIH | Environmental change | Acute | Feeding | Predictive validity<br>No food restriction |
| CUS | Varied mild | Chronic | Anhedonia | Construct and<br>Predictive validity |
| ELS | Deprivation | Chronic | Endocrine or behavioral | Etiologic validity |
| Social defeat | Conflict/defeat | Acute or chronic | Social avoidance | Construct and<br>Predictive validity |

LH: learned helplessness, FST: forced swim test, TST: tail suspension test, NSF: novelty-suppressed feeding, NIH: novelty-induced hypophagia, CUS: chronic unpredictable mild stress, ELS: early-life stress.

# IV. Models

## A. Learned helplessness

The LH paradigm uses a stress-exposure period in which rats or mice are exposed to inescapable stress (e.g., electrical footshock) in one or more sessions. In a subsequent session, the animals are tested for their performance in an active avoidance test. In a typical active avoidance test, animals are confined to one side of a shuttle box chamber where footshocks are delivered but the animal has the opportunity of actively escaping the footshock. Animals previously exposed to inescapable stress show reduced abilities to escape in this model. The reduced ability to escape is restored by different forms of antidepressant treatment, including tricyclic antidepressants, selective serotonin reuptake inhibitors, monoamine oxidase inhibitors, and electroconvulsive shock therapy (Martin *et al.*, 1990; Sherman *et al.*, 1982). This model has good validity for predicting antidepressant efficacy (Willner, 1984), and there are few reports on false positives. This model has been used to demonstrate the importance of controllability of the stressor as a key psychological component in inducing the behavioral deficit (Anisman and Matheson, 2005). Animals that are helpless in this model also show several features that have similarity with human depression, including decreased motor activity, weight loss, altered sleep, decreased motivation, and increases in stress hormones (Maier, 1984). Despite the presence of these features along with behavioral helplessness, there is no way of knowing if the LH model reproduces the pathophysiology of depression.

The time course for induction and for treatment effect in LH paradigms is improved (subchronic) compared with the acute responsiveness of the FST and TST (explained later). LH models can identify subgroups of stress-exposed animals that are more prone to becoming helpless. Using LH to identify vulnerable and resistant subgroups can be a useful strategy for investigating mechanisms underlying differential susceptibility and has analogy to the human situation. Limitations of LH models include low reproducibility and the relevance of the induction methods has been questioned (Nestler *et al.*, 2002).

## B. Forced swim test

The FST involves placing a rat or mouse in a cylinder with enough water so that it cannot touch the bottom with its hind paws (Porsolt *et al.*, 1977a,b, 1978). A normal animal will show an immediate burst of activity, try to escape, and then eventually adopt an "immobile" posture, where it will make only those movements necessary to keep its head above water. The development of immobility may be facilitated by prior exposure to the test and a 24-h prior preexposure to the test is often used (Porsolt *et al.*, 1978). Immobility is quantified during brief test periods and classical antidepressants such as the monoamine oxidase inhibitors, tricyclics, and atypical antidepressants all decrease the duration of immobility in rats and mice in a dose-dependent manner (Borsini and Meli, 1988; Porsolt *et al.*, 1977a,b). A modified FST procedure is often used in rats that allows behavior in response to norepinephrine-selective drugs to be distinguished from behavior after treatment with serotonin-selective antidepressants. The modification involves separately quantifying the predominant active behaviors as either swimming or climbing. Swimming behavior predominates for serotonergic antidepressants and climbing predominates for drugs that are primarily noradrenergic, allowing the FST to detect this distinction (Detke *et al.*, 1995; Lucki, 1997). The FST may yield false positive results with drugs that increase locomotor activity, and correspondingly, decrease immobility (e.g., amphetamine). In addition, the FST does not uniformly distinguish acute from chronic antidepressant effects. The FST is sensitive to genetic variation as indicated by strain differences in performance and drug effects in rats and mice (López-Rubalcava and Lucki, 2000; Porsolt *et al.*, 1978).

## C. Tail suspension test

The TST is conceptually similar to the FST and is suggested to have greater sensitivity. A mouse is suspended by the tail in this test and observed for the extent of immobility versus active movement (Steru *et al.*, 1985). Similar to the FST, the TST is also based on the adoption of a passive response in a stress situation. Acute antidepressant treatment given prior to the test reduces immobility time in the TST and it is considered to have good predictive validity (Cryan *et al.*, 2005; Perrault *et al.*, 1992; Steru *et al.*, 1985). Although conceptually similar, the TST and FST do not show identical sensitivities to pharmacologic agents or to strain differences, suggesting that responding in these tests may be determined by nonidentical substrates (Bai *et al.*, 2001). Different mouse strains respond differently to basal immobility in the TST, indicating that this test is sensitive to genetic influence (Ripoll *et al.*, 2003). The strain response profile for antidepressant response is different from the profile for basal response, indicating that the determinants for basal and antidepressant responding are not identical (Liu and Gershenfeld, 2001; Trullas *et al.*, 1989). Differential sensitivity between strains for antidepressant response is suggested to be related to variations in monoamine levels (Ripoll *et al.*, 2003).

The TST is used only in mice and not in rats due to their larger size and weight. The TST has similar limitations to the FST, including a false positive response to psychostimulants and acute drug response. The high reliability of the FST and TST has also contributed to their use and they are both considered useful for investigating differences between strains in reactivity to stress.

LH, the FST, and TST do not reproduce the pathophysiology of depression but they are useful in that they induce changes that are sensitive to therapeutic agents in a manner predictive of their effects in humans. The FST and TST have been used extensively for this purpose, but the selectivity of these tests for monoamine-based mechanisms may limit their ability to detect novel mechanisms (Lucki, 1997; Thiébot *et al.*, 1992; Weiss and Kilts, 1995; Willner, 1990).

## D. Hyponeophagia paradigms

Examples of hyponeophagia tests that are used in rats and mice are novelty-induced hypophagia (NIH) and novelty-suppressed feeding (NSF) paradigms. They are anxiety based and compare feeding behavior in an anxiogenic versus a nonanxiogenic environment. The stress employed in these models is very mild relative to most other tests for antidepressant action, and consists of placing the experimental animal in a novel environment to induce anxiety during testing. The animal experiences conflict between the desire to approach and feed or drink, and the anxiety-induced avoidance of the novel environment.

The NIH procedure described by Dulawa *et al.* (2004), measures latency and volume for consumption in a familiar (home cage) versus a novel environment. Mice are habituated to drink a palatable liquid (sweetened milk) and then given the opportunity to approach and consume it in two test sessions. The first session occurs in the home cage and serves as a control for potential treatment effects on appetite. The subsequent test session occurs in a similar cage except that additional parameters (location, lighting) are chosen to be mildly anxiogenic. Consumption measures from the novel cage are compared with the same measures obtained in the home cage and the difference score is the measure of hyponeophagia. The inclusion of the home cage control that utilizes equivalent measures in a home cage detects and controls for potential alterations in consumption-related variables. Using palatable food or drink as the test substance avoids the use of food deprivation, which can complicate interpretation. These hyponeophagia models have good predictive validity. They respond to the anxiolytic effects of benzodiazepines and barbiturates and respond to antidepressants which are anxiolytic (Dulawa and Hen, 2005). Importantly, the predictive validity also applies to the temporal course of response. The hyponeophagia paradigms detect the clinically relevant acute and chronic anxiolytic effect of benzodiazepines (Bodnoff *et al.*, 1988). They detect anxiolytic effects of antidepressants only after chronic treatment, which agrees with the clinical profile for this effect in humans (Bodnoff *et al.*, 1989; Dulawa and Hen, 2005). Additionally strengthening the predictive validity is the fact that hyponeophagia paradigms can detect increased anxiety, including that resulting from acute SSRI treatment, an effect that is not reliably detected in other models but is clinically relevant (Dulawa and Hen, 2005). While being sensitive to chronic influences, the paradigms benefit from having no significant training requirement. Hyponeophagia models are sensitive to genetic influences on anxiety (Dulawa and Hen, 2005).

The anxiety component in hyponeophagia models provides a degree of face validity. There is significant comorbidity of major depression and anxiety disorders (Fava *et al.*, 2000; Kaufman and Charney, 2000). Anxiety and depression may result from closely related mechanisms as suggested by their common symptomology, family and population studies that suggest that vulnerability may result from common genetic factors, imaging studies that indicate ADT-responsive abnormalities in similar brain areas for the two disorders, and comparable efficacy of ADT in the treatment of both disorders (Drevets *et al.*, 2008; Kaufman and Charney, 2000; Ressler and Nemeroff, 2000; Serretti *et al.*, 2009). These considerations suggest vulnerability of common neural substrates and support the growing body of evidence for overlap in the neural circuitry that modulates anxiety and mood. They further validate the appropriate use of anxiety components in antidepressant models.

## E. Chronic unpredictable mild stress

In comparison to LH and FST/TST procedures that rely on relatively short-term aversive stress exposure, the chronic unpredictable mild stress (CUS) paradigm was developed to study neural changes that result from stress of a more chronic nature. CUS paradigms aim to model a chronic depressive-like state that develops gradually over time in response to stress, and is thus considered more naturalistic in the induction. A CUS paradigm was first studied by Katz and colleagues, and this idea was further developed by Willner (Katz *et al.*, 1981a,b; Willner, 1997; Willner *et al.*, 1987) providing the basis for most of the currently used paradigms. Most of the procedures employed for CUS share certain common features such as use of a stressor variety, use of stressors that are mild in severity, and use of stress exposure schedules that are semirandom and unpredictable. Rats or mice are exposed to a series of different stress conditions over a period of several weeks. Several stressors (6–8) are applied (1 or 2 per day) for several hours each day. Typical stressors include overnight illumination, periods of food or water restriction, cage tilt, and isolation or crowded housing. The sequential and unpredictable stress exposure decreases the likelihood of the animals habituating to any one reoccurring condition (Aguilera, 1998; Magariños and McEwen, 1995; Tannenbaum *et al.*, 2002).

The gradual development of a decrease in reward sensitivity or anhedonia is a central focus of CUS paradigms. Decreased ability to experience reward is a characteristic common to all forms of depression and it is amenable to repeated measurement as a quantifiable endpoint for assessing the effectiveness of CUS. Exposure to CUS can result in several other behavioral and physiologic changes that have analogy with symptoms of depression, such as decreased reward-related behavior, decreased self-care, and changes in sleep that respond to antidepressant treatment. These and other abnormalities, including increased hypothalamic–pituitary–adrenal (HPA) axis activation and immune system abnormalities, support face validity of this model (Willner, 2005). These changes develop gradually over time with CUS exposure and suggest improved face validity of this compared with the more acute stress models. Construct validity for CUS is largely based on the development of reduced sucrose preference, which is interpreted to reflect anhedonia, a core symptom of depression (Willner, 1997).

Anhedonia in the CUS model responds to chronic but not acute treatment with several classes of antidepressant drugs, indicating good predictive validity (Papp *et al.*, 1996). Hedonic measures are not altered by antidepressant treatment in the control animals in the model, in agreement with the lack of effect of antidepressants in altering hedonic response in humans (Willner *et al.*, 1987). False positive responses are reported but predictive validity for CUS is strengthened by the time course and the general lack of effectiveness of nonantidepressants. Reliability had been questioned for the CUS model, but is considered significantly improved, as reviewed by Willner (1997, 2005). The CUS paradigms are time-consuming to perform, but the growing reliability and the temporal characteristics as well as the validity of using anhedonia as an endpoint has resulted in increasing use of CUS models.

## F. Hedonic sensitivity

Methods for quantifying hedonic sensitivity include conditioned place preference procedures in which animals learn to associate a particular environment with reward experience, brain-stimulation reward (BSR) paradigms, and quantifying consumption of sweet solutions. Quantifying consumption of sweetened fluids (sucrose or saccharin) is the most commonly employed endpoint for assessing CUS effectiveness. Rats previously habituated to sucrose are typically given a choice of drinking sucrose versus water in a two-bottle test. While control rats typically show a preference for drinking weak sucrose solutions, rats exposed to CUS loose this preference. The development of this effect can be demonstrated by repeated sucrose preference testing during the course of CUS exposure. The time-dependent reversal of this effect with chronic antidepressant treatment can also be demonstrated by repeated testing. The dependence of sucrose consumption/preference on several experimental variables has been investigated, including sucrose concentration, temporal parameters, test duration, and body weight (D'Aquila *et al.*, 1997; Muscat and Willner, 1992; Willner *et al.*, 1996). Compared to rats, stress-sensitive changes in sucrose preference are more difficult to establish in mice. Decreases in sucrose consumption in a one-bottle test rather than preference can be demonstrated in mice after CUS, but this measure does not have the advantage of the simultaneous control measure for water drinking, which is a component of preference tests (Monleon *et al.*, 1995; Pothion *et al.*, 2004). Anhedonic measures after CUS are sensitive to strain effects for both rats and mice (Nielsen *et al.*, 2000; Pothion *et al.*, 2004).

In BSR paradigms, animals with implanted electrodes are trained to perform a specific operant response that results in the administration of rewarding electrical stimulation into a specific area of the brain. Operant responding relative to the stimulation intensity is quantified as a measure of reward system sensitivity and the motivation to obtain reward. BSR paradigms can be used to investigate stress effects on reward function, including effects of CUS and ADT, although results can be variable and subject to individual differences (Moreau *et al.*, 1992; Nielsen *et al.*, 2000). BSR paradigms are useful in that they can distinguish elevations in reward function from aversion, and a BSR paradigm has been developed in which alterations in reward function can be distinguished from treatment-induced influences on response performance (Markou *et al.*, 1992; Todtenkopf *et al.*, 2004; Carlezon and Chartoff, 2007). BSR paradigms avoid the appetitive and satiation effects that can complicate interpretation of behavioral responding for consummatory rewards such as food. Despite the positive features of BSR paradigms, they require surgery, training, and are less widely used than sucrose testing.

## G. Early-life stress

Early-life adverse experience is an important predisposing factor for psychopathology in humans. Several human studies indicate that exposure to stress or adversity early in life increases the risk for depression, and that stress exposure may interact with genetic risk factors (Agid *et al.*, 1999, 2000; Caspi *et al.*, 2003; Kaufman *et al.*, 2006; Weiss *et al.*, 1999). Experimental paradigms have been developed in an effort to model ELS, and are used as models in which to investigate determinants of experience-dependent susceptibility to depressive illness. The ELS models typically employ stress exposure during critical periods of development and result in stable phenotypic changes. ELS-induced changes that have been particularly replicable involve alterations in neural systems that regulate or respond to stress such as the HPA axis and include endocrine, neurochemical, and behavioral alterations. Changes in stress-responsive systems after ELS could be relevant to consequences in humans after ELS and may suggest mechanisms that predispose to depressive illness. Improved construct/etiological validity of ELS models will depend on further phenotyping of ELS animals in depression models and improvements in the validity of research models or assessing depression-related traits after ELS (Heim *et al.*, 2004; Pryce *et al.*, 2005).

*Maternal separation.* Parental care is increasingly implicated as an important modifier of stress effects during development in humans, and maternal deprivation paradigms are useful as developmental animal models of predisposition to affective disorders/depression (Heim and Nemeroff, 2001; Holmes *et al.*, 2005; Kendler *et al.*, 2002; Newport *et al.*, 2002). A role of maternal behavior in programming emotion-related behavior in the offspring has been suggested to have evolutionary/survival value by "translating" the stress level in the environment into offspring emotional reactivity that is suited to that environment (Zhang *et al.*, 2006).

A number of maternal deprivation paradigms exist that utilize repeated periods of separation of preweanling rats from the mother. Preweanling rats are exposed to daily episodes of 3–6 h separation during a critical period in the first 2 postnatal weeks and the separation can include the intact litter from the mother, or individual pups can be separated from littermates and the mother. Previously separated animals are then allowed to develop under normal conditions through adulthood, when phenotypic characteristics are evaluated. As adults, previously separated rats show behavioral abnormalities, including increased anxiety and fear responses, reduced motor activity, reduced social motivation, reduced hedonic responding, sleep and appetite disturbances, and endocrine and neurochemical alterations in stress-relevant systems (Ladd *et al.*, 2000; Levine, 1957; Mintz *et al.*, 2005; Plotsky and Meaney, 1993; Rüedi-Bettschen *et al.*, 2005, 2006). Stress responsiveness of the HPA axis is a consistent finding, but traits in depression tests are more variable and appear to depend on rat strain (Pryce *et al.*, 2005). Many of the behavioral changes in maternal separation models have analogy with symptoms of depression and the neuroendocrine changes are consistent with depression (Heim *et al.*, 2004; Pryce *et al.*, 2001, 2005). The stability of the phenotypic changes allows these models to be useful for investigation of mechanisms, including gene expression, related to mood disorders (Law *et al.*, 2009). Some effects of maternal separation are counteracted by chronic antidepressant drug treatment or ECT (Leventopoulos *et al.*, 2009), but establishment of predictive validity requires further studies. Maternal separation paradigms in rodents target critical periods of postnatal development when the brain is very susceptible to experience-dependent alterations. These models are suggested to have face validity for disrupted parenting behavior in humans that can result from a number of situations, including parental depression, also likely to occur during critical periods of development (Newport *et al.*, 2002).

A variation on maternal separation paradigms is a related paradigm that uses the quantification of levels of maternal care as it occurs naturally rather than experimentally manipulating maternal care (Francis *et al.*, 1999; Liu *et al.*, 1997). In this work, licking/grooming of pups and arched back nursing have been identified as important features of maternal behavior in female rats. Naturally occurring variations in these maternal behaviors are quantified and low levels are considered to represent a stress condition for the offspring. Levels of maternal care have been demonstrated to correlate with levels of stress-reactivity (HPA activity and anxiety-related phenotype) in the adult offspring (Liu *et al.*, 1997). This result has analogy to the maternal separation paradigms described above and with the correlation of HPA reactivity in humans with prior ELS (Heim *et al.*, 2001). Alterations in HPA axis responsiveness in humans after ELS may predispose to later depression (Heim *et al.*, 2001, 2004; Kendler *et al.*, 2002). This paradigm thus represents a way to identify a population of depression/anxiety-susceptible individuals in an experimental setting and the stable phenotype allows for the study of mechanisms that may underlie features of the phenotype.

ELS paradigms are sensitive and time-consuming and results obtained depend on choice of comparison control groups used. But, unlike some of the shorter-term stress models, the ELS paradigms produce animals with lasting depression-related features and can therefore inform the study of stress contributions in predisposing individuals to chronic anxiety and depressive illness. Familial transmission of depression is likely to involve both genetic and environmental components, and individuals who inherit genetic susceptibility to depression may also be exposed to adversity in the early environment as a result of depression-related behavior in the parents (Kendler *et al.*, 2002; Newport *et al.*, 2002). The work of Meaney and colleagues, with their paradigm of variation in maternal care and cross-fostering, has made significant advances in describing detailed molecular mechanisms whereby stress-related phenotype can be transmitted from mother to offspring via stable changes in gene expression (Kaffman and Meaney, 2007).

*Prenatal stress.* The impact of stress has also been modeled in prenatal stress paradigms. Maternal stress of various types, for example, noise exposure or restraint during gestation results in alterations in the offspring, including increased anxiety, increased indices of depression in depression models, and altered HPA axis activity (Alonso *et al.*, 1991; Maccari *et al.*, 2003; McCormick *et al.*, 1995; Morilak and Frazer, 2004; Morley-Fletcher *et al.*, 2003; Secoli and Teixeira, 1998; Smith *et al.*, 2004; Weinstock *et al.*, 1992). Behavioral alterations in the FST depression model, anxiety, and HPA axis changes that result from prenatal stress are reversible with chronic antidepressant treatment (Morley-Fletcher *et al.*, 2004; Poltyrev and Weinstock, 2004). Alterations in HPA axis are similar to those caused by prenatal stress in humans (Weinstock, 1997). Prenatal stress paradigms have construct and face validity, but the anxiety and depression-related changes that are induced in the mothers complicate the interpretation as to the relative contributions of gestational versus postnatal care effects.

## H. Social defeat

Social stress represents a significant type of adversity in many species and is thought to play a role in the development of depression and other psychopathology in humans (Agid *et al.*, 2000; Bjorkqvist, 2001; Huhman, 2006). The use of social conflict as a stressor and the use of social interaction as a quantifiable endpoint both have validity for depression (Heim and Nemeroff, 2001). Experimental models in rodents frequently utilize a conflict situation that results in one animal becoming or retaining dominant status and another ending up subordinate or "defeated". A phenotypic trait produced in these models is social avoidance, which can be quantified and is suggested to model social withdrawal in human depression (Berton *et al.*, 2006; Koolhaas *et al.*, 1997; Van Kampen *et al.*, 2002).

Social stress models are suggested to represent an induction of a depressive-like state that may be more relevant to human depression compared with models that employ acute or severe stressors. In these models, social conflict is created between male animals. This can be done by introducing an intruder animal into the home cage of another resident. The experiments are generally designed taking into account factors such as strain, body weight, and social status to ensure an outcome in which a defeated animal is produced. Paradigms are used which vary the number of conflict sessions and the nature of the conflict (psychological vs. physical). Physical attack and threat of attack (exposure to sensory contact with another animal but with a barrier to physical attack) can be used separately or combined within a paradigm. Control animals for these experiments should also be exposed to social contact but without conflict or defeat. Two important depression-related features that occur in defeated animals are anhedonia, measured as reduced preference for sweet solutions, and social avoidance in the presence of an unfamiliar animal (Meerlo *et al.*, 1996; Rygula *et al.*, 2005; Von Frijtag *et al.*, 2002). Other behavioral or physiologic changes include decreased sexual behavior and increased defensive behavior, increased anxiety, decreased locomotor or exploratory activity, changes in circadian rhythmicity, alterations in feeding and body weight, sleep disturbances, and impaired immune function (Bohus *et al.*, 1993; Koolhaas *et al.*, 1997; Martinez *et al.*, 1998; Meerlo *et al.*, 1996). The HPA axis is activated in defeated animals, which is similar to other stress models (Buwalda *et al.*, 1999).

Social avoidance and anhedonia that result from social defeat are long-lasting and are sensitive to chronic but not acute treatment with antidepressant drugs (Berton *et al.*, 2006; Huhman, 2006; Meerlo *et al.*, 1996, 2002;

Von Frijtag *et al.*, 2002). This indicates the utility of social defeat models in studying time-dependent neural processes relevant to depression. Animals can be identified as susceptible or resistant to the effects of social defeat, indicating further value of social defeat models for investigating substrates of individual vulnerability (Krishnan *et al.*, 2007). Social defeat has proven useful in identifying molecular mechanisms that can induce stable changes in phenotype (Krishnan *et al.*, 2007).

## V. Concluding Remarks

The more rapid and acute antidepressant-responsive assays such as the FST and TST have been useful in identifying new drugs that share mechanisms with the older known drugs. While these assays are sensitive to the identified antidepressant mechanisms, it is possible that they are not sensitive to other mechanisms that could be of therapeutic value in treating depression. The identification of novel and improved antidepressant mechanisms depends on models that can recapitulate critical processes operative in depression. Until critical processes are identified, models that incorporate stress exposure, time-dependent induction and treatment response, and individual differences in susceptibility will be the most valuable in facilitating the study of mechanisms underlying depression and its treatment.

## *REFERENCES*

Agid, O., Shapira, B., Zislin, J., Ritsner, M., Hanin, B., Murad, H., Troudart, T., Bloch, M., Heresco-Levy, U., and Lerer, B. (1999). Environment and vulnerability to major psychiatric illness: A case control study of early parental loss in major depression, bipolar disorder and schizophrenia. *Mol. Psychiatry* **4**(2), 163–172.

Agid, O., Kohn, Y., and Lerer, B. (2000). Environmental stress and psychiatric illness. *Biomed. Pharmacother.* **54**(3), 135–141.

Aguilera, G. (1998). Corticotropin releasing hormone receptor regulation and the stress response. *Trends Endocrinol. Metab.* **9**(8), 329–336.

Alonso, S. J., Arevalo, R., Afonso, D., and Rodriguez, M. (1991). Effects of maternal stress during pregnancy on forced swimming test behavior of the offspring. *Physiol. Behav.* **50,** 511–517.

Anisman, H., and Matheson, K. (2005). Stress, depression, and anhedonia: Caveats concerning animal models. *Neurosci. Biobehav. Rev.* **29**(4–5), 525–546.

Bai, F., Li, X., Clay, M., Lindstrom, T., and Skolnick, P. (2001). Intra- and interstrain differences in models of "behavioral despair". *Pharmacol. Biochem. Behav.* **70,** 187–192.

Berton, O., McClung, C. A., Dileone, R. J., Krishnan, V., Renthal, W., Russo, S. J., Graham, D., Tsankova, N. M., Bolanos, C. A., Rios, M., Monteggia, L. M., Self, D. W., and Nestler, E. J. (2006). Essential role of BDNF in the mesolimbic dopamine pathway in social defeat stress. *Science* **311**(5762), 864–868.

Bjorkqvist, K. (2001). Social defeat as a stressor in humans. *Physiol. Behav.* **73,** 435–442.

Bodnoff, S. R., Suranyi-Cadotte, B., Aitken, D. H., Quirion, R., and Meaney, M. J. (1988). The effects of chronic antidepressant treatment in an animal model of anxiety. *Psychopharmacology (Berl.)* **95**(3), 298–302.
Bodnoff, S. R., Suranyi-Cadotte, B., Quirion, R., and Meaney, M. J. (1989). A comparison of the effects of diazepam versus several typical and atypical anti-depressant drugs in an animal model of anxiety. *Psychopharmacology (Berl.)* **97**(2), 277–279.
Bohus, B., Koolhaas, J. M., Heijnen, C. J., and de Boer, O. (1993). Immunological responses to social stress: Dependence on social environment and coping abilities. *Neuropsychobiology* **28**(1–2), 95–99.
Borsini, F., and Meli, A. (1988). Is the forced swimming test a suitable model for revealing antidepressant activity? *Psychopharmacology (Berl.)* **94,** 147–160.
Buwalda, B., de Boer, S. F., Schmidt, E. D., Felszeghy, K., Nyakas, C., Sgoifo, A., Van der Vegt, B. J., Tilders, F. J., Bohus, B., and Koolhaas, J. M. (1999). Long-lasting deficient dexamethasone suppression of hypothalamic-pituitary-adrenocortical activation following peripheral CRF challenge in socially defeated rats. *J. Neuroendocrinol.* **11**(7), 513–520.
Carlezon, W. A. Jr, and Chartoff, E. H. (2007). Intracranial self-stimulation (ICSS) in rodents to study the neurobiology of motivation. *Nat Protoc.* **2**(11), 2987–2995.
Caspi, A., Sugden, K., Moffitt, T. E., Taylor, A., Craig, I. W., Harrington, H., McClay, J., Mill, J., Martin, J., Braithwaite, A., and Poulton, R. (2003). Influence of life stress on depression: Moderation by a polymorphism in the 5-HTT gene. *Science* **301**(5631), 386–389.
Crowley, J. J., and Lucki, I. (2005). Opportunities to discover genes regulating depression and antidepressant response from rodent behavioral genetics. *Curr. Pharm. Des.* **11**(2), 157–169.
Cryan, J. F., Mombereau, C., and Vassout, A. (2005). The tail suspension test as a model for assessing antidepressant activity: Review of pharmacological and genetic studies in mice. *Neurosci. Biobehav. Rev.* **29**(4–5), 571–625.
D'Aquila, P. S., Newton, J., and WIllner, P. (1997). Diurnal variation in the effect of chronic mild stress on sucrose intake and preference. *Physiol. Behav.* **62,** 421–426.
Detke, M. J., Rickels, M., and Lucki, I. (1995). Active behaviors in the rat forced swimming test differentially produced by serotonergic and noradrenergic antidepressants. *Psychopharmacology (Berl.)* **121**(1), 66–72.
Diagnostic Statistical Manual IV American Psychiatric Press, Washington, DC.
Drevets, W. C., Price, J. L., and Furey, M. L. (2008). Brain structural and functional abnormalities in mood disorders: Implications for neurocircuitry models of depression. *Brain Struct. Funct.* **213**(1–2), 93–118.
Dulawa, S. C., and Hen, R. (2005). Effects of chronic fluoxetine in animal models of anxiety and depression. *Neurosci. Biobehav. Rev.* **29**(4–5), 771–783.
Dulawa, S. C., Holick, K. A., Gundersen, B., and Hen, R. (2004). Recent advances in animal models of chronic antidepressant effects: The novelty-induced hypophagia test. *Neuropsychopharmacology* **29**(7), 1321–1330.
Dunn, A. J., Swiergiel, A. H., and de Beaurepaire, R. (2005). Cytokines as mediators of depression: What can we learn from animal studies? *Neurosci. Biobehav. Rev.* **29**(4–5), 891–909.
El Yacoubi, M., and Vaugeois, J. M. (2007). Genetic rodent models of depression. *Curr. Opin. Pharmacol.* **7**(1), 3–7.
Fava, M., Rankin, M. A., Wright, E. C., Alpert, J. E., Nierenberg, A. A., Pava, J., and Rosenbaum, J. F. (2000). Anxiety disorders in major depression. *Compr. Psychiatry* **41**(2), 97–102.
Francis, D. D., Champagne, F. A., Liu, D., and Meaney, M. J. (1999). Maternal care, gene expression, and the development of individual differences in stress reactivity. *Ann. NY Acad. Sci.* **896,** 66–84.

Heim, C., and Nemeroff, C. B. (2001). The role of childhood trauma in the neurobiology of mood and anxiety disorders: Preclinical and clinical studies. *Biol. Psychiatry* **49,** 1023–1039.

Heim, C., Newport, D. J., Bonsall, R., Miller, A. H., and Nemeroff, C. B. (2001). Altered pituitry–adrenal axis responses to provocative challenge tests in adult survivors of childhood abuse. *Am. J. Psychiatry* **158,** 575–581.

Heim, C., Plotsky, P. M., and Nemeroff, C. B. (2004). Importance of studying the contributions of early adverse experience to neurobiological findings in depression. *Neuropsychopharmacology* **29**(4), 641–648.

Henn, F. A., and Vollmayr, B. (2005). Stress models of depression: Forming genetically vulnerable strains. *Neurosci. Biobehav. Rev.* **29**(4–5), 799–804.

Herman, J. P., and Cullinan, W. E. (1997). Neurocircuitry of stress: Central control of the hypothalamo-pituitary-adrenocortical axis. *Trends Neurosci.* **20**(2), 78–84.

Holmes, A., le Guisquet, A. M., Vogel, E., Millstein, R. A., Leman, S., and Belzung, C. (2005). Early life genetic, epigenetic and environmental factors shaping emotionality in rodents. *Neurosci. Biobehav. Rev.* **29**(8), 1335–1346.

Huhman, K. L. (2006). Social conflict models: Can they inform us about human psychopathology? *Horm. Behav.* **50**(4), 640–646.

Kaffman, A., and Meaney, M. J. (2007). Neurodevelopmental sequelae of postnatal maternal care in rodents: Clinical and research implications of molecular insights. *J. Child Psychol. Psychiatry* **48**(3–4), 224–244.

Katz, R. J., Roth, K. A., and Carroll, B. J. (1981a). Acute and chronic stress effects on open field activity in the rat: Implications for a model of depression. *Neurosci. Biobehav. Rev.* **5**(2), 247–251.

Katz, R. J., Roth, K. A., and Schmaltz, K. (1981b). Amphetamine and tranylcypromine in an animal model of depression: Pharmacological specificity of the reversal effect. *Neurosci. Biobehav. Rev.* **5**(2), 259–264.

Kaufman, J., and Charney, D. (2000). Comorbidity of mood and anxiety disorders. *Depress. Anxiety* **12**(Suppl. 1), 69–76.

Kaufman, J., Yang, B. Z., Douglas-Palumberi, H., Grasso, D., Lipschitz, D., Houshyar, S., Krystal, J. H., and Gelernter, J. (2006). Brain-derived neurotrophic factor-5-HTTLPR gene interactions and environmental modifiers of depression in children. *Biol. Psychiatry* **59**(8), 673–680.

Keller, M. C., Neale, M. C., and Kendler, K. S. (2007). Association of different adverse life events with distinct patterns of depressive symptoms. *Am. J. Psychiatry* **164**(10), 1521–1529.

Kendler, K. S., Karkowski, L. M., and Prescott, C. A. (1999). Causal relationship between stressful life events and the onset of major depression. *Am. J. Psychiatry* **156,** 837–841.

Kendler, K. S., Gardner, C. O., and Prescott, C. A. (2002). Toward a comprehensive developmental model for major depression in women. *Am. J. Psychiatry* **159**(7), 1133–1145.

Kessler, R. C. (1997). The effects of stressfull life events on depression. *Annu. Rev. Psychol.* **48,** 191–214.

Kessler, R. C., Berglund, P., Demler, O., Jin, R., Merikangas, K. R., and Walters, E. E. (2005). Lifetime prevalence and age-of-onset distributions of DSM-IV disorders in the National Comorbidity Survey Replication. *Arch. Gen. Psychiatry* **62**(6), 593–602.

Koolhaas, J. M., De Boer, S. F., De Rutter, A. J., Meerlo, P., and Sgoifo, A. (1997). Social stress in rats and mice. *Acta Physiol. Scand. Suppl.* **640,** 69–72.

Krishnan, V., Han, M. H., Graham, D. L., Berton, O., Renthal, W., Russo, S. J., Laplant, Q., Graham, A., Lutter, M., Lagace, D. C., Ghose, S., Reister, R., *et al.* (2007). Molecular adaptations underlying susceptibility and resistance to social defeat in brain reward regions. *Cell* **131**(2), 391–404.

Ladd, C. O., Huot, R. L., Thrivikraman, K. V., Nemeroff, C. B., Meaney, M. J., and Plotsky, P. M. (2000). Long-term behavioral and neuroendocrine adaptations to adverse early experience. *Prog. Brain Res.* **122,** 81–103.
Law, A. J., Pei, Q., Walker, M., Gordon-Andrews, H., Weickert, C. S., Feldon, J., Pryce, C. R., and Harrison, J. (2009). Early parental deprivation in the marmoset monkey produces long-term changes in hippocampal expression of genes involved in synaptic plasticity and implicated in mood disorder. *Neuropsychopharmacology* **34**(6), 1381–1394.
Leventopoulos, M., Russig, H., Feldon, J., Pryce, C. R., and Opacka-Juffry, J. (2009). Early deprivation leads to long-term reductions in motivation for reward and 5-HT1A binding and both effects are reversed by fluoxetine. *Neuropharmacology* **56**(3), 692–701.
Levine, S. (1957). Infantile experience and resistance to physiological stress. *Science* **126,** 405–406.
Liu, X., and Gershenfeld, H. K. (2001). Genetic differences in the tail-suspension test and its relationship to imipramine response among 11 inbred strains of mice. *Biol. Psychiatry* **49**(7), 575–581.
Liu, D., Diorio, J., Tannenbaum, B., Caldji, C., Francis, D., Freedman, A., Sharma, S., Pearson, D., Plotsky, P. M., and Meaney, M. J. (1997). Maternal care, hippocampal glucocorticoid receptors, and hypothalamic-pituitary-adrenal responses to stress. *Science* **277**(5332), 1620–1621.
López-Rubalcava, C., and Lucki, I. (2000). Strain differences in the behavioral effects of antidepressant drugs in the rat forced swimming test. *Neuropsychopharmacology* **22**(2), 191–199.
Lucki, I. (1997). The forced swimming test as a model for core and component behavioral effects of antidepressant drugs. *Behav. Pharmacol.* **8,** 523–532.
Maccari, S., Darnaudery, M., Morley-Fletcher, S., *et al.* (2003). Prenatal stress and long-term consequences: Implications of glucocorticoid hormones. *Neurosci. Biobehav. Rev.* **27,** 119–127.
Magariños, A. M., and McEwen, B. S. (1995). Stress-induced atrophy of apical dendrites of hippocampal CA3c neurons: Comparison of stressors. *Neuroscience* **69,** 83–88.
Maier, S. F. (1984). Learned helplessness and animal models of depression. *Prog. Neuropsychopharmacol. Biol. Psychiatry* **8**(3), 435–446.
Markou, A., Hauger, R. L., and Koob, G. F. (1992). Desmethylimipramine attenuates cocaine withdrawal in rats. *Psychopharmacology (Berl.)* **109**(3), 305–314.
Martin, P., Soubrié, P., and Puech, A. J. (1990). Reversal of helpless behavior by serotonin uptake blockers in rats. *Psychopharmacology (Berl.)* **101**(3), 403–407.
Martinez, M. M., Calvo-Torrent, M. A., and Pico-Alfonso, M. A. (1998). Social defeat and subordination as models of social stress in laboratory rodents: A review. *Aggress. Behav.* **24,** 241–256.
McCormick, C. M., Smythe, J. W., Sharma, S., and Meaney, M. J. (1995). Sex-specific effects of prenatal stress on hypothalamic-pituitary-adrenal responses to stress and brain glucocorticoid receptor density in adult rats. *Brain Res. Dev. Brain Res.* **84,** 55–61.
McKinney, W. T. (2001). Overview of the past contributions of animal models and their changing place in psychiatry. *Semin. Clin. Neuropsychiatry* **6**(1), 68–78.
Meerlo, P., Overkamp, G. J., Daan, S., Van Den Hoofdakker, R. H., and Koolhaas, J. M. (1996). Changes in behaviour and body weight following a single or double social defeat in rats. *Stress* **1,** 21–32.
Meerlo, P., Sgoifo, A., and Turek, F. W. (2002). The effects of social defeat and other stressors on the expression of circadian rhythms. *Stress* **5,** 15–22.
Mintz, M., Rüedi-Bettschen, D., Feldon, J., and Pryce, C. (2005). Early social and physical deprivation leads to reduced social motivation in adulthood in Wistar rats. *Behav. Brain Res.* **156,** 311–320.

Monleon, S., D'Aquila, P., Parra, A., Simon, V. M., Brain, P. F., and Willner, P. (1995). Attenuation of sucrose consumption in mice by chronic mild stress and its restoration by imipramine. *Psychopharmacology (Berl.)* **117**(4), 453–457.

Moreau, J. L., Jenck, F., Martin, J. R., Mortas, P., and Haefely, W. E. (1992). Antidepressant treatment prevents chronic unpredictable mild stress-induced anhedonia as assessed by ventral tegmentum self-stimulation behavior in rats. *Eur. Neuropsychopharmacol.* **2**(1), 43–49.

Morilak, D. A., and Frazer, A. (2004). Antidepressants and brain monoaminergic systems: A dimensional approach to understanding their behavioural effects in depression and anxiety disorders. *Int. J. Neuropsychopharmacol.* **7**(2), 193–218.

Morley-Fletcher, S., Rea, M., Maccari, S., and Laviola, G. (2003). Environmental enrichment during adolescence reverses the effects of prenatal stress on play behaviour and HPA axis reactivity in rats. *Eur. J. Neurosci.* **18,** 3367–3374.

Morley-Fletcher, S., Darnaudéry, M., Mocaer, E., Froger, N., Lanfumey, L., Laviola, G., Casolini, P., Zuena, A. R., Marzano, L., Hamon, M., and Maccari, S. (2004). Chronic treatment with imipramine reverses immobility behaviour, hippocampal corticosteroid receptors and cortical 5-HT(1A) receptor mRNA in prenatally stressed rats. *Neuropharmacology* **47**(6), 841–847.

Muscat, R., and Willner, P. (1992). Suppression of sucrose drinking by chronic mild unpredictable stress: A methodological analysis. *Neurosci. Biobehav. Rev.* **16**(4), 507–517.

Nestler, E. J., Gould, E., Manji, H., Buncan, M., Duman, R. S., Greshenfeld, R. K., Hen, R., Koester, S., Lederhendler, I., Meaney, M., Robbins, T., Winsky, L., and Zalcman, S. (2002). Preclinical models: Status of basic research in depression. *Biol. Psychiatry* **52,** 503–528.

Newport, D. J., Stowe, Z. N., and Nemeroff, C. B. (2002). Parental depression: Animal models of an adverse life event. *Am. J. Psychiatry* **159**(8), 1265–1283.

Nielsen, C. K., Arnt, J., and Sánchez, C. (2000). Intracranial self-stimulation and sucrose intake differ as hedonic measures following chronic mild stress: Interstrain and interindividual differences. *Behav. Brain Res.* **107**(1–2), 21–33.

Overstreet, D. H., Friedman, E., Mathé, A. A., and Yadid, G. (2005). The Flinders Sensitive Line rat: A selectively bred putative animal model of depression. *Neurosci. Biobehav. Rev.* **29**(4–5), 739–759.

Papp, M., Moryl, E., and Willner, P. (1996). Pharmacological validation of the chronic mild stress model of depression. *Eur. J. Pharmacol.* **296**(2), 129–136.

Perrault, G., Morel, E., Zivkovic, B., and Sanger, D. J. (1992). Activity of litoxetine and other serotonin uptake inhibitors in the tail suspension test in mice. *Pharmacol. Biochem. Behav.* **42**(1), 45–47.

Pittenger, C., and Duman, R. S. (2008). Stress, depression, and neuroplasticity: A convergence of mechanisms. *Neuropsychopharmacology* **33**(1), 88–109.

Plotsky, P. M., and Meaney, M. J. (1993). Early, postnatal experience alters hypothalamic corticotropin-releasing factor (CRF) mRNA, median eminence CRF content and stress-induced release in adult rats. *Brain Res. Mol. Brain Res.* **18**(3), 195–200.

Poltyrev, T., and Weinstock, M. (2004). Gender difference in the prevention of hyperanxiety in adult prenatally stressed rats by chronic treatment with amitriptyline. *Psychopharmacology (Berl.)* **171,** 270–276.

Porsolt, R. D., Bertin, A., and Jalfre, M. (1977a). Behavioral despair in mice: A primary screening test for antidepressants. *Arch. Int. Pharmacodyn. Ther.* **229**(2), 327–336.

Porsolt, R. D., Le Pichon, M., and Jalfre, M. (1977b). Depression: A new animal model sensitive to antidepressant treatments. *Nature* **266**(5604), 730–732.

Porsolt, R. D., Bertin, A., and Jalfre, M. (1978). "Behavioural despair" in rats and mice: Strain differences and the effects of imipramine. *Eur. J. Pharmacol.* **51**(3), 291–294.

Pothion, S., Bizot, J. C., Trovero, F., and Belzung, C. (2004). Strain differences in sucrose preference and in the consequences of unpredictable chronic mild stress. *Behav. Brain Res.* **155**(1), 135–146.

Pryce, C. R., Bettschen, D., Bahr, N. I., and Feldon, J. (2001). Comparison of the effects of infant handling, isolation, and nonhandling on acoustic startle, prepulse inhibition, locomotion, and HPA activity in the adult rat. *Behav. Neurosci.* **115**(1), 71–83.

Pryce, C. R., Rüedi-Bettschen, D., Dettling, A. C., Weston, A., Russig, H., Ferger, B., and Feldon, J. (2005). Long-term effects of early-life environmental manipulations in rodents and primates: Potential animal models in depression research. *Neurosci. Biobehav. Rev.* **29** (4–5), 649–674.

Ressler, K. J., and Nemeroff, C. B. (2000). Role of serotonergic and noradrenergic systems in the pathophysiology of depression and anxiety disorders. *Depress. Anxiety* **12**(Suppl. 1), 2–19.

Ripoll, N., David, D. J., Dailly, E., Hascoët, M., and Bourin, M. (2003). Antidepressant-like effects in various mice strains in the tail suspension test. *Behav. Brain Res.* **143**(2), 193–200.

Rüedi-Bettschen, D., Pedersen, E., Feldon, J., and Pryce, C. (2005). Rat early deprivation under specific conditions leads to reduced interest in reward in adulthood in Wistar rats. *Behav. Brain Res.* **156,** 297–310.

Rüedi-Bettschen, D., Zhang, W., Russig, H., Ferger, B., Weston, A., Pedersen, E. M., Feldon, J., and Pryce, C. R. (2006). Early deprivation leads to altered behavioural, autonomic and endocrine responses to environmental challenge in adult Fischer rats. *Eur. J. Neurosci.* **24**(10), 2879–2893.

Rygula, R., Abumaria, N., Flügge, G., Fuchs, E., Rüther, E., and Havemann-Reinecke, U. (2005). Anhedonia and motivational deficits in rats: Impact of chronic social stress. *Behav. Brain Res.* **162**(1), 127–134.

Secoli, S. R., and Teixeira, N. A. (1998). Chronic prenatal stress affects development and behavioral depression in rats. *Stress* **2,** 273–280.

Serretti, A., Chiesa, A., Calati, R., Perna, G., Bellodi, L., and De Ronchi, D. (2009). Common genetic, clinical, demographic and psychosocial predictors of response to pharmacotherapy in mood and anxiety disorders. *Int. Clin. Psychopharmacol.* **24**(1), 1–18.

Sherman, A. D., Sacquitne, J. L., and Petty, F. (1982). Specificity of the learned helplessness model of depression. *Pharmacol. Biochem. Behav.* **16**(3), 449–454.

Smith, J. W., Seckl, J. R., Evans, A. T., Costall, B., and Smythe, J. W. (2004). Gestational stress induces post-partum depression-like behaviour and alters maternal care in rats. *Psychoneuroendocrinology* **29**(2), 227–244.

Sonawalla, S. B., and Fava, M. (2001). Severe depression: Is there a best approach? *CNS Drugs* **15**(10), 765–776.

Song, C., and Leonard, B. E. (2005). The olfactory bulbectomised rat as a model of depression. *Neurosci. Biobehav. Rev.* **29**(4–5), 627–647.

Steru, L., Chermat, R., Thierry, B., and Simon, P. (1985). The tail suspension test: A new method for screening antidepressants in mice. *Psychopharmacology* **85**(3), 367–370.

Tannenbaum, B., Tannenbaum, G., and Anisman, H. (2002). Neurochemical and behavioral alterations elicited by a chronic intermittent stressor regimen: Implications for allostatic load. *Brain Res.* **953,** 82–92.

Thiébot, M.-H., Martin, P., and Puech, A. J. (1992). Animal behavioural studies in the evaluation of antidepressant drugs. *Br. J. Psychiatry* **160,** 44–50.

Todtenkopf, M. S., Marcus, J. F., Portoghese, P. S., and Carlezon, W. A. Jr. (2004). Effects of kappa-opioid receptor ligands on intracranial self-stimulation in rats. *Psychopharmacology (Berl).* **172**(4), 463–470.

Trullas, R., Jackson, B., and Skolnick, P. (1989). Genetic differences in a tail suspension test for evaluating antidepressant activity. *Psychopharmacology (Berl.)* **99**(2), 287–288.

Van Kampen, M., Kramer, M., Hiemke, C., Flugge, G., and Fuchs, E. (2002). The chronic psychosocial stress paradigm in male tree shrews: Evaluation of a novel animal model for depressive disorders. *Stress* **5,** 37–46.

Von Frijtag, J. C., Van den Bos, R., and Spruijt, B. M. (2002). Imipramine restores the long-term impairment of appetitive behavior in socially stressed rats. *Psychopharmacology (Berl.)* **162**(3), 232–238.

Wang, P. S., Simon, G., and Kessler, R. C. (2003). The economic burden of depression and the cost-effectiveness of treatment. *Int. J. Methods Psychiatr. Res.* **12**(1), 22–33.

Weinstock, M. (1997). Does prenatal stress impair coping and regulation of hypothalamic-pituitary-adrenal axis? *Neurosci. Biobehav. Rev.* **21**(1), 1–10.

Weinstock, M., Matlina, E., Maor, G. I., Rosen, H., and McEwen, B. S. (1992). Prenatal stress selectively alters the reactivity of the hypothalamic-pituitary adrenal system in the female rat. *Brain Res.* **595,** 195–200.

Weiss, J., and Kilts, C. D. (1995). Animal models of depression and schizophrenia pp. 89–131. American Psychiatric Press, Washington, DC.

Weiss, E. L., Longhurst, J. G., and Mazure, C. M. (1999). Childhood sexual abuse as a risk factor for depression in women: Psychosocial and neurobiological correlates. *Am. J. Psychiatry* **156**(6), 816–828.

Willner, P. (1984). The validity of animal models of depression. *Psychopharmacology (Berl.)* **83**(1), 1–16.

Willner, P. (1990). Animal models of depression: An overview. *Pharmacol. Ther.* **45,** 425–455.

Willner, P. (1997). Validity, reliability and utility of the chronic mild stress model of depression: A 10-year review and evaluation. *Psychopharmacology* **134,** 319–329.

Willner, P. (2005). Chronic mild stress (CMS) revisited: Consistency and behavioural-neurobiological concordance in the effects of CMS. *Neuropsychobiology* **52**(2), 90–110.

Willner, P., Towell, A., Sampson, D., Sophokleous, S., and Muscat, R. (1987). Reduction of sucrose preference by chronic unpredictable mild stress, and its restoration by a tricyclic antidepressant. *Psychopharmacology (Berl.)* **93**(3), 358–364.

Willner, P., Moreau, J. L., Nielsen, C. K., Papp, M., and Sluzewska, A. (1996). Decreased hedonic responsiveness following chronic mild stress is not secondary to loss of body weight. *Physiol. Behav.* **60**(1), 129–134.

Zhang, T. Y., Bagot, R., Parent, C., Nesbitt, C., Bredy, T. W., Caldji, C., Fish, E., Anisman, H., Szyf, M., and Meaney, M. J. (2006). Maternal programming of defensive responses through sustained effects on gene expression. *Biol. Psychiatry* **73**(1), 72–89.

CHAPTER TWO

# Astrocytes in the Amygdala

Ryan T. Johnson,* S. Marc Breedlove,*,† *and* Cynthia L. Jordan*,†

## Contents



## Abstract

The amygdala has received considerable attention because of its established role in specific behaviors and disorders such as anxiety, depression, and autism. Studies have revealed that the amygdala is a complex and dynamic brain region that is highly connected with other areas of the brain. Previous works have focused on neurons, demonstrating that the amygdala in rodents is highly plastic and sexually dimorphic. However, our more recent work explores sex differences in nonneuronal cells, joining a rich literature concerning glia in the amygdala. Prior investigation of glia in the amygdala can generally be divided into disease-related and hormone-related categories, with both areas of research producing interesting findings concerning glia in this important brain region. Despite a wide range of research topics, the collected findings make it clear that glia in the amygdala are sensitive and plastic cells that respond and develop in a highly region specific manner. 

* Neuroscience Program, Michigan State University, East Lansing, Michigan, USA
† Psychology Department, Michigan State University, East Lansing, Michigan, USA

*Vitamins and Hormones*, Volume 82
ISSN 0083-6729, DOI: 10.1016/S0083-6729(10)82002-3

# I. The Amygdala

Located in the temporal lobes, the cluster of nuclei comprising the amygdala is involved in several important behaviors, including social behavior, fear, and anxiety (Adolphs *et al.*, 2005; Bishop *et al.*, 2004; LeDoux, 1998; Mah *et al.*, 2007; Spezio *et al.*, 2007; Truitt *et al.*, 2007; Williams *et al.*, 2006). Tract-tracing work along with recent resting-state fMRI data has demonstrated that the amygdala is reciprocally connected via direct and indirect connections to important homeostatic and cognitive centers such as the hypothalamus, hippocampus, and much of the sensory cortex (Kilpatrick *et al.*, 2006). Given these extensive connections, it is not surprising that the amygdala has been implicated in many behavioral disorders and diseases, including schizophrenia, drug addiction, anxiety disorders, depression, and autism.

Much of the knowledge regarding amygdala function stems from its role in fear and anxiety. Conditioning experiments have identified the amygdala as crucial to the fear response, and additional methodologies have mapped the amygdala's expanded role in emotional memory and motivation (Bechara *et al.*, 1995; Cahill, 2000; reviewed in Cardinal *et al.*, 2002). Within these fields, it is generally acknowledged that cellular plasticity within the amygdala plays a vital role in shaping behavior. This plasticity has most often been investigated in the context of long-term potentiation (LTP) associated with fear conditioning (Rogan and LeDoux, 1995; Yu *et al.*, 2008). However, neuroendocrine researchers have also been intrigued by amygdala plasticity because of the region's dense concentrations of steroid hormone receptors (Simerly *et al.*, 1990), its well-documented collection of functional and morphological sex differences (Cahill *et al.*, 2004; Hines *et al.*, 1992; Johnson *et al.*, 2008; Kilpatrick *et al.*, 2006; Rubinow and Juraska, 2009), and its responsiveness to circulating gonadal hormones throughout the lifespan (Cooke and Woolley, 2005; Cooke *et al.*, 1999, 2003; Martinez *et al.*, 2006; Morris *et al.*, 2008). These characteristics strongly suggest that the sex differences seen in amygdala anatomy may underlie sex biases in behavior and disease susceptibility, and that amygdala plasticity may be a key player in the regulation of such attributes.

The exploration of amygdala anatomy has mostly been with rodent models although feline and nonhuman primates have contributed significantly (Rees and Michael, 1984; reviewed in McDonald, 1998). In common laboratory species and in humans, the amygdala is divided into central, medial, cortical, lateral, basal, and accessory basal divisions which are strongly interconnected, although the direction and strength of these connections does not appear uniform among the subregions (Alheid *et al.*, 1995; Grove, 1988a,b; Shammah-Lagnado *et al.*, 2000).

Plasticity within the various amygdala subregions and in the amygdala as a whole has been examined through many paradigms and has been covered in numerous reviews (Cooke, 2006; Sah *et al.*, 2008; Samson *et al.*, 2005; Kinsley and Lambert, 2008). The human literature has examined plasticity in amygdala volume and functionality (Giedd *et al.*, 1997; Killgore and Yurgelun-Todd, 2004) and the rodent literature has repeatedly demonstrated plasticity of amygdala neurons based on fear-related LTP (reviewed in Sigurdsson *et al.*, 2007, see also Sah *et al.*, 2008) and by manipulating hormones (Cooke *et al.*, 1999; Morris *et al.*, 2008). Undoubtedly, the examination of neuronal plasticity in the amygdala, whether in the context of hormones or LTP, has greatly expanded our understanding of the function of this region. However, recently another component of amygdala anatomy, glial cells, has been explored, which may fill many of the gaps in our understanding of this dynamic region. This review examines the current findings concerning glia in the amygdala.

## II. Glia

As in most of the central nervous system, several types of glia are present in the amygdala, including astrocytes, oligodendrocytes, and microglia. However, much of the exploration of glia in the limbic system has focused either on astrocytes specifically or on glia in general. Thus, reports examining microglia or oligodendrocytes in the amygdala are rare, but these cells are undoubtedly included in counts of total glial within the amygdala. This circumstance makes attributing specific effects or phenomenon to specific glial subtypes sometimes difficult, but perhaps future investigations will alleviate some of this confusion.

Astrocytes are perhaps the most thoroughly investigated glia subtype in the central nervous system and certainly in the amygdala. Astrocytes were first described in detail in 1895 when Cajal gave one type of glia the name "astrocytes," based upon their extensive star-like arbors. Cajal (1909) suggested that astrocytes were functioning to insulate neural connections while Golgi (1885), who was also investigating glia, proposed that astrocytes provide a link between neurons and the blood supply based upon their predisposition to contact blood vessels. Unfortunately, after these initial descriptions and hypotheses, astrocytes were mostly ignored. It was not until the identification of the astrocyte inflammatory response that these cells once again received attention.

Numerous reports demonstrate that astrocytes rapidly respond to all manner of brain trauma, including injury, disease, and genetic disorders with a rapid synthesis of glial fibrillary acidic protein (GFAP) at the site of injury. This reliable, complex response has been carefully examined and reviewed in detail (Eddleston and Mucke, 1993; Eng and Ghirnikar, 1994). The information derived from studies of reactive gliosis has been vital in understanding brain injury and should not be minimized, but as Cotter *et al.* (2001) described it, "glia were typecast, their crucial role in other cortical function overlooked."

Fortunately, the past two decades have seen a dramatic increase in attention to glia. Much of the interest stems from findings that astrocytes exhibit rapidly propagating calcium waves, suggesting that they may form a second path of communication working in conjunction with the neuronal system (Guthrie *et al.*, 1999). More recent work suggests that propagating calcium waves are extremely rare *in vivo* (Wang *et al.*, 2006), but astrocytes are still considered active and dynamic cells. Mature astrocytes generally have large arbors extending from their cell bodies, and in the cortex, it has been estimated that each astrocyte connects on average to approximately four neurons, 300–600 dendrites, and 1,000,000 synapses (Halassa *et al.*, 2007). Astrocyte processes are closely involved in the formation of new synapses and these processes are highly mobile and extend or retract to modulate contact between neurons (Hatton, 2002; Nishida and Okabe, 2007; Ullian *et al.*, 2001). Interestingly, increased branch complexity in astrocytes has been associated with an increase in functional synapses in some regions (Elmariah *et al.*, 2005; Pfrieger and Barres, 1997) but has also been associated with decreases in dendritic spines (Mong *et al.*, 2001). Thus, it may be that astrocytes have opposing roles under different conditions, promoting and maintaining synapses in some cases, while promoting their elimination in others. In many ways, a dramatic sensitivity to varying conditions has become the hallmark of glia, and their responsiveness to various disease states, which we review next, clearly illustrates this point.

## III. Amygdala Glia and Disease

Glia are known to be important metabolic factories for normal brain functioning, responding quickly and topographically to stimuli with rapid increases in metabolic activity (Schummers *et al.*, 2008). The nature of the glial response to disease has been puzzling, with long-standing questions about whether this response is beneficial and neuroprotective (Thippeswamy *et al.*, 2005; Yamamuro *et al.*, 2003) or is actually detrimental to repair (Mukhopadhyay *et al.*, 1994; also see Mena and García de Yébenes, 2008). Furthermore, in several of the diseases and disorders, we describe next, it is not at all certain if the effects seen in glia are merely a response to disease or a contributing cause. As in most cases, that dichotomy may be an illusion, with glia likely playing both roles at various times and in various instances. However, what is clear is that the extreme sensitivity of glia to brain dysfunction makes them powerful indicators of abnormal brain function and has guided the examination of amygdala glia in several disease models.

## A. Epileptic states

Investigation of patients with temporal lobe epilepsy demonstrates glial satellitosis, the accumulation of glia around neurons, in the lateral amygdala (Faber-zuschratter *et al.*, 2009) and interesting glial phenomena have been detailed in the mouse amygdala as well. Kainic acid, an excitotoxin regularly used to investigate the brain's response to seizure-like conditions, induces several responses in glia in rodents. After kainic acid injection, microglia become immunopositive for cyclin-dependent kinase 4 and cyclin D1, important regulators of apoptotic cell death (Ino and Chiba, 2001). In amygdala astrocytes, kainic acid injection results in prolonged expression of metallothioneins, a family of proteins involved in neuroprotection (Kim *et al.*, 2003). The apolipoproteins, a family including apolipoprotein E, are primarily expressed in astrocytes under normal conditions. Grootendorst *et al.* (2000) found, however, that in cases when damage induced by kainic acid extended beyond the injection site into regions such as the amygdala, apolipoprotein E expression was also found in neurons. In cases of mild damage, this effect was not seen, but increases in GFAP-immunoreactivity and microgliosis were seen. The authors conclude that in severe cases, damage to the glial network is large enough to allow accumulation of apolipoproteins from damaged glia in the surrounding neurons, while in mild cases the reactive glia response may ameliorate this outcome (Grootendorst *et al.*, 2000).

Amygdala-kindled seizures also promote astrocyte proliferation in the piriform cortex of rats (Vessal *et al.*, 2004), and kindling of the olfactory bulb, a major source of input for the amygdala, results in astrocyte proliferation in the basolateral amygdala in addition to the piriform cortex (Woldbye *et al.*, 1996). These two studies together suggest that over stimulation of the amygdala, be it through direct kindling or though kindling of a primary input source, can have long-reaching effects on other parts of the brain. The authors suggest that astrocytes become activated in response to the formation of a new synaptic pathway brought on by the hyperexcitability of the epileptic discharge and conclude that astrocytes may support the formation of the kindling pathway (Vessal *et al.*, 2004, 2005). Jung *et al.* (2009) also found amygdala astrocyte proliferation in a lithium–pilocarpine injection model, and surgical specimens from patients suffering from temporal lobe epilepsy demonstrate an inverse correlation between the extent of astrocytic-reactive gliosis and inhibitory synapses on GAD-positive projection neurons (Yilmazer-Hanke *et al.*, 2007). Again, these results suggest that astrocyte activity may serve to promote a "rewiring" of seizure sensitive regions.

Astrocytes in the piriform cortex become immunoreactive for nestin, an embryonic intermediate neurofilament protein, after amygdaloid kindling, providing further evidence that astrocyte morphology is affected by seizures (Umeoka *et al.*, 2001). Relatedly, while astrocyte reactivity in the hippocampus and amygdala after amygdala-kindled seizure was equivalent in S-100$\beta$ knockouts compared to controls, the knockout animals kindled more rapidly and exhibited more severe seizures, suggesting changes in astrocytic structural proteins are related to seizure progression (Dyck *et al.*, 2002). However, as S-100$\beta$ is expressed only in a subtype of adult astrocytes around blood vessels (Hachem *et al.*, 2005), only some astrocytes would be involved. Investigators have also begun to explore whether astrocytes help to preserve neurons after seizures. Vascular endothelial growth factor (VEGF) protects neurons from cell death (Newton *et al.*, 2003), and astrocytes in the amygdala upregulate VEGF protein expression approximately 24 h after seizure induction. This VEGF upregulation in astrocytes may benefit surrounding neurons (Nicoletti *et al.*, 2008).

Finally, seizure-like states are also examined for their beneficial effects. Electroconvulsive seizures are used for the treatment of major depression, and Wennström *et al.* (2004) examined Wistar rats given five electroconvulsive treatments and injected with bromodeoxyuridine (BrdU) to monitor cell proliferation. As the seizure literature would predict, the authors report a proliferation of glial cells in the amygdala after electroconvulsive treatment. Specifically, the most dramatic increase was in oligodendrocytes, a proliferation that lasted 3 weeks after treatment and led to the establishment of new mature oligodendrocytes. Interestingly, this effect was region and cell-type specific with oligodendrocyte proliferation being found in the central, lateral, and basal subregions and microglia proliferation found in the medial subregion. The authors speculate that the formation of these new cells may play a role in regulating synaptic function (Wennström *et al.*, 2004).

## B. Depression

As the previous report suggests, glia are also implicated in major depressive disorder and bipolar disorder. For example, social exploratory behavior is sometimes used as a model of depression, and Lee *et al.* (2007) found that blocking glutamate uptake in the amygdala resulted in a dose-dependent reduction in social exploratory behavior and altered circadian activity patterns, reminiscent of depressive states in humans. Importantly, selectively blocking glutamate uptake by astrocytes with dihydrokainic acid in the basolateral amygdala also reduced social exploratory behavior, an effect that could be reversed by simultaneous injection of the NMDA receptor antagonist AP5. These results led the authors to conclude that impaired uptake and metabolism of glutamate by astrocytes may be critically involved in depression (Lee *et al.*, 2007).

Glial numbers in the amygdala are also affected in people with major depressive disorder. Bowley *et al.* (2002) examined the amygdala from patients with major depressive and bipolar disorder postmortem and found that glia density and glia/neuron ratio were greatly reduced compared to controls. Interestingly, this reduction was mostly seen in the left hemisphere and was not due to a change in the number of neurons. This reduction in glia was also found in untreated bipolar cases but not in cases treated with either lithium or valproate, suggesting these medications may reduce the loss of glia in the amygdala (Bowley *et al.*, 2002). While a 4-week lithium treatment of adult males rats reduced oligodendrocyte proliferation compared to controls (Orre *et al.*, 2009), the authors note that they did not look for effects of lithium on the survival of cells.

Since the identification of reduced glial numbers in the amygdala in major depressive disorder, other reports have examined this phenomenon in more detail. One study examined the expression of glial markers in the basolateral amygdala in Wister Kyoto (WKY) rats, often used as a model for anxiety and depression due to their exaggerated responses to stressors and depressive-like performance in behavioral tests (Armario *et al.*, 1995; De La Garza and Mahoney, 2004). Gosselin *et al.* (2009) found a significant deficit in the number of GFAP-immunopositive cells (a common marker for astrocytes) in the basolateral amygdala of WKY rats compared with Sprague-Dawley control animals, a result confirmed by Western analysis of GFAP expression levels. However, the use of a second astrocyte marker, S-100$\beta$, revealed no differences in expression between controls and WKY rats. Furthermore, while in control animals GFAP and S-100$\beta$ were colocalized, in WKY rats a population of S-100$\beta$-positive cells was found that were devoid of GFAP immunoreactivity. This led the authors to suggest that in the WKY model, GFAP expression by astrocytes is altered, not the number or density of astrocytes *per se* (Gosselin *et al.*, 2009).

Another report found similar results using a maternal deprivation paradigm, a model of early-life stress, which is a major risk factor in the etiology of depressive disorders. The maternally deprived animals exhibited a significant reduction in GFAP-immunoreactive astrocyte density in the basolateral amygdala and several other brain regions. This reduction in immunoreactivity was greater than the overall reduction in cell density, which suggests, as in the previous report, that GFAP *expression*, rather than the *number* of astrocytes, was altered in this animal model (Leventopoulos *et al.*, 2007). However, it is also clear that early stress can alter cell numbers in the amygdala. Using a prenatal stress paradigm, Kraszpulski *et al.* (2006) found altered glial numbers in different amygdala subregions at different developmental time points. For example, the number of glia in the lateral amygdala nucleus was reduced in prenatally stressed animals at postnatal day 7 only, whereas glia numbers were reduced in the basolateral and central nuclei at postnatal day 25. Neuron number exhibited the same pattern and the authors speculate that these reductions in cell number may underlie the enhanced anxiety seen in prenatally stressed animals in adulthood (Kraszpulski *et al.*, 2006). Unfortunately, how reductions of GFAP expression and/or reduced glia number in the amygdala relate to depression is still unclear.

Despite several reports that astrocytes are affected in the amygdala of both human cases and in animal models of major depressive disorder, other evidence suggests that astrocytes may not be altered during depression. Hamidi *et al.* (2004) labeled tissue from patients diagnosed with major depressive disorder for S-100$\beta$, human leukocyte antigen, and stained for Nissl. Using S-100$\beta$, the group reported no significant difference in astrocyte density in depressed patients compared to controls. Moreover, human leukocyte antigen was used to identify microglia and revealed no difference. However, oligodendrocyte density (identified by their compact deeply stained nuclei in Nissl staining) was significantly lower in this population of depressed patients. The authors concluded that the reduction of glia previously reported in the amygdala of patients suffering from major depressive disorder was due to a reduction in oligodendrocytes. These results are difficult to interpret because Gosselin *et al.* (2009) found a dissociation of S-100$\beta$ and GFAP-labeled cells in their WKY rat model of depression. Perhaps, if Hamidi and colleagues had also utilized GFAP immunoreactivity as their astrocytic marker, they might have seen a reduction in the number of immunopositive astrocytes as well. So, it is possible that *both* astrocytes and oligodendrocytes are reduced in the amygdala of depressive disorder cases. Adding to the complexity is an interesting report examining S-100A1, a protein closely related to S-100$\beta$ that also colocalizes with GFAP. S-100A1 knockout mice appear to develop normally and show no obvious differences in brain development. However, S-100A1 knockout males exhibit *reduced* anxiety based on an approach-avoidance paradigm (Ackermann *et al.*, 2006). Reduced anxiety in knockouts suggests S-100A1 may somehow facilitate fear and anxiety, two behaviors often elevated in models of major depressive disorder. Regardless of what changes occur to amygdala glia, a clear cause–effect relationship remains elusive, and the question of whether changes in glial numbers in the amygdala are a response to or a cause of depression remains unanswered (see Hercher *et al.*, 2009).

## C. Proteinopathies

Glia are also being investigated in other diseases. In Parkinson's disease, $\alpha$-synuclein fibrils aggregate and eventually lead to cellular dysfunction, and the presence of $\alpha$-synuclein-immunoreactive inclusions in neurons has been an important disease marker in research on this disease

(Mukaetova-Ladinska and McKeith, 2006). Astrocytic α-synuclein-immunoreactive inclusions have also been described in the human amygdala (Terada *et al.*, 2003), and these astrocytic inclusions appear in the amygdala at stage 3 of the disease, after their initial appearance in neurons, and their immunoreactivity increases in the latter stages of the disease (Braak *et al.*, 2007). In fact, astrocyte inclusion bodies generally follow the progression of neuronal inclusions throughout the brain, including the amygdala, as the disease progresses (Braak *et al.*, 2007). Astrocytes are not thought to produce α-synuclein themselves, so they may take up the altered α-synuclein molecules from afflicted neurons, likely explaining why glia are also eventually affected by disease. Whether these α-synuclein-immunoreactive glia influence the progression of disease symptoms is unclear, but based upon how astrocytes respond to trauma in the undiseased brain, diseased astrocytes in Parkinson's disease may cause a loss of neuronal connections, effectively severing the amygdala from other brain regions in the latter stages of the disease (Braak *et al.*, 2007). Given that α-synuclein-immunoreactive astrocytes and astrocytes positive for other disease markers (tau and ubiquitin inclusion bodies) have been found in the amygdala of non-Parkinson's patients who were diagnosed with diffuse neurofibrillary tangles with calcification (DNTC) dementia (Hashimoto *et al.*, 2003; Yokota *et al.*, 2002), this disease may share pathophysiological mechanisms with Parkinson's and other proteinopathies (Yokota *et al.*, 2002; also see Jellinger, 2008).

Glia also appear to be affected in Alzheimer's disease (AD), and astrogliosis in the amygdala has been reported in many AD cases (Brockhaus, 1938; Corsellis, 1970). Scott *et al.* (1992) investigated the cortical and basal amygdala subregions in brains of AD patients postmortem. They counted glia, dividing them into two categories: large, which the authors presumed to be astrocytes, and small, presumed to be oligodendrocytes and microglia, based on size. When examining glial cell density (cells per $mm^2$), they found that the overall glia density was greater in AD cases in both the basal and cortical regions. However, this was attributable mostly to increased density of large glia. When the data were corrected for structural atrophy, total numbers of glia were *reduced* in AD cases in both size categories in the basal amygdala. In the cortical region, only the number of small glia was reduced (Scott *et al.*, 1992). Additional reports found an increase in the number of astrocytes expressing peroxiredoxin 6, an antioxidant enzyme, in the amygdala of AD cases compared to controls. Peroxiredoxin 6 staining, which is not found in oligodendrocytes or microglia and only at very low levels in neurons, was found mostly in astrocytes associated with amyloid-$\beta$ plaques, suggesting the plaques produce reactive oxygen species via astrocytes (Power *et al.*, 2008). Clearance of plaques in AD may not be the responsibility solely of astrocytes, however. Kaku *et al.* (2003) found that a mutant mouse (osteopetrotic mice), known to have reduced number of microglia, also have greater fibrillary plaques in the amygdala and other regions compared to controls, suggesting microglia regulate plaque formation and/or clearance. Additional reports confirm that activated microglia follow plaque formation in the amygdala, and activated astrocytes form around these plaques later on (Dudal *et al.*, 2004).

One difficulty in measuring astrocyte proliferation in any age-related disease such as AD is the well-documented increase in GFAP production as the brain ages (Kohmana *et al.*, 1995; Linnemann and Skarsfelt, 1994; Schipper, 1996). Interestingly, this increase in GFAP can be attenuated in several brain regions, including the amygdala, by systemic administration of the steroid pregnenolone in aging animals (Legrand and Alonso, 1998). This result implies that reduced hormone levels during aging may cause widespread increases in GFAP production, and this may contribute to the increased incidence of many neural disorders with age. This finding might not be surprising because pregnenolone is a precursor to several other steroid hormones, and glia are known to be highly sensitive to steroid hormones as discussed below.

## D. Other pathologies

In addition to what is known about glia in epilepsy, depression, and various proteinopathies, glia also figure into other areas of disease and disorder research. For example, *myo*-inositol, a glial marker and second messenger in intracellular calcium regulation, is reduced in the amygdala of narcoleptic patients compared to controls, suggesting glia involvement in sleep disorders (Poryazova *et al.*, 2009). Furthermore, Yokota *et al.* (2008) found several interesting glial abnormalities in the amygdala during their investigation of cases of neurofibromatosis type 1, a disease in which nervous tissue grows to form tumors. Yokota and colleagues conclude that the glial clusters and satellitosis may be due to altered astrocyte growth regulation. Finally, interleukin-$\beta$, which is released during infection and is responsible for inducing fever and behavioral changes, promotes the phosphorylation of Erk1/2 in astrocytes in the amygdala and other regions (Nadjar *et al.*, 2005).

These reports suggest that glia in the amygdala may be involved in, or at least be a sensitive indicator of, a broad range diseases and disorders. Due to their sensitivity and prominent role in maintaining brain homeostasis, the use of glia as indicators of neurological trauma has great potential for disease research. Furthermore, given their role in the formation of synapses and facilitation of neuronal communication, exploration of glia in disease models may lead to novel treatments focusing on restoring lost communication pathways.

## IV. Amygdala Glia and Hormones

Klüver and Bucy (1939) were perhaps the first to identify the amygdala's involvement in sexual behavior. Given that sex is intimately linked with gonadal hormones, neuroendocrinologists have been investigating how gonadal hormones shape this region both in adulthood and during development. This line of research has revealed an array of sexual dimorphisms within the amygdala (reviewed in Hamann, 2005; Stefanova and Ovtscharoff, 2000) and these sex differences have important implications beyond sexual behavior, as most of the diseases associated with the amygdala, such as depression and autism, exhibit strong sex biases in the population. While it is abundantly clear that gonadal hormones are key components in the early establishment of sex differences and of CNS plasticity, the mechanisms behind hormone-induced alterations in neuroanatomy are still not clear. Glia are responsive to gonadal hormones (Mong *et al.*, 1999, 2001) and investigators are now asking how glia may be involved in the ontogeny of sex differences, and may mediate hormone-induced plasticity in adulthood. This line of questioning is beginning to provide a much more detailed and rich picture of the amygdala.

The evidence for gonadal hormone influence on neurons is well established (reviewed in Woolley, 2007). However, glia also respond to gonadal hormones signals. Estrogen receptors are upregulated in reactive astrocytes after injury (Blurton-Jones and Tuszynski, 2001) and estrogen influences astrocyte morphology in the hypothalamus (Mong *et al.*, 1996), hippocampus (Milner *et al.*, 2000), and in primary culture (Garcia-Segura *et al.*, 1989). Estrogen appears to influence glia in the amygdala as well. The scaffold protein, MNAR/PELP1, is important coactivator in estrogen's nongenomic activity and has been found in glia in the amygdala (Khan *et al.*, 2006), suggesting glia may respond directly to estrogen through nongenomic means.

### A. Gonadal hormones and amygdala glia during developmental

Since the initial discovery gonadal hormone-sensitive glia, neuroendocrine research has begun to explore glia–hormone interactions across the lifespan, and evidence suggests that hormones may influence glia in the amygdala early in development. For example, a single dose of estrogen during neonatal development increased the number of proliferating glia in the basolateral nucleus 3 days later in male rats but not in females (Dmitar *et al.*, 1995). A slightly more complex pattern of results was found in the medial, cortical, and central nuclei. In estrogen-treated male rat pups, the percentage of BrdU-labeled glia was increased in all three nuclei, based upon light microscopy identification, whereas in females, estrogen increased the number of labeled glia only in the medial nucleus, while decreasing their numbers in the cortical nucleus and having no effect in the central nucleus (Drekic *et al.*, 1995). Glia also appear responsive to hormones during puberty, another critical period in development. Using BrdU labeling, Ahmed *et al.* (2008) demonstrated that new cells, including GFAP-positive cells, are added to the medial amygdala during puberty in rats. This pubertal addition was sexually dimorphic with males having more BrdU-positive cells in the medial amygdala than females, a finding that fits with the larger adult volume of the male medial amygdala in rats. Interestingly, the addition of BrdU-positive cells in males was eliminated by prepubertal castration, strongly suggesting that the addition of new cells in the amygdala during puberty, including GFAP-positive cells, is modulated by gonadal hormones. Thus, while gliogenesis in the amygdala appears to be regulated by gonadal hormones during development, the level of responsiveness is both sex- and subregion-specific. This specificity is also found in the adult amygdala in response to gonadal hormones.

## B. Gonadal hormones and amygdala glia during adulthood

In addition to influencing amygdala glia during development, several reports have documented glial sensitivity to fluctuations in adult hormone levels within the amygdala of adult animals. For example, Blutstein *et al.* (2006) report that estradiol treatment of mice produces an increase in glutamine synthase gene expression in the medial amygdala of adults, an interesting effect given that neurons are incapable of synthesizing glutamate and are dependent upon glia to replenish their glutamine supply. Additionally, treatment of meadow voles with either testosterone propionate or estradiol benzoate, but not 5alpha-dihydrotestosterone, resulted in a significant increase in BrdU-labeled cells, 35% of which were glia, in the amygdala but not in the dentate gyrus or ventromedial hypothalamus (VMH) (Fowler *et al.*, 2003). Interestingly, estradiol benzoate seems to have this effect in meadow voles but not in prairie voles (Fowler *et al.*, 2005). The presence of an effect on glia in the amygdala, a region closely linked to sexual behavior, in a polygamous rodent species (meadow voles) but not in the monogamous rodent species (prairie voles), suggest hormone-dependent changes in the number of amygdala glia may be involved in reproductive behaviors, which may contribute to changes in female mating behavior.

The medial amygdala is involved in mating behavior in several species and glia within this region seem particularly responsive to changes in adult hormone levels. For example, multiparous females given postpartum contact with pups exhibit reduced numbers of GFAP-positive cells in the medial amygdala compared to pup-exposed primiparous females (Featherstone *et al.*, 2000), a difference which may be due to greater estrogen exposure in postpartum dams. In addition, the density of GFAP-immunoreactivity is higher in the medial subregion during the proestrus phase compared to the other phases of the estrous cycle in rats (Martinez *et al.*, 2006), and in ovariectomized females, injection of estradiol, alone or with progesterone, increased GFAP-immunoreactive density in portions of the medial amygdala (Martinez *et al.*, 2006).

Similarly, testosterone increases glial cell numbers (based on their appearance in Nissl staining) in the medial posterodorsal amygdala (MePD) of ovariectomized adult females compared to ovariectomized controls (Morris *et al.*, 2008), a pattern of results strongly suggesting estrogenic metabolites of testosterone may lead to glial proliferation in some parts of the adult medial amygdala. In the anteroventral medial amygdala (MePV), by contrast, Carrillo *et al.* (2007) found no changes in glial number across the estrous cycle, despite changes in volume and other elements. Thus, even within subregions of the medial amygdala (MePD compared to MePV), glia may vary in their responsiveness to changes in steroid hormone levels.

In addition to hormone sensitivity in adulthood, dramatic sex differences can be found in adult amygdala glia. For example, total glia numbers based on Nissl staining are sexually dimorphic in the MePD of adult rats with males having more glia than females (Morris *et al.*, 2008). GFAP-ir density measures also suggest a sex difference in astrocytes within the MePD and MePV regions with females having greater GFAP-ir density than males (Rasia-Filho *et al.*, 2002), which may be related to the smaller volume of this nucleus in females.

Recent work in our lab has further explored sex differences in astrocytes and the role gonadal hormones play in shaping these cells. Using unbiased stereology, we found that male rats have more astrocytes than females in the MePD and that testicular feminized mutant (TFM) males, which have dysfunctional androgen receptors, were feminine in this regard. These results suggest that MePD astrocytes are dependent upon functional androgen receptors to reach masculine numbers (Johnson *et al.*, 2008). Interestingly, this sex difference was found only in the right hemisphere, suggesting androgens may act on MePD astrocytes in a hemisphere-dependent manner. In the same groups of animals (wild-type males, females, and TFM males), we also found that MePD astroctyes in males have more complex arbors than do MePD astrocytes in either females or TFM males, but in this case, only in the left hemisphere. Again, these results suggest an androgen receptor-dependent masculinization of MePD astrocyte morphology in a hemisphere-dependent manner.

Of course, gonadal hormones can influence astrocytes through several possible pathways. In the VMH, data suggest that estrogens influence the complexity of astrocyte arbors via indirect actions through neurons (Mong *et al.*, 2002). Neurons in the amygdala, especially the medial regions, are rich in steroid hormone receptors, suggesting that a similar indirect pathway may occur in the MePD. Alternatively, steroid hormones may act directly on glial cells. Astrocytes are known to express both androgen and estrogen receptors, although this is highly region- and species-specific (Azcoitia *et al.*, 1999; DonCarlos *et al.*, 2006; Finley and Kritzer, 1999). However, as far as we are aware, no report has confirmed the presence of either estrogen or androgen receptors in amygdala astrocytes *in vivo*, leaving one to speculate that the majority of hormone effects on astrocytes in the amygdala are through indirect pathways—that steroids directly affect neurons, which then affect astrocytes. It is important to note that the lack of a direct pathway for hormone action does not diminish the importance of investigating steroid hormone influence on glia. If anything, an indirect mechanism, with gonadal hormones affecting neurons which then affect astrocytes, offers a means of studying neuron–astrocyte interactions. Although the mechanisms are unclear, these reports clearly indicate that hormones are involved in the modulation of amygdala glia throughout the lifespan and only continued investigation will lead to an understanding of the mechanism through which such modulation may influence amygdala-based behaviors.

## C. Other agents and amygdala glia

In addition to gonadal hormones, glia in the amygdala are responsive to a broad range of nongonadal hormones, neurotransmitters and toxins. For example, Zhang *et al.* (2009) found that a tryptophan restrictive diet resulted in increased astrocyte arbor size and branching in the mouse amygdala, suggesting sensitivity to serotonin. Given that serotonin receptors have been found in GFAP-positive cells in the amygdala of rats (Xu and Pandey, 2000), serotonin may influence astrocyte morphology directly.

In addition, Glass *et al.* (2002) noted that although adrenergic receptors are mostly found on neuronal processes, some are present on glial processes in the rat central amygdala. Similarly, although principally neuron-based, glucocorticoid receptors have also been identified in glial processes of the lateral amygdala in rats (Johnson *et al.*, 2005), and Banisadr *et al.* (2002) note that CCR2, the receptor for chemokine monocyte chemoattractant protein-1 (MCP-1/CCL2) is found on glia in the amygdala, suggesting these cells are involved in the immunological response at inflammation sites.

Other reports note intense glial reactivity in the amygdala in response to soman, a potent and destructive nerve agent (Collombet *et al.*, 2005a,b). A second study demonstrates that some regrowth/migration of cells appears to occur in the amygdala following soman poisoning and that treatment with cytokines causes these cells to differentiate into astrocytes (Collombet *et al.*, 2005a,b). Additionally, the hyperalgesic effects of bradykinin administration into the amygdala can be blocked by the glial metabolic inhibitor fluorocitrate among other agents (Dalmolin *et al.*, 2007), suggesting glia metabolism contributes to the hyperalgesia. Finally, while we know that glia are involved in the formation of interamygdala connections in the rat brain during early development (Cooke and Simerly, 2005), glial cell division in the amygdala is also very sensitive during development in some species. Kent and Harman (1998) found that slight elevations in body temperature in developing Wallabies result in reduced glial cell division in the amygdala. Although it is not clear what the behavioral consequences of these early changes are, given other work in the field, they may lead to lasting psychosocial changes in the adult animals.

Collectively, these studies hint at the involvement of glia in numerous processes beyond the well-known diseases, disorders and hormone-based findings in which they are currently being aggressively investigated. Perhaps with continued investigation, it will become clear how these results can be integrated into a better understanding of the amygdala and the dynamic cells within it.

## V. Conclusions

While the above findings present a scattering of evidence that make a cohesive report of glia function in the amygdala difficult to portray, it is clear that glia in general have offered a wealth of intriguing and informative findings within the past several decades. Several excellent reviews have summarized glia-based findings in various brain regions and in the brain as a whole (Barres, 2008; Eng, Ghirnikar and Lee, 2000; Freeman and Doherty, 2006; Garcia-Segura and Melcangi, 2006). In this review, we have gathered findings regarding glia in the amygdala and it is our hope that this collection of findings will promote continued investigation of glia in this critical brain region. A deeper understanding of glia will be informative for measures of amygdala functionality such as fMRI, a technique closely linked to astrocyte activity (Schummers *et al.*, 2008), which is often used to investigate this region. As this review suggest, it is clear that these cells respond and develop in a highly region specific manner, making further study of glia a necessity. Likewise, the sensitivity and plasticity of glia may make them abundantly useful in diagnosis and treatment of various disorders and diseases, especially those exhibiting strong sex biases in the population. Glia are clearly involved in several disorders and diseases and the amygdala's prominent sex differences suggest glia in this region may be linked to sex differences in the vulnerability to disorders such as depression, schizophrenia, anxiety disorders, and autism.

## ACKNOWLEDGMENTS

Supported by NIH grants NS28421 and NS045195.

## REFERENCES

Ackermann, G. E., Marenholz, I., Wolfer, D. P., Chan, W. Y., Schäfer, B., Erne, P., and Heizmann, C. W. (2006). S100A1-deficient male mice exhibit increased exploratory activity and reduced anxiety-related responses. *Biochim. Biophys. Acta* **1763,** 1307–1319.

Adolphs, R., Gosselin, F., Buchanan, T. W., Tranel, D., Schyns, P., and Damasio, A. R. (2005). A mechanism for impaired fear recognition after amygdala damage. *Nature* **433,** 68–72.

Ahmed, E. I., Zehr, J. L., Schulz, K. M., Lorenz, B. H., DonCarlos, L. L., and Sisk, C. L. (2008). Pubertal hormones modulate the addition of new cells to sexually dimorphic brain regions. *Nat. Neurosci.* **11,** 995–997.

Alheid, G. F., Beltramino, C. A., De Olmos, J. S., Forbes, M. S., Swanson, D. J., and Heimer, L. (1995). The neuronal organization of the supracapsular part of the stria terminalis in the rat: The dorsal component of the extended amygdala. *Neuroscience* **84,** 967–996.

Armario, A., Gavalda, A., and Marti, J. (1995). Comparison of the behavioural and endocrine response to forced swimming stress in five inbred strains of rats. *Psychoneuroendocrinology* **20,** 879–890.

Azcoitia, I., Sierra, A., and García-Segura, L. M. (1999). Localization of estrogen receptor beta-immunoreactivity in astrocytes of the adult rat brain. *Glia* **26,** 260–267.

Banisadr, G., Quéraud-Lesaux, F., Boutterin, M. C., Pélaprat, D., Zalc, B., Rostène, W., Haour, F., and Parsadaniantz, S. M. (2002). Distribution, cellular localization and functional role of CCR2 chemokine receptors in adult rat brain. *J. Neurochem.* **81,** 257–269.

Barres, B. A. (2008). The mystery and magic of glia: A perspective on their roles in health and disease. *Neuron* **60,** 430–440.

Bechara, A., Tranel, D., Damasio, H., Adolphs, R., Rockland, C., and Damasio, A. R. (1995). Double dissociation of conditioning and declarative knowledge relative to the amygdala and hippocampus in humans. *Science* **25,** 1115–1118.

Bishop, S. J., Duncan, J., and Lawrence, A. D. (2004). State anxiety modulation of the amygdala response to unattended threat-related stimuli. *J. Neurosci.* **17,** 10364–10368.

Blurton-Jones, M., and Tuszynski, M. H. (2001). Reactive astrocytes express estrogen receptors in the injured primate brain. *J. Comp. Neurol.* **433,** 115–123.

Blutstein, T., Devidze, N., Choleris, E., Jasnow, A. M., Pfaff, D. W., and Mong, J. A. (2006). Oestradiol up-regulates glutamine synthetase mRNA and protein expression in the hypothalamus and hippocampus: Implications for a role of hormonally responsive glia in amino acid neurotransmission. *J. Neuroendocrinol.* **18,** 692–702.

Bowley, M. P., Drevets, W. C., Ongur, D., and Price, J. L. (2002). Low glial numbers in the amygdala in major depressive disorder. *Biol. Psychiatry* **52,** 404–412.

Braak, H., Sastre, M., and Del Tredici, K. (2007). Development of alpha-synuclein immunoreactive astrocytes in the forebrain parallels stages of intraneuronal pathology in sporadic Parkinson's disease. *Acta Neuropathol.* **114,** 231–241.

Brockhaus, H. (1938). Zur Normalen und Pathologischen Anatomic des Mandelkerngebietes. *J. Psychol. Neurol.* **49,** 1–136.

Cahill, L. (2000). Neurobiological mechanisms of emotionally influenced, long-term memory. *Prog Brain Res.* **126,** 29–37, Review.

Cahill, L., Uncapher, M., Kilpatrick, L., Alkire, M. T., and Turner, J. (2004). Sex-related hemispheric lateralization of amygdala function in emotionally influenced memory: An FMRI investigation. *Learn. Mem.* **11,** 261–266.

Cajal, S. R. (1909). Histologie du Systeme Nerveuz de l'Homme et des Vertebres. A Maloine, Paris.

Cardinal, R. N., Parkinson, J. A., Hall, J., and Everitt, B. J. (2002). Emotion and motivation: The role of the amygdala, ventral striatum, and prefrontal cortex. *Neurosci. Biobehav. Rev.* **26,** 321–352, Review.

Carrillo, B., Pinos, H., Guillamón, A., Panzica, G., and Collado, P. (2007). Morphometrical and neurochemical changes in the anteroventral subdivision of the rat medial amygdala during estrous cycle. *Brain Res.* **1150,** 83–93.

Collombet, J. M., Four, E., Bernabé, D., Masqueliez, C., Burckhart, M. F., Baille, V., Baubichon, D., and Lallement, G. (2005a). Soman poisoning increases neural progenitor proliferation and induces long-term glial activation in mouse brain. *Toxicology* **208,** 319–334.

Collombet, J. M., Four, E., Burckhart, M. F., Masqueliez, C., Bernabé, D., Baubichon, D., Hérodin, F., and Lallement, G. (2005b). Effect of cytokine treatment on the neurogenesis process in the brain of soman-poisoned mice. *Toxicology* **210,** 9–23.

Cooke, B. M. (2006). Steroid-dependent plasticity in the medial amygdala. *Neuroscience* **138,** 997–1005, Review.

Cooke, B. M., and Simerly, R. B. (2005). Ontogeny of bidirectional connections between the medial nucleus of the amygdala and the principal bed nucleus of the stria terminalis in the rat. *J. Comp. Neurol.* **489,** 42–58.

Cooke, B. M., and Woolley, C. S. (2005). Sexually dimorphic synaptic organization of the medial amygdala. *J. Neurosci.* **16,** 10759–10767.

Cooke, B. M., Tabibnia, G., and Breedlove, S. M. (1999). A brain sexual dimorphism controlled by adult circulating androgens. *Proc. Natl. Acad. Sci. USA* **22,** 7538–7540.

Cooke, B. M., Breedlove, S. M., and Jordan, C. L. (2003). Both estrogen receptors and androgen receptors contribute to testosterone-induced changes in the morphology of the medial amygdala and sexual arousal in male rats. *Horm. Behav.* **43,** 336–346.

Corsellis, C. L. (1970). The limbic areas in Alzheimer's disease and in other conditions associated with dementia. *In* Alzheimer's disease and related conditions, (G. E. W. Wolstenholme, Ed.), pp. 37–50. A Ciba Foundation symposium, London: Churchill.

Cotter, D. R., Pariante, C. M., and Everall, I. P. (2001). Glial cell abnormalities in major psychiatric disorders: The evidence and implications. *Brain Res. Bull.* **55,** 585–595, Review.

Dalmolin, G. D., Silva, C. R., Bellé, N. A., Rubin, M. A., Mello, C. F., Calixto, J. B., and Ferreira, J. (2007). Bradykinin into amygdala induces thermal hyperalgesia in rats. *Neuropeptides* **41,** 263–270.

De La Garza, R., and Mahoney, J. J. (2004). A distinct neurochemical profile in WKY rats at baseline and in response to acute stress: implications for animal models of anxiety and depression. *Brain Res.* **1021,** 209–218.

Dmitar, D., Slobodan, M., Bojan, S., Cvetković, D., and Lozance, O. (1995). Study of neurons and glial cells of basolateral amygdala in male and female rats neonatally treated with estrogen. *Int. J. Neurosci.* **83,** 145–151.

DonCarlos, L. L., Sarkey, S., Lorenz, B., Azcoitia, I., Garcia-Ovejero, D., Huppenbauer, C., and Garcia-Segura, L. M. (2006). Novel cellular phenotypes and subcellular sites for androgen action in the forebrain. *Neuroscience* **138,** 801–807, Review.

Dudal, S., Morissette, C., Lacombe, D., Tremblay, P., and Gervais, F. (2004). Differences in the amyloid-beta-induced inflammatory response in microglia from C57BL/6 and A/J strains of mice. *J. Neuroimmunol.* **153,** 26–35.

Drekic, D., Malobabic, S., Gledic, D., and Cvetkovic, D. (1995). Different neuronal and glial cell groups in corticomedial amygdala react differently to neonatally administered estrogen. *Neuroscience* **66,** 475–481.

Dyck, R. H., Bogoch, I. I., Marks, A., Melvin, N. R., and Teskey, G. C. (2002). Enhanced epileptogenesis in S100B knockout mice. *Brain Res. Mol. Brain Res.* **106,** 22–29.

Eddleston, M., and Mucke, L. (1993). Molecular profile of reactive astrocytes; implications for their role in neurologic diseases. *Neurosci.* **54,** 15–36.

Elmariah, S. B., Hughes, E. G., Oh, E. J., and Balice-Gordon, R. J. (2005). Neurotrophin signaling among neurons and glia during formation of tripartite synapses. *Neuron Glia Biol.* **1,** 1–11.

Eng, L., and Ghirnikar, R. S. (1994). GFAP and astrogliosis. *Brain Pathol.* **4.**

Eng, L. F., Ghirnikar, R. S., and Lee, Y. L. (2000). Glial fibrillary acidic protein: GFAP-thirty-one years (1969–2000). *Neurochem. Res.* **25,** 1439–1451.

Faber-Zuschratter, H., Hüttmann, K., Steinhäuser, C., Becker, A., Schramm, J., Okafo, U., Shanley, D., and Yilmazer-Hanke, D. M. (2009). Ultrastructural and functional characterization of satellitosis in the human lateral amygdala associated with Ammon's horn sclerosis. *Acta Neuropathol.* **117,** 545–555.

Featherstone, R. E., Fleming, A. S., and Ivy, G. O. (2000). Plasticity in the maternal circuit: Effects of experience and partum condition on brain astrocyte number in female rats. *Behav. Neurosci.* **114,** 158–172.

Finley, S. K., and Kritzer, M. F. (1999). Immunoreactivity for intracellular androgen receptors in identified subpopulations of neurons, astrocytes and oligodendrocytes in primate prefrontal cortex. *J. Neurobiol.* **40,** 446–457.

Fowler, C. D., Freeman, M. E., and Wang, Z. (2003). Newly proliferated cells in the adult male amygdala are affected by gonadal steroid hormones. *J. Neurobiol.* **57,** 257–269.

Fowler, C. D., Johnson, F., and Wang, Z. (2005). Estrogen regulation of cell proliferation and distribution of estrogen receptor-alpha in the brains of adult female prairie and meadow voles. *J. Comp. Neurol.* **489,** 166–179.

Freeman, M. R., and Doherty, J. (2006). Glial cell biology in Drosophila and vertebrates. *Trends Neurosci.* **29,** 82–90, Review.

Garcia-Segura, L. M., and Melcangi, R. C. (2006). Steroids and glial cell function. *Glia* **54,** 485–498.

Garcia-Segura, L. M., Torres-Aleman, I., and Naftolin, F. (1989). Astrocytic shape and glial fibrillary acidic protein immunoreactivity are modified by estradiol in primary rat hypothalamic cultures. *Brain Res. Dev. Brain Res.* **47,** 298–302.

Giedd, J. N., Castellanos, F. X., Rajapakse, J. C., Vaituzis, A. C., and Rapoport, J. L. (1997). Sexual dimorphism of the developing human brain. *Prog. Neuropsychopharmacol. Biol. Psychiatry* **21,** 1185–1201.

Glass, M. J., Colago, E. E., and Pickel, V. M. (2002). Alpha-2A-adrenergic receptors are present on neurons in the central nucleus of the amygdala that project to the dorsal vagal complex in the rat. *Synapse* **46,** 258–268.

Golgi, C. (1885). Sulla fina anatomia degli organi centrali del sistema nervoso. *Riv. Sper. Fremiat. Med. Leg. Alienazioni Ment.* **11,** 72–123.

Gosselin, R. D., Gibney, S., O'Malley, D., Dinan, T. G., and Cryan, J. F. (2009). Region specific decrease in glial fibrillary acidic protein immunoreactivity in the brain of a rat model of depression. *Neuroscience* **159,** 915–925.

Grootendorst, J., Mulder, M., Haasdijk, E., de Kloet, E. R., and Jaarsma, D. (2000). Presence of apolipoprotein E immunoreactivity in degenerating neurones of mice is dependent on the severity of kainic acid-induced lesion. *Brain Res.* **868,** 165–175.

Grove, E. A. (1988a). Efferent connections of the substantia innominata in the rat. *J. Comp. Neurol.* **277,** 347–364.

Grove, E. A. (1988b). Neural associations of the substantia innominata in the rat: Afferent connections. *J. Comp. Neurol.* **277,** 315–346.

Guthrie, P. B., Knappenberger, J., Segal, M., Bennett, M. V., Charles, A. C., and Kater, S. B. (1999). ATP released from astrocytes mediates glial calcium waves. *J. Neurosci.* **19,** 520–528.

Hachem, S., Aguirre, A., Vives, V., Marks, A., Gallo, V., and Legraverend, C. (2005). Spatial and temporal expression of S100B in cells of oligodendrocyte lineage. *Glia* **51,** 81–97.

Halassa, M. M., Fellin, T., Takano, H., Dong, J. H., and Haydon, P. G. (2007). Synaptic islands defined by the territory of a single astrocyte. *J. Neurosci.* **27,** 6473–6477.

Hamann, S. (2005). Sex differences in the responses of the human amygdala. *Neuroscientist* **11,** 288–293, Review.

Hamidi, M., Drevets, W. C., and Price, J. L. (2004). Glial reduction in amygdala in major depressive disorder is due to oligodendrocytes. *Biol. Psychiatry* **55,** 563–569.

Hashimoto, N., Takeuchi, T., Ishihara, R., Ukai, K., Kobayashi, H., Iwata, H., Iwai, K., Mizuno, Y., Yamaguchi, H., and Shibayama, H. (2003). Glial fibrillary tangles in diffuse neurofibrillary tangles with calcification. *Acta Neuropathol.* **106,** 150–156.

Hatton, G. I. (2002). Glial-neuronal interactions in the mammalian brain. *Adv. Physiol. Educ.* **26,** 225–237, Review.

Hercher, C., Turecki, G., and Mechawar, N. (2009). Through the looking glass: Examining neuroanatomical evidence for cellular alterations in major depression. *J. Psychiatr Res.* **43,** 947–961.

Hines, M., Allen, L. S., and Gorski, R. A. (1992). Sex differences in subregions of the medial nucleus of the amygdala and the bed nucleus of the stria terminalis of the rat. *Brain Res.* **8,** 321–326.

Ino, H., and Chiba, T. (2001). Cyclin-dependent kinase 4 and cyclin D1 are required for excitotoxin-induced neuronal cell death in vivo. *J. Neurosci.* **21,** 6086–6094.

Jellinger, K. A. (2008). Neuropathological aspects of Alzhemier disease, Parkinson disease and frontotemporal dementia. *Neurodegener Dis.* **5,** 118–121.

Johnson, L. R., Farb, C., Morrison, J. H., McEwen, B. S., and LeDoux, J. E. (2005). Localization of glucocorticoid receptors at postsynaptic membranes in the lateral amygdala. *Neuroscience* **136,** 289–299.

Johnson, R. T., Breedlove, S. M., and Jordan, C. L. (2008). Sex differences and laterality in astrocyte number and complexity in the adult rat medial amygdala. *J. Comp. Neurol.* **511,** 599–609.

Jung, K. H., Chu, K., Lee, S. T., Kim, J. H., Kang, K. M., Song, E. C., Kim, S. J., Park, H. K., Kim, M., Lee, S. K., and Roh, J. K. (2009). Region-specific plasticity in the epileptic rat brain: A hippocampal and extrahippocampal analysis. *Epilepsia* **50,** 537–549.

Kaku, M., Tsutsui, K., Motokawa, M., Kawata, T., Fujita, T., Kohno, S., Tohma, Y., Ohtani, J., Tenjoh, K., and Tanne, K. (2003). Amyloid beta protein deposition and neuron loss in osteopetrotic (op/op) mice. *Brain Res. Brain Res. Protoc.* **12,** 104–108.

Kent, A. R., and Harman, A. M. (1998). The effects of a transient increase in temperature on cell generation and cell death in the hippocampus and amygdala of the wallaby, *Setonix brachyurus* (quokka). *Exp. Brain Res.* **122,** 301–308.

Khan, M. M., Hadman, M., De Sevilla, L. M., Mahesh, V. B., Buccafusco, J., Hill, W. D., and Brann, D. W. (2006). Cloning, distribution, and colocalization of MNAR/PELP1 with glucocorticoid receptors in primate and nonprimate brain. *Neuroendocrinology* **84,** 317–329.

Killgore, W. D., and Yurgelun-Todd, D. A. (2004). Sex-related developmental differences in the lateralized activation of the prefrontal cortex and amygdala during perception of facial affect. *Percept. Mot. Skills* **99,** 371–391.

Kilpatrick, L. A., Zald, D. H., Pardo, J. V., and Cahill, L. F. (2006). Sex-related differences in amygdala functional connectivity during resting conditions. *Neuroimage* **30,** 452–461.

Kim, D., Kim, E. H., Kim, C., Sun, W., Kim, H. J., Uhm, C. S., Park, S. H., and Kim, H. (2003). Differential regulation of metallothionein-I, II, and III mRNA expression in the rat brain following kainic acid treatment. *NeuroReport* **14,** 679–682.

Kinsley, C. H., and Lambert, K. G. (2008). Reproduction-induced neuroplasticity: Natural behavioural and neuronal alterations associated with the production and care of offspring. *J. Neuroendocrinol.* **20,** 515–525, Review.

Klüver, H., and Bucy, P. C. (1939). Preliminary analysis of functions of the temporal lobes in monkeys. *J. Neuropsychiatry Clin. Neurosci.* **9,** 606–620.

Kohmana, S. G., Goss, J. R., Finch, C. E., and McNeill, T. H. (1995). Increases of glial fibrillary acidic protein in the aging female mouse brain. *Neurobiol. Aging* **16,** 59–67.

Kraszpulski, M., Dickerson, P. A., and Salm, A. K. (2006). Prenatal stress affects the developmental trajectory of the rat amygdala. *Stress* **9,** 85–95.

LeDoux, J. (1998). Fear and the brain: Where have we been and where are we going? *Biol. Psychiatry* **44,** 1229–1238, Review.

Lee, Y., Gaskins, D., and Shekhar, A. (2007). Glia mechanisms in mood regulation: A novel model of mood disorders. *Psychopharmacology* **191,** 55–65.

Legrand, A., and Alonso, G. (1998). Pregnenolone reverses the age-dependent accumulation of glial fibrillary acidic protein within astrocytes of specific regions of the rat brain. *Brain Res.* **802,** 125–133.

Leventopoulos, M., Ruedi-Bettschen, D., Knuesel, I., Feldon, J., Pryce, C. R., and Opacka-Juffry, J. (2007). Long-term effects of early life deprivation on brain glia in Fischer rats. *Brain Res.* **1142,** 119–126.

Linnemann, D., and Skarsfelt, T. (1994). Regional changes in expression of NCAM, GFAP, and S100 in aging rat brain. *Neurobiol. Aging* **15,** 651–665.

Mah, L., Zarate, C. A. Jr., Singh, J., Duan, Y. F., Luckenbaugh, D. A., Manji, H. K., and Drevets, W. C. (2007). Regional cerebral glucose metabolic abnormalities in bipolar II depression. *Biol. Psychiatry* **61,** 765–775.

Martinez, F. G., Hermel, E. E., Xavier, L. L., Viola, G. G., Riboldi, J., Rasia-Filho, A. A., and Achaval, M. (2006). Gonadal hormone regulation of glial fibrillary acidic protein immunoreactivity in the medial amygdala subnuclei across the estrous cycle and in castrated and treated female rats. *Brain Res.* **1108,** 117–126.

Mena, M. A., and García de Yébenes, J. (2008). Glial cells as players in parkinsonism: The "good", the "bad", and the "mysterious" glia. *Neuroscientist.* **14**, 544–60, Review.

McDonald, A. J. (1998). Cortical pathways to the mammalian amygdala. *Prog. Neurobiol.* **55,** 257–332.

Milner, T. A., Shah, P., and Pierce, J. P. (2000). Beta-adrenergic receptors primarily are located on the dendrites of granule cells and interneurons but also are found on astrocytes and a few presynaptic profiles in the rat dentate gyrus. *Synapse* **36,** 178–193.

Mong, J. A., Glaser, E., and McCarthy, M. M. (1999). Gonadal steroids promote glial differentiation and alter neuronal morphology in the developing hypothalamus in a regionally specific manner. *J. Neurosci.* **19,** 1464–1472.

Mong, J. A., Kurzweil, R. L., Davis, A. M., Rocca, M. S., and McCarthy, M. M. (1996). Evidence for sexual differentiation of glia in rat brain. *Horm Behav.* **30,** 553–562.

Mong, J. A., Roberts, R. C., Kelly, J. J., and McCarthy, M. M. (2001). Gonadal steroids reduce the density of axospinous synapses in the developing rat arcuate nucleus: An electron microscopy analysis. *J. Comp. Neurol.* **432,** 259–267.

Mong, J. A., Nuñez, J. L., and McCarthy, M. M. (2002). GABA mediates steroid-induced astrocyte differentiation in the neonatal rat hypothalamus. *J. Neuroendocrinol.* **14,** 45–55.

Morris, J. A., Jordan, C. L., and Breedlove, S. M. (2008). Sexual dimorphism in neuronal number of the posterodorsal medial amygdala is independent of circulating androgens and regional volume in adult rats. *J. Comp. Neurol.* **506,** 851–859.

Mukaetova-Ladinska, E. B., and McKeith, I. G. (2006). Pathophysiology of synuclein aggregation in Lewy body disease. *Mech. Ageing Dev.* **127,** 188–202.

Mukhopadhyay, G., Doherty, P., Walsh, F. S., Crocker, P. R., and Filbin, M. T. (1994). A novel role for myelin-associated glycoprotein as an inhibitor of axonal regeneration. *Neuron* **13,** 757–767.

Nadjar, A., Combe, C., Busquet, P., Dantzer, R., and Parnet, P. (2005). Signaling pathways of interleukin-1 actions in the brain: Anatomical distribution of phospho-ERK1/2 in the brain of rat treated systemically with interleukin-1beta. *Neuroscience* **134,** 921–932.

Newton, S. S., Collier, E. F., Hunsberger, J., Adams, D., Terwilliger, R., Selvanayagam, E., and Duman, R. S. (2003). Gene profile of electroconvulsive seizures: Induction of neurotrophic and angiogenic factors. *J. Neurosci.* **23,** 10841–10851.

Nicoletti, J. N., Shah, S. K., McCloskey, D. P., Goodman, J. H., Elkady, A., Atassi, H., Hylton, D., Rudge, J. S., Scharfman, H. E., and Croll, S. D. (2008). Vascular endothelial growth factor is up-regulated after status epilepticus and protects against seizure-induced neuronal loss in hippocampus. *Neuroscience* **151,** 232–241.

Nishida, H., and Okabe, S. (2007). Direct astrocytic contacts regulate local maturation of dendritic spines. *J. Neurosci.* **27,** 331–340.

Orre, K., Wennström, M., and Tingström, A. (2009). Chronic lithium treatment decreases NG2 cell proliferation in rat dentate hilus, amygdala and corpus callosum. *Prog. Neuropsychopharmacol. Biol. Psychiatry* **33,** 503–510.

Pfrieger, F. W., and Barres, B. A. (1997). Synaptic efficacy enhanced by glial cells in vitro. *Science* **277,** 1684–1687.

Poryazova, R., Schnepf, B., Werth, E., Khatami, R., Dydak, U., Meier, D., Boesiger, P., and Bassetti, C. L. (2009). Evidence for metabolic hypothalamo-amygdala dysfunction in narcolepsy. *Sleep* **32,** 607–613.

Power, J. H., Asad, S., Chataway, T. K., Chegini, F., Manavis, J., Temlett, J. A., Jensen, P. H., Blumbergs, P. C., and Gai, W. P. (2008). Peroxiredoxin 6 in human brain: Molecular forms, cellular distribution and association with Alzheimer's disease pathology. *Acta Neuropathol.* **115,** 611–622.

Rasia-Filho, A. A., Xavier, L. L., dos Santos, P., Gehlen, G., and Achaval, M. (2002). Glial fibrillary acidic protein immunodetection and immunoreactivity in the anterior and posterior medial amygdala of male and female rats. *Brain Res. Bull.* **58,** 67–75.

Rees, H. D., and Michael, R. P. (1984). Estrogen target neurons in the forebrain of the cat during postnatal development. *Brain Res.* **315,** 73–79.

Rogan, M. T., and LeDoux, J. E. (1995). LTP is accompanied by commensurate enhancement of auditory-evoked responses in a fear conditioning circuit. *Neuron* **15,** 127–136.

Rubinow, M. J., and Juraska, J. M. (2009). Neuron and glia numbers in the basolateral nucleus of the amygdala from preweaning through old age in male and female rats: A stereological study. *J. Comp. Neurol.* **512,** 717–725.

Sah, P., Westbrook, R. F., and Lüthi, A. (2008). Fear conditioning and long-term potentiation in the amygdala: What really is the connection? *Ann. NY Acad. Sci.* **1129,** 88–95, Review.

Samson, R. D., Duvarci, S., and Pare, D. (2005). Synaptic Plasticity in the central nucleus of the amygdala. *Rev. Neurosci.* **16,** 287–302.

Schipper, H. M. (1996). Astrocytes, brain aging, and neurodegeneration. *Neurobiol. Aging* **17,** 467–480.

Schummers, J., Yu, H., and Sur, M. (2008). Tuned responses of astrocytes and their influence on hemodynamic signals in the visual cortex. *Science* **320,** 1628–1643.

Scott, S. A., Dekosky, S. T., Sparks, D. L., Knox, C. A., and Scheff, S. W. (1992). Amygdala cell loss and atrophy in Alzheimers-disease. *Ann. Neurol.* **32,** 555–563.

Shammah-Lagnado, S. J., Beltramino, C. A., McDonald, A. J., Miselis, R. R., Yang, M., de Olmos, J., Heimer, L., and Alheid, G. F. (2000). Supracapsular bed nucleus of the stria terminalis contains central and medial extended amygdala elements: Evidence from anterograde and retrograde tracing experiments in the rat. *J. Comp. Neurol.* **422,** 533–555.

Sigurdsson, T., Doyère, V., Cain, C. K., and LeDoux, J. E. (2007). Long-term potentiation in the amygdala: A cellular mechanism of fear learning and memory. *Neuropharmacology* **52,** 215–227, Review.

Simerly, R. B., Chang, C., Muramatsu, M., and Swanson, L. W. (1990). Distribution of androgen and estrogen receptor mRNA-containing cells in the rat brain: An in situ hybridization study. *J. Comp. Neurol.* **294,** 76–95.

Spezio, M. L., Huang, P. Y., Castelli, F., and Adolphs, R. (2007). Amygdala damage impairs eye contact during conversations with real people. *J. Neurosci.* **27,** 3994–3997.

Stefanova, N., and Ovtscharoff, W. (2000). Sexual dimorphism of the bed nucleus of the stria terminalis and the amygdala. *Adv. Anat. Embryol. Cell Biol.* **158,** 1–78, Review.

Terada, S., Ishizu, H., Yokota, O., Tsuchiya, K., Nakashima, H., Ishihara, T., Fujita, D., Uéda, K., Ikeda, K., and Kuroda, S. (2003). Glial involvement in diffuse Lewy body disease. *Acta Neuropathol.* **105,** 163–169.

Thippeswamy, T., McKay, J. S., Morris, R., Quinn, J., Wong, L. F., and Murphy, D. (2005). Glial-mediated neuroprotection: Evidence for the protective role of the NO-cGMP pathway via neuron-glial communication in the peripheral nervous system. *Glia* **49,** 197–210.

Truitt, W. A., Sajdyk, T. J., Dietrich, A. D., Oberlin, B., McDougle, C. J., and Shekhar, A. (2007). From anxiety to autism: Spectrum of abnormal social behaviors modeled by progressive disruption of inhibitory neuronal function in the basolateral amygdala in Wistar rats. *Psychopharmacology* **191,** 107–118.

Ullian, E. M., Sapperstein, S. K., Christopherson, K. S., and Barres, B. A. (2001). Control of synapse number by glia. *Science* **291,** 657–661.

Umeoka, S., Miyamoto, O., Nakagawa, T., Janjua, N. A., Nagao, S., and Itano, T. (2001). Expression of an embryonic intermediate filament protein in amygdaloid kindled rats. *Epilepsy Res.* **43,** 249–253.

Vessal, M., Dugani, C. B., Solomon, D. A., Burnham, W. M., and Ivy, G. O. (2004). Astrocytic proliferation in the piriform cortex of amygdala-kindled subjects: A quantitative study in partial versus fully kindled brains. *Brain Res.* **1022,** 47–53.

Vessal, M., Dugani, C. B., Solomon, D. A., McIntyre Burnham, W., and Ivy, G. O. (2005). Might astrocytes play a role in maintaining the seizure-prone state? *Brain Res.* **1044,** 190–196.

Wang, T. F., Zhou, C., Tang, A. H., Wang, S. Q., and Chai, Z. (2006). Cellular mechanism for spontaneous calcium oscillations in astrocytes. *Acta Pharmacol. Sin.* **27,** 861–868.

Wennström, M., Hellsten, J., and Tingström, A. (2004). Electroconvulsive seizures induce proliferation of NG2-expressing glial cells in adult rat amygdala. *Biol. Psychiatry* **55,** 464–471.

Williams, L. M., Kemp, A. H., Felmingham, K., Barton, M., Olivieri, G., Peduto, A., Gordon, E., and Bryant, R. A. (2006). Trauma modulates amygdala and medial prefrontal responses to consciously attended fear. *Neuroimage* **29,** 347–357.

Woldbye, D. P., Bolwig, T. G., Kragh, J., and Jørgensen, O. S. (1996). Synaptic degeneration and remodelling after fast kindling of the olfactory bulb. *Neurochem. Res.* **21,** 585–593.

Woolley, C. S. (2007). Acute effects of estrogen on neuronal physiology. *Annu. Rev. Pharmacol. Toxicol.* **47,** 657–680, Review.

Xu, T., and Pandey, S. C. (2000). Cellular localization of serotonin(2A) (5HT(2A)) receptors in the rat brain. *Brain Res. Bull.* **51,** 499–505.

Yamamuro, A., Ago, Y., Takuma, K., Maeda, S., Sakai, Y., Baba, A., and Matsuda, T. (2003). Possible involvement of astrocytes in neuroprotection by the cognitive enhancer T-588. *Neurochem. Res.* **28,** 1779–1783.

Yilmazer-Hanke, D. M., Faber-Zuschratter, H., Blümcke, I., Bickel, M., Becker, A., Mawrin, C., and Schramm, J. (2007). Axo-somatic inhibition of projection neurons in the lateral nucleus of amygdala in human temporal lobe epilepsy: An ultrastructural study. *Exp. Brain Res.* **177,** 384–399.

Yokota, O., Terada, S., Ishizu, H., Tsuchiya, K., Kitamura, Y., Ikeda, K., Uéda, K., and Kuroda, S. (2002). NACP/alpha-synuclein immunoreactivity in diffuse neurofibrillary tangles with calcification (DNTC). *Acta Neuropathol.* **104,** 333–341.

Yokota, O., Tsuchiya, K., Hayashi, M., Kakita, A., Ohwada, K., Ishizu, H., Takahashi, H., and Akiyama, H. (2008). Glial clusters and perineuronal glial satellitosis in the basal ganglia of neurofibromatosis type 1. *Acta Neuropathol.* **116,** 57–66.

Yu, S. Y., Wu, D. C., Liu, L., Ge, Y., and Wang, Y. T. (2008). Role of AMPA receptor trafficking in NMDA receptor-dependent synaptic plasticity in the rat lateral amygdala. *J. Neurochem.* **106,** 889–899.

Zhang, L., Corona-Morales, A. A., Vega-González, A., García-Estrada, J., and Escobar, A. (2009). Dietary tryptophan restriction in rats triggers astrocyte cytoskeletal hypertrophy in hippocampus and amygdala. *Neurosci. Lett.* **450,** 242–245.

CHAPTER THREE

# Limbic Effects of High-Frequency Stimulation of the Subthalamic Nucleus

Yasin Temel[*,†]

## Contents



## Abstract

The use of stimulation electrodes implanted in the brain to control severely disabling neurological and psychiatric conditions is a fast emerging area of clinical neuroscience. For instance, high-frequency stimulation of the subthalamic nucleus (STN) has become the surgical therapy of choice for advanced Parkinson's disease. This therapy improves motor disability substantially and also the quality of life, but some patients show postoperative behavioral changes such as depression and mania. These behavioral effects can be explained on the basis of the anatomical data. The STN is interconnected not only with motor areas, but also with associative and limbic regions. In this chapter, the author discusses relevant articles, provides anatomical details, and presents an integrated view. 

* Departments of Neuroscience and Neurosurgery, Maastricht University Medical Center, Maastricht, The Netherlands
† European Graduate School of Neuroscience (EURON), Maastricht, The Netherlands

*Vitamins and Hormones*, Volume 82
ISSN 0083-6729, DOI: 10.1016/S0083-6729(10)82003-5

# I. Introduction

The use of stimulation electrodes implanted in the brain to control severely disabling neurological and psychiatric conditions is an exciting and fast emerging area of clinical neuroscience (Wichmann and Delong, 2006). As an example, high-frequency stimulation (HFS) of the subthalamic nucleus (STN) has become the surgical therapy of choice for advanced Parkinson's disease (PD) and to date thousands of patients worldwide have benefited from this procedure. This therapy, also referred as "deep brain stimulation" (DBS), has long-lasting beneficial effects on the motor disability of PD patients, not only demonstrated with case series (Krack *et al.*, 2003; Rodriguez-Oroz *et al.*, 2005; Visser-Vandewalle *et al.*, 2005), but also with controlled trials (Deuschl *et al.*, 2006; Weaver *et al.*, 2009). The most substantial improvements are in rigidity, hypokinesia and tremor, and to a lesser extent in axial (including gait) and some autonomic symptoms. Another important advantage of DBS of the STN is that the anti-Parkinsonian medication (dopamine replacement therapy) can be reduced substantially and therefore medication-induced complications such as daily motor fluctuations (on–off fluctuations) are decreased significantly.

## A. Stereotaxy and HFS

The electrical stimuli are delivered continuously by an electrode placed in the STN by a neurosurgical technique called stereotaxy. With stereotaxy, it is possible to direct an electrode to an intracerebral target with extremely high precision without causing substantial damage to overlying structures. The patient receives a stereotactic frame mounted on his head and this allows to apply the three-dimensional Cartesian coordinate system, which is then used to define a target on a stereotactic CT or MRI image (Temel *et al.*, 2008) (Fig. 3.1).

The surgery for DBS of the STN is generally under local anesthesia (awake surgery) to allow intraoperative testing for the therapeutic- and side effects, although some surgeons prefer general anesthesia. The stimuli are produced by an internal pulse generator implanted either beneath the clavicle or in the abdomen subcutaneously or submuscularly, which is connected with the electrodes by internalized cables. In PD, only stimulations at higher frequencies are therapeutic, whereas low-frequency stimulation deteriorates the motor symptoms (Timmermann *et al.*, 2004).

DBS has generally the same clinical effects as a lesion with respect to the improvement of motor disability in movement disorders, but has more advantages such as its adjustability and reversibility (Benabid, 2003; Temel *et al.*, 2004a). Nevertheless, the cellular effects of DBS and lesions are

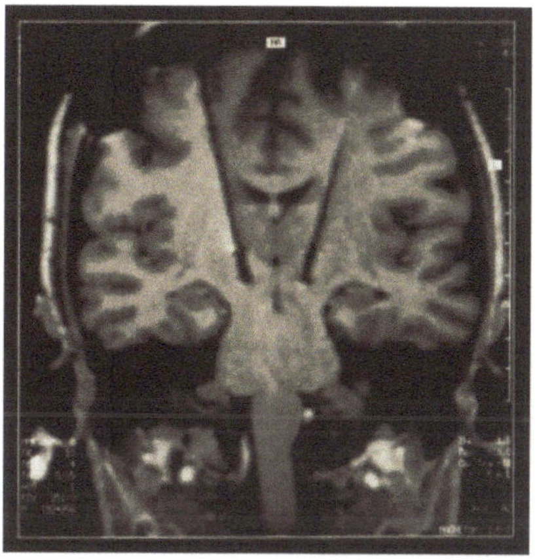

**Figure 3.1** T1-weighed postoperative magnetic resonance image of a patient with advanced PD who has received stimulating electrodes in the STN. This image illustrates how "deep" the electrodes are implanted in the brain for stimulation. The STN cannot be seen on a T1-weighed postoperative magnetic resonance image.

considerably different: a lesion destroys and DBS modulates the activity of neuronal elements (Dostrovsky and Lozano, 2002; Grill and McIntyre, 2001; Grill *et al.*, 2004; McIntyre *et al.*, 2004). To this day, fundamental knowledge regarding the application of electrical currents to deep brain structures is still not complete. A number of possible mechanisms have, however, been proposed (Dostrovsky and Lozano, 2002). One of the more popular hypotheses was that DBS caused a reduction of neuronal activity by means of a depolarization block. This proposed mechanism involved the suppression of voltage-gated sodium, and T-type calcium currents leading to an interruption of spontaneous activity within the neurons (Beurrier *et al.*, 2001). However, this theory is being abandoned. A second mechanism is that silencing of target nuclei by DBS is achieved by stimulation of GABAergic afferents to the target cells and the consequent hyperpolarization of postsynaptic terminals by release of the inhibitory neurotransmitter GABA (Moser *et al.*, 2003). Recordings made in the stimulated nucleus show inhibition or decreased activity during and after the stimulus train (Benazzouz *et al.*, 2004; Filali *et al.*, 2004; Tai *et al.*, 2003). However, electrical recordings from the efferent nuclei from the stimulated nucleus indicate that DBS can not only decrease but in certain conditions also increase the output of the stimulated nucleus (Hashimoto *et al.*, 2003; Vlamings *et al.*, 2009; Windels *et al.*, 2003). Quantitative models have revealed that DBS inhibits the cell bodies of neurons around the electrode by activation of presynaptic terminals, while stimulating the output of local neurons, by initiation of action potentials in the axon remote from the cell body (Grill and McIntyre, 2001; McIntyre *et al.*, 2004). Furthermore, this dual effect appeared with high frequencies only (Grill *et al.*, 2004). Another line of evidence suggests that HFS of the STN suppresses pathological oscillatory activity in the basal ganglia (Kuhn *et al.*, 2008). The exaggerated synchronization of basal ganglia neurons has been suggested to be responsible for some of the motor deficits in PD. In other words, lines of evidences are not entirely in agreement, but there is a more or less a generally accepted view that HFS of the STN decreases its neuronal activity.

## B. Limbic effects of high-frequency stimulation of the subthalamic nucleus

The STN is believed to be one of the main regulators of motor function through its fundamental role within the basal ganglia–thalamocortical motor circuit (Albin *et al.*, 1989; Alexander and Crutcher, 1990; Temel *et al.*, 2005a). As HFS of the STN has become a widely applied surgical procedure, more data has appeared on its nonmotor effects (Gervais-Bernard *et al.*, 2009; Hariz, 2002; Hariz and Johansson, 2001; Lyons *et al.*, 2001; Oh *et al.*, 2002; Voon *et al.*, 2008). Concerning the latter, clinicians observed behavioral effects such as stimulation-dependent cognitive alterations (Saint-Cyr *et al.*, 2000; Temel *et al.*, 2004b) and changes in emotional behavior (Berney *et al.*, 2002; Kulisevsky *et al.*, 2002; Sensi *et al.*, 2004). Basic researchers applying STN HFS in animal models to study the underlying neuronal mechanisms found clear-cut effects on cognitive and limbic functions (Baunez *et al.*, 2007; Desbonnet *et al.*, 2004; Klavir *et al.*, 2009; Temel *et al.*, 2005b). Anatomical tracing studies in rodents and primates have shown that the STN is composed of parts that project to associative and limbic areas of the pallidal complex and substantia nigra pars reticulata (SNr) and that the basal ganglia associative and limbic circuits are processed through the STN (Alexander and Crutcher, 1990; Alexander *et al.*, 1990a; Parent and Hazrati, 1995a,b). In the next sections, I will discuss the limbic effects of HFS of the STN in PD patients and animal models of PD. However, it is not the aim of this chapter to provide a complete review of the literature, and therefore I have not cited each article describing limbic effects of HFS of the STN. Here, I aimed to cite some relevant articles, provide conceptual and anatomical details, and present an integrated view.

One of the first studies reporting limbic effects following HFS of the STN in patients suffering from advanced PD was by Kumar *et al.* (1998). One patient out of seven experienced a mild personality change in combination with disinhibition. Rodriquez and colleagues reported in the same year one patient with a transient dysthymia and another patient with a severe depressive state, after HFS of the STN. One year later, Kumar *et al.* (1999) described a patient with acute depression when the most effective electrode contact was activated, and Moro *et al.* (1999) reported depression in one patient out of seven. In 2000, Houeto *et al.* (2000) reported four cases of depression out of 23 operated patients and Molinuevo *et al.* (2000) reported one depressed patient in a series of 15 patients.

In the next years, more studies appeared on the limbic effects of HFS of the STN. Volkmann *et al.* (2001) described the results of STN HFS in 16 patients and 6 patients developed signs of depression postoperatively, and Dujardin *et al.* (2001) reported that two patients became depressed after DBS of the STN.

In 2002, Valldeoriola *et al.* (2002) observed that 5 out of their 26 patients became depressed with HFS of the STN. In another study in the same year, the adjustment disorders, psychiatric disorders, and personality changes were evaluated in 24 patients with STN HFS (Houeto *et al.*, 2002). Four patients experienced a depression, 18 patients had generalized anxiety, 15 patients showed emotional hyperreactivity, and in 8 patients, a personality disorder was diagnosed. One patient with hypersexuality after surgery was reported by Krause *et al.* (2001).

Krack *et al.* (2001) reported two patients with hypomania as a result of STN HFS. Similar observations of manic episodes following STN HFS were also reported by Kulisevsky *et al.* (2002). In other studies, depression was found in 2 out of 26 patients (Ostergaard *et al.*, 2002), 3 out of 20 patients (Vingerhoets *et al.*, 2002), 3 out of 17 patients (Martinez-Martin *et al.*, 2002), 6 out of 24 patients (Berney *et al.*, 2002), 3 out of 31 patients (Doshi *et al.*, 2002), and in 6 out of 18 patients (Thobois *et al.*, 2002). Romito *et al.* (2002a) reported the long-term results of STN stimulation in 22 patients with advanced PD. Motor performance and activities of daily living improved significantly. However, the most frequently observed complication was behavioral; depression in two patients, manic psychosis in two patients, and hypersexuality in four patients.

In 2003, Daniele *et al.* (2003) published their long-term results on the cognitive and behavioral effects of HFS of the STN in patients with PD. Three out of 20 patients had transient manic symptoms with hypersexuality. Kleiner-Fisman *et al.* (2003) observed a variety of changes in their patients. In a study with 25 patients, four patients developed mood changes, one patient had a suicide attempt, and one patient developed hypersexuality. At the end of the same year, Krack *et al.* (2003) published their five-year follow-up results of STN HFS. In total, 49 patients were examined. As compared with the preoperative condition, the patients' scores at 5 years for motor function were significantly improved. Nevertheless, seven patients became depressed and eight patients developed signs of hypomania. One preoperatively depressed patient committed suicide postoperatively. Depression and hypomania following STN HFS were also observed by others in the same year (Herzog *et al.*, 2003; Romito *et al.*, 2003). In more recent years, these early observations have been confirmed by other publications (Gervais-Bernard *et al.*, 2009; Hershey *et al.*, 2008; Soulas *et al.*, 2008; Voon *et al.*, 2008; York *et al.*, 2008).

We have listed above some clinical studies observing behavioral changes in their patients following STN HFS. There are also reports that have not found any negative effect on limbic behavior (Ardouin *et al.*, 1999; Funkiewiez *et al.*, 2003; Jahanshahi *et al.*, 2000; Perozzo *et al.*, 2001; Schneider *et al.*, 2003; Witt *et al.*, 2008). We have carried out a meta-analysis on these data with the inclusion of all available reports on the behavioral effects of STN HFS, including case series and reports (Temel and Visser-Vandewalle, 2005). In about 50% of the patients, a change in nonmotor behavior was observed (Table 3.1). The most frequently observed change was in cognitive functions, and in ~10% of the patients, limbic changes (i.e., mood disorders) were reported. The majority of these behavioral side effects were mild or even subclinical, and severe only in some cases.

In the following paragraphs, I will describe some relevant animal studies on the behavioral effects of HFS of the STN. In 2003, the first study in rats appeared on the effects of STN stimulation on behavioral parameters. Darbaky *et al.* (2003) performed unilateral STN stimulations in rats with unilateral striatal injections of 6-hydroxytryptamine (6-OHDA). The neurotoxin 6-OHDA is used for chronic dopamine depletion in animals to mimic the situation in PD. The stimulation was applied during a choice reaction time task. Unilateral STN HFS decreased the apomorphine-induced circling behavior and reduced catalepsy induced by haloperidol, which are indicative of a beneficial effect of STN HFS on 6-OHDA-induced motor deficits. STN HFS did not result in any effect on nonmotor functions. One year later, Desbonnet *et al.* (2004) reported the first study in which bilateral STN HFS was applied in rats. Results showed a significant linear decrease in premature responding with decreasing amplitudes and at high frequencies only. The same group conducted a second study in which bilateral STN HFS was applied in bilateral striatal 6-OHDA depleted rats (Temel *et al.*, 2005b). Bilateral STN HFS significantly decreased

**Table 3.1** The results of the analysis of the postoperative behavioral changes in Parkinson's disease patients with HFS of the STN

| Behavioral complications | % |
|---|---|
| Cognitive alterations | 41 (19–63) |
| Depression | 8 (0–19) |
| (Hypo)mania | 4 (0–12) |
| Anxiety disorders | <2 |
| Personality changes, hypersexuality, apathy, and aggressiveness | <0.5 |

The values are expressed as mean percentages and 95% confidence intervals (CI) and show the percentage of patients within the study population experiencing a certain form of a behavioral change. Adopted from Temel *et al.* (2006).

6-OHDA-induced deficit in premature responding and motor deficit. These data showed that bilateral STN HFS could acutely and separately influence the 6-OHDA-induced motor deficit and impulsivity-related behavior. In 2007, Baunez *et al.* (2007) reported on the effect of bilateral HFS of the STN in dopamine-depleted and naïve animals in a visual attentional task. Their results showed that STN HFS did not alleviate the deficits induced by the dopamine lesion such as omissions and latency to make correct responses, but induced perseverative approaches to the food magazine, an indicator of enhanced motivation. In sham-operated controls, STN HFS significantly reduced accuracy and induced perseverative behavior. In the same year, Temel *et al.* (2007) found that bilateral STN HFS induced acute and reversible depression-like behavior in a rat model and linked this to a decrease in firing rate of 5-hydrotryptamine neurons in the dorsal raphe nucleus (DRN).

## C. Anatomy

The primate and rat STN is a relatively small densely populated biconvex-shaped nucleus located between the zona incerta dorsally and substantia nigra posteriorly. Embryologically, the STN is part of the diencephalon, but in the adult brain its anatomical position is at the diencephalon–midbrain junction. It is surrounded by dense bundles of myelinated fibers such as the internal capsule (Hamani *et al.*, 2004; Yelnik and Percheron, 1979). The STN has approximately 560,000 cells in humans and 25,000 cells in rats (Hardman *et al.*, 1997, 2002; Oorschot, 1996). The volume of the STN is circa 240 $mm^3$ in humans and 0.8 $mm^3$ in rats (Hamani *et al.*, 2004; Hardman *et al.*, 2002). There are some considerable differences between primates and rats with respect to the dendritic organization of STN neurons. Since the STN is much larger in the primate than in the rat, the dendrites extend to a small part of the primate STN whereas in the rat the dendrites can extend across almost the whole nucleus (Heimer *et al.*, 1995). In the primate and cat, the STN is considered to be a closed nucleus except for his medial border with the lateral hypothalamic area (Parent and Hazrati, 1995b; Yelnik and Percheron, 1979). This means that the dendritic fields are confined within the nucleus except for the medial border. In the rat, the distal dendrites occasionally cross the dorsal, medial, and ventral borders of the nucleus (Hammond and Yelnik, 1983; Smith *et al.*, 1998). These data suggest that the specificity regarding the afferent input might be higher in the primate than in the rat (Heimer *et al.*, 1995). Concerning the axons of this nucleus, it is thought that the most STN neurons are long-axoned projection neurons (Rafols and Fox, 1976). Most axons emerge directly from the cell body, and some of them form intrinsic collaterals. This collateral system is more frequent in rats, but relatively rare in primates (Sato *et al.*, 2000; Shink *et al.*, 1996).

The functional subdivision of the basal ganglia into the different basal ganglia-thalamocortical circuits has also been applied to the primate STN (Hamani *et al.*, 2004). These functional domains are largely defined by the afferents and to a lesser extent by the efferent connections. The primate STN has anatomically three subdivisions: the dorsolaterally located somatomotor part (two-thirds of the STN), the ventromedially located associative part (one-third of the STN), and the medial tip represents the limbic part (Fig. 3.2) (Hamani *et al.*, 2004; Parent and Hazrati, 1995a,b). In the rodent STN, two major anatomical domains can be distinguished. The medial part is reciprocally connected with the ventral pallidum (VP) (associative and limbic functions), and the lateral part with the dorsal parts of the basal ganglia (sensorimotor functions) and related sensorimotor cortical areas (Groenewegen and Berendse, 1990; Heimer and Zahm, 1995). Because of the difference in dendritic architecture in the rodent and primate STN, the segregation of the STN into functional parts seems to be more evident in primates than in rodents.

The STN receives and projects to a number of different regions inside and outside the basal ganglia. The main afferents arise from the cortex (glutamatergic), globus pallidus externus (GPe) (GABAergic), parafascicular (PF) (glutamatergic) and centromedian nuclei (CM) (glutamatergic) of the thalamus, substantia nigra pars compacta (SNc) (dopaminergic), pedunculopontine nucleus (PPN) (cholinergic and glutamatergic), and the DRN (serotonergic) (Canteras *et al.*, 1990; Hamani *et al.*, 2004; Orieux *et al.*, 2000; Parent and Hazrati, 1995b; Steinbusch, 1981). In turn, STN efferents

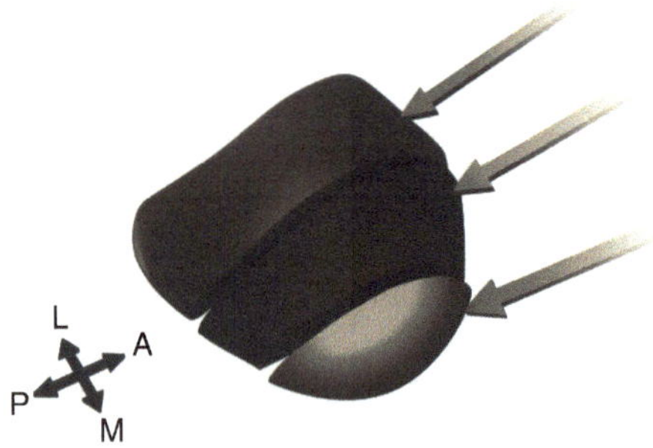


**Figure 3.2** This figure illustrates the functional subdivisions of the primate STN and shows the mediolateral orientation of this nucleus. The functional subdivisions of the basal ganglia into the different basal ganglia-thalamocortical circuits have also been applied to the primate STN. The STN has anatomically three subdivisions: the dorsolaterally located somatomotor part (upper part), the ventromedially located associative part (middle part), and the medial tip (lower part) represents the limbic part. Each circuit is being processed within the corresponding functional part of the STN. Adopted from Temel *et al.* (2005a).

(glutamate mediated neurotransmission) are predominantly directed to globus pallidus internus (GPi), GPe, SNr, SNc, and the PPN (Hamani *et al.*, 2004; Parent and Hazrati, 1995b).

Motor and cognitive information from the cortex to the basal ganglia and further, is processed through the corticobasal ganglia-thalamocortical motor and associative circuit, respectively (see for review Temel *et al.*, 2005a). Emotional, motivational, and affective processes (limbic information) related to the basal ganglia are represented by the basal ganglia-thalamocortical limbic circuit (Alexander *et al.*, 1986, 1990b; Nakano, 2000) (Fig. 3.3). Here, I will outline the limbic circuit. In the primate, projections from the hippocampus, the amygdala, limbic and paralimbic

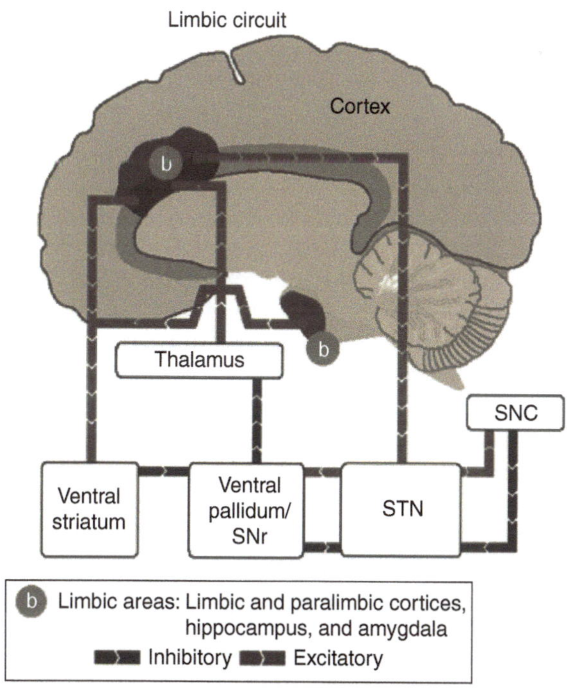


**Figure 3.3** Schematic illustration of the primate basal ganglia-thalamocortical limbic circuit. Projections from the hippocampus, the amygdala, limbic, and paralimbic cortices are primarily concentrated at the level of the ventral striatum. The ventral striatum consists basically of the nucleus accumbens, ventromedial part of the caudate–putamen and the medium-celled portion of the olfactory tubercle. The ventral striatum projects then to the ventral pallidum. From here the limbic circuit is directed to the mediodorsal nucleus of the thalamus. This circuit is closed by a thalamocortical pathway to the anterior cingulate area and medial orbitofrontal cortex. The STN has reciprocal connections with the ventral pallidum. The ventral pallidum is considered to be the major limbic circuit output region. Adopted from Temel *et al.* (2005a). SNC: substantia nigra pars compacta; SNr: substantia nigra pars reticulata; STN: subthalamic nucleus.

cortices are primarily concentrated at the level of the ventral striatum (Alexander *et al.*, 1990b; Parent and Hazrati, 1995a). The ventral striatum consists basically of the nucleus accumbens, ventromedial part of the caudate-putamen and the medium-celled portion of the olfactory tubercle (Nakano, 2000; Nauta, 1979; Parent and Hazrati, 1995a). The ventral striatum projects in turn to the VP. From here, the limbic circuit is directed to the mediodorsal (MD) nucleus of the thalamus (Alexander *et al.*, 1990b). This circuit is closed by a thalamocortical pathway to the anterior cingulated area and medial orbitofrontal cortex. The STN has reciprocal connections with the VP (Nauta and Cole, 1978; Haber *et al.*, 1985). The VP is considered to be the major limbic circuit output region. Modulation of the neuronal activity of the STN directly influences the activity of both NMDA and non-NMDA expressing neurons in the VP (Turner *et al.*, 2001). Within this concept of the limbic circuit, the STN again has an important role as it is directly connected with the output region of this circuit. Some authors made a subdivision in the limbic circuit into three subcircuits according to the cortical origin (Haber *et al.*, 1990; Nakano, 2000). However, the functionality of these subcircuits remains elusive.

In rats, the limbic circuit originates from the limbic areas (infralimbic, ventral prelimbic, and ventral agranular insular areas) and projects to the shell of the nucleus accumbens (Berendse *et al.*, 1992; Brog *et al.*, 1993). From here, the projections are directed toward the subcommissural part of the VP and from here further processed to the MD nucleus of the thalamus (Heimer and Zahm, 1995). The STN projects to the VP, as explained earlier. Finally, after passing the MD nucleus of the thalamus, the circuit is directed back to the cortical areas (Groenewegen *et al.*, 1990).

As well in humans as rats, a strong glutamatergic corticosubthalamic projection exists (Fogelson *et al.*, 2005; Magill *et al.*, 2005). It still remains unknown how this projection fits into the existing scheme of the direct and indirect pathway and further research is warranted to establish the clinical importance of this pathway.

## II. Discussion

The clinical and experimental data demonstrates that modulating the neuronal activity of the STN by DBS results in substantial improvement of the motor disability, but can be accompanied by behavioral changes. In patients, different aspects of behavior were changed, varying from a decline in executive functions to mood disorders. Concerning the limbic changes, depression was seen postoperatively (Berney *et al.*, 2002; Doshi *et al.*, 2002; Dujardin *et al.*, 2001; Houeto *et al.*, 2000, 2002; Krack *et al.*, 2003; Kumar *et al.*, 1999; Martinez-Martin *et al.*, 2002; Molinuevo *et al.*,

2000; Moro *et al.*, 1999; Ostergaard *et al.*, 2002; Rodriguez *et al.*, 1998; Romito *et al.*, 2002a; Thobois *et al.*, 2002; Valldeoriola *et al.*, 2002; Vingerhoets *et al.*, 2002; Volkmann *et al.*, 2001), in some cases leading to suicide (attempts) (Doshi *et al.*, 2002; Kleiner-Fisman *et al.*, 2003; Krack *et al.*, 2003), and (hypo)mania as well (Daniele *et al.*, 2003; Krack *et al.*, 2001; Kulisevsky *et al.*, 2002; Romito *et al.*, 2002b). Furthermore, changes in personality (Houeto *et al.*, 2002; Kumar *et al.*, 1998) and anxiety (Dujardin *et al.*, 2001; Houeto *et al.*, 2002), and hypersexuality (Kleiner-Fisman *et al.*, 2003; Krause *et al.*, 2001), were also observed following STN DBS, albeit only in single cases. Patients with preoperative cognitive deficits and affective disorders seem to be at risk for further deterioration after surgery (Kumar *et al.*, 1998; Rodriguez *et al.*, 1998). In animal studies, specific behavioral changes after STN stimulation were observed.

Anatomically, motor, cognitive and limbic information related to the basal ganglia is processed by the motor, and associative and limbic circuits, respectively. The STN has anatomically a central position within these circuits. In addition, these circuits are processed within the primate STN by specific groups of neurons located dorsolaterally (motor), medially (limbic), and ventromedially (associative). In the rodent STN, anatomical data suggest that the medial part receives information from associative and limbic parts of the VP (Groenewegen and Berendse, 1990; Heimer and Zahm, 1995). Because of the difference in dendritic architecture in the rodent and primate STN as explained earlier, the segregation of the STN into functional parts seems to be more evident in primates than in rodents.

In almost all clinical and experimental studies, improvement in motor disability was accompanied by unexpected changes in behavior. This suggests that stimulations of the STN have generally been nonselective. With DBS in patients, it is very unlikely that with the current technique of chronic stimulation (macrostimulation) and current dimensions of electrodes, to selectively influence the motor part of the STN. The challenge is to technically improve the surgical procedure to modulate only the motor part of the STN. The application of micro-, instead of macroelectrodes, for chronic stimulation, together with the introduction of high-resolution imaging techniques resulting in accurate visualization of the STN, may lead to more regionally selective modulation.

## III. Conclusion

DBS of the STN is a frequently performed surgical procedure to treat PD patients in the advanced stage. This therapy improves motor disability substantially and also the quality of life, but some patients have shown unexpected postoperative behavioral changes such as depression and mania. These behavioral effects can be explained on the basis of the anatomical data. The STN is interconnected not only with motor areas, but also with associative and limbic regions. The next step in DBS therapy is to refine the technology in such a way that only the desired effect is obtained, for instance with more selective stimulation, thereby preventing the "unwanted" side effects.

## ACKNOWLEDGMENTS

The scientific work of YT on "HFS of the STN" has received support from the Netherlands Organisation for Scientific Research (NWO) and the Dutch Brain Foundation (Hersenstichting Nederland).

## REFERENCES

Albin, R. L., Young, A. B., and Penney, J. B. (1989). The functional anatomy of basal ganglia disorders. *Trends Neurosci.* **12,** 366–375.

Alexander, G. E., and Crutcher, M. D. (1990). Functional architecture of basal ganglia circuits: Neural substrates of parallel processing. *Trends Neurosci.* **13,** 266–271.

Alexander, G. E., DeLong, M. R., and Strick, P. L. (1986). Parallel organization of functionally segregated circuits linking basal ganglia and cortex. *Annu. Rev. Neurosci.* **9,** 357–381.

Alexander, G. E., Crutcher, M. D., and DeLong, M. R. (1990). Basal ganglia-thalamocortical circuits: Parallel substrates for motor, oculomotor, "prefrontal" and "limbic" functions. *In* "Progress in Brain Research", (H. B. M. Uylings, C. G. Van Eden, and J. P. C. De Bruin, *et al.*, Eds.), Vol. 85, pp. 119–146. Elsevier Science Publishers, Amsterdam.

Alexander, G. E., Crutcher, M. D., and DeLong, M. R. (1990). Basal ganglia-thalamocortical circuits: Parallel substrates for motor, oculomotor, "prefrontal" and "limbic" functions. *Prog. Brain Res.* **85,** 119–146.

Ardouin, C., Pillon, B., Peiffer, E., *et al.* (1999). Bilateral subthalamic or pallidal stimulation for Parkinson's disease affects neither memory nor executive functions: A consecutive series of 62 patients. *Ann. Neurol.* **46,** 217–223.

Baunez, C., Christakou, A., Chudasama, Y., *et al.* (2007). Bilateral high-frequency stimulation of the subthalamic nucleus on attentional performance: Transient deleterious effects and enhanced motivation in both intact and parkinsonian rats. *Eur. J. NeuroSci.* **25,** 1187–1194.

Benabid, A. L. (2003). Deep brain stimulation for Parkinson's disease. *Curr. Opin. Neurobiol.* **13,** 696–706.

Benazzouz, A., Tai, C. H., Meissner, W., *et al.* (2004). High-frequency stimulation of both zona incerta and subthalamic nucleus induces a similar normalization of basal ganglia metabolic activity in experimental parkinsonism. *FASEB J.* **18,** 528–530.

Berendse, H. W., Galis-de Graaf, Y., and Groenewegen, H. J. (1992). Topographical organization and relationship with ventral striatal compartments of prefrontal corticostriatal projections in the rat. *J. Comp. Neurol.* **316,** 314–347.

Berney, A., Vingerhoets, F., Perrin, A., *et al.* (2002). Effect on mood of subthalamic DBS for Parkinson's disease: A consecutive series of 24 patients. *Neurology* **59,** 1427–1429.

Beurrier, C., Bioulac, B., Audin, J., and Hammond, C. (2001). High-frequency stimulation produces a transient blockade of voltage-gated currents in subthalamic neurons. *J. Neurophysiol.* **85,** 1351–1356.
Brog, J. S., Salyapongse, A., Deutch, A. Y., and Zahm, D. S. (1993). The patterns of afferent innervation of the core and shell in the "accumbens" part of the rat ventral striatum: Immunohistochemical detection of retrogradely transported fluoro-gold. *J. Comp. Neurol.* **338,** 255–278.
Canteras, N. S., Shammah-Lagnado, S. J., Silva, B. A., and Ricardo, J. A. (1990). Afferent connections of the subthalamic nucleus: A combined retrograde and anterograde horse-radish peroxidase study in the rat. *Brain Res.* **513,** 43–59.
Daniele, A., Albanese, A., Contarino, M. F., *et al.* (2003). Cognitive and behavioural effects of chronic stimulation of the subthalamic nucleus in patients with Parkinson's disease. *J. Neurol. Neurosurg. Psychiatry* **74,** 175–182.
Darbaky, Y., Forni, C., Amalric, M., and Baunez, C. (2003). High frequency stimulation of the subthalamic nucleus has beneficial antiparkinsonian effects on motor functions in rats, but less efficiency in a choice reaction time task. *Eur. J. NeuroSci.* **18,** 951–956.
Desbonnet, L., Temel, Y., Visser-Vandewalle, V., *et al.* (2004). Premature responding following bilateral stimulation of the rat subthalamic nucleus is amplitude and frequency dependent. *Brain Res.* **1008,** 198–204.
Deuschl, G., Schade-Brittinger, C., Krack, P., *et al.* (2006). A randomized trial of deep-brain stimulation for Parkinson's disease. *N. Engl. J. Med.* **355,** 896–908.
Doshi, P. K., Chhaya, N., and Bhatt, M. H. (2002). Depression leading to attempted suicide after bilateral subthalamic nucleus stimulation for Parkinson's disease. *Mov. Disord.* **17,** 1084–1085.
Dostrovsky, J. O., and Lozano, A. M. (2002). Mechanisms of deep brain stimulation. *Mov. Disord.* **17**(Suppl 3), S63–S68.
Dujardin, K., Defebvre, L., Krystkowiak, P., *et al.* (2001). Influence of chronic bilateral stimulation of the subthalamic nucleus on cognitive function in Parkinson's disease. *J. Neurol.* **248,** 603–611.
Filali, M., Hutchison, W. D., Palter, V. N., *et al.* (2004). Stimulation-induced inhibition of neuronal firing in human subthalamic nucleus. *Exp. Brain Res.* **156,** 274–281.
Fogelson, N., Williams, D., Tijssen, M., *et al.* (2005). Different functional loops between cerebral cortex and the subthalmic area in Parkinson's disease. *Cereb. Cortex* .
Funkiewiez, A., Ardouin, C., Krack, P., *et al.* (2003). Acute psychotropic effects of bilateral subthalamic nucleus stimulation and levodopa in Parkinson's disease. *Mov. Disord.* **18,** 524–530.
Gervais-Bernard, H., Xie-Brustolin, J., Mertens, P., *et al.* (2009). Bilateral subthalamic nucleus stimulation in advanced Parkinson's disease: Five year follow-up. *J. Neurol.* **256,** 225–233.
Grill, W. M., and McIntyre, C. (2001). Extracellular excitation of central neurons: Implications for the mechanism of deep brain stimulation. *Thalamus Relat. Syst.* **1,** 269–277.
Grill, W. M., Snyder, A. N., and Miocinovic, S. (2004). Deep brain stimulation creates an informational lesion of the stimulated nucleus. *NeuroReport* **15,** 1137–1140.
Groenewegen, H. J., and Berendse, H. W. (1990). Connections of the subthalamic nucleus with ventral striatopallidal parts of the basal ganglia in the rat. *J. Comp. Neurol.* **294,** 607–622.
Groenewegen, H. J., Berendse, H. W., Wolters, J. G., and Lohman, A. H. (1990). The anatomical relationship of the prefrontal cortex with the striatopallidal system, the thalamus and the amygdala: Evidence for a parallel organization. *Prog. Brain Res.* **85,** 95–116.
Haber, S. N., Groenewegen, H. J., Grove, E. A., and Nauta, W. J. (1985). Efferent connections of the ventral pallidum: Evidence of a dual striato pallidofugal pathway. *J. Comp. Neurol.* **235,** 322–335.

Haber, S. N., Wolfe, D. P., and Groenewegen, H. J. (1990). The relationship between ventral striatal efferent fibers and the distribution of peptide-positive woolly fibers in the forebrain of the rhesus monkey. *Neuroscience* **39,** 323–338.

Hamani, C., Saint-Cyr, J. A., Fraser, J., *et al.* (2004). The subthalamic nucleus in the context of movement disorders. *Brain* **127,** 4–20.

Hammond, C., and Yelnik, J. (1983). Intracellular labelling of rat subthalamic neurones with horseradish peroxidase: Computer analysis of dendrites and characterization of axon arborization. *Neuroscience* **8,** 781–790.

Hardman, C. D., Halliday, G. M., McRitchie, D. A., and Morris, J. G. (1997). The subthalamic nucleus in Parkinson's disease and progressive supranuclear palsy. *J. Neuropathol. Exp. Neurol.* **56,** 132–142.

Hardman, C. D., Henderson, J. M., Finkelstein, D. I., *et al.* (2002). Comparison of the basal ganglia in rats, marmosets, macaques, baboons, and humans: Volume and neuronal number for the output, internal relay, and striatal modulating nuclei. *J. Comp. Neurol.* **445,** 238–255.

Hariz, M. I. (2002). Complications of deep brain stimulation surgery. *Mov. Disord.* **17**(Suppl 3), S162–S166.

Hariz, M. I., and Johansson, F. (2001). Hardware failure in parkinsonian patients with chronic subthalamic nucleus stimulation is a medical emergency. *Mov. Disord.* **16,** 166–168.

Hashimoto, T., Elder, C. M., Okun, M. S., *et al.* (2003). Stimulation of the subthalamic nucleus changes the firing pattern of pallidal neurons. *J. Neurosci.* **23,** 1916–1923.

Heimer, L., Alheid, G. F., and Zahm, D. S. (1995). Basal ganglia. *In* "The Rat Nervous System" (G. Paxinos, Ed.), Second edition, pp. 579–628. Academic Press Inc, San Diego.

Hershey, T., Wu, J., Weaver, P. M., *et al.* (2008). Unilateral vs. bilateral STN DBS effects on working memory and motor function in Parkinson disease. *Exp. Neurol.* **210,** 402–408.

Herzog, J., Volkmann, J., Krack, P., *et al.* (2003). Two-year follow-up of subthalamic deep brain stimulation in Parkinson's disease. *Mov. Disord.* **18,** 1332–1337.

Houeto, J. L., Damier, P., Bejjani, P. B., *et al.* (2000). Subthalamic stimulation in Parkinson disease: A multidisciplinary approach. *Arch. Neurol.* **57,** 461–465.

Houeto, J. L., Mesnage, V., Mallet, L., *et al.* (2002). Behavioural disorders, Parkinson's disease and subthalamic stimulation. *J. Neurol. Neurosurg. Psychiatry* **72,** 701–707.

Jahanshahi, M., Ardouin, C. M., Brown, R. G., *et al.* (2000). The impact of deep brain stimulation on executive function in Parkinson's disease. *Brain* **123**(Pt 6), 1142–1154.

Klavir, O., Flash, S., Winter, C., and Joel, D. (2009). High frequency stimulation and pharmacological inactivation of the subthalamic nucleus reduces 'compulsive' lever-pressing in rats. *Exp. Neurol.* **215,** 101–109.

Kleiner-Fisman, G., Fisman, D. N., Sime, E., *et al.* (2003). Long-term follow up of bilateral deep brain stimulation of the subthalamic nucleus in patients with advanced Parkinson disease. *J. Neurosurg.* **99,** 489–495.

Krack, P., Kumar, R., Ardouin, C., *et al.* (2001). Mirthful laughter induced by subthalamic nucleus stimulation. *Mov. Disord.* **16,** 867–875.

Krack, P., Batir, A., Van Blercom, N., *et al.* (2003). Five-year follow-up of bilateral stimulation of the subthalamic nucleus in advanced Parkinson's disease. *N. Engl. J. Med.* **349,** 1925–1934.

Krause, M., Fogel, W., Heck, A., *et al.* (2001). Deep brain stimulation for the treatment of Parkinson's disease: Subthalamic nucleus versus globus pallidus internus. *J. Neurol. Neurosurg. Psychiatry* **70,** 464–470.

Kuhn, A. A., Kempf, F., Brucke, C., *et al.* (2008). High-frequency stimulation of the subthalamic nucleus suppresses oscillatory beta activity in patients with Parkinson's disease in parallel with improvement in motor performance. *J. Neurosci.* **28,** 6165–6173.

Kulisevsky, J., Berthier, M. L., Gironell, A., *et al.* (2002). Mania following deep brain stimulation for Parkinson's disease. *Neurology* **59,** 1421–1424.

Kumar, R., Lozano, A. M., Kim, Y. J., *et al.* (1998). Double-blind evaluation of subthalamic nucleus deep brain stimulation in advanced Parkinson's disease. *Neurology* **51,** 850–855.

Kumar, R., Lozano, A. M., Sime, E., *et al.* (1999). Comparative effects of unilateral and bilateral subthalamic nucleus deep brain stimulation. *Neurology* **53,** 561–566.

Lyons, K. E., Koller, W. C., Wilkinson, S. B., and Pahwa, R. (2001). Surgical and device-related events with deep brain stimulation. *Neurology* **56,** S19.004.

Magill, P. J., Sharott, A., Harnack, D., *et al.* (2005). Coherent spike-wave oscillations in the cortex and subthalamic nucleus of the freely moving rat. *Neuroscience* **132,** 659–664.

Martinez-Martin, P., Valldeoriola, F., Tolosa, E., *et al.* (2002). Bilateral subthalamic nucleus stimulation and quality of life in advanced Parkinson's disease. *Mov. Disord.* **17,** 372–377.

McIntyre, C. C., Grill, W. M., Sherman, D. L., and Thakor, N. V. (2004). Cellular effects of deep brain stimulation: Model-based analysis of activation and inhibition. *J. Neurophysiol.* **91,** 1457–1469.

Molinuevo, J. L., Valldeoriola, F., Tolosa, E., *et al.* (2000). Levodopa withdrawal after bilateral subthalamic nucleus stimulation in advanced Parkinson disease. *Arch. Neurol.* **57,** 983–988.

Moro, E., Scerrati, M., Romito, L. M., *et al.* (1999). Chronic subthalamic nucleus stimulation reduces medication requirements in Parkinson's disease. *Neurology* **53,** 85–90.

Moser, A., Gieselberg, A., Ro, B., *et al.* (2003). Deep brain stimulation: Response to neuronal high frequency stimulation is mediated through GABA(A) receptor activation in rats. *Neurosci. Lett.* **341,** 57–60.

Nakano, K. (2000). Neural circuits and topographic organization of the basal ganglia and related regions. *Brain Dev.* **22**(Suppl 1), S5–S16.

Nauta, H. J. (1979). A proposed conceptual reorganization of the basal ganglia and telencephalon. *Neuroscience* **4,** 1875–1881.

Nauta, H. J., and Cole, M. (1978). Efferent projections of the subthalamic nucleus: An autoradiographic study in monkey and cat. *J. Comp. Neurol.* **180,** 1–16.

Oh, M. Y., Abosch, A., Kim, S. H., *et al.* (2002). Long-term hardware-related complications of deep brain stimulation. *Neurosurgery* **50,** 1268–1274discussion 1274–1266.

Oorschot, D. E. (1996). Total number of neurons in the neostriatal, pallidal, subthalamic, and substantia nigral nuclei of the rat basal ganglia: A stereological study using the cavalieri and optical disector methods. *J. Comp. Neurol.* **366,** 580–599.

Orieux, G., Francois, C., Feger, J., *et al.* (2000). Metabolic activity of excitatory parafascicular and pedunculopontine inputs to the subthalamic nucleus in a rat model of Parkinson's disease. *Neuroscience* **97,** 79–88.

Ostergaard, K., Sunde, N., and Dupont, E. (2002). Effects of bilateral stimulation of the subthalamic nucleus in patients with severe Parkinson's disease and motor fluctuations. *Mov. Disord.* **17,** 693–700.

Parent, A., and Hazrati, L. N. (1995a). Functional anatomy of the basal ganglia. I. The cortico-basal ganglia-thalamo-cortical loop. *Brain Res. Brain Res. Rev.* **20,** 91–127.

Parent, A., and Hazrati, L. N. (1995b). Functional anatomy of the basal ganglia. II. The place of subthalamic nucleus and external pallidum in basal ganglia circuitry. *Brain Res. Brain Res. Rev.* **20,** 128–154.

Perozzo, P., Rizzone, M., Bergamasco, B., *et al.* (2001). Deep brain stimulation of the subthalamic nucleus in Parkinson's disease: Comparison of pre- and postoperative neuropsychological evaluation. *J. Neurol. Sci.* **192,** 9–15.

Rafols, J. A., and Fox, C. A. (1976). The neurons in the primate subthalamic nucleus: A Golgi and electron microscopic study. *J. Comp. Neurol.* **168,** 75–111.

Rodriguez, M. C., Guridi, O. J., Alvarez, L., *et al.* (1998). The subthalamic nucleus and tremor in Parkinson's disease. *Mov. Disord.* **13**(Suppl 3), 111–118.

Rodriguez-Oroz, M. C., Obeso, J. A., Lang, A. E., *et al.* (2005). Bilateral deep brain stimulation in Parkinson's disease: A multicentre study with 4 years follow-up. *Brain* **128,** 2240–2249.

Romito, L. M., Scerrati, M., Contarino, M. F., *et al.* (2002a). Long-term follow up of subthalamic nucleus stimulation in Parkinson's disease. *Neurology* **58,** 1546–1550.

Romito, L. M., Raja, M., Daniele, A., *et al.* (2002b). Transient mania with hypersexuality after surgery for high frequency stimulation of the subthalamic nucleus in Parkinson's disease. *Mov. Disord.* **17,** 1371–1374.

Romito, L. M., Scerrati, M., Contarino, M. F., *et al.* (2003). Bilateral high frequency subthalamic stimulation in Parkinson's disease: Long-term neurological follow-up. *J. Neurosurg. Sci.* **47,** 119–128.

Saint-Cyr, J. A., Trepanier, L. L., Kumar, R., *et al.* (2000). Neuropsychological consequences of chronic bilateral stimulation of the subthalamic nucleus in Parkinson's disease. *Brain* **123**(Pt 10), 2091–2108.

Sato, F., Lavallee, P., Levesque, M., and Parent, A. (2000). Single-axon tracing study of neurons of the external segment of the globus pallidus in primate. *J. Comp. Neurol.* **417,** 17–31.

Schneider, F., Habel, U., Volkmann, J., *et al.* (2003). Deep brain stimulation of the subthalamic nucleus enhances emotional processing in Parkinson disease. *Arch. Gen. Psychiatry* **60,** 296–302.

Sensi, M., Eleopra, R., Cavallo, M. A., *et al.* (2004). Explosive-aggressive behavior related to bilateral subthalamic stimulation. *Parkinsonism Relat. Disord.* **10,** 247–251.

Shink, E., Bevan, M. D., Bolam, J. P., and Smith, Y. (1996). The subthalamic nucleus and the external pallidum: Two tightly interconnected structures that control the output of the basal ganglia in the monkey. *Neuroscience* **73,** 335–357.

Smith, Y., Bevan, M. D., Shink, E., and Bolam, J. P. (1998). Microcircuitry of the direct and indirect pathways of the basal ganglia. *Neuroscience* **86,** 353–387.

Soulas, T., Gurruchaga, J. M., Palfi, S., *et al.* (2008). Attempted and completed suicides after subthalamic nucleus stimulation for Parkinson's disease. *J. Neurol. Neurosurg. Psychiatry* **79,** 952–954.

Steinbusch, H. W. (1981). Distribution of serotonin-immunoreactivity in the central nervous system of the rat-cell bodies and terminals. *Neuroscience* **6,** 557–618.

Tai, C. H., Boraud, T., Bezard, E., *et al.* (2003). Electrophysiological and metabolic evidence that high-frequency stimulation of the subthalamic nucleus bridles neuronal activity in the subthalamic nucleus and the substantia nigra reticulata. *FASEB J.* **17,** 1820–1830.

Temel, Y., and Visser-Vandewalle, V. (2005). Behavioural effects of subthalamic stimulation in advanced Parkinson disease: A systematic review. *In* Proceedings of the World Society for Stereotactic and Functional Neurosurgery pp. 11–18.

Temel, Y., Ackermans, L., Celik, H., *et al.* (2004a). Management of hardware infections following deep brain stimulation. *Acta Neurochir. (Wien)* **146,** 355–361.

Temel, Y., Blokland, A., Ackermans, L., *et al.* (2004b). Improved motor responding but central slowing following bilateral subthalamic stimulation in patients with advanced Parkinson disease. *Acta Neurochir. (Wien)* **146,** 865–866.

Temel, Y., Blokland, A., Steinbusch, H. W., and Visser-Vandewalle, V. (2005a). The functional role of the subthalamic nucleus in cognitive and limbic circuits. *Prog. Neurobiol.* **76,** 393–413.

Temel, Y., Visser-Vandewalle, V., Aendekerk, B., *et al.* (2005b). Acute and separate modulation of motor and cognitive performance in parkinsonian rats by bilateral stimulation of the subthalamic nucleus. *Exp. Neurol.* **193,** 43–52.

Temel, Y., Kessels, A., Tan, S., *et al.* (2006). Behavioural changes after bilateral subthalamic stimulation in advanced Parkinson disease: A systematic review. *Parkinsonism Relat. Disord.* **12,** 265–272.

Temel, Y., Boothman, L. J., Blokland, A., *et al.* (2007). Inhibition of 5-HT neuron activity and induction of depressive-like behavior by high-frequency stimulation of the subthalamic nucleus. *Proc. Natl. Acad. Sci. USA* **104,** 17087–17092.

Temel, Y., Prinsenberg, T., and Visser-Vandewalle, V. (2008). Imaging of the subthalamic nucleus for deep brain stimulation: A systematic review. *Neuromodulation* **11,** 8–12.

Thobois, S., Mertens, P., Guenot, M., *et al.* (2002). Subthalamic nucleus stimulation in Parkinson's disease: Clinical evaluation of 18 patients. *J. Neurol.* **249,** 529–534.

Timmermann, L., Wojtecki, L., Gross, J., *et al.* (2004). Ten-Hertz stimulation of subthalamic nucleus deteriorates motor symptoms in Parkinson's disease. *Mov. Disord.* **19,** 1328–1333.

Turner, M. S., Lavin, A., Grace, A. A., and Napier, T. C. (2001). Regulation of limbic information outflow by the subthalamic nucleus: Excitatory amino acid projections to the ventral pallidum. *J. Neurosci.* **21,** 2820–2832.

Valldeoriola, F., Pilleri, M., Tolosa, E., *et al.* (2002). Bilateral subthalamic stimulation monotherapy in advanced Parkinson's disease: Long-term follow-up of patients. *Mov. Disord.* **17,** 125–132.

Vingerhoets, F. J., Villemure, J. G., Temperli, P., *et al.* (2002). Subthalamic DBS replaces levodopa in Parkinson's disease: Two-year follow-up. *Neurology* **58,** 396–401.

Visser-Vandewalle, V., van der Linden, C., Temel, Y., *et al.* (2005). Long-term effects of bilateral subthalamic nucleus stimulation in advanced Parkinson disease: A four year follow-up study. *Parkinsonism Relat. Disord.* **11,** 157–165.

Vlamings, R., Visser-Vandewalle, V., Kozan, R., *et al.* (2009). Bilateral high frequency stimulation of the subthalamic nucleus normalizes COX activity in the substantia nigra of Parkinsonian rats. *Brain Res.* **1288,** 143–148.

Volkmann, J., Allert, N., Voges, J., *et al.* (2001). Safety and efficacy of pallidal or subthalamic nucleus stimulation in advanced PD. *Neurology* **56,** 548–551.

Voon, V., Krack, P., Lang, A. E., *et al.* (2008). A multicentre study on suicide outcomes following subthalamic stimulation for Parkinson's disease. *Brain* **131,** 2720–2728.

Weaver, F. M., Follett, K., Stern, M., *et al.* (2009). Bilateral deep brain stimulation vs best medical therapy for patients with advanced Parkinson disease: A randomized controlled trial. *JAMA* **301,** 63–73.

Wichmann, T., and Delong, M. R. (2006). Deep brain stimulation for neurologic and neuropsychiatric disorders. *Neuron* **52,** 197–204.

Windels, F., Bruet, N., Poupard, A., *et al.* (2003). Influence of the frequency parameter on extracellular glutamate and gamma-aminobutyric acid in substantia nigra and globus pallidus during electrical stimulation of subthalamic nucleus in rats. *J. Neurosci. Res.* **72,** 259–267.

Witt, K., Daniels, C., Reiff, J., *et al.* (2008). Neuropsychological and psychiatric changes after deep brain stimulation for Parkinson's disease: A randomised, multicentre study. *Lancet Neurol.* **7,** 605–614.

Yelnik, J., and Percheron, G. (1979). Subthalamic neurons in primates: A quantitative and comparative analysis. *Neuroscience* **4,** 1717–1743.

York, M. K., Dulay, M., Macias, A., *et al.* (2008). Cognitive declines following bilateral subthalamic nucleus deep brain stimulation for the treatment of Parkinson's disease. *J. Neurol. Neurosurg. Psychiatry* **79,** 789–795.

CHAPTER FOUR

# Hippocampal Mossy Fiber Synaptic Transmission and Its Modulation

Katsunori Kobayashi*,†

## Contents



## Abstract

Signal transmission between the dentate gyrus and hippocampal CA3 region is mediated by the mossy fibers (MFs), the axons of dentate granule cells. The MF- CA3 synaptic transmission is regulated by various neurotransmitters and neuromodulators. Since single MF inputs activated in bursts can generate postsynaptic action potentials and play an instructive role in associative synaptic plasticity at CA3 recurrent excitatory synapses, modulations of this input are supposed to have a substantial impact on activity of the hippocampal neuronal network. Intrinsic properties of the MF synapse and its modulatory system can be changed by environment and experience. Some of these extrinsic influences on the MF synapse may be mediated by hormones. The modulatory system at the MF synapse could also be a potential target for pharmacological treatments of psychiatric disorders. Here I review some of characteristic properties of the MF synapse and its modulatory system. © 2010 Elsevier Inc.

## I. Introduction

The dentate gyrus is positioned at the entrance of the hippocampal excitatory trisynaptic circuit. Signal transmission from the dentate gyrus to the hippocampal proper is mediated by the axons of dentate granule cells, the

* Department of Pharmacology, Nippon Medical School, Tokyo, Japan
† Japan Science and Technology Agency, CREST, Saitama, Japan

*Vitamins and Hormones*, Volume 82
ISSN 0083-6729, DOI: 10.1016/S0083-6729(10)82004-7

mossy fibers (MFs). The MFs form large synaptic terminals on proximal dendrites of CA3 pyramidal cells, mainly in apical dendritic regions in the stratum lucidum (Amaral and Dent, 1981). The MF system has been postulate to serve as a strong excitatory input to CA3 pyramidal cells, and the anatomical distribution of MFs has been correlated with some of hippocampus-dependent behaviors (Hausheer-Zarmakupi *et al.*, 1996; Schwegler *et al.*, 1981). Although the large MF bouton can form tens of synaptic contacts with a postsynaptic pyramidal cell (Acsády *et al.*, 1998; Chicurel and Harris, 1992), influence of firing of single MFs on postsynaptic cells is generally weak during low-frequency transmission. However, during high-frequency firing of MFs, the synaptic efficacy at the MF-CA3 pyramidal cell synapse is strongly enhanced and single MF inputs can generate action potentials in postsynaptic neurons (Henze *et al.*, 2002; Kobayashi and Poo, 2004; Lawrence *et al.*, 2004; Sachidhanandam *et al.*, 2009). Such postsynaptic activity induced by high-frequency activation of single MF inputs plays an instructive role in the induction of associative synaptic plasticity at the CA3 recurrent excitatory synapses (Kobayashi and Poo, 2004; Sachidhanandam *et al.*, 2009). Thus, the MF input indeed has a strong impact on the CA3 neuronal circuit, when activated at high frequencies. In addition to the activity-dependent modifications, various transmitters and neuromodulators can regulate MF synaptic transmission. Neuromodulators could control activity of the CA3 neuronal circuit by changing the efficacy of the instructive MF inputs to CA3. Some of synaptic modulations can directly contribute to the activity-dependent modifications of MF synaptic transmission. Thus, the modulatory system at the MF synapse is supposed to be critically involved in physiological functions of the MF synapse in the hippocampal neuronal circuit. In this chapter, characteristic physiological properties of the MF synapse and its modulatory system are reviewed, with some attention to influences of extrinsic factors, such as environment, experience, and drug treatments, on them.

## A. Physiological characteristics of mossy fiber synapse

The MF-CA3 pyramidal cell synapse has been characterized by prominent activity-dependent short-term synaptic plasticity. Reversible short-term synaptic modifications in response to changes in the presynaptic firing rates can be commonly seen at both excitatory and inhibitory synapses. However, the magnitude of changes in the synaptic efficacy is exceptionally large at the hippocampal MF synapse (Kobayashi *et al.*, 1996; Langdon *et al.*, 1995; Salin *et al.*, 1996). At the MF synapse, the efficacy of synaptic transmission is augmented upon an increase in firing frequencies of MFs, a phenomenon called frequency facilitation. Activity-dependent synaptic facilitation is generally ascribed to enhanced transmitter release caused by a rise in residual $Ca^{2+}$ concentrations in presynaptic terminals, and synapses with low initial transmitter release probability exhibit large synaptic facilitation in general (Zucker and Regehr, 2002). The prominent facilitation at the MF synapse can be at least partly explained by low basal transmitter release probability. Strong tonic synaptic inhibition by ambient adenosine has been shown to maintain the low release probability (Moore *et al.*, 2003; but see Kukley *et al.*, 2005). Release of $Ca^{2+}$ from intracellular $Ca^{2+}$ stores has been shown to contribute to facilitation of MF synaptic transmission and of $Ca^{2+}$ transients in MF terminals (Lauri *et al.*, 2003; Liang *et al.*, 2002; Scott and Rusakov, 2006; Scott *et al.*, 2008; Shimizu *et al.*, 2008; but see Breustedt and Schmitz, 2004; Carter *et al.*, 2002). Feedback regulation by facilitatory glutamate autoreceptors also seems to play an important role in the MF synaptic facilitation (see below).

The prominent frequency facilitation at the MF synapse is crucial for dentate-to-CA3 spike transmission and induction of associative synaptic plasticity in the CA3 recurrent circuit (Kobayashi and Poo, 2004; Sachidhanandam *et al.*, 2009). This frequency facilitation is established with postnatal development (Marchal and Mulle, 2004). In mice heterozygous for α-calcium/calmodulin-dependent protein kinase II (CaMKII), the frequency facilitation at the MF synapse is strongly reduced as compared with wild-type mice in adults (Yamasaki *et al.*, 2008). Although CaMKII seems to have a direct contribution to the frequency facilitation (Salin *et al.*, 1996), the reduced facilitation in CaMKII mutant mice is largely ascribed to failure of maturation of the dentate granule cells (Yamasaki *et al.*, 2008). A similar defect of the frequency facilitation has been demonstrated in mice lacking the kainate receptor (KAR) GluK1 (Breustedt and Schmitz, 2004; Contractor *et al.*, 2001; Sachidhanandam *et al.*, 2009; but see Kwon and Castillo, 2008b). Both of these mutant mice show profound behavioral abnormalities (Shaltiel *et al.*, 2008; Yamasaki *et al.*, 2008). The magnitude of paired-pulse facilitation (PPF), another form of short-term plasticity induced by paired stimulation, at the MF synapse has been correlated with behaviors of mice (Kobayashi *et al.*, 2006). These lines of evidence suggest potential importance of the strong synaptic facilitation in physiological functions of the MF system and behavioral regulation.

Long-term plasticity has also been demonstrated at the MF-CA3 pyramidal cell synapse. Unlike most other excitatory synapses in the hippocampus, long-term potentiation (LTP) and long-term depression (LTD) at the MF synapse are independent of *N*-methyl-D-aspartate (NMDA) receptors (Harris and Cotman, 1986; Kobayashi *et al.*, 1996; Zalutsky and Nicoll, 1990). Both LTP and LTD can be induced even in the absence of fast glutamatergic synaptic transmission (Castillo *et al.*, 1994; Chen *et al.*, 2001; Ito and Sugiyama, 1991; Kobayashi *et al.*, 1996; Yeckel *et al.*, 1999), suggesting that the induction of long-term plasticity at the MF synapse is mainly regulated by activity of presynaptic cells. However, postsynaptic factors have also been shown to be involved in the induction of MF LTP (Contractor *et al.*, 2002; Jaffe and Johnston, 1990; Urban and Barrionuevo, 1996; Yeckel *et al.*, 1999), while the expression of LTP has been consistently shown to be mediated by presynaptic processes (Maeda *et al.*, 1997; Staubli *et al.*, 1990; Weisskopf and Nicoll, 1995; Xiang *et al.*, 1994; Zalutsky and Nicoll, 1990). The Eph receptor–ephrin system has been suggested to play a role as retrograde signaling in the generation of MF LTP (Contractor *et al.*, 2002). In young animals, distinct forms of MF LTD that require $Ca^{2+}$ elevation in postsynaptic pyramidal cells have also been demonstrated (Domenici *et al.*, 1998; Lei *et al.*, 2003).

In addition to pyramidal cells, MFs also form synapses onto excitatory interneurons called mossy cells in the hilar region of the dentate gyrus and inhibitory interneurons. MFs form large synaptic terminals onto mossy cells (Acsády *et al.*, 1998). Physiological properties of the MF–mossy cell synapse are generally similar to that onto pyramidal cells, but the magnitude and frequency dependence of the synaptic facilitation appears to be slightly different (Lysetskiy *et al.*, 2005). Inhibitory interneurons have been shown to be major targets of MFs (Acsády *et al.*, 1998). MFs form small synaptic terminals onto these neurons. At the MF–inhibitory interneuron synapses, high-frequency stimulation causes either synaptic depression or facilitation that is much smaller in magnitude than that at the MF–pyramidal cell synapses (Toth *et al.*, 2000). In cultured slices, recordings from synaptically connected pairs of granule and pyramidal cells revealed that single action potentials in a granule cell evoked a net inhibitory response in a pyramidal cell due to feed forward synaptic inhibition, but high-frequency firing caused net excitatory responses (Mori *et al.*, 2004). This activity-dependent switching can be explained by differences in the number of synaptic contacts and in the activity-dependent short-term synaptic modifications between synapses made by MFs onto inhibitory interneurons and those onto pyramidal cells described above. However, results of minimal-stimulation studies in acute hippocampal slices do not support the finding in the cultured slice (Kobayashi and Poo, 2004; Sachidhanandam *et al.*, 2009). In this review, the MF synapse refers to the MF–pyramidal cell synapse unless otherwise specified.

## B. Modulation of mossy fiber synaptic transmission

MF synaptic transmission is regulated by various neurotransmitters and neuromodulators. Most of them can act on presynaptic terminals and regulate release of transmitters. Therefore, these modulators are supposed to influence both short- and long-term plasticity at the MF synapse. Some of presynaptic modulations have been suggested to directly contribute to the activity-dependent nature of the MF synaptic transmission. As described above, tonic presynaptic inhibition by ambient adenosine has been shown to be required for maintaining the low transmission efficacy and wide dynamic range of activity-dependent synaptic modifications (Moore *et al.*, 2003; but see Kukley *et al.*, 2005). Experience and drug treatments can change the efficacy and expression levels of neuromodulators acting on the

MF synapse, thereby possibly regulating hippocampal functions. Here, some of synaptic modulations that are essential for physiological functions of the MF synapse and/or would be of potential importance in regulation of hippocampal functions by external factors are reviewed.

## 1. Glutamate

At the MF synapse, fast excitatory postsynaptic currents (EPSCs) are mainly mediated by α-amino-3-hydroxyl-5-methyl-4-isoxazole-propionate (AMPA) receptors as in other excitatory synapses in the brain. Glutamate can also activate NMDA and KARs at the MF synapse. Although NMDA receptors are not required for the long-term plasticity of MF synaptic transmission mediated by AMPA receptors, they are involved in the induction of LTP that is mechanistically distinct from conventional MF LTP and is selectively expressed by NMDA receptor-mediated synaptic responses (Kwon and Castillo, 2008a). High-density kainate binding has been demonstrated along the MF pathway (Monaghan and Cotman, 1982; Mulle *et al.*, 1998; Tremblay *et al.*, 1985). Activation of postsynaptic KARs can generate slow depolarizing synaptic currents (Castillo *et al.*, 1997; Vignes and Collingridge 1997). This slow synaptic current has a minor contribution to synaptic potentials during low-frequency transmission, but plays an important role in spike generation during repetitive high-frequency MF firing (Kwon and Castillo, 2008b; Sachidhanandam *et al.*, 2009). Much attention has been paid to the potential involvement of presynaptic KARs in the activity-dependent modifications of the MF synaptic transmission. Application of kainate at low concentrations (50–100 nM) can potentiate MF synaptic transmission probably via presynaptic mechanisms (Breustedt and Schmitz, 2004; Contractor *et al.*, 2003; Ji and Stäubli, 2002; Lauri *et al.*, 2001; Rodríguez-Moreno and Sihra, 2004; Schmitz *et al.*, 2001; but see Kwon and Castillo, 2008b; Pinheiro *et al.*, 2007). Kainate can depolarize MFs (Kamiya and Ozawa, 2000; Schmitz *et al.*, 2000, 2001), and subthreshold depolarization of MF terminals can enhance action potential-dependent release of transmitters (Alle and Geiger, 2006). Activation of KARs by endogenous glutamate has also been shown to enhance MF synaptic transmission. Thus, facilitation of NMDA EPSCs recorded in the presence of an AMPA receptor-specific antagonist can be reduced by AMPA/KAR antagonists (Schmitz *et al.*, 2001; but see Kwon and Castillo, 2008b). Paired-pulse stimulation of MFs can induce facilitation of presynaptic $Ca^{2+}$ transients that is also blocked by AMPA/KAR antagonists (Kamiya *et al.*, 2002; but see Mori-Kawakami *et al.*, 2003). KAR-specific antagonists can also reduce facilitation of MF synaptic transmission and/or presynaptic $Ca^{2+}$ transients during repetitive MF firing (Dargan *et al.*, 2009; Lauri *et al.*, 2001; Scott *et al.*, 2008; but see Breustedt and Schmitz, 2004; Kwon and Castillo, 2008b). These lines of evidence suggest that presynaptic KARs activated by glutamate released from MF terminals contribute to the large facilitation at the MF synapse probably via augmentation of action potential-driven $Ca^{2+}$ elevation in MF terminals. In addition to postsynaptic KARs, these presynaptic KARs have also been shown to play an important role in spike transmission between MFs and pyramidal cells (Sachidhanandam *et al.*, 2009). On the other hand, at the MF synapse formed onto inhibitory interneurons and hilar mossy cells, blockade of KARs had no significant effects on paired-pulse ratios of synaptic currents or presynaptic $Ca^{2+}$ transients (Scott *et al.*, 2008). Thus, the enhancement of synaptic transmission by kainate autoreceptors appears to be specific to the MF–pyramidal cell synapse.

Some of mutant mice lacking KAR subunits show impaired MF synaptic facilitation. Kainate-induced potentiation of MF synaptic transmission is impaired in GluK2- (Breustedt and Schmitz, 2004) and GluK5-deficient mice (Contractor *et al.*, 2003). Mice lacking GluK2 have been shown to exhibit reduced MF PPF and frequency facilitation in some studies (Breustedt and Schmitz, 2004; Contractor *et al.*, 2001; Sachidhanandam *et al.*, 2009), but not in other studies (Kwon and Castillo, 2008b; Mulle *et al.*, 1998). GluK3-deficient mice also showed reduced MF synaptic facilitation (Pinheiro *et al.*, 2007; Sachidhanandam *et al.*, 2009; but see Kwon and Castillo, 2008b). In mice lacking GluK4 and GluK5, which do not form homomeric receptors, MF PPF was reduced, but frequency facilitation was intact (Fernandes *et al.*, 2009). These subunits may be involved in localization of the receptor complex near the transmitter release sites (Fernandes *et al.*, 2009). Although GluK1-deficient mice showed intact MF synaptic facilitation (Breustedt and Schmitz, 2004; Contractor *et al.*, 2001), pharmacological studies suggested a contribution of the GluK1 subunit to the MF synaptic facilitation, kainate-induced potentiation, and presynaptic $Ca^{2+}$ transients during repetitive MF stimulation (Dargan *et al.*, 2009; Lauri *et al.*, 2001; but see Breustedt and Schmitz, 2004; Kwon and Castillo, 2008b). Thus, the subtype of presynaptic KARs involved in the synaptic facilitation remains controversial. The discrepancies in results of mutant studies could be due to differences in experimental conditions such as genetic background of mice, age, or housing environments. Since the large frequency facilitation at the MF synapse is established with postnatal development (Marchal and Mulle, 2004), any defects in maturation of the MF synapse caused by the gene knockout combined with other factors would potentially impair the MF synaptic facilitation in adults (Kobayashi, 2009; Yamasaki *et al.*, 2008).

Higher concentrations of kainate ($\geq$200 nM) inhibit MF synaptic transmission (Contractor *et al.*, 2000; Kamiya and Ozawa, 2000; Kwon and Castillo, 2008b; Pinheiro *et al.*, 2007; Schmitz *et al.*, 2000). In GluK2-deficient mice, kainate has no inhibitory effects on MF synaptic transmission (Contractor *et al.*, 2000). Large MF depolarization could reduce the presynaptic $Ca^{2+}$ influx, thereby suppressing the transmitter release from

MF terminals (Kamiya and Ozawa, 2000). Therefore, the bidirectional effects of KAR activation on MF synaptic transmission may be due to differences in the magnitude of kainate-induced MF depolarization. It has also been reported that the kainate-induced synaptic modifications are mediated by cAMP cascades. While the synaptic inhibition by kainate requires G protein activation (Negrete-Díaz *et al.*, 2006), the synaptic potentiation does not (Rodríguez-Moreno and Sihra, 2004). Thus, differential coupling to cAMP cascades may be also involved in the bidirectional effects of KAR activation.

KARs have been implicated in the induction of long-term plasticity as well. In mice lacking GluK2 or GluK3, but not GluK1, MF LTP has been shown to be impaired (Contractor *et al.*, 2001; Pinheiro *et al.*, 2007). On the other hand, pharmacological studies have demonstrated suppression of MF LTP by GluK1 receptor antagonists (Bortolotto *et al.*, 1999; Dargan *et al.*, 2009; Lauri *et al.*, 2001, 2003). Since MF LTP can be induced in the presence of high concentrations of the nonselective ionotropic glutamate receptor antagonist kynurenic acid (10–20 mM) (Castillo *et al.*, 1994; Chen *et al.*, 2001; Yeckel *et al.*, 1999; but see Bortolotto *et al.*, 1999), KARs does not seem to be indispensable for the induction of MF LTP, but may facilitate the LTP induction (Schmitz *et al.*, 2003). MF LTP has been shown to be mediated by cAMP cascades (Huang *et al.*, 1994; Weisskopf *et al.*, 1994). KARs may contribute to the induction of MF LTP by activating cAMP cascades (Rodríguez-Moreno and Sihra, 2004) and/or by enhancing $Ca^{2+}$ elevation in MF terminals (Lauri *et al.*, 2003).

MF synaptic transmission is regulated by several subtypes of metabotropic glutamate receptors (mGluRs). Immunohistochemical investigation revealed that groups II and III mGluRs are predominantly localized at presynaptic elements of the hippocampal synapses including the MF synapse (Shigemoto *et al.*, 1997). A low concentration of 2-amino-4-phosphonobutyric acid (AP4), a potent agonist of mGluR4 of the group III mGluRs, has been known to block MF synaptic transmission in guinea pigs (Lanthorn *et al.*, 1984; Yamamoto *et al.*, 1983), but not in rats (Lanthorn *et al.*, 1984). Application of group II mGluR agonists strongly suppresses MF synaptic transmission in guinea pigs (Manzoni *et al.*, 1995), rats (Kamiya *et al.*, 1996) and mice (Yokoi *et al.*, 1996), but has minimal effects at the synapse between the recurrent excitatory fibers and CA3 pyramidal cells (Kamiya *et al.*, 1996; Kobayashi and Poo, 2004). Daumas *et al.* (2009) have shown that infusions of a group II mGluR agonist into the hippocampal CA3 region *in vivo* impaired contextual fear conditioning. Given the specific effect of the group II mGluR agonist on the MF synapse, this finding suggests the major role for the MF system in the formation of the contextual fear memory. Blockade of the presynaptic mGluRs enhances the frequency facilitation of the MF synaptic transmission (Kwon and Castillo, 2008b; Scanziani *et al.*, 1997; Toth *et al.*, 2000; Vogt and Nicoll, 1999), suggesting that endogenous glutamate released

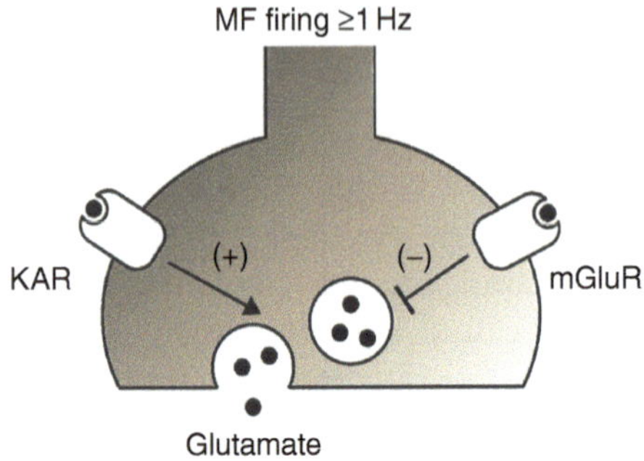


**Figure 4.1** Bidirectional modulation of MF synaptic transmission by presynaptic glutamate receptors. When MFs are firing at moderate to high frequencies, glutamate can facilitate and inhibit the release of transmitters from MFs by activating presynaptic kainate receptors (KARs) and mGluRs, respectively. Strong activation of KARs by high concentrations of agonists can inhibit the transmitter release.

from MF terminals during repetitive simulation can activate presynaptic mGluRs and inhibit the release of glutamate. Therefore, during repetitive firing of MFs, the release of glutamate from MF terminals can be bidirectionally modulated by KARs and mGluRs (Fig. 4.1).

Prolonged repetitive stimulation of MFs at 1 Hz induces MF LTD that can be blocked by mGluR antagonists (Kobayashi *et al.*, 1996; Tzounopoulos *et al.*, 1998) and is impaired in mice lacking mGluR2, one of the group II mGluRs (Yokoi *et al.*, 1996). Therefore, group II mGluRs may play an important role in the induction of MF LTD (but see Wostrack and Dietrich, 2009). One of the group III mGluRs, mGluR7, is localized at the MF terminals onto interneurons, but not onto CA3 pyramidal cells (Shigemoto *et al.*, 1997). High-frequency stimulation of MFs can induce LTD at the MF–interneuron synapses presumably via activation of mGluR7 (Maccaferri *et al.*, 1998; Pelkey *et al.*, 2005). Some studies have demonstrated the involvement of group I mGluRs in MF LTP (Bashir *et al.*, 1993; Conquet *et al.*, 1994; Yeckel *et al.*, 1999; but see Fitzjohn *et al.*, 1998; Hsia *et al.*, 1995; Manzoni *et al.*, 1994).

## 2. Neuropeptides

MFs contain high amounts of neuropeptides such as enkephalin and dynorphin (Gall *et al.*, 1990). The expression level of these neuropeptides is known to be altered by seizures (Gall *et al.*, 1990; Hong *et al.*, 1993; Marksteiner *et al.*, 1990), stress (Shirayama *et al.*, 2004), and drug treatments (Rattan and Tejwani, 1997). Thus, peptidergic modulations might mediate some of the changes in hippocampal functions caused by extrinsic influences.

Dynorphin, one of opioid peptides, is highly concentrated at the MF pathway (McGinty *et al.*, 1983). In guinea pigs, dynorphin can presynaptically inhibit MF synaptic transmission via activation of $\kappa$-opioid receptors (KORs) (Salin *et al.*, 1995; Weisskopf *et al.*, 1993). Dendrotoxin-sensitive potassium channels have been shown to be involved in KOR-dependent synaptic inhibition (Simmons and Chavkin, 1996). Tetanic stimulation of MFs causes synaptic depression at neighboring MF synapses that can be blocked by naloxone, a broad-spectrum antagonist of opioid receptors (Weisskopf *et al.*, 1993). Naloxone or KOR-specific antagonists can also enhance MF LTP induced by tetanic stimulation (Harrison *et al.*, 2002; Weisskopf *et al.*, 1993; but see Williams and Johnston, 1996). Therefore, endogenous opioid peptide, most probably dynorphin, released during the tetanic stimulation has inhibitory effects on synaptic transmission and the induction of LTP (Fig. 4.2). Since synaptic facilitation induced by 1 Hz stimulation was not affected by naloxone, low-frequency MF firing does not seem to be sufficient for release of dynorphin (Weisskopf *et al.*, 1993). In rats, effects of dynorphin on MF synaptic transmission depend on the strains (Salin *et al.*, 1995), and naloxone has been shown to block the induction of MF LTP probably by antagonizing $\mu$-opioid receptors (Derrick *et al.*, 1991; Jin and Chavkin, 1999; Williams and Johnston, 1996; but see Salin *et al.*, 1995). In guinea pigs chronically treated with morphine, the magnitude of MF LTP was increased and naloxone did not enhance LTP (Harrison *et al.*, 2002). Application of exogenous dynorphin could still inhibit MF synaptic transmission in these guinea pigs. Therefore, KORs are functional after the morphine treatment, but do not seem to be activated by endogenous dynorphin. Consistent with this finding, chronic morphine treatments have been shown to reduce dynorphin levels in the hippocampus (Rattan and Tejwani, 1997).

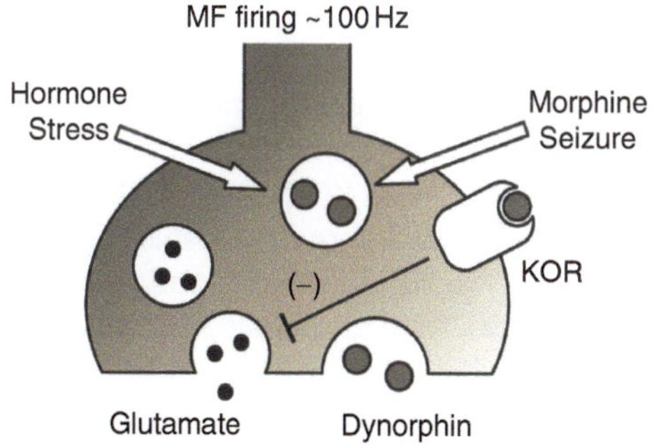


**Figure 4.2** Regulation of MF synaptic transmission mediated by dynorphin. High-frequency firing of MFs causes release of opioid peptide dynorphin from MFs, and released dynorhpin inhibits synaptic transmission and the induction of LTP by activating $\kappa$-opioid receptors (KORs). Extrinsic factors and hormones can change expression levels of dynorhpin, thereby potentially regulating MF synaptic transmission.

Seizures are known to modulate expression of neuropeptides in MFs (Gall *et al.*, 1990; Hong *et al.*, 1993). Dynorphin is supposed to play a role as an anticonvulsant by suppressing synaptic transmission. Dynorphin immunoreactivity decreases transiently after seizures, probably due to release of dynorphin during the seizures (Hong *et al.*, 1993; Simonato and Romualdi, 1996). Since high-frequency stimulation is required for efficient release of dynorphin (Weisskopf *et al.*, 1993), dynorphin would be suitable for suppressing excess neuronal activity during seizures. Prodynorphin mRNA levels are either increased or decreased depending on the type of seizures and time after the seizure period (Gall *et al.*, 1990; Hong *et al.*, 1993; Simonato and Romualdi, 1996). Seizures seem to cause overproduction of dynorphin and a subsequent rebound decrease. Seizures also affect expression levels of other neuropeptides. Neuropeptide Y (NPY) is expressed at low levels in MFs in control conditions, but is markedly upregulated after seizures or during epilepsy (Gall *et al.*, 1990; Marksteiner *et al.*, 1990; Tu *et al.*, 2005). Since NPY can also inhibit synaptic transmission (Klapstein and Colmers, 1993), the upregulation of NPY would protect hippocampal neurons from excess neuronal activity, especially in the reorganized network in the epileptic brain (Tu *et al.*, 2005).

## 3. Monoamines

Noradrenaline has been shown to enhance MF LTP without significantly affecting the basal synaptic efficacy in rats (Hopkins and Johnston, 1984, 1988). This effect of noradrenaline is mediated by activation of $\beta$-adrenergic receptors and intracellular cAMP in postsynaptic neurons (Hopkins and Johnston, 1988). In cultured slices, noradrenaline can suppress MF synaptic transmission via activation of $\alpha_1$ receptors (Scanziani *et al.*, 1993). Recently, other monoamines have also been shown to modulate MF synaptic transmission. Dopamine can potentiate MF synaptic transmission in mice (Kobayashi and Suzuki, 2007; Kobayashi *et al.*, 2006). The dopamine-induced potentiation was mediated by $D_1$-like receptors and cAMP (Kobayashi and Suzuki, 2007). Serotonin can also potentiate MF synaptic transmission via activation of 5-$HT_4$ receptors in mice (Kobayashi *et al.*, 2008). The 5-$HT_4$ receptor is also coupled to cAMP cascades, and dopamine and serotonin have been shown to share intracellular pathways (Kobayashi *et al.*, 2008). Both dopamine and serotonin reduced PPF (Kobayashi and Suzuki, 2007; Kobayashi *et al.*, 2008), suggesting the presynaptic site for the expression of potentiation induced by these modulators. This also indicates that the effects of these monoaminergic modulations depend on activity of MFs, with higher frequency transmission being less sensitive to the modulations. The brain monoaminergic system is thought to be the primary target of antidepressant drugs. Chronic treatment of mice with the serotonergic antidepressant fluoxetine bidirectionally changed the 5-$HT_4$-receptor-mediated synaptic modulation at the MF synapse:

Fluoxetine increased the magnitude of MF synaptic potentiation induced by lower concentrations of serotonin, but decreased that by higher concentrations (Kobayashi *et al.*, 2008). Therefore, chronic fluoxetine can stabilize the serotonergic modulation of MF synaptic transmission. These synaptic changes caused by fluoxetine were associated with reduction of activity of mice in a novel open field (Kobayashi *et al.*, 2008). The magnitude of the dopamine-induced potentiation was correlated with open-field activity (Kobayashi *et al.*, 2006). Thus, the monoaminergic modulation of MF synaptic transmission seems to be associated with regulation of reactivity of animals to novelty. Depletion of serotonin by *p*-chloroamphetamine causes a decrease in hippocampal expression of prodynorphin mRNAs (Di Benedetto *et al.*, 2004), raising the possibility that the serotonergic system can also indirectly modulate MF synaptic transmission by regulating dynorphin levels.

## C. Hormone and mossy fiber synapse

Corticosteroid hormones are essential for the integrity of the dentate gyrus and MF system (Joëls, 2007; Sousa *et al.*, 1999). Both high- and low-affinity corticosterone receptors, mineralocorticoid, and glucocorticoid receptors, respectively, are expressed in the dentate gyrus (Joëls, 2007). Forebrain mineralocorticoid receptor knockout mice exhibit abnormalities of the MF projection (Berger *et al.*, 2006). Adrenalectomy of adult rats causes a substantial loss of dentate granule cells 3–4 months after the surgery that can be prevented by corticosterone treatment (Sloviter *et al.*, 1989). Adrenolectomy in rats also causes a loss of MF synapses and reduces the size of MF terminals and postsynaptic excrescences on which MFs form synapses (Sousa *et al.*, 1999). The decrease in the MF synapse number was completely reversed by low doses of corticosetrone, while the MF terminal morphology was partially restored. Corticosterone applied at a higher dose can cause decreases in the MF synapse number and size in intact rats (Sousa *et al.*, 2000), suggesting that the corticosetrone level should be maintained at optimal levels for the anatomical and structural integrity of the MF system. Adrenalectomy also causes decreases in dynorphin mRNA levels in the dentate gyrus and kainate binding in the hilus and CA3 region. These changes can be detected in 7 days after the surgery and prevented by the mineralocorticoid aldosterone (Watanabe *et al.*, 1995). Chronic corticosterone applied alone can increase dynorphin mRNA levels (Watanabe *et al.*, 1995). Dynorphin levels have been reported to be altered by gonadal steroids as well (Torres-Reveron *et al.*, 2009). These lines of evidence raise the possibility that corticosteroids and other hormones can modify MF synaptic transmission and plasticity via changes in intrinsic properties of the synapse and/or dynorphin-mediated modulations. It has been reported that adrenalectomy had no clear effects on MF synaptic potentiation induced by tetanic stimulation *in vivo* (Pavlides and McEwen, 1999). In this study, however, the tetanic stimulation induced strong short-term depression instead of potentiation or facilitation, and the magnitude of lasting potentiation was very small. Further *in vivo* or *ex vivo* analyses should be carried out to clarify the role of corticosteroids in regulating functional properties of the MF synapse.

In slices, some hormones can acutely modify synaptic transmission at the MF synapse. Thyrotropin-releasing hormone can augments LTP of population spikes evoked by MF stimulation (Ishihara *et al.*, 1991). A brief application of insulin induces MF LTD that is distinct from that induced by repetitive electrical stimulation of MFs (Huang *et al.*, 2003). These findings suggest that some hormones can act as direct modulators of MF synaptic transmission.

## D. Effects of stress and experience on mossy fiber synapse

The central nervous system is modified by experience and environment. Stress can cause structural changes and synaptic dysfunctions in the hippocampus (Garcia, 2001; McEwen, 1999). Since stress can generally precipitate psychiatric disorders, stress-induced neuronal dysfunctions may represent the cellular pathophysiology of psychiatric disorders.

Neonatal maternal separation has been shown to reduce the density of infrapyramidal bands of MFs and increase anxiety-related behaviors (Huot *et al.*, 2002). Chronic unpredictable stress reduces the number and size of MF terminals (Sousa *et al.*, 2000). Stress can increase serum corticosterone levels, and corticosterone administration mimics the effects of unpredictable stress on the MF system (Sousa *et al.*, 2000). In rats exposed to chronic restraint stress, synaptic vesicles in the MF terminals were more densely localized in the vicinity of active zones and the area occupied by mitochondria is increased (Magariños *et al.*, 1997). Chronic restraint stress also causes shrinkage of postsynaptic thorny excrescences in the dendrite of CA3 pyramidal cells (Stewart *et al.*, 2005). On the other hand, environmental enrichment increases complexities of large MF terminals and the number of synapses in the stratum lucidum (Galimberti *et al.*, 2006; Gogolla *et al.*, 2009).

Much less is known about effects of stress or environment on physiological properties of the MF synapse. Exposure of rats to acute tail suspension stress has been shown to reduce the magnitude of MF LTP and extracellular zinc levels (Takeda *et al.*, 2009a). Zinc is contained in MFs in high levels and has been shown to have multiple effects on the induction of MF LTP (Huang *et al.*, 2008; Li *et al.*, 2001; Takeda *et al.*, 2008). The decrease in extracellular zinc levels upon exposure to acute stress seemed to be due to influx of zinc into cells, and corticosterone administration failed to mimic this effect of stress (Takeda *et al.*, 2009b). Environmental enrichment can reduce PPF at the MF synapse without affecting basal synaptic efficacy and concomitantly reduce open-field activity, while social isolation had no significant effects on either of them (Kobayashi *et al.*, 2006).

Exposure to acute stress or learned helplessness, an animal model of depression, induces robust increases in dynorphin-like immunoreactivity at the MF pathway (Shirayama *et al.*, 2004). Since dynorphin inhibits the induction of MF LTP (Weisskopf *et al.*, 1993), increased dynorphin levels may also contribute to the stress-induced suppression of MF LTP. Infusion of $\kappa$-opioid antagonist into the dentate gyrus or CA3 region causes antidepressant-like effects in learned helplessness (Shirayama *et al.*, 2004). Repeated electroconvulsive stimulation, an experimental model of electroconvulsive therapy for depression, causes a strong decrease in dynorphin-like immunoreactivity along MFs (Kanamatsu *et al.*, 1986). Therefore, the dynorphin-mediated inhibition of the MF synaptic transmission seems to be associated with depression-related behaviors in learned helplessness and might also be related to therapeutic treatments of depression in humans. This idea is in agreement with the recent finding described above that modifications of the serotonergic modulation at the MF synapse are associated with some of behavioral effects of antidepressant drugs (Kobayashi, 2009; Kobayashi *et al.*, 2008).

## II. Conclusion

In this review, physiological properties of the hippocampal MF synapse and its modulatory system were described. Glutamate-mediated modulations have been shown to be directly involved in shaping the prominent activity dependence of MF synaptic transmission. Neuropeptides can be released from the MF terminals during high-frequency transmission and regulate LTP induced by high-frequency stimulation of MFs. Monoamines can potentiate MF synaptic transmission, with rather preferential effects on low-frequency transmission, but also facilitate LTP induced by high-frequency stimulation. Experience and drug treatments can affect these modulatory systems either directly or via the hormonal system, as exemplified in dynorphin-mediated modulations (Fig. 4.2), and could modify MF synaptic transmission in different ways depending on the target modulators that they act on. However, it largely remains to be elucidated how such regulation of the MF system contributes to a variety of modifications of the brain functions by extrinsic factors. There is an emerging idea that the dentate gyrus and MF synapse are critically involved in some psychiatric disorders and their treatments (Kobayashi, 2009; Reif *et al.*, 2007; Sahay and Hen, 2007; Yamasaki *et al.*, 2008). Changes in the MF system induced by stress could be a candidate cellular basis for the pathophysiology of psychiatric disorders. Although there are many lines of evidence for stress-induced structural changes of the MF synapse and CA3 pyramidal cells, it is poorly understood how functional properties of this synapse, including its modulatory system, are modified. There has been a substantial accumulation of knowledge about molecular mechanisms of synaptic modulation and plasticity. In future studies, more efforts should be made to elucidate the association of synaptic modulations with adaptive and pathological changes of brain functions and behaviors.

## REFERENCES


Acsády, L., Kamondi, A., Sík, A., Freund, T., and Buzsáki, G. (1998). GABAergic cells are the major postsynaptic targets of mossy fibers in the rat hippocampus. *J. Neurosci.* **18,** 3386–3403.

Alle, H., and Geiger, J. R. P. (2006). Combined analog and action potential coding in hippocampal mossy fibers. *Science* **311,** 1290–1293.

Amaral, D. G., and Dent, J. A. (1981). Development of the mossy fibers of the dentate gyrus: I A light and electron microscopic study of the mossy fibers and their expansions. *J. Comp. Neurol.* **195,** 51–86.

Bashir, Z. I., Bortolotto, Z. A., Davies, C. H., Berretta, N., Irving, A. J., Seal, A. J., Henley, J. M., Jane, D. E., Watkins, J. C., and Collingridge, G. L. (1993). Induction of LTP in the hippocampus needs synaptic activation of glutamate metabotropic receptors. *Nature* **363,** 347–350.

Berger, S., Wolfer, D. P., Selbach, O., Alter, H., Erdmann, G., Reichardt, H. M., Chepkova, A. N., Welzl, H., Haas, H. L., Lipp, H.-P., and Schütz, G. (2006). Loss of the limbic mineralocorticoid receptor impairs behavioral plasticity. *Proc. Natl. Acad. Sci. USA* **103,** 195–200.

Bortolotto, Z. A., Clarke, V. R. J., Delany, C. M., Parry, M. C., Smolders, I., Vignes, M., Ho, K. H., Miu, P., Brinton, B. T., Fantaske, R., Ogden, A., Gates, M., *et al.* (1999). Kainate receptors are involved in synaptic plasticity. *Nature* **402,** 297–301.

Breustedt, J., and Schmitz, D. (2004). Assessing the role of GLUK5 and GLUK6 at hippocampal mossy fiber synapses. *J. Neurosci.* **24,** 10093–10098.

Carter, A. G., Vogt, K. E., Foster, K. A., and Regehr, W. G. (2002). Assessing the role of calcium-induced calcium release in short-term presynaptic plasticity at excitatory central synapses. *J. Neurosci.* **22,** 21–28.

Castillo, P. E., Weisskopf, M. G., and Nicoll, R. A. (1994). The role of $Ca^{2+}$ channels in hippocampal mossy fiber synaptic transmission and long-term potentiation. *Neuron* **12,** 261–269.

Castillo, P. E., Malenka, R. C., and Nicoll, R. A. (1997). Kainate receptors mediate a slow postsynaptic current in hippocampal CA3 neurons. *Nature* **388,** 182–186.

Chen, Y.-L., Huang, C.-C., and Hsu, K.-S. (2001). Time-dependent reversal of long-term potentiation by low-frequency stimulation at the hippocampal mossy fiber-CA3 synapses. *J. Neurosci.* **21,** 3705–3714.

Chicurel, M. E., and Harris, K. M. (1992). Three-dimensional analysis of the structure and composition of CA3 branched dendritic spines and their synaptic relationships with mossy fiber boutons in the rat hippocampus. *J. Comp. Neurol.* **325,** 169–182.

Conquet, F., Bashir, Z. I., Davies, C. H., Daniel, H., Ferraguti, F., Bordi, F., Franz-Bacon, K., Reggiani, A., Matarese, V., Condé, F., Collingridge, G. L., and Crépel, F. (1994). Motor deficit and impairment of synaptic plasticity in mice lacking mGluR1. *Nature* **372,** 237–243.

Contractor, A., Swanson, G. T., Sailer, A., O'Gorman, S., and Heinemann, S. F. (2000). Identification of the kainate receptor subunits underlying modulation of excitatory synaptic transmission in the CA3 region of the hippocampus. *J. Neurosci.* **20,** 8269–8278.
Contractor, A., Swanson, G., and Heinemann, S. F. (2001). Kainate receptors are involved in short- and long-term plasticity at mossy fiber synapses in the hippocampus. *Neuron* **29,** 209–216.
Contractor, A., Rogers, C., Maron, C., Henkemeyer, M., Swanson, G. T., and Heinemann, S. F. (2002). Trans-synaptic Eph receptor-ephrin signaling in hippocampal mossy fiber LTP. *Science* **296,** 1864–1869.
Contractor, A., Sailer, A. W., Darstein, M., Maron, C., Xu, J., Swanson, G. T., and Heinemann, S. F. (2003). Loss of kainate receptor-mediated heterosynaptic facilitation of mossy-fiber synapses in KA2$^{-/-}$ mice. *J. Neurosci.* **23,** 422–429.
Dargan, S. L., Clarke, V. R. J., Alushin, G. M., Sherwood, J. L., Nisticò, R., Bortolotto, Z. A., Ogden, A. M., Bleakman, D., Doherty, A. J., Lodge, D., Mayer, M. L., Fitzjohn, S. M., *et al.* (2009). ACET is a highly potent and specific kainate receptor antagonist: Characterisation and effects on hippocampal mossy fibre function. *Neuropharmacology* **56,** 121–130.
Daumas, S., Ceccom, J., Halley, H., Francés, B., and Lassalle, J. M. (2009). Activation of metabotropic glutamate receptor type 2/3 supports the involvement of the hippocampal mossy fiber pathway on contextual fear memory consolidation. *Learn. Mem.* **16,** 504–507.
Derrick, B. E., Weinberger, S. B., and Martinez, J. L. Jr. (1991). Opioid receptors are involved in an NMDA receptor-independent mechanism of LTP induction at hippocampal mossy fiber-CA3 synapses. *Brain Res. Bull.* **27,** 219–223.
Di Benedetto, M., D'Addario, C., Collins, S., Izenwasser, S., Candeletti, S., and Romualdi, P. (2004). Role of serotonin on cocaine-mediated effects on prodynorphin gene expression in the rat brain. *J. Mol. Neurosci.* **22,** 213–222.
Domenici, M. R., Berretta, N., and Cherubini, E. (1998). Two distinct forms of long-term depression coexist at the mossy fiber-CA3 synapse in the hippocampus during development. *Proc. Natl. Acad. Sci. USA* **95,** 8310–8315.
Fernandes, H. B., Catches, J. S., Petralia, R. S., Copits, B. A., Xu, J., Russell, T. A., Swanson, G. T., and Contractor, A. (2009). High-affinity kainate receptor subunits are necessary for ionotropic but not metabotropic signaling. *Neuron* **63,** 818–829.
Fitzjohn, S. M., Bortolotto, Z. A., Palmer, M. J., Doherty, A. J., Ornstein, P. L., Schoepp, D. D., Kingston, A. E., Lodge, D., and Collingridge, G. L. (1998). The potent mGlu receptor antagonist LY341495 identifies roles for both cloned and novel mGlu receptors in hippocampal synaptic plasticity. *Neuropharmacology* **37,** 1445–1458.
Galimberti, I., Gogolla, N., Alberi, S., Santos, A. F., Muller, D., and Caroni, P. (2006). Long-term rearrangements of hippocampal mossy fiber terminal connectivity in the adult regulated by experience. *Neuron* **50,** 749–763.
Gall, C., Lauterborn, J., Isackson, P., and White, J. (1990). Seizures, neuropeptide regulation, and mRNA expression in the hippocampus. *Prog. Brain Res.* **83,** 371–390.
Garcia, R. (2001). Stress, hippocampal plasticity, and spatial learning. *Synapse* **40,** 180–183.
Gogolla, N., Galimberti, I., Deguchi, Y., and Caroni, P. (2009). Wnt signaling mediates experience-related regulation of synapse numbers and mossy fiber connectivities in the adult hippocampus. *Neuron* **62,** 510–525.
Harris, E. W., and Cotman, C. W. (1986). Long-term potentiation of guinea pig mossy fiber responses is not blocked by N-methyl D-aspartate antagonists. *Neurosci. Lett.* **70,** 132–137.
Harrison, J. M., Allen, R. G., Pellegrino, M. J., Williams, J. T., and Manzoni, O. J. (2002). Chronic morphine treatment alters endogenous opioid control of hippocampal mossy fiber synaptic transmission. *J. Neurophysiol.* **87,** 2464–2470.

Hausheer-Zarmakupi, Z., Wolfer, D. P., Leisinger-Trigona, M.-C., and Lipp, H.-P. (1996). Selective breeding for extremes in open-field activity of mice entails a differentiation of hippocampal mossy fibers. *Behav. Genet.* **26,** 167–176.

Henze, D. A., Wittner, L., and Buzsáki, G. (2002). Single granule cells reliably discharge targets in the hippocampal CA3 network *in vivo*. *Nat. Neurosci.* **5,** 790–795.

Hong, J. S., McGinty, J. F., Lee, P. H. K., Xie, C. W., and Mitchell, C. L. (1993). Relationship between hippocampal opioid peptides and seizures. *Prog. Neurobiol.* **40,** 507–528.

Hopkins, W. F., and Johnston, D. (1984). Frequency-dependent noradrenergic modulation of long-term potentiation in the hippocampus. *Science* **226,** 350–352.

Hopkins, W. F., and Johnston, D. (1988). Noradrenergic enhancement of long-term potentiation at mossy fiber synapses in the hippocampus. *J. Neurophysiol.* **59,** 667–687.

Hsia, A. Y., Salin, P. A., Castillo, P. E., Aiba, A., Abeliovich, A., Tonegawa, S., and Nicoll, R. A. (1995). Evidence against a role for metabotropic glutamate receptors in mossy fiber LTP: The use of mutant mice and pharmacological antagonists. *Neuropharmacology* **34,** 1567–1572.

Huang, Y.-Y., Li, X.-C., and Kandel, E. R. (1994). cAMP contributes to mossy fiber LTP by initiating both a covalently mediated early phase and macromolecular synthesis-dependent late phase. *Cell* **79,** 69–79.

Huang, C.-C., You, J.-L., Lee, C.-C., and Hsu, K.-S. (2003). Insulin induces a novel form of postsynaptic mossy fiber long-term depression in the hippocampus. *Mol. Cell. Neurosci.* **24,** 831–841.

Huang, Y. Z., Pan, E., Xiong, Z.-Q., and McNamara, J. O. (2008). Zinc-mediated transactivation of TrkB potentiates the hippocampal mossy fiber-CA3 pyramid synapse. *Neuron* **57,** 546–558.

Huot, R. L., Plotsky, P. M., Lenox, R. H., and McNamara, R. K. (2002). Neonatal maternal separation reduces hippocampal mossy fiber density in adult Long Evans rats. *Brain Res.* **950,** 52–63.

Ishihara, K., Katsuki, H., Kawabata, A., Sasa, M., Satoh, M., and Takaori, S. (1991). Effects of thyrotropin-releasing hormone and a related analog, CNK-602A, on long-term potentiation in the mossy fiber-CA3 pathway of guinea pig hippocampal slices. *Brain Res.* **554,** 203–208.

Ito, I., and Sugiyama, H. (1991). Roles of glutamate receptors in long-term potentiation at hippocampal mossy fiber synapses. *NeuroReport* **2,** 333–336.

Jaffe, D., and Johnston, D. (1990). Induction of long-term potentiation at hippocampal mossy-fiber synapses follows a Hebbian rule. *J. Neurophysiol.* **64,** 948–960.

Ji, Z., and Stäubli, U. (2002). Presynaptic kainate receptors play different physiological roles in mossy fiber and associational-commissural synapses in CA3 of hippocampus from adult rats. *Neurosci. Lett.* **331,** 71–74.

Jin, W., and Chavkin, C. (1999). Mu opioids enhance mossy fiber synaptic transmission indirectly by reducing $GABA_B$ receptor activation. *Brain Res.* **821,** 286–293.

Joëls, M. (2007). Role of corticosteroid hormones in the dentate gyrus. *Prog. Brain Res.* **163,** 355–370.

Kamiya, H., and Ozawa, S. (2000). Kainate receptor-mediated presynaptic inhibition at the mouse hippocampal mossy fibre synapse. *J. Physiol.* **523,** 653–665.

Kamiya, H., Shinozaki, H., and Yamamoto, C. (1996). Activation of metabotropic glutamate receptor type 2/3 suppresses transmission at rat hippocampal mossy fibre synapses. *J. Physiol.* **493,** 447–455.

Kamiya, H., Ozawa, S., and Manabe, T. (2002). Kainate receptor-dependent short-term plasticity of presynaptic $Ca^{2+}$ influx at the hippocampal mossy fiber synapses. *J. Neurosci.* **22,** 9237–9243.

Kanamatsu, T., McGinty, J. F., Mitchell, C. L., and Hong, J. S. (1986). Dynorphin- and enkephalin-like immunoreactivity is altered in limbic-basal ganglia regions of rat brain after repeated electroconvulsive shock. *J. Neurosci.* **6,** 644–649.

Klapstein, G. J., and Colmers, W. F. (1993). On the sites of presynaptic inhibition by neuropeptide Y in rat hippocampus *in vitro*. *Hippocampus* **3,** 103–111.

Kobayashi, K. (2009). Targeting the hippocampal mossy fiber synapse for the treatment of psychiatric disorders. *Mol. Neurobiol.* **39,** 24–36.

Kobayashi, K., and Poo, M.-M. (2004). Spike train timing-dependent associative modification of hippocampal CA3 recurrent synapses by mossy fibers. *Neuron* **41,** 445–454.

Kobayashi, K., and Suzuki, H. (2007). Dopamine selectively potentiates hippocampal mossy fiber to CA3 synaptic transmission. *Neuropharmacology* **52,** 552–561.

Kobayashi, K., Manabe, T., and Takahashi, T. (1996). Presynaptic long-term depression at the hippocampal mossy fiber-CA3 synapse. *Science* **273,** 648–650.

Kobayashi, K., Ikeda, Y., and Suzuki, H. (2006). Locomotor activity correlates with modifications of hippocampal mossy fibre synaptic transmission. *Eur. J. Neurosci.* **24,** 1867–1873.

Kobayashi, K., Ikeda, Y., Haneda, E., and Suzuki, H. (2008). Chronic fluoxetine bidirectionally modulates potentiating effects of serotonin on the hippocampal mossy fiber synaptic transmission. *J. Neurosci.* **28,** 6272–6280.

Kukley, M., Schwan, M., Fredholm, B. B., and Dietrich, D. (2005). The role of extracellular adenosine in regulating mossy fiber synaptic plasticity. *J. Neurosci.* **25,** 2832–2837.

Kwon, H. B., and Castillo, P. E. (2008a). Long-term potentiation selectively expressed by NMDA receptors at hippocampal mossy fiber synapses. *Neuron* **57,** 108–120.

Kwon, H.-B., and Castillo, P. E. (2008b). Role of glutamate autoreceptors at hippocampal mossy fiber synapses. *Neuron* **60,** 1082–1094.

Langdon, R. B., Johnson, J. W., and Barrionuevo, G. (1995). Posttetanic potentiation and presynaptically induced long-term potentiation at the mossy fiber synapse in rat hippocampus. *J. Neurobiol.* **26,** 370–385.

Lanthorn, T. H., Ganong, A. H., and Cotman, C. W. (1984). 2-Amino-4-phosphonobutyrate selectively blocks mossy fiber-CA3 responses in guinea pig but not rat hippocampus. *Brain Res.* **290,** 174–178.

Lauri, S. E., Bortolotto, Z. A., Bleakman, D., Ornstein, P. L., Lodge, D., Isaac, J. T. R., and Collingridge, G. L. (2001). A critical role of a facilitatory presynaptic kainate receptor in mossy fiber LTP. *Neuron* **32,** 697–709.

Lauri, S. E., Bortolotto, Z. A., Nistico, R., Bleakman, D., Ornstein, P. L., Lodge, D., Isaac, J. T. R., and Collingridge, G. L. (2003). A role for $Ca^{2+}$ stores in kainate receptor-dependent synaptic facilitation and LTP at mossy fiber synapses in the hippocampus. *Neuron* **39,** 327–341.

Lawrence, J. J., Grinspan, Z. M., and McBain, C. J. (2004). Quantal transmission at mossy fibre targets in the CA3 region of the rat hippocampus. *J. Physiol.* **554,** 175–193.

Lei, S., Pelkey, K. A., Topolnik, L., Congar, P., Lacaille, J. C., and McBain, C. J. (2003). Depolarization-induced long-term depression at hippocampal mossy fiber-CA3 pyramidal neuron synapses. *J. Neurosci.* **23,** 9786–9795.

Li, Y., Hough, C. J., Frederickson, C. J., and Sarvey, J. M. (2001). Induction of mossy fiber→Ca3 long-term potentiation requires translocation of synaptically released $Zn^{2+}$. *J. Neurosci.* **21,** 8015–8025.

Liang, Y., Yuan, L.-L., Johnston, D., and Gray, R. (2002). Calcium signaling at single mossy fiber presynaptic terminals in the rat hippocampus. *J. Neurophysiol.* **87,** 1132–1137.

Lysetskiy, M., Földy, C., and Soltesz, I. (2005). Long- and short-term plasticity at mossy fiber synapses on mossy cells in the rat dentate gyrus. *Hippocampus* **15,** 691–696.

Maccaferri, G., Tóth, K., and McBain, C. J. (1998). Target-specific expression of presynaptic mossy fiber plasticity. *Science* **279,** 1368–1370.

Maeda, T., Kaneko, S., Akaike, A., and Satoh, M. (1997). Direct evidence for increase in excitatory amino acids release during mossy fiber LTP in rat hippocampal slices as revealed by the patch sensor methods. *Neurosci. Lett.* **224,** 103–106.
Magariños, A. M., Verdugo, J. M. G., and McEwen, B. S. (1997). Chronic stress alters synaptic terminal structure in hippocampus. *Proc. Natl. Acad. Sci. USA* **94,** 14002–14008.
Manzoni, O. J., Weisskopf, M. G., and Nicoll, R. A. (1994). MCPG antagonizes metabotropic glutamate receptors but not long-term potentiation in the hippocampus. *Eur. J. Neurosci.* **6,** 1050–1054.
Manzoni, O. J., Castillo, P. E., and Nicoll, R. A. (1995). Pharmacology of metabotropic glutamate receptors at the mossy fiber synapses of the guinea pig hippocampus. *Neuropharmacology* **34,** 965–971.
Marchal, C., and Mulle, C. (2004). Postnatal maturation of mossy fibre excitatory transmission in mouse CA3 pyramidal cells: A potential role for kainate receptors. *J. Physiol.* **561,** 27–37.
Marksteiner, J., Ortler, M., Bellmann, R., and Sperk, G. (1990). Neuropeptide Y biosynthesis is markedly induced in mossy fibers during temporal lobe epilepsy of the rat. *Neurosci. Lett.* **112,** 143–148.
McEwen, B. S. (1999). Stress and hippocampal plasticity. *Annu. Rev. Neurosci.* **22,** 105–122.
McGinty, J. F., Henriksen, S. J., Goldstein, A., Terenius, L., and Bloom, F. E. (1983). Dynorphin is contained within hippocampal mossy fibers: Immunochemical alterations after kainic acid administration and colchicine-induced neurotoxicity. *Proc. Natl. Acad. Sci. USA* **80,** 589–593.
Monaghan, D. T., and Cotman, C. W. (1982). The distribution of [$^3$H]kainic acid binding sites in rat CNS as determined by autoradiography. *Brain Res.* **252,** 91–100.
Moore, K. A., Nicoll, R. A., and Schmitz, D. (2003). Adenosine gates synaptic plasticity at hippocampal mossy fiber synapses. *Proc. Natl. Acad. Sci. USA* **100,** 14397–14402.
Mori, M., Abegg, M. H., Gähwiler, B. H., and Gerber, U. (2004). A frequency-dependent switch from inhibition to excitation in a hippocampal unitary circuit. *Nature* **431,** 453–456.
Mori-Kawakami, F., Kobayashi, K., and Takahashi, T. (2003). Developmental decrease in synaptic facilitation at the mouse hippocampal mossy fibre synapse. *J. Physiol.* **553,** 37–48.
Mulle, C., Sailer, A., Pérez-Otaño, I., Dickinson-Anson, H., Castillo, P. E., Bureau, I., Maron, C., Gage, F. H., Mann, J. R., Bettler, B., and Heinemann, S. F. (1998). Altered synaptic physiology and reduced susceptibility to kainate-induced seizures in GluR6-deficient mice. *Nature* **392,** 601–605.
Negrete-Díaz, J. V., Sihra, T. S., Delgado-García, J. M., and Rodríguez-Moreno, A. (2006). Kainate receptor-mediated inhibition of glutamate release involves protein kinase A in the mouse hippocampus. *J. Neurophysiol.* **96,** 1829–1837.
Pavlides, C., and McEwen, B. S. (1999). Effects of mineralocorticoid and glucocorticoid receptors on long-term potentiation in the CA3 hippocampal field. *Brain Res.* **851,** 204–214.
Pelkey, K. A., Lavezzari, G., Racca, C., Roche, K. W., and McBain, C. J. (2005). mGluR7 is a metaplastic switch controlling bidirectional plasticity of feedforward inhibition. *Neuron* **46,** 89–102.
Pinheiro, P. S., Perrais, D., Coussen, F., Barhanin, J., Bettler, B., Mann, J. R., Malva, J. O., Heinemann, S. F., and Mulle, C. (2007). GluR7 is an essential subunit of presynaptic kainate autoreceptors at hippocampal mossy fiber synapses. *Proc. Natl. Acad. Sci. USA* **104,** 12181–12186.
Rattan, A. K., and Tejwani, G. A. (1997). Effect of chronic treatment with morphine, midazolam and both together on dynorphin(1-13) levels in the rat. *Brain Res.* **754,** 239–244.
Reif, A., Schmitt, A., Fritzen, S., and Lesch, K.-P. (2007). Neurogenesis and schizophrenia: Dividing neurons in a divided mind? *Eur. Arch. Psychiatry Clin. Neurosci.* **257,** 290–299.

Rodríguez-Moreno, A., and Sihra, T. S. (2004). Presynaptic kainate receptor facilitation of glutamate release involves protein kinase A in the rat hippocampus. *J. Physiol.* **557,** 733–745.

Sachidhanandam, S., Blanchet, C., Jeantet, Y., Cho, Y. H., and Mulle, C. (2009). Kainate receptors act as conditional amplifiers of spike transmission at hippocampal mossy fiber synapses. *J. Neurosci.* **29,** 5000–5008.

Sahay, A., and Hen, R. (2007). Adult hippocampal neurogenesis in depression. *Nat. Neurosci.* **10,** 1110–1115.

Salin, P. A., Weisskopf, M. G., and Nicoll, R. A. (1995). A comparison of the role of dynorphin in the hippocampal mossy fiber pathway in guinea pig and rat. *J. Neurosci.* **15,** 6939–6945.

Salin, P. A., Scanziani, M., Malenka, R. C., and Nicoll, R. A. (1996). Distinct short-term plasticity at two excitatory synapses in the hippocampus. *Proc. Natl. Acad. Sci. USA* **93,** 13304–13309.

Scanziani, M., Gähwiler, B. H., and Thompson, S. M. (1993). Presynaptic inhibition of excitatory synaptic transmission mediated by $\alpha$ adrenergic receptors in area CA3 of the rat hippocampus *in vitro*. *J. Neurosci.* **13,** 5393–5401.

Scanziani, M., Salin, P. A., Vogt, K. E., Malenka, R. C., and Nicoll, R. A. (1997). Use-dependent increases in glutamate concentration activate presynaptic metabotropic glutamate receptors. *Nature* **385,** 630–634.

Schmitz, D., Frerking, M., and Nicoll, R. A. (2000). Synaptic activation of presynaptic kainate receptors on hippocampal mossy fiber synapses. *Neuron* **27,** 327–338.

Schmitz, D., Mellor, J., and Nicoll, R. A. (2001). Presynaptic kainate receptor mediation of frequency facilitation at hippocampal mossy fiber synapses. *Science* **291,** 1972–1976.

Schmitz, D., Mellor, J., Breustedt, J., and Nicoll, R. A. (2003). Presynaptic kainate receptors impart an associative property to hippocampal mossy fiber long-term potentiation. *Nat. Neurosci.* **6,** 1058–1063.

Schwegler, H., Lipp, H. P., Van der Loos, H., and Buselmaier, W. (1981). Individual hippocampal mossy fiber distribution in mice correlates with two-way avoidance performance. *Science* **214,** 817–819.

Scott, R., and Rusakov, D. A. (2006). Main determinants of presynaptic $Ca^{2+}$ dynamics at individual mossy fiber-CA3 pyramidal cell synapses. *J. Neurosci.* **26,** 7071–7081.

Scott, R., Lalic, T., Kullmann, D. M., Capogna, M., and Rusakov, D. A. (2008). Target-cell specificity of kainate autoreceptor and $Ca^{2+}$-store-dependent short-term plasticity at hippocampal mossy fiber synapses. *J. Neurosci.* **28,** 13139–13149.

Shaltiel, G., Maeng, S., Malkesman, O., Pearson, B., Schloesser, R. J., Tragon, T., Rogawski, M., Gasior, M., Luckenbaugh, D., Chen, G., and Manji, H. K. (2008). Evidence for the involvement of the kainate receptor subunit GluR6 (GRIK2) in mediating behavioral displays related to behavioral symptoms of mania. *Mol. Psychiatry* **13,** 858–872.

Shigemoto, R., Kinoshita, A., Wada, E., Nomura, S., Ohishi, H., Takada, M., Flor, P. J., Neki, A., Abe, T., Nakanishi, S., and Mizuno, N. (1997). Differential presynaptic localization of metabotropic glutamate receptor subtypes in the rat hippocampus. *J. Neurosci.* **17,** 7503–7522.

Shimizu, H., Fukaya, M., Yamasaki, M., Watanabe, M., Manabe, T., and Kamiya, H. (2008). Use-dependent amplification of presynaptic $Ca^{2+}$ signaling by axonal ryanodine receptors at the hippocampal mossy fiber synapse. *Proc. Natl. Acad. Sci. USA* **105,** 11998–12003.

Shirayama, Y., Ishida, H., Iwata, M., Hazama, G.-I., Kawahara, R., and Duman, R. S. (2004). Stress increases dynorphin immunoreactivity in limbic brain regions and dynorphin antagonism produces antidepressant-like effects. *J. Neurochem.* **90,** 1258–1268.

Simmons, M. L., and Chavkin, C. (1996). κ-Opioid receptor activation of a dendrotoxin-sensitive potassium channel mediates presynaptic inhibition of mossy fiber neurotransmitter release. *Mol. Pharmacol.* **50,** 80–85.

Simonato, M., and Romualdi, P. (1996). Dynorphin and epilepsy. *Prog. Neurobiol.* **50,** 557–583.

Sloviter, R. S., Valiquette, G., Abrams, G. M., Ronk, E. C., Sollas, A. L., Paul, L. A., and Neubort, S. (1989). Selective loss of hippocampal granule cells in the mature rat brain after adrenalectomy. *Science* **243,** 535–538.

Sousa, N., Madeira, M. D., and Paula-Barbosa, M. M. (1999). Corticosterone replacement restores normal morphological features to the hippocampal dendrites, axons and synapses of adrenalectomized rats. *J. Neurocytol.* **28,** 541–558.

Sousa, N., Lukoyanov, N. V., Madeira, M. D., Almeida, O. F. X., and Paula-Barbosa, M. M. (2000). Reorganization of the morphology of hippocampal neurites and synapses after stress-induced damage correlates with behavioral improvement. *Neuroscience* **97,** 253–266.

Staubli, U., Larson, J., and Lynch, G. (1990). Mossy fiber potentiation and long-term potentiation involve different expression mechanisms. *Synapse* **5,** 333–335.

Stewart, M. G., Davies, H. A., Sandi, C., Kraev, I. V., Rogachevsky, V. V., Peddie, C. J., Rodriguez, J. J., Cordero, M. I., Donohue, H. S., Gabbott, P. L. A., and Popov, V. I. (2005). Stress suppresses and learning induces plasticity in CA3 of rat hippocampus: A three-dimensional ultrastructural study of thorny excrescences and their postsynaptic densities. *Neuroscience* **131,** 43–54.

Takeda, A., Kanno, S., Sakurada, N., Ando, M., and Oku, N. (2008). Attenuation of hippocampal mossy fiber long-term potentiation by low micromolar concentrations of zinc. *J. Neurosci. Res.* **86,** 2906–2911.

Takeda, A., Ando, M., Kanno, S., and Oku, N. (2009a). Unique response of zinc in the hippocampus to behavioral stress and attenuation of subsequent mossy fiber long-term potentiation. *Neurotoxicology* **30,** 712–717.

Takeda, A., Sakurada, N., Ando, M., Kanno, S., and Oku, N. (2009b). Facilitation of zinc influx via AMPA/kainate receptor activation in the hippocampus. *Neurochem. Int.* **55,** 376–382.

Torres-Reveron, A., Khalid, S., Williams, T. J., Waters, E. M., Jacome, L., Luine, V. N., Drake, C. T., McEwen, B. S., and Milner, T. A. (2009). Hippocampal dynorphin immunoreactivity increases in response to gonadal steroids and is positioned for direct modulation by ovarian steroid receptors. *Neuroscience* **159,** 204–216.

Toth, K., Suares, G., Lawrence, J. J., Philips-Tansey, E., and McBain, C. J. (2000). Differential mechanisms of transmission at three types of mossy fiber synapse. *J. Neurosci.* **20,** 8279–8289.

Tremblay, E., Represa, A., and Ben-Ari, Y. (1985). Autoradiographic localization of kainic acid binding sites in the human hippocampus. *Brain Res.* **343,** 378–382.

Tu, B., Timofeeva, O., Jiao, Y., and Nadler, J. V. (2005). Spontaneous release of neuropeptide Y tonically inhibits recurrent mossy fiber synaptic transmission in epileptic brain. *J. Neurosci.* **25,** 1718–1729.

Tzounopoulos, T., Janz, R., Südhof, T. C., Nicoll, R. A., and Malenka, R. C. (1998). A role for cAMP in long-term depression at hippocampal mossy fiber synapses. *Neuron* **21,** 837–845.

Urban, N. N., and Barrionuevo, G. (1996). Induction of Hebbian and non-Hebbian mossy fiber long-term potentiation by distinct patterns of high-frequency stimulation. *J. Neurosci.* **16,** 4293–4299.

Vignes, M., and Collingridge, G. L. (1997). The synaptic activation of kainate receptors. *Nature* **388,** 179–182.

Vogt, K. E., and Nicoll, R. A. (1999). Glutamate and $\gamma$-aminobutyric acid mediate a heterosynaptic depression at mossy fiber synapses in the hippocampus. *Proc. Natl. Acad. Sci. USA* **96,** 1118–1122.

Watanabe, Y., Weiland, N. G., and McEwen, B. S. (1995). Effects of adrenal steroid manipulations and repeated restraint stress on dynorphin mRNA levels and excitatory amino acid receptor binding in hippocampus. *Brain Res.* **680,** 217–225.

Weisskopf, M. G., and Nicoll, R. A. (1995). Presynaptic changes during mossy fibre LTP revealed by NMDA receptor-mediated synaptic responses. *Nature* **376,** 256–259.

Weisskopf, M. G., Zalutsky, R. A., and Nicoll, R. A. (1993). The opioid peptide dynorphin mediates heterosynaptic depression of hippocampal mossy fibre synapses and modulates long-term potentiation. *Nature* **362,** 423–427.

Weisskopf, M. G., Castillo, P. E., Zalutsky, R. A., and Nicoll, R. A. (1994). Mediation of hippocampal mossy fiber long-term potentiation by cyclic AMP. *Science* **265,** 1878–1882.

Williams, S. H., and Johnston, D. (1996). Actions of endogenous opioids on NMDA receptor-independent long-term potentiation in area CA3 of the hippocampus. *J. Neurosci.* **16,** 3652–3660.

Wostrack, M., and Dietrich, D. (2009). Involvement of Group II mGluRs in mossy fiber LTD. *Synapse* **63,** 1060–1068.

Xiang, Z., Greenwood, A. C., Kairiss, E. W., and Brown, T. H. (1994). Quantal mechanism of long-term potentiation in hippocampal mossy-fiber synapses. *J. Neurophysiol.* **71,** 2552–2556

Yamamoto, C., Sawada, S., and Takada, S. (1983). Suppressing action of 2-amino-4-phosphonobutyric acid on mossy fiber-induced excitation in the guinea pig hippocampus. *Exp. Brain Res.* **51,** 128–134.

Yamasaki, N., Maekawa, M., Kobayashi, K., Kajii, Y., Maeda, J., Soma, M., Takao, K., Tanda, K., Ohira, K., Toyama, K., Kanzaki, K., Fukunaga, K., *et al.* (2008). Alpha-CaMKII deficiency causes immature dentate gyrus, a novel candidate endophenotype of psychiatric disorders. *Mol. Brain* **1,** 6.

Yeckel, M. F., Kapur, A., and Johnston, D. (1999). Multiple forms of LTP in hippocampal CA3 neurons use a common postsynaptic mechanism. *Nat. Neurosci.* **2,** 625–633.

Yokoi, M., Kobayashi, K., Manabe, T., Takahashi, T., Sakaguchi, I., Katsuura, G., Shigemoto, R., Ohishi, H., Nomura, S., Nakamura, K., Nakao, K., Katsuki, M., *et al.* (1996). Impairment of hippocampal mossy fiber LTD in mice lacking mGluR2. *Science* **273,** 645–647.

Zalutsky, R. A., and Nicoll, R. A. (1990). Comparison of two forms of long-term potentiation in single hippocampal neurons. *Science* **248,** 1619–1624.

Zucker, R. S., and Regehr, W. G. (2002). Short-term synaptic plasticity. *Annu. Rev. Physiol.* **64,** 355–405.

CHAPTER FIVE

# Xenobiotics in the Limbic System—Affecting Brain's Network Function

Ralf P. Meyer,* Georgios Pantazis,† Nina Killer,‡ Carolin Bürck,‡ Ricarda Schwab,‡ Monika Brandt,‡ Rolf Knoth,§ *and* Marcel Gehlhaus§

## Contents



## Abstract

Xenobiotic compounds enter the brain through nutrition, environmentals, and drugs. In order to maintain intrinsic homeostasis, the brain has to adapt to xenobiotic influx. Among others, steroid hormones appear as crucial mediators in this process. However, especially in the therapy of neurological diseases or

* Medizinische Fakultät der Universität Freiburg, Breisacherstraße 64, Freiburg, Germany
† Institut für Hirnforschung, Universitätsklinikum Tübingen, Calwerstraße 3, Tübingen, Germany
‡ Hochschule Furtwangen University, Department Mechanical & Process Engineering, Jakob-Kienzle-Straße 17, Villingen-Schwenningen, Germany
§ Pathologisches Institut, Abt. Neuropathologie, Neurozentrum, Universität Freiburg, Breisacherstraße 64, Freiburg, Germany

*Vitamins and Hormones*, Volume 82
ISSN 0083-6729, DOI: 10.1016/S0083-6729(10)82005-9

brain tumors, long-term application of neuroactive drugs is advised. Several clinically important malignancies based on hormonal dysbalance rise up after treatment with neuroactive drugs, for example, sexual and mental disorders or severe cognitive changes. A drug–hormone cross talk proceeding over drug-mediated cytochrome P450 induction predominantly in the limbic system and the blood–brain barrier, consequently altered steroid hormone metabolism, and P450-mediated change of steroid hormone receptor expression and signaling may serve as an explanation for such disorders. Especially, the interplay between the expression of AR and P450 at the blood–brain barrier and in structures of the limbic system is of considerable interest in understanding brain's reaction on xenobiotic treatment. This chapter summarizes present models and concepts on brain's reaction after xenobiotics crossing the blood–brain barrier and invading the limbic system. 

## I. Introduction

The limbic system represents complex core structures in the phylogenetically old part of the mammalian brain. Based on the initial anatomical and functional definitions and descriptions of Broca in 1878, Papez in 1937, and MacLean (Maclean, 1952), the concept of what the limbic system really signifies has changed again and again by the scientific community. Consisting of a set of brain structures including the hippocampus, amygdala, anterior thalamic nuclei, hypothalamus, and limbic cortex (Conn and Freeman, 2000), the limbic system nowadays can be regarded as a balancing structure in processing input from and to external and internal environment (McLachlan, 2009). Emotional, autonomic, motor, and cognitive responses are determined considerably through memory and motivation or relayed from or to cortical structures like the frontal cortex. Especially, memory and cognition, sexual behavior, and also stress and fear are processed, weighed, and balanced by the limbic system in subtle manner (McLachlan, 2009; Roozendaal *et al.*, 2009).

Many of these neurophysiologic parameters are regulated by so-called neuroactive steroids of either gonadal or brain origin, like testosterone, estradiol, and glucocorticoids (Hojo *et al.*, 2004; Janowsky *et al.*, 1994; McEwen, 1994). The limbic system is considered as a prominent area of neurosteroid biosynthesis and action (Leranth *et al.*, 2003; Yau *et al.*, 2003). Neuroactive steroids are associated with memory, behavior, mood, neuroprotection, aging, and neurotransmission. In addition, steroid hormones were credited with playing an important role in brain development, function, and plasticity (Melcangi and Panzica, 2006). Their corresponding receptors AR, ER$\alpha$ and $\beta$, and glucocorticoid receptor (GR) show high concentrations and considerable overlap of expression in the structures of the limbic system. This predominantly applies to the hippocampus, hypothalamus, and amygdala, and also the thalamic region (Kawata, 1995; McEwen *et al.*, 2001; Meyer *et al.*, 2009).

It becomes evident that disturbances in steroid hormone level or action more or less influence limbic system function. Such a disturbance can either originate from endogenous factors like defects in gonadal or cerebral steroid hormone production or from exogenous delivered compounds derived from environment, nutrition, or medication. Several recent studies demonstrate that especially xenobiotics entering the brain via diffusion or transport over the blood–brain barrier can affect steroid hormone metabolism in brain and alter steroid hormone receptor-mediated downstream signaling (Gehlhaus *et al.*, 2007; Killer *et al.*, 2009; Meyer *et al.*, 2006). The resulting effects on brain physiology are suspected to include clinically relevant endocrine disorders with dramatic impairment of quality of life, for example, bad or depressive mood state, sexual deficits, or detrimental effects on cognition (Herzog and Fowler, 2005).

This chapter summarizes the effects of xenobiotics on steroid hormone action in the limbic system and gives an outlook of the situation at the blood–brain barrier when xenobiotics reach the brain. This so-called drug–hormone cross talk appears to affect brain's network function.

## II. Xenobiotic Uptake and the Limbic System

### A. Sources of xenobiotic uptake

A xenobiotic compound reaching the brain first has to cross the blood–brain barrier. This structure is brain's most powerful protection barrier and therefore, the penetration of xenobiotics into the brain is highly regulated (Gragera *et al.*, 1993). Three groups of xenobiotics can be distinguished in their ability to cross the blood–brain barrier (Seelig *et al.*, 1994): a region of very hydrophobic compounds that fail to enter the central nervous system because they remain adsorbed to the membrane, a central area of less hydrophobic compounds that can cross the blood–brain barrier, and a region of relatively hydrophilic compounds that do not cross the blood–brain barrier unless applied at high concentrations. Most of the substances that act in the central nervous system belong to the second, central area. They cross the blood–brain barrier mainly by diffusion or selective transport systems like the drug transporters of the ATP-binding cassette type (ABC-type transporters) or the organic anion transporters (Bauer *et al.*, 2004; Handschin *et al.*, 2003; Meyer *et al.*, 2001b; Stanley *et al.*, 2009).

The vast majority of xenobiotics entering the brain emanates from drugs, nutrition, or environmentals. The treatment of neurological disorders, brain tumors, or headache attacks with neuroactive drugs is a major challenge predominantly in industrial countries (Diener *et al.*, 1999; Wittchen and

Jacobi, 2004). Around 10% of the European population is affected by neurological disorders like neurodegeneration, dementia, or schizophrenia (source: 16th Meeting of the European Neurological Society (ENS), Lausanne, Switzerland, 2006), 10–15% of patients suffer from depression (Bermejo *et al.*, 2005). Epilepsy as one of the most common neurological disorders affects approximately 1% of the world's population (Forsgren *et al.*, 2005); by far the most of these patients are treated with antiepileptic drugs (AEDs) (Prilipko *et al.*, 2005). The therapy of brain tumors with high-dose chemotherapy (CHT) using cytarabine (AraC) or methotrexate (MTX) is steadily increasing for otherwise-incurable CNS tumors like gliomas or lymphomas (Newton, 2005).

Herbal remedies in self-medication or nutrition are trendy, as it is a part of life style and most people consider them as healthy and safe because of their natural origin (Fricker, 2008). In the United States, consumers spend about $5.1 billion on herbal remedies and approximately 24% of the US population (44.6 million) are regular users of herbal remedies, often in combination with prescription medications (Hepner *et al.*, 2002).

Such a large amount of neuroactive drug or nutrition intake unequivocally implicates interactions with cells, cellular components, or signaling pathways at the blood–brain barrier or inside the brain, being either benign or leading to adverse drug reactions or unintentional side effects.

## B. Side effects in the limbic system

Endocrine disorders and reproductive dysfunction as well as mood and cognitive changes are unusually common among women and men with epilepsy, as compared to the average population. A common feature of these disorders is, that the level of the androgen hormone testosterone was reduced while estrogen level was enhanced, in both, serum and brain (Herzog, 2002; Isojarvi *et al.*, 2005; Meyer *et al.*, 2006; Molleken *et al.*, 2009). There is ongoing discussion on whether these changes in steroid hormone levels emerge from epilepsy itself or from the treatment with AEDs (Herzog, 2008; Luef and Rauchenzauner, 2009). Recent studies imply a crucial role for AEDs in modulating steroid hormone levels in the brain (Frye, 2006; Killer *et al.*, 2009; Meyer *et al.*, 2006). Furthermore, there is a certain risk of suicidal ideation and behavior as adverse effect of AED treatment in epilepsy (Bell *et al.*, 2009). Therefore, the benefits of seizure control by the use of AEDs have to be carefully weighed against possible adverse effects like behavioral problems and psychiatric disorders.

CHT-induced cognitive changes are a severe complication after successful therapy in long-term cancer survivors, for example, of breast cancer and primary CNS lymphoma. Persistent changes of cognitive function emerge in a significant proportion of successfully treated cancer patients after high-dose CHT with substances like AraC or MTX, accompanied by a dramatic impairment of quality of life. Disintegration of drug transporters and disturbance of androgen and estrogen steroid hormone signaling have been discussed as possible candidate mechanisms (Ahles and Saykin, 2002, 2007; Gervasini, 2009; Minisini *et al.*, 2004; Tannock *et al.*, 2004). Interestingly, AEDs like carbamazepine, oxcarbazepine, and phenytoin were found to interact with anticancer drugs in the manner that these AEDs can cause an increase or decrease in anticancer drug concentration, giving rise to adverse drug effects (Yap *et al.*, 2008).

## III. The Main Players

The (bio) chemical properties of the substances in consideration and the nature of possible candidate mechanisms, as discussed above, imply a particular contribution of cytochrome P450 enzymes and steroid hormone receptors as major parts of neurochemical circuits occurring after a xenobiotic substance has crossed the blood–brain barrier. The following paragraph will enlighten these main players in more detail.

### A. Cytochrome P450

Cytochrome P450 (P450, CYP) is the collective term for a large superfamily of heme-containing enzymes that play an important role in the oxidative metabolism of numerous endogenous and foreign compounds like steroid hormones and xenobiotics, namely pharmaceuticals (Flockhart, 2009; Nelson, 2009). The P450 families 1–4 are involved in drug and steroid hormone metabolism, the families 5–52 participate preferentially in steroid hormone or vitamin synthesis (Meyer *et al.*, 2007).

Although the main site for P450-dependent drug metabolism in humans and mammals is the liver, several P450 isoforms are expressed in defined regions and cellular populations of the brain. One major side of brain P450 expression are the glial cells of the brain barrier regions, like the blood–brain barrier or the cerebrospinal fluid–brain barrier. These "barrier" P450s, namely CYP1A1, 1A2, 2B1, 2C11, 2C29, 2E1, and some CYP4Fs, are thought to act as initial filter system when drugs enter the brain (Delaporte and Renton, 1997; Koehler *et al.*, 2009; Meyer *et al.*, 2004, 2007; Tindberg, 2003; Wang *et al.*, 2008). Neuronal P450s are discussed to metabolize odorants, xenobiotics, and steroid hormones specifically attending the affected cell populations. They deactivate steroid hormones or fatty acids in a subtle regulated manner and prepare them for exclusion from the brain. Sites of expression are the olfactory mucosa (CYP1A, 1B, 2A, 2G) (Gu *et al.*, 1998; Huang *et al.*, 2000) and olfactory bulb, striatum, hypothalamus, and cortex (CYP1A1) (Meyer *et al.*, 2002; Schilter and Omiecinski, 1993).

CYP3A is present in steroid hormone-sensitive areas such as hippocampus, hypothalamus, olfactory bulb, and cerebellum (Hagemeyer *et al.*, 2003; Meyer *et al.*, 2009; Wang *et al.*, 2000), while CYP2B6 and 2E1 are found in diverse cortical neuron populations, the olfactory bulb, and the cerebellum (Howard *et al.*, 2003; Lee *et al.*, 2006).

Besides their physiological role in fatty acid and steroid metabolism, various brain P450 isoforms are subject to induction by drugs or other xenobiotics crossing the blood–brain barrier. Examples for potent inducers are alcohol, some neuroleptics, anticonvulsants, and endocrine factors. The main P450 isoforms affected in brain are, among others, CYP1A1/2, 2B1/6, 2C9/19/29, several 2Ds, 3A4/11, and 4Fs (Gervasini *et al.*, 2004; Hedlund *et al.*, 2001; Kalsotra *et al.*, 2006; Lee *et al.*, 2006; Meyer *et al.*, 2007). It is estimated, that about 70% of all clinically used drugs interact with the P450 system (Ingelman-Sundberg, 2004a,b). Induction is considered as a major part of "what happens" after drug influx, as it will lead to altered steroid hormone and/or drug metabolism. Consequently, the physiological balance of drug filtering or steroid hormone metabolism is deranged, which may directly affect downstream signaling pathways.

## B. Steroid hormones and their receptors

Neuroactive steroids are main regulators of brain functions and affect and modulate a wide range of neurophysiologic parameters like mood, behavior, sexuality, memory, cognition, stress (Hojo *et al.*, 2004; Janowsky *et al.*, 1994; McEwen, 1994). Especially, the structures of the limbic system are main targets within the brain for gonadal steroid hormones like testosterone, oestradiol, and glucocorticoids and a relevant region in the brain's capacity for neurosteroid synthesis (Leranth *et al.*, 2003; McEwen, 1996; Yau *et al.*, 2003).

Dysbalance of androgen, estrogen, or glucocorticoid hormone level and action affects hormone signaling, by which the resulting effects may get multiplied in either benign or malign direction. This becomes increasingly clear by the fact, that the corresponding steroid hormone receptors AR, estrogen receptor $\alpha$ and $\beta$, and GR show considerable overlap of expression in the limbic system, supposing synergism in hormone action (Kawata, 1995; Meyer *et al.*, 2009). An example may be the top–down regulation of limbic stress circuits. Psychogenic stimuli, for example, evoked by antipsychotic or stimulating drugs, are processed in limbic forebrain structures, including the amygdala, the hippocampus, and the prefrontal cortex. The output from these limbic structures converges on crucial subcortical relay sites, allowing downstream processing of limbic information (Ulrich-Lai and Herman, 2009). Stress-induced plasticity in the amygdala affects changes in other brain regions and has long-term consequences in cognitive performance and pathological anxiety (Roozendaal *et al.*,

2009). This demonstrates the complex network of steroid hormone-induced information flow in the brain.

But what happens, if xenobiotics interfere with these networks? It is known that, for example, food intake can remodel or, on the opposite, being remodeled by hypothalamic neuronal networks (Magni *et al.*, 2009). Eestrogen or estrogenic drugs interact with cholinergic and serotonergic systems and affect hippocampal and forebrain areas, thereby enhancing memory and cognition (Maki and Dumas, 2009). Epileptic patients treated with certain AEDs demonstrate decreased testosterone and enhanced estrogen levels in serum, and exert sexual or psychiatric disorders to significant extent (Herzog *et al.*, 2006; Mula and Monaco, 2009). In summary, xenobiotic drug action can obviously interact with and modulate steroid hormone-based molecular circuits in the limbic system. This will be discussed concisely in paragraph IV.

## IV. Molecular Circuits in the Limbic System

### A. Induction of P450 → steroid hormone metabolism → androgen receptor regulation

Several recent studies describe the interplay between xenobiotic drug uptake, P450 induction, and steroid hormone regulation as a so-called "drug–hormone cross talk" preferentially occurring in the limbic system. The classical AEDs phenytoin, carbamazepine, and oxcarbazepine were used as model drugs to investigate this cross talk in human and mouse brain (Killer *et al.*, 2009; Meyer *et al.*, 2007). The drug–hormone cross talk model is based on the generally accepted rule, that the conversion of neuroactive steroids to endocrine inactive hydroxylated metabolites by P450s can be regarded as one of the major pathways to inactivate androgens in the brain. This belongs to the main functions of brain P450s like CYP2B, 2C, and 3A under physiological conditions (Hedlund *et al.*, 2001; Meyer *et al.*, 2007; Miksys and Tyndale, 2004; Warner and Gustafsson, 1995).

Neuroactive drugs entering the brain very often have the potency to induce these androgen-metabolizing P450s by binding to specific P450-regulating nuclear receptors, so-called xenosensors, like pregnane X receptor (PXR) or constitutive active receptor (CAR) (Anakk *et al.*, 2003; Handschin and Meyer, 2003). It was shown that in addition to the induction of P450, the expression of AR was concomitantly increased in neurons, both on mRNA and protein level. Furthermore, androgen receptor dependent hormonal signaling gets elevated (Gehlhaus *et al.*, 2007; Killer *et al.*, 2009; Meyer *et al.*, 2006). Testosterone was increasingly metabolized to physiologically inactive hydroxytestosterones in a CYP3A-dependent reaction (Meyer *et al.*, 2006, 2009). Figure 5.1 summarizes these findings

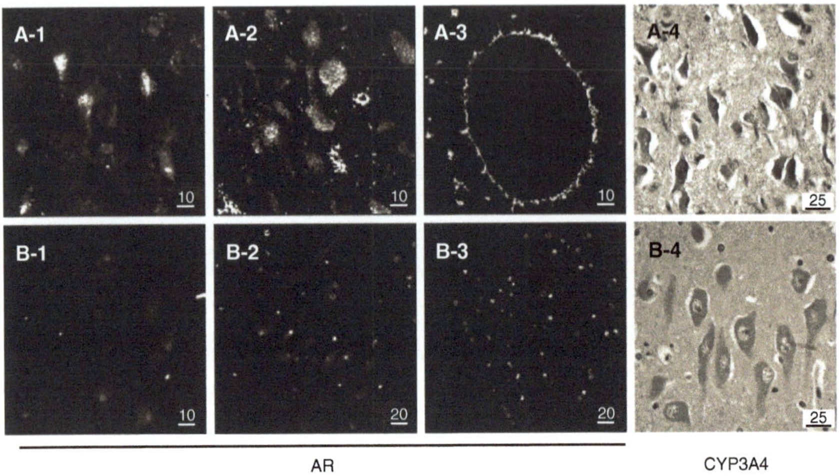


**Figure 5.1** Expression of AR and CYP3A4 in the limbic system of human brain. Series A-1–A-4. Samples from patients with temporal lobe epilepsy, treated with carbamazepine. (A-1, A-2): Representative images of AR expression in hippocampal pyramidal neurons of the CA-2 region of two different individuals. (A-3): AR expression surrounding a capillary vessel in the hilar region of hippocampus. (A-4): CYP3A4 expression in CA2 pyramidal neurons of hippocampus. Series B1–B4: Sections from deceased persons without brain pathology, no AED therapy. (B-1, B-2): Representative images of AR expression in hippocampal pyramidal neurons of the CA-2 region of two different individuals. (B-3): AR expression in the subventricular zone of hippocampus. (B-4): CYP3A4 expression in CA2 pyramidal neurons of hippocampus. (A-1–A3 and B-1–B-3): High-power confocal fluorescence images of sections from paraffin embedded samples from surgical or autoptical resected hippocampus using an anti-AR antibody (Santa Cruz Biotechnology, polyclonal rabbit anti-AR, N-20, sc-816, and C-19, sc-815; Santa Cruz, CA, USA). (A-4, B-4): Immunohistochemical detection of CYP3A4 (dark) in paraffin embedded samples (see above) using an anti-CYP3A4 antibody (polyclonal antirat CYP3A2, cross-reactive to human CYP3A4, BD Gentest, Heidelberg, Germany) (Killer *et al.*, 2009). As previously reported, written informed consent had been obtained from the patients or their relatives in charge for the use of resection material for research, as approved by the institutional Ethics Committee (EK Freiburg 169/05) (Killer *et al.*, 2009). More and detailed methodology information is given elsewhere (Killer *et al.*, 2009; Meyer *et al.*, 2009). Scale bars ($\mu$m) as indicated.

demonstrating elevated AR and CYP3A4 expression in an epileptic patient, treated with the P450-inducing AED carbamazepine as compared to a deceased person without brain pathology or AED treatment.

These data point to a signaling loop beginning with drug-mediated induction of CYP3A, subsequent CYP3A-dependent depletion of testosterone level in hippocampus leading to enhanced expression of AR in these target cells, which in turn gives rise to altered steroid hormone signaling cascades in the brain. The crucial role for brain P450 in this process was demonstrated by the use of cell line models. Specific inhibition of CYP3A isoforms in these cell lines leads to cohesive downregulation of AR expression and signaling indicating that AEDs affect neuronal androgen signaling via a P450-dependent pathway (Gehlhaus *et al.*, 2007; Killer *et al.*, 2009; Meyer *et al.*, 2009).

The drug-hormone cross talk within the limbic system appears at least in part causative for several clinical important affective, cognitive, and reproductive malignancies observed in patients with epilepsy or with certain brain tumors, as mentioned above (paragraph II). Epilepsy patients who had been treated with P450-inducing AEDs suffer from such disorders to much higher amount than such, who had been treated with noninducers (Isojarvi, 2008; Stores, 1975). Reduced bioavailable testosterone could be measured in serum of these patients and it has been demonstrated, that their androgen levels are altered mainly in limbic system rather than in blood serum (Frye, 2006). Severe cognitive changes also occur in a significant subset of brain tumor patients, for example, with primary CNS lymphoma, although they had been successfully treated with high-dose CHT. These cognitive changes are accompanied by a dramatic impairment of quality of life and emerge as an increasingly important clinical problem (Correa and Ahles, 2008). The reduction of androgen and estrogen levels within the brain as a response to the treatment with chemotherapeutic drugs interacting with certain P450s is discussed as a possible candidate mechanism (Ahles and Saykin, 2007).

However, an interesting finding when analyzing the human samples was the occurrence of a strong AR immunosignal around the capillaries and vessels of AED-treated patients (Figs. 5.1A–5.3). Such an immunosignal, highly reminiscent of a "lion's fire ring" from circus attraction, could not be found in control patients. The interesting question is, if this AR expression is part of downstream signaling cascades proceeding from the blood–brain barrier to neuronal structures of the limbic system.

## V. Downstream Effects from the Blood–Brain Barrier to the Limbic System

### A. Epilepsy and tumors: reaction on xenobiotic treatment

Epilepsy and brain tumors give rise to psychopathological defects, as mentioned above. A key role of steroid hormone dysbalance, possibly evoked by the uptake of antiepileptic or chemotherapeutic drugs is discussed for quite some time. An interesting case supporting this hypothesis is demonstrated in Fig. 5.2. Figure 5.2A and B illustrates AR and CYP3A4 expression in a patient with initial diagnosis of astrocytoma (WHO grade III). Investigated after the first surgical intervention, only moderate constitutive expression of both, AR and CYP3A4 were detectable in cortical structures (Fig. 5.2A and B). In hippocampus, only moderate background staining of AR could

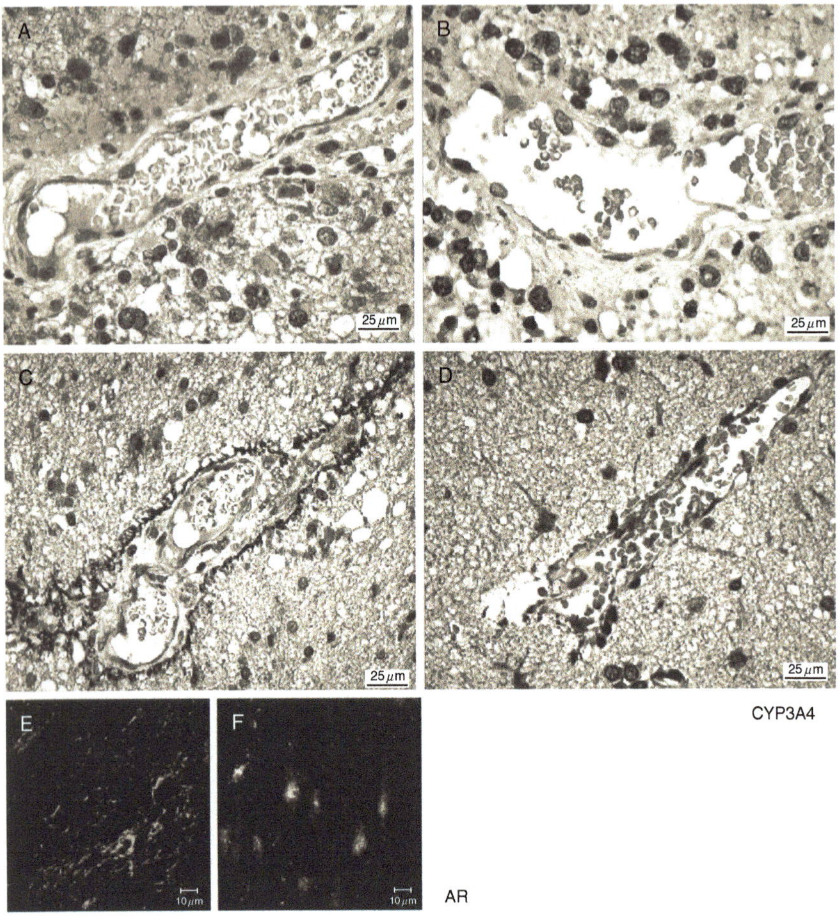


**Figure 5.2** Development of AR and CYP3A4 expression after AED therapy. (A, B): Sample from cortex of a patient with astrocytoma (WHO grade III). Only moderate expression of AR (A) and CYP3A4 (B). (C, D): Recurrence of astrocytoma 4 month later in the same patient. The patient had developed epilepsy after the first surgical intervention and was treated with carbamazepine for 4 months. Excessive expression of AR (C) and CYP3A4 (D) (both visible as dark fall out) around a cortical capillary. (A–D): Immunohistochemical detection of antigens in paraffin embedded material as mentioned in legend of Fig. 5.1. (E, F): AR expression in the CA2 pyramidal cell layer of hippocampus before (E) and after (F) treatment with carbamazepine. High-power confocal fluorescence images of sections from paraffin embedded samples. Scale bars (μm) as indicated.

be observed (Fig. 5.2E). However, after this operation, the patient developed epilepsy, which could be efficiently controlled by the application of carbamazepine. Nevertheless, 4 month later the patient was subject to relapse of the tumor, resulting in a second surgical intervention.

Anew investigation of cortex and hippocampus revealed a dramatic increase of AR and CYP3A4 expression, with the AR immunosignal remaining to the "fire ring" structure found in other AED-treated patients (Figs. 5.1A, 5.2C, D, F, and 5.3). Ascribed to the case history of tumor onset and beginning of AED treatment, it appears conclusive, that most likely the AED treatment led to the observed severe alterations occurring at the blood–brain barrier, in addition to the incidents in the limbic system.

Similar strong expression of AR and CYP3A4 was found in a patient which was treated with high-dose MTX due to acute leukemia (Fig. 5.3). All investigated structures of the limbic system, the ependymal cells of the lateral ventricle (Fig. 5.3A and B), the vessels in stratum oriens (Fig. 5.3A and B), the pyramidal neurons of hippocampal CA2 region (Fig. 5.3E and F), and cortical capillaries (Fig. 5.3C and D) demonstrate strong expression of AR and CYP3A4. These findings support the hypothesis, that treatment with CHT drugs can also influence the hormonal balance in brain via a P450-dependent pathway in terms of the drug–hormone cross talk.

As deduced from recent studies, the demonstrated increase of expression of AR and CYP3A4 could most likely be drawn back to AED or CHT treatment rather than to the disease alone (Gehlhaus *et al.*, 2007; Killer *et al.*, 2009; Meyer *et al.*, 2006).

## B. Androgen receptor "fire ring" at the blood–brain barrier

The occurrence of the AR-"fire ring" around the capillaries of AED- or CHT-treated patients (Fig. 5.4A) led us speculate to what cellular structures this expression of AR may relate to. Therefore, samples from AED-treated epileptic patients were investigated by double staining fluorescence technique (Killer *et al.*, 2009; Meyer *et al.*, 2009) (Fig. 5.4). The capillaries and vessels of the hilar region (Fig. 5.4B), the granule cell layer (Fig. 5.4C), and the inner molecular layer (Fig. 5.4D) showed considerable overlap of the glial marker S-100 and AR expression. The granule neurons themselves however, did not show any costaining (Fig. 5.4E). We conclude from these findings, that AR is expressed in glial structures around the capillaries, besides the neuronal expression found in granule cells and pyramidal neurons of hippocampus. The interplay of AR and drug-metabolizing P450 isoforms is further corroborated by the strong expression of CYP2C19 found at the blood–brain barrier (Fig. 5.4F and G). CYP2C19 is the human analog of murine CYP2C29, which is known as a barrier P450-metabolizing invading xenobiotics (Meyer *et al.*, 2001a,b, 2004).

The present findings conclude, that in addition to the drug–hormone cross talk within the neuronal structures of the limbic system, reactive glial cells at the blood–brain barrier obviously cooperate or coreact to the influx of xenobiotics in the brain.

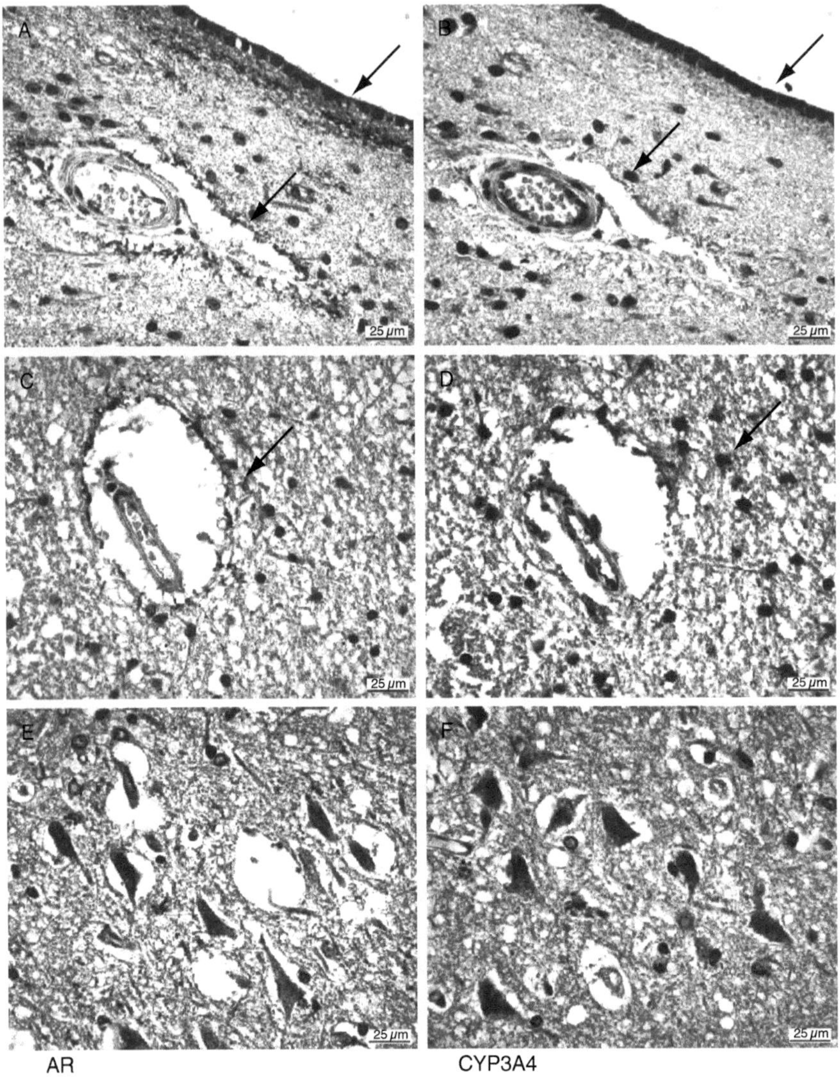


**Figure 5.3** AR and CYP3A4 expression in a tumor patient treated with high-dose MTX. (A, C, E): Expression of AR in the ependymal cells of the lateral ventricle (A), capillaries of hippocampal stratum oriens (A) and cortex (C), and in pyramidal cells of CA2 region in hippocampus (E). (B, D, F): Direct consecutive slices demonstrating CYP3A4 expression in the ependymal cells of the lateral ventricle (B), capillaries of hippocampal stratum oriens (B) and cortex (D), and in pyramidal cells of CA2 region in hippocampus (F). Immunohistochemical detection of antigens in paraffin embedded material as mentioned in legend of Fig. 5.1. The arrows point to strong sites of AR or CYP3A4 expression. Scale bars (μm) as indicated.

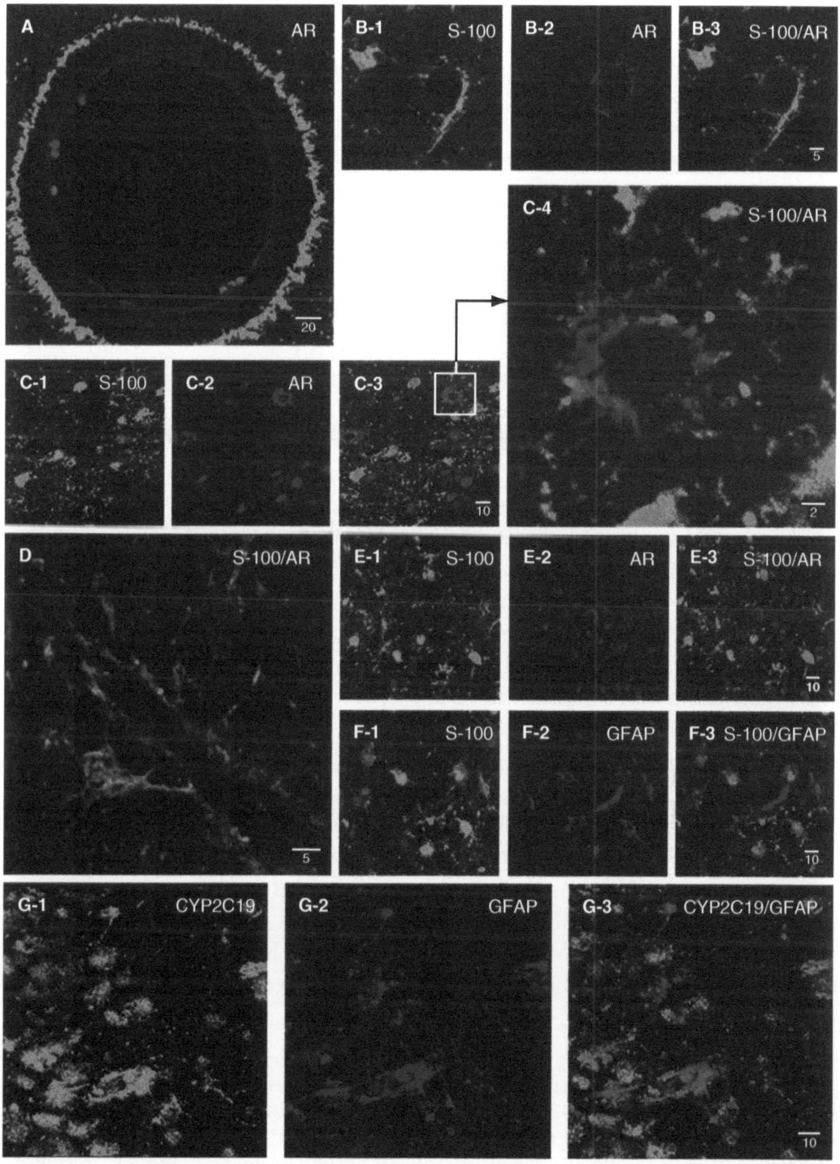


**Figure 5.4** Characterization of AR expression at the capillaries of the limbic system. High-power confocal images of double-labeled sections illustrating expression and colocalization of AR, the glial markers S-100 and GFAP, and CYP2C19. Colocalization of the antigens appears yellow in the overlay mode (B-3, C-3, C-4, D, E-3, F-3, and G-3). All sections were derived from paraffin embedded biopsy or autopsy samples of epilepsy patients treated with P450-inducing AEDs (carbamazepine, oxcarbazepine, or phenytoin) (see legend of Fig. 5.1 for details). (A): AR expression (green) around a capillary vessel in the subventricular zone of hippocampus. (B-1–B-3): S-100 (green,

## C. Disintegrated brain network function?

Several pathologies of the brain involve a disturbance of blood–brain barrier function, and, in many of these, astrocyte function is directly affected. This includes stroke and trauma, infectious processes, multiple sclerosis, but also brain tumors, epilepsy, and neurodegeneration. In most of these cases, the drug transporter ABCB1, a member of the ATP-binding cassette subfamily B, shows dysfunction or reduced efficacy (Abbott *et al.*, 2006). It is known, that ABCB1 is regulated by the xenosensors CAR and PXR and thereby reacts on the induction of P450s of the CYP2C and 3A subfamilies (Bauer *et al.*, 2004; Handschin and Meyer, 2003; Stanley *et al.*, 2009). During the last years, the function of astrocytes at the blood–brain barrier has received a reappraisal, emerging from "brain glue" to important communication elements. This cell–cell communication is mainly driven by gliotransmitters like glutamate or steroid hormones (Volterra and Meldolesi, 2005). Steroid hormones are attributed to have a protective function at the blood–brain barrier in that they cause barrier tightening and improve its function (Abbott *et al.*, 2006). These findings lead us to speculate that the observed expression of AR and CYP3A4 in glial structures at the blood–brain barrier is part of a network type reaction of the brain after the influx of xenobiotics in the brain. This includes the drug–hormone cross talk in the limbic system, leading to alterations in AR-mediated signaling.

The localization of AR in fibrillary structures of glial cells at the blood–brain barrier gives rise to speculation about the function. Recent studies demonstrate a significant portion of AR present in glial cells of the limbic forebrain, the hippocampus, and several sites of the cortex (Conejo *et al.*, 2005; DonCarlos *et al.*, 2006; Finley and Kritzer, 1999). It was further shown that AR immunoreaction is not limited to nuclear or cytoplasmic structures. It was additionally found in glial dendrites, and its immunoreactivity was not appreciably different in males and females, as observed in rhesus monkeys (Conejo *et al.*, 2005; Finley and Kritzer, 1999). Different sites of cellular expression of steroid hormone receptors obviously come

B-1) and AR (red, B-2) expression at a capillary vessel in the hilar region of hippocampus. (C-1–C-4): S-100 (green, C-1) and AR (red, C-2) expression in the granule cell layer of dentate gyrus. (D) S-100 (green) and AR (red) expression at a vessel in the inner molecular layer of the hippocampus. (E-1–E-3): S-100 (green, E-1) and AR (red, E-2) in the granule cell layer of the hippocampus. (F-1–F-3): S-100 (green, F-1) and GFAP (red, F-2) expression in the CA-2 region of the hippocampus. (G-1–G-3): CYP2C19 (green, G-1) and GFAP (red, G-2) expression in the granule cell layer of the hippocampus. Antibodies used: AR as mentioned above (Fig. 5.1), S-100 (RDI-S100abm, Fitzgerald, concord, MA, USA, and anti-S100, Z0311, Dako, Glostrup, Denmark), GFAP (antimouse 556329, BD Pharmingen, Heidelberg, Germany), and CYP19 (polyclonal rabbit antimouse CYP2C29 (K-6, purified IgG-fraction), own manufacturing (Hagemeyer *et al.*, 2009)). Scale bars (μm) as indicated. (See Color Insert.)

along with different function. The intracellular localization of estrogen receptors may serve as example. When in the nucleus, these proteins are estradiol-activated transcription factors, but when trafficked to the cell membrane, ERα and ERβ rapidly activate protein kinase pathways, alter membrane electrical properties, modulate ion flux, and can mediate long-term effects through gene expression. Furthermore, they strongly interact with glutamate receptors to regulate intracellular calcium (Micevych *et al.*, 2009).

The strong reaction of AR and CYP3A4 in both the glial cells of the blood–brain barrier and the neurons of the limbic system leads us to conclude that after influx of xenobiotics either consecutive or parallel processes proceed, which may lead to an imbalance of brain network function. An important and most interesting question arising from these findings is, how these reactions are connected to each other and if they are part of a possible neuroprotective scenario in the brain exerted by steroid hormone action or image of progressive breakdown of cerebral functions activated by the xenobiotic compound.

## VI. Conclusions and Future Directions

Network homeostasis is a prerequisite for maintaining brain's functional stability. Biochemical and neurophysiologic circuits in brain undergo rearrangements throughout life: development, experience and behavior constantly modify synaptic strength, and network connectivity (Maffei and Fontanini, 2009). Cognitive and psychological functions have to adapt to steadily alternating situations. Xenobiotic uptake to the brain is in first line a perturbance of homeostasis, independent of its intent. The structures of the brain have to react to this perturbance in order to remain homeostasis. The blood–brain barrier and the limbic system can be seen as a first line of reaction or defense in processes where foreign compounds invading the brain. The drug–hormone cross talk gives insight in neurobiochemical pathways within the brain after such a xenobiotic influx. Drug-metabolizing P450s like CYP1A, 2B, 2C, and 3A have the potency to direct neuroactive drugs to endocrine signaling. The interplay between drugs and endocrine systems in the brain appears increasingly important in understanding several malignancies like cognitive, mental, and sexual disorders after drug therapy, for example, in epilepsy and other neurological and neoplastic diseases.

The mechanism by which this "feed forward" of drug signal information proceeds has to be fully elucidated. Especially, the interplay between the reaction at the blood–brain barrier and that in the limbic structures is of considerable interest. How are they related to each other? Is it a part of neuroprotection? Or is it just progressive breakdown of brain structures?

The blood–brain barrier is increasingly regarded as an important target to fight disease. The underlying logic is that if blood–brain barrier dysfunction can be reduced, halted, or reversed, this could be valuable therapy in conditions in which neuronal damage is secondary to, or exacerbated by, blood–brain barrier damage (Abbott *et al.*, 2006). New and other players have to be investigated for their role in this process. This includes in particular the drug transporters and xenosensors. Some reports demonstrate effective regulation of brain P450s by drug transporters and xenosensors at the blood–brain barrier (Anakk *et al.*, 2004; Bauer *et al.*, 2004, 2006).

We conclude that the understanding of the neurochemical correlation between drug application and steroid hormone signaling may help us to reduce potential toxicology of neuroactive drugs by altering their design or redirecting their signaling cascades proceeding from the blood–brain barrier to structures of the limbic system or the cortex. The drug-hormone cross talk may guide us to novel strategies for the improvement of therapy in neurological diseases and brain tumors.

## ACKNOWLEDGMENTS

This work was supported by grants from the Deutsche Forschungsgemeinschaft (DFG Me 1544/4) and from the Forschungskommission of the University Hospital Freiburg (MEY 55/01, MEY 322/04, MEY 375/05 and MEY 465/06). We thank Margarethe Ditter (Department of Neuropathology, University of Freiburg) for her extraordinary technical assistance. This work is dedicated to Benedikt Volk, University of Freiburg, to honor his continuous discussion, advice, and support.

## REFERENCES

Abbott, N. J., Ronnback, L., and Hansson, E. (2006). Astrocyte–endothelial interactions at the blood–brain barrier. *Nat. Rev. Neurosci.* **7,** 41–53.

Ahles, T. A., and Saykin, A. J. (2002). Breast cancer chemotherapy-related cognitive dysfunction. *Clin. Breast Cancer* **3**(Suppl. 3), S84–S90.

Ahles, T. A., and Saykin, A. J. (2007). Candidate mechanisms for chemotherapy-induced cognitive changes. *Nat. Rev. Cancer* **7,** 192–201.

Anakk, S., Kalsotra, A., Shen, Q., Vu, M. T., Staudinger, J. L., Davies, P. J., and Strobel, H. W. (2003). Genomic characterization and regulation of CYP3a13: Role of xenobiotics and nuclear receptors. *FASEB J.* **17,** 1736–1738.

Anakk, S., Kalsotra, A., Kikuta, Y., Huang, W., Zhang, J., Staudinger, J. L., Moore, D. D., and Strobel, H. W. (2004). CAR/PXR provide directives for Cyp3a41 gene regulation differently from Cyp3a11. *Pharmacogenomics J.* **4,** 91–101.

Bauer, B., Hartz, A. M., Fricker, G., and Miller, D. S. (2004). Pregnane X receptor up-regulation of P-glycoprotein expression and transport function at the blood–brain barrier. *Mol. Pharmacol.* **66,** 413–419.

Bauer, B., Yang, X., Hartz, A. M., Olson, E. R., Zhao, R., Kalvass, J. C., Pollack, G. M., and Miller, D. S. (2006). In vivo activation of human pregnane X receptor tightens the blood–brain barrier to methadone through P-glycoprotein up-regulation. *Mol. Pharmacol.* **70,** 1212–1219.

Bell, G. S., Mula, M., and Sander, J. W. (2009). Suicidality in people taking antiepileptic drugs: What is the evidence? *CNS Drugs* **23,** 281–292.

Bermejo, I., Kratz, S., Gaebel, W., Berger, M., Schneider, F., Pfeiffer-Gerschel, T., Hegerl, U., and Haerter, M. (2005). Self-evaluation of depression by patients and condition identification by primary care physicians—A comparison. *Die. Med. Welt.* **56,** 73–78.

Conejo, N. M., Gonzalez-Pardo, H., Cimadevilla, J. M., Arguelles, J. A., Diaz, F., Vallejo-Seco, G., and Arias, J. L. (2005). Influence of gonadal steroids on the glial fibrillary acidic protein-immunoreactive astrocyte population in young rat hippocampus. *J. Neurosci. Res.* **79,** 488–494.

Conn, P. M., and Freeman, M. E. (2000). Neuroendocrinology in Physiology and Medicine. Humana Press, Totowa, NJ.

Correa, D. D., and Ahles, T. A. (2008). Neurocognitive changes in cancer survivors. *Cancer J.* **14,** 396–400.

Delaporte, E., and Renton, K. W. (1997). Cytochrome P4501A1 and cytochrome P4501A2 are downregulated at both transcriptional and post-transcriptional levels by conditions resulting in interferon-alpha/beta induction. *J. Pharm. Biomed. Anal.* **60,** 787–796.

Diener, H. C., Kaube, H., and Limmroth, V. (1999). Antimigraine drugs. *J. Neurol.* **246,** 515–519.

DonCarlos, L. L., Sarkey, S., Lorenz, B., Azcoitia, I., Garcia-Ovejero, D., Huppenbauer, C., and Garcia-Segura, L. M. (2006). Novel cellular phenotypes and subcellular sites for androgen action in the forebrain. *Neuroscience* **138,** 801–807.

Finley, S. K., and Kritzer, M. F. (1999). Immunoreactivity for intracellular androgen receptors in identified subpopulations of neurons, astrocytes and oligodendrocytes in primate prefrontal cortex. *J. Neurobiol.* **40,** 446–457.

Flockhart, D. A. (2009). P450-Drug interaction table. Available at http://medicine.iupui.edu/clinpharm/ddis/last accessed date: Dec 6th, 2009.

Forsgren, L., Beghi, E., Oun, A., and Sillanpaa, M. (2005). The epidemiology of epilepsy in Europe—A systematic review. *Eur. J. Neurol.* **12,** 245–253.

Fricker, G. (2008). Drug interactions with natural products at the blood brain barrier. *Curr. Drug Metab.* **9,** 1019–1026.

Frye, C. A. (2006). Role of androgens in epilepsy. *Expert Rev. Neurother.* **6,** 1061–1075.

Gehlhaus, M., Schmitt, N., Volk, B., and Meyer, R. P. (2007). Antiepileptic drugs affect neuronal androgen signaling via a cytochrome P450-dependent pathway. *J. Pharm. Exp. Ther.* **322,** 550–559.

Gervasini, G. (2009). Polymorphisms in methotrexate pathways: What is clinically relevant, what is not, and what is promising. *Curr. Drug Metab.* **10**(6), 547–566.

Gervasini, G., Carrillo, J. A., and Benitez, J. (2004). Potential role of cerebral cytochrome P450 in clinical pharmacokinetics: Modulation by endogenous compounds. *Clin. Pharmacokinet.* **43,** 693–706.

Gragera, R. R., Muñiz, E., and Martinez-Rodriguez, R. (1993). Molecular and ultrastructural basis of the blood–brain barrier function. Immunohistochemical demonstration of Na + /K + ATPase, à-actin, phosphocreatine and clathrin in the capillary wall and its microenvironment. *Cell. Mol. Biol.* **39,** 819–828.

Gu, J., Zhang, Q. Y., Genter, M. B., Lipinskas, T. W., Negishi, M., Nebert, D. W., and Ding, X. (1998). Purification and characterization of heterologously expressed mouse CYP2A5 and CYP2G1: Role in metabolic activation of acetaminophen and 2,6-dichlorobenzonitrile in mouse olfactory mucosal microsomes. *J. Pharm. Exp. Ther.* **285,** 1287–1295.

Hagemeyer, C. E., Rosenbrock, H., Ditter, M., Knoth, R., and Volk, B. (2003). Predominantly neuronal expression of cytochrome P450 isoforms CYP3A11 and CYP3A13 in mouse brain. *Neuroscience* **117,** 521–529.

Hagemeyer, C. E., Burck, C., Schwab, R., Knoth, R., and Meyer, R. P. (2009). 7-Benzyloxyresorufin-O-dealkylase activity as a marker for measuring cytochrome P450 CYP3A induction in mouse liver. *Anal. Biochem.* 10.1016/j.ab.2009.11.004.

Handschin, C., and Meyer, U. A. (2003). Induction of drug metabolism: The role of nuclear receptors. *Pharmacol. Rev.* **55,** 649–673.

Handschin, C., Podvinec, M., and Meyer, U. A. (2003). In silico approaches, and in vitro and in vivo experiments to predict induction of drug metabolism. *Drug News Perspect.* **16,** 423–434.

Hedlund, E., Gustafsson, J. A., and Warner, M. (2001). Cytochrome P450 in the brain; A review. *Curr. Drug Metab.* **2,** 245–263.

Hepner, D. L., Harnett, M., Segal, S., Camann, W., Bader, A. M., and Tsen, L. C. (2002). Herbal medicine use in parturients. *Anesth. Analg.* **94,** 690–693, table of contents.

Herzog, A. G. (2002). Altered reproductive endocrine regulation in men with epilepsy: Implications for reproductive function and seizures. *Ann. Neurol.* **51,** 539–542.

Herzog, A. G. (2008). Disorders of reproduction in patients with epilepsy: Primary neurological mechanisms. *Seizure* **17,** 101–110.

Herzog, A. G., and Fowler, K. M. (2005). Sexual hormones and epilepsy: Threat and opportunities. *Curr. Opin. Neurol.* **18,** 167–172.

Herzog, A. G., Drislane, F. W., Schomer, D. L., Pennell, P. B., Bromfield, E. B., Dworetzky, B. A., Farina, E. L., and Frye, C. A. (2006). Differential effects of antiepileptic drugs on neuroactive steroids in men with epilepsy. *Epilepsia* **47,** 1945–1948.

Hojo, Y., Hattori, T. A., Enami, T., Furukawa, A., Suzuki, K., Ishii, H. T., Mukai, H., Morrison, J. H., Janssen, W. G., Kominami, S., Harada, N., Kimoto, T., and Kawato, S. (2004). Adult male rat hippocampus synthesizes estradiol from pregnenolone by cytochromes P45017alpha and P450 aromatase localized in neurons. *Proc. Natl. Acad. Sci. USA* **101,** 865–870.

Howard, L. A., Miksys, S., Hoffmann, E., Mash, D., and Tyndale, R. F. (2003). Brain CYP2E1 is induced by nicotine and ethanol in rat and is higher in smokers and alcoholics. *Br. J. Pharmacol.* **138,** 1376–1386.

Huang, P., Rannug, A., Ahlbom, E., Hakansson, H., and Ceccatelli, S. (2000). Effect of 2,3,7,8-tetrachlorodibenzo-p-dioxin on the expression of cytochrome P450 1A1, the aryl hydrocarbon receptor, and the aryl hydrocarbon receptor nuclear translocator in rat brain and pituitary. *Toxicol. Appl. Pharmacol.* **169,** 159–167.

Ingelman-Sundberg, M. (2004a). Human drug metabolising cytochrome P450 enzymes: Properties and polymorphisms. *Naunyn. Schmiedebergs Arch. Pharmacol.* **369,** 89–104.

Ingelman-Sundberg, M. (2004b). Pharmacogenetics of cytochrome P450 and its applications in drug therapy: The past, present and future. *Trends Pharmacol. Sci.* **25,** 193–200.

Isojarvi, J. (2008). Disorders of reproduction in patients with epilepsy: Antiepileptic drug related mechanisms. *Seizure* **17,** 111–119.

Isojarvi, J. I., Tauboll, E., and Herzog, A. G. (2005). Effect of antiepileptic drugs on reproductive endocrine function in individuals with epilepsy. *CNS Drugs* **19,** 207–223.

Janowsky, J. S., Oviatt, S. K., and Orwoll, E. S. (1994). Testosterone influences spatial cognition in older men. *Behav. Neurosci.* **108,** 325–332.

Kalsotra, A., Zhao, J., Anakk, S., Dash, P. K., and Strobel, H. W. (2007). Brain trauma leads to enhanced lung inflammation and injury: Evidence for role of P4504Fs in resolution. *J. Cereb. Blood Flow Metab.* **27,** 963–974.

Kawata, M. (1995). Roles of steroid hormones and their receptors in structural organization in the nervous system. *Neurosci. Res.* **24,** 1–46.

Killer, N., Hock, M., Gehlhaus, M., Capetian, P., Knoth, R., Pantazis, G., Volk, B., and Meyer, R. P. (2009). Modulation of androgen and estrogen receptor expression by antiepileptic drugs and steroids in hippocampus of patients with temporal lobe epilepsy. *Epilepsia* **50,** 1875–1890.

Koehler, R. C., Roman, R. J., and Harder, D. R. (2009). Astrocytes and the regulation of cerebral blood flow. *Trends Neurosci.* **32,** 160–169.

Lee, A. M., Miksys, S., Palmour, R., and Tyndale, R. F. (2006). CYP2B6 is expressed in African Green monkey brain and is induced by chronic nicotine treatment. *Neuropharmacology* **50,** 441–450.

Leranth, C., Petnehazy, O., and MacLusky, N. J. (2003). Gonadal hormones affect spine synaptic density in the CA1 hippocampal subfield of male rats. *J. Neurosci.* **23,** 1588–1592.

Luef, G., and Rauchenzauner, M. (2009). Epilepsy and hormones: A critical review. *Epilepsy Behav.* **15,** 73–77.

Maclean, P. D. (1952). Some psychiatric implications of physiological studies on frontotemporal portion of limbic system (visceral brain). *Electroencephalogr. Clin. Neurophysiol.* **4,** 407–418.

Maffei, A., and Fontanini, A. (2009). Network homeostasis: A matter of coordination. *Curr. Opin. Neurobiol.* **19,** 168–173.

Magni, P., Dozio, E., Ruscica, M., Celotti, F., Masini, M. A., Prato, P., Broccoli, M., Mambro, A., More, M., and Strollo, F. (2009). Feeding behavior in mammals including humans. *Ann. NY Acad. Sci.* **1163,** 221–232.

Maki, P. M., and Dumas, J. (2009). Mechanisms of action of estrogen in the brain: Insights from human neuroimaging and psychopharmacologic studies. *Semin. Reprod. Med.* **27,** 250–259.

McEwen, B. S. (1994). How do sex and stress hormones affect nerve cells? *Ann. NY Acad. Sci.* **743,** 1–16.

McEwen, B. S. (1996). Gonadal and adrenal steroids regulate neurochemical and structural plasticity of the hippocampus via cellular mechanisms involving NMDA receptors. *Cell. Mol. Neurobiol.* **16,** 103–116.

McEwen, B., Akama, K., Alves, S., Brake, W. G., Bulloch, K., Lee, S., Li, C., Yuen, G., and Milner, T. A. (2001). Tracking the estrogen receptor in neurons: Implications for estrogen-induced synapse formation. *Proc. Natl. Acad. Sci. USA* **98,** 7093–7100.

McLachlan, R. S. (2009). A brief review of the anatomy and physiology of the limbic system. *Can. J. Neurol. Sci.* **36**(Suppl. 2), S84–S87.

Melcangi, R. C., and Panzica, G. C. (2006). Neuroactive steroids: Old players in a new game. *Neuroscience* **138,** 733–739.

Meyer, R. P., Hagemeyer, C. E., Knoth, R., Kurz, G., and Volk, B. (2001a). Oxidative hydrolysis of scoparone by cytochrome P450 Cyp2c29 reveals a novel metabolite. *Biochem. Biophys. Res. Commun.* **285,** 32–39.

Meyer, R. P., Knoth, R., Schiltz, E., and Volk, B. (2001b). Possible function of astrocyte cytochrome P450 in control of xenobiotic phenytoin in the brain: In vitro studies on murine astrocyte primary cultures. *Exp. Neurol.* **167,** 376–384.

Meyer, R. P., Podvinec, M., and Meyer, U. A. (2002). Cytochrome P450 CYP1A1 accumulates in the cytosol of kidney and brain and is activated by heme. *Mol. Pharmacol.* **62,** 1061–1067.

Meyer, R. P., Volk, B., Knoth, R., Aschner, M., and Costa, L. G. (2004). Function of astrocyte cytochrome P450 in control of xenobiotic metabolism. The Role of Glia in Neurotoxicity. Vol. 2, pp. 61–72. CRC Press, Boca Raton.

Meyer, R. P., Hagemeyer, C. E., Knoth, R., Kaufmann, M. R., and Volk, B. (2006). Anti-epileptic drug phenytoin enhances androgen metabolism and androgen receptor expression in murine hippocampus. *J. Neurochem.* **96,** 460–472.

Meyer, R. P., Gehlhaus, M., Knoth, R., and Volk, B. (2007). Expression and function of cytochrome P450 in brain drug metabolism. *Curr. Drug Metab.* **8,** 297–306.

Meyer, R. P., Gehlhaus, M., Schwab, R., Burck, C., Knoth, R., and Hagemeyer, C. E. (2009). Concordant up-regulation of cytochrome P450 Cyp3a11, testosterone oxidation and androgen receptor expression in mouse brain after xenobiotic treatment. *J. Neurochem.* **109,** 670–681.

Micevych, P., Kuo, J., and Christensen, A. (2009). Physiology of membrane oestrogen receptor signalling in reproduction. *J. Neuroendocrinol.* **21,** 249–256.

Miksys, S., and Tyndale, R. F. (2004). The unique regulation of brain cytochrome P450 2 (CYP2) family enzymes by drugs and genetics. *Drug Metab. Rev.* **36,** 313–333.

Minisini, A., Atalay, G., Bottomley, A., Puglisi, F., Piccart, M., and Biganzoli, L. (2004). What is the effect of systemic anticancer treatment on cognitive function? *Lancet Oncol.* **5,** 273–282.

Molleken, D., Richter-Appelt, H., Stodieck, S., and Bengner, T. (2009). Sexual quality of life in epilepsy: Correlations with sex hormone blood levels. *Epilepsy Behav.* **14,** 226–231.

Mula, M., and Monaco, F. (2009). Antiepileptic drugs and psychopathology of epilepsy: An update. *Epileptic Disord.* **11,** 1–9.

Nelson, D. R. (2009). P450 nomenclature and overview. Available at http://drnelson.utmem.edu/CytochromeP450.html. last accessed date: Dec 6th, 2009.

Newton, H. B. (2005). Clinical pharmacology of brain tumor chemotherapy. *In* "Handbook ofBrain Tumor Chemotherapy" (H. B. Newton, Ed.), pp. 21–42. Academic Press, Oxford.

Prilipko, L., deBoer, H. M., Saxena, S., and Who, S. (2005). Atlas: Epilepsy Care in the World. World Health Organisation, Geneva, Switzerland.

Roozendaal, B., McEwen, B. S., and Chattarji, S. (2009). Stress, memory and the amygdala. *Nat. Rev. Neurosci.* **10,** 423–433.

Schilter, B., and Omiecinski, C. J. (1993). Regional distribution and expression modulation of cytochrome P-450 and epoxide hydrolase mRNAs in the rat brain. *Mol. Pharmacol.* **44,** 990–996.

Seelig, A., Gottschlich, R., and Devant, R. M. (1994). A method to determine the ability of drugs to diffuse through the blood–brain barrier. *Proc. Natl. Acad. Sci. USA* **91,** 68–72.

Stanley, L. A., Horsburgh, B. C., Ross, J., Scheer, N., and Wolf, C. R. (2009). Drug transporters: Gatekeepers controlling access of xenobiotics to the cellular interior. *Drug Metab. Rev.* **41,** 27–65.

Stores, G. (1975). Behavioural effects of anti-epileptic drugs. *Dev. Med. Child Neurol.* **17,** 647–658.

Tannock, I. F., Ahles, T. A., Ganz, P. A., and Van Dam, F. S. (2004). Cognitive impairment associated with chemotherapy for cancer: Report of a workshop. *J. Clin. Oncol.* **22,** 2233–2239.

Tindberg, N. (2003). Phorbol ester induces CYP2E1 in astrocytes, through a protein kinase C- and tyrosine kinase-dependent mechanism. *J. Neurochem.* **86,** 888–895.

Ulrich-Lai, Y. M., and Herman, J. P. (2009). Neural regulation of endocrine and autonomic stress responses. *Nat. Rev. Neurosci.* **10,** 397–409.

Volterra, A., and Meldolesi, J. (2005). Astrocytes, from brain glue to communication elements: The revolution continues. *Nat. Rev. Neurosci.* **6,** 626–640.

Wang, H., Napoli, K. L., and Strobel, H. W. (2000). Cytochrome P450 3A9 catalyzes the metabolism of progesterone and other steroid hormones. *Mol. Cell. Biochem.* **213,** 127–135.

Wang, Y., Zhao, J., Kalsotra, A., Turman, C. M., Grill, R. J., Dash, P. K., and Strobel, H. W. (2008). CYP4Fs expression in rat brain correlates with changes in LTB4 levels after traumatic brain injury. *J. Neurotrauma.* **25,** 1187–1194.

Warner, M., and Gustafsson, J. A. (1995). Cytochrome P450 in the brain: Neuroendocrine functions. *Front. Neuroendocrinol.* **16,** 224–236.

Wittchen, H. U., and Jacobi, F. (2004). Angstst"rungen. Gesundheitsberichterstattung des Bundes Robert Koch-Institut, Berlin, Germanypp. 1–16.

Yap, K. Y., Chui, W. K., and Chan, A. (2008). Drug interactions between chemotherapeutic regimens and antiepileptics. *Clin. Ther.* **30,** 1385–1407.

Yau, J. L., Rasmuson, S., Andrew, R., Graham, M., Noble, J., Olsson, T., Fuchs, E., Lathe, R., and Seckl, J. R. (2003). Dehydroepiandrosterone 7-hydroxylase CYP7B: Predominant expression in primate hippocampus and reduced expression in Alzheimer's disease. *Neuroscience* **121,** 307–314.

CHAPTER SIX

# Brain Plasticity after Ischemic Episode

Galyna G. Skibo *and* Alexander G. Nikonenko

## Contents



## Abstract

Brain plasticity describes the potential of the organ for adaptive changes involved in various phenomena in health and disease. A substantial amount of experimental evidence, received in animal and cell models, shows that a cascade of plastic changes at the molecular, cellular, and tissue levels, is initiated in different regions of the postischemic brain. Underlying mechanisms include neurochemical alterations, functional changes in excitatory and inhibitory synapses, axonal and dendritic sprouting, and reorganization of sensory and motor central maps. Multiple lines of evidence indicate numerous points in which the process of postischemic recovery may be influenced with the aim to restore the full capacity of the brain tissue injured by an ischemic episode. 

## I. Mechanisms of Brain Ischemic Injury

Cerebral ischemia is caused by major insufficiencies in brain blood supply. Its primary effect is a loss of oxygen and energy substrates, causing dramatic shortages in ATP production and an electron leakage generating reactive oxygen species (ROS) (Lo, 2008; Nikonenko *et al.*, 2009). ROS, together with other reactive molecules, promote lipid, protein, and DNA oxidation affecting normal cell physiology. In addition, ROS may activate specific signal transduction pathways, leading to cell death (Perez-Pinzon *et al.*, 2005).

Department of Cytology, Bogomoletz Institute of Physiology, Kiev, Ukraine

*Vitamins and Hormones*, Volume 82
ISSN 0083-6729, DOI: 10.1016/S0083-6729(10)82006-0

Transient hypoxia slightly decreases respiration in mitochondria, but together with a rise in intracellular $Ca^{2+}$, it causes dramatic effects which can amount to the disintegration of mitochondrial membranes (Bonanni *et al.*, 2006). Persisting glycolysis leads to a significant drop in tissue pH. Some of the acid-sensing, proton-gated cation channels widely expressed in CNS neurons, are highly permeable for $Ca^{2+}$ ions and play important roles in acid-induced cell death (Gao *et al.*, 2005). The described events are most prominent in the focus of ischemic injury. The latter is surrounded by a moderately hypoperfused region, called penumbra, where energy production is compromised to a lesser extent, but ischemia-related events here may eventually evolve into cell death (Lo, 2008).

Cerebral ischemia causes a dramatic loss of extracellular fluid and concomitant swelling of brain cells (Nicholson and Sykova, 1998). Progressive ischemia-related swelling of perivascular astrocytic feet is associated with the narrowing of microvascular lumen (Naganuma, 1990). Early brain swelling may be influenced by the degranulation of mast cells, the numbers of which dramatically increase in the ischemic hemisphere (Vexler and Yenari, 2009).

Soon after ischemia, the malfunction of ATP-dependent processes leads to a dramatic change in ion gradients associated with a sustained abnormal release of glutamate from neurons and glial cells (Benveniste *et al.*, 1984). Excessive glutamate levels stimulate N-methyl-D-aspartate (NMDA) receptors expressed in neurons and certain glial cell types inducing a rapid opening of $Ca^{2+}$-permeable channels across cell membranes (Nakamichi *et al.*, 2004). Resulting high cytoplasmic levels of $Ca^{2+}$ initiate a cascade of intracellular events that may cause cell damage. This process, called excitotoxicity, involves the synergistic action of proteases, ROS, and $Ca^{2+}$ overload (Choi and Rothman, 1990).

Mild ischemia was reported to potentiate synaptic transmission, while more severe ischemic insults were found to suppress it (Hammond *et al.*, 1994; Jourdain *et al.*, 2002). Postischemic long-term potentiation (LTP) indicates that pyramidal neurons in the hippocampal CA1 area become dysfunctional long before their death (Miyazaki *et al.*, 1993). Changes in synaptic transmission implicate alterations in the composition and function of various receptors in excitatory and inhibitory synapses. NMDA receptors contribute little to basal synaptic transmission, but are crucial for the induction of certain forms of synaptic plasticity. Antagonists selectively targeting NMDA receptors containing the NR2B subunit blocked postischemic LTP and reduced ischemic damage (Picconi *et al.*, 2006). Abnormal influx of $Ca^{2+}$ through $\alpha$-amino-3-hydroxy-5-methyl-4-isoxazolepropionic acid (AMPA) receptors is thought to contribute to the neuronal death associated with ischemia. Recent evidence indicates that the expression of GluR2 subunit, which presence makes AMPA receptors $Ca^{2+}$-impermeable, is downregulated in vulnerable neurons prior to cell death (Tanaka *et al.*, 2000). Ischemic insults promote internalization of GluR2-containing AMPA receptors from synaptic sites via clathrin-dependent endocytosis and facilitate delivery of GluR2-lacking AMPA receptors to synaptic sites via exocytosis (Liu *et al.*, 2006).

Cerebral ischemia is followed by an increase of neuronal excitability and a decrease of gamma-aminobutyric acid (GABA)-related inhibition in non-ischemic, structurally connected brain areas. This is associated with a downregulation of GABA receptor binding, and an altered composition of GABA receptors (Witte and Stoll, 1997). Ischemia-related deficits of GABAergic transmission occur between days and weeks after the insult being associated with structural degeneration of GABAergic interneurons in a tissue surrounding the infarct (Neumann-Haefelin *et al.*, 1998).

A variety of substances are released into the extracellular space in the brain in response to acute ischemic stress, and their effects are exclusively mediated by receptor-linked second messenger systems. One of them, adenylate cyclase, is activated in the initial phase of acute ischemia, and its activity gradually decreases in the late phase of ischemia (Tanaka, 2001).

Focal cerebral ischemia was shown to inhibit protein synthesis in the infarct core (Popa-Wagner *et al.*, 1999) and later in the penumbra (Zhang *et al.*, 2006). A marked increase in the immunoreactivity of ubiquitin, used as a tag to label proteins committed to be hydrolyzed, was shown in postischemic hippocampal neurons (Gubellini *et al.*, 1997). The modification of kainate receptor by its conjugation with SUMO protein, ubiquitine relative, may influence the function of this receptor. A dramatic increase in such conjugation was observed in the striatal infarct area. Levels of kainate receptors were decreased only in the infarct zone, whereas AMPA receptor levels were decreased in other brain areas as well (Cimarosti *et al.*, 2008).

Intense proteolysis of cytoskeletal proteins occurs in the brain within minutes of transient ischemia because of the activation of calcium-sensitive proteases (calpains), and fodrin (Yokota *et al.*, 2003), α-spectrin (Neumar *et al.*, 2001), microtubule-associated protein-2 (MAP2) (Buddle *et al.*, 2003), and other cytoskeletal proteins are being degraded. MAP tau protein, influencing the dynamics of microtubule assembly, is rapidly dephosphorylated after transient cerebral ischemia (Shackelford and Yeh, 1998). Calpain-dependent cleavage of collapsin response mediator protein (CRMP), one of the direct modulators of microtubules, signals axonal retraction and neuronal death, while blocking this signal transduction pathway prevents these events from happening after cerebral ischemia (Hou *et al.*, 2008).

Cell death associated with a short ischemic episode tends to be delayed and selective for neurons, whereas prolonged ischemia leads to a broader and more rapid cell death dynamics. The lag between an ischemic episode and neuronal death depends upon the severity and/or duration of ischemia (Rosenblum, 1997). The ischemia-related degeneration of pyramidal cells in the hippocampal CA1 area increased progressively leading to injury and death in 79.5% of neurons within 7 days (Kovalenko *et al.*, 2006). Hippocampal interneurons are generally more resistant than pyramidal cells to excitotoxic insults possibly due to differences in NMDA receptor expression and/or subunit composition (Avignone *et al.*, 2005).

Brain cells affected by ischemia die by necrosis, apoptosis, autophagy, or hybrid forms of cell death. Apoptotic cells represent a minor fraction of dying postischemic neurons, while the rest follow a necrosis-like pathway (Martin *et al.*, 1998; Müller *et al.*, 2004). Upregulation of c-Jun N-terminal kinases is viewed as an early sign of apoptosis in neurons and caspase-3 is upregulated in the majority of apoptotic cells (Müller *et al.*, 2004; Rami, 2003). Caspases are a family of cysteine proteases that undergo proteolytic maturation in a sequential manner. Effector caspases cleave various molecules, including proteins, inhibiting apoptosis, and the activation of effector caspases is regarded as a terminal event of apoptosis (McLaughlin, 2004). Caspase activation occurs as early as 1 h after ischemia in discrete cell populations of the adult brain (Chen *et al.*, 2005).

Some evidence suggests that ischemia-related neuronal death may involve the elements of autophagy—a homeostatic process required for the recycling of proteins and damaged organelles (Rami and Kögel, 2008). In addition, some cells in the penumbra were found to display hybrid cell death features combining various apoptotic and necrotic alterations (Wei *et al.*, 2004). The neuronal vulnerability is region-specific and the dorsolateral striatum, hippocampal CA1 and CA2 areas, particular cortical areas are among the most susceptible to ischemic damage. On the other hand, anterior thalamus, paramedian neocortex, and dentate gyrus demonstrate noticeable resistance to ischemic insults.

A short ischemic episode induces a progressive decrease in synaptic numbers in the rat CA1 area exceeding 30%—1 day and 65%—7 days after the occlusion. Nearly half of the remaining synapses display degenerative features (Kovalenko *et al.*, 2006). The so-called postsynaptic density (PSD), the important constituent of a synapse, is one of the most affected. Rapid ischemia-related enlargement of PSD was shown in excitatory spine synapses in the CA1 area, both *in vivo* and *in vitro* (Kovalenko *et al.*, 2006). This enlargement was more pronounced and longer-term in the CA1 area than in the dentate gyrus. PSD is composed of a cytoskeletal frame, in which proteins, including neurotransmitter receptors, ion channels, and adaptor proteins, are anchored. The ischemia-related activation of proteases results in a partial degradation of certain cross-linker proteins in PSD, such as spectrin and MAP2, leading to a more diffuse and irregular PSD structure (Martone *et al.*, 1999). The increased amount of material associated with PSD in the postischemic brain can represent denatured and aggregated proteins that are heavily ubiquitinized (Hu *et al.*, 2000). These changes may cause persistent alteration in synaptic transmission.

Synaptic vesicles (SVs) constitute an integral component of a chemical synapse specializing in the storage of a neurotransmitter and their availability for release determines some presynaptic properties. SVs can be divided into three pools characterized by different release-competence: the readily releasable pool (RRP), reserve pool, and resting pool (Sudhof, 2000). In excitatory synapses of the CA1 area, SVs numbers tend to decrease progressively after ischemia (Kovalenko *et al.*, 2006). It is consistent with the data showing that cerebral ischemia activates $Ca^{2+}$-dependent exocytosis of SVs (Hammond *et al.*, 1994). Data indicate that during days after ischemia SVs exocytosis is not balanced by the adequate levels of SVs endocytosis.

Spatial rearrangement of SVs was observed as early as 15 min after ischemia both *in vivo* and *in vitro*, the average distance from these organelles to active zones being increased significantly. During the next 7 days, spatial patterns formed by SVs became normal. It has to be noted that these changes reflect mainly the disappearance of those organelles which are in close proximity to active zones (Kovalenko *et al.*, 2006; Nikonenko and Skibo, 2004).

The intervesicle distance may be used as a measure of vesicle clustering potential. It is of interest that the value of this parameter experiences a marked increase after ischemia. The change is more pronounced for RRP and part of the reserve pool cluster facing an active zone. The increase in intervesicle spacings indicates the impaired integrity of the spatial clusters of SVs and may point to abnormal interactions between SVs proteins after ischemia (Fig. 6.1). The expression of SNAP-25 and synaptophysin implicated in the regulation of SVs cycling has been modified following transient ischemia (Ishimaru *et al.*, 2001). A postischemic increase in calpain proteolytic activity and related degradation of cytoskeletal proteins may be responsible for the impairments of SVs transport.

Cerebral ischemia is associated with the activation of resident glial cells (astrocytes and microglial cells), production of inflammatory cytokines, and leukocyte infiltration into the brain (Kriz and Lalancette-Hebert, 2009). Leukocyte entry into the brain is normally restricted but it may take place after ischemia if the integrity of the blood–brain barrier is disrupted (Vexler and Yenari, 2009). The early ischemia-related upregulation of matrix metalloproteinases (MMPs), modulators of extracellular matrix, contributes to the disruption of this barrier. The inhibition of MMPs results in a significant reduction of ischemic brain injury (Amantea *et al.*, 2007).

Cyclooxygenase-2 (COX-2), an enzyme converting arachidonic acid to prostaglandins, is expressed both in neurons and astrocytes. COX-2 is induced in the postischemic brain, its gene being early regulated by NMDA receptor activity (Savonenko *et al.*, 2009). On the other hand, reaction products of COX-2 influence excitatory glutamatergic neurotransmission (Yang and Chen, 2008). COX-2 expression is upregulated in

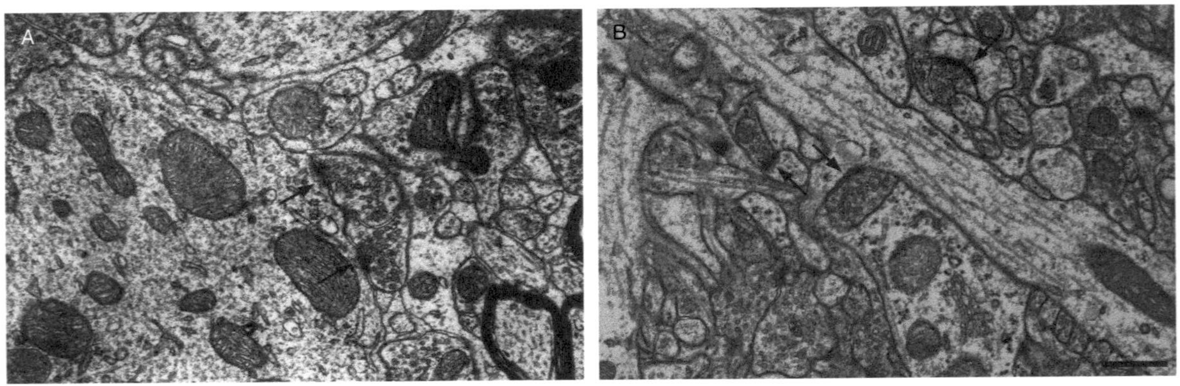


**Figure 6.1** Micrographs of hippocampal tissue in the CA1 area *str. radiatum* of intact mongolian gerbil (A) and gerbil subjected to a short episode of transient global cerebral ischemia (B). Scale bar = 0.5 $\mu$m.

neurons committed to survive, in a pattern suggesting differential sensitivity to COX-2 upregulation (Choi *et al.*, 2006). The complement system also plays a role in ischemic brain injury. Lack of C3 or inhibition of C3a receptor are associated with decreased granulocyte infiltration and reduced oxidative stress (Vexler and Yenari, 2009).

Microglial cells are resident macrophages that, among other activities, monitor the functional status of synapses and contribute to the ischemia-related increase in the turnover of synaptic contacts in the brain tissue (Wake *et al.*, 2009). After cerebral ischemia, microglial cells proliferate, migrate toward neuronal perikarya and phagocytize degenerated neuronal bodies. The question whether ischemia-related activation of microglial cells is beneficial or detrimental is still open. Activated microglial cells can damage various cell types by producing toxic products, inflammatory cytokines, and chemokines (Vexler and Yenari, 2009).

Several cytokines and cell adhesion molecules are implicated in early ischemia-related deterioration of the brain tissue. Cytokines are induced both at the site of infarction and in remote nonaffected brain areas. Expression of interleukin 1$\beta$ (IL-1$\beta$), an inflammatory cytokine driving neutrophils into the brain, is elevated during ischemia (Serou *et al.*, 1999). The inactivation of IL-1$\alpha/\beta$ gene significantly reduces ischemia-related neuronal loss. There is some evidence that IL-1 may act through the release of nitric oxide (NO) by the form of nitric oxide synthase (NOS), which is usually not expressed in the brain except in the inflammatory state—inducible NOS (Mizushima *et al.*, 2002). IL-6 is largely thought of as a proinflammatory cytokine, but whether it plays a significant role in ischemic injury is far from clear (Vexler and Yenari, 2009). For example, it was shown that IL-6 exerts trophic effects on postischemic hippocampal neurons (Matsuda *et al.*, 1996).

NO can have contrasting roles in ischemic injury, protective when generated by endothelial NOS and detrimental when produced by neuronal NOS or inducible NOS. The latter is capable of producing large amounts of NO over extended periods, which, in turn, directly damages cellular structures, produces a number of toxic species, such as peroxynitrate, activates COX-2, or inflammatory cells. The increase in neuronal NOS expression was prominent in brain areas adjacent to severe neuronal damage (Ishida *et al.*, 2001). After initial stimulation by cytokines, large amounts of NO produced by inducible NOS in the microglia may cause cellular damage (Paakkari and Lindsberg, 1995).

Astrocytes respond to ischemia by increasing their numbers and size together with elongation of cytoplasmic processes (Kajihara *et al.*, 2001), increase in the expression of glial fibrillary acidic protein (GFAP) of the intermediate filaments (Rossi *et al.*, 2007), reorganization of their gap junctions supporting a functional syncytium (Li *et al.*, 1998), and accumulation of glycogen (Kajihara *et al.*, 2001). Severe ischemia can induce apoptosis in astrocytes (Yu *et al.*, 2001), however other data suggest that these cells mostly die by a nonapoptotic mechanism (Chu *et al.*, 2007). Astrocytes are thought to produce several neurotrophic factors to protect neurons from delayed ischemia-related death. One of them, brain-derived neurotrophic factor (BDNF), is known to promote neuronal survival, guide axonal pathfinding, and participate in activity-dependent synaptic plasticity (Endo, 2005).

Ischemia-related changes in the brain tissue include a transient accumulation of amyloid-$\beta$ precursor protein and amyloid-$\beta$ peptides adjacent to the ischemic lesion. Soluble amyloid-$\beta$ species are known to cause synaptic dysfunction (Hu *et al.*, 2008). Selective phospholipids of synaptic membranes are reservoirs for lipid second messengers, including platelet-activating factor (PAF). Under physiological conditions, PAF modulates neurotransmitter release, but at high concentrations, it becomes neurotoxic. PAF activates the expression of immediate-early genes encoding transcription factors (e.g., c-fos) and enzymes (e.g., COX-2 and MMPs) (Bazan and Allan, 1996).

It can be concluded that ischemic injury of the brain unrolls at different levels and has both functional and structural implications. The deficiency in energy metabolism both in neurons and glial cells is an initiating factor. Temporal and spatial dynamics of ischemic injury are dictated by specific properties of neuronal populations and suggest the action of multiple factors, including excitotoxicity, ion imbalance, oxidative stress, inflammation, etc. The reaction of brain tissue depends on the type (global or focal) and duration of cerebral ischemia. If the latter is relatively short, then cell death tends to be delayed and selective for neurons, whereas longer episodes of ischemia cause broader and faster destructive changes.

## II. Mechanisms of Adult Brain Plasticity

The adult brain demonstrates a considerable degree of structural and functional plasticity. The concept of brain plasticity covers all the mechanisms involved in the capacity of the organ to remodel itself in response to various demands in health and disease. Synaptic plasticity, the ability of a synapse to change in strength, is possibly the most important aspect of brain plasticity. It is known that the function of a synapse can be modulated by prior activity and a wide spectrum of neuromodulators changing the amount of a neurotransmitter released into a synapse and responsiveness of the postsynaptic cell to that neurotransmitter (Song and Huganir, 2002).

At the molecular level, synaptic plasticity is driven by the modification of the existing synaptic proteins and the activation of specific genes, leading to the production of new synaptic proteins. Long-lasting changes in synaptic connectivity between neurons can involve the generation and elimination of synapses. Synaptic strength is shown to depend on the number of ion channels residing in synaptic membranes (Debanne *et al.*, 2003). It is well known that neurons change their own excitability in response to various stimuli by changing the density of receptors in the synaptic membranes, receptors being added dynamically by exocytosis and removed by endocytosis. AMPA receptors are delivered to the membrane due to repetitive NMDA receptor activation (Song and Huganir, 2002). Regulatory forms of plasticity, changing the numbers of NMDA receptors at the synapse, also exist to provide a negative feedback (Pérez-Otaño and Ehlers, 2005).

Phenomena of brain plasticity include also cell renewal. Persistent neurogenesis was shown to occur in discrete regions of the adult brain, including the hippocampal dentate gyrus and subventricular zone (SVZ) of lateral ventricles. New cells born in SVZ migrate via the rostral migratory stream toward the olfactory bulb replacing its neurons throughout the life of an animal. New cells originating in the dentate gyrus migrate into the granule cell layer, where they differentiate, become physiologically mature and integrate in the neuronal circuits (Iwai *et al.*, 2002). Neurogenesis in adult brain increases in various conditions, including ischemia (Morris *et al.*, 2007) while decreasing under stress (Gould *et al.*, 1998). The question whether adult brain neurogenesis is sufficient to replace died neurons and influence the functional recovery of the brain tissue after ischemia is still a matter of discussion.

Neuronal regeneration is limited in the brain may be due to the presence of inhibitory factors and lack of growth factors. However, in some regions of the adult brain, for example, in hippocampus, neurons retain their capacity for axonal growth (Frotscher *et al.*, 1996). Axonal rearrangement is believed to be an important mechanism underlying learning and memory. On the other hand, it is crucially important for lesion-related plasticity in the form of axonal sprouting in which undamaged axons grow new nerve endings to reconnect neurons whose links were injured. Experimental evidence suggests that neurons compete for neurotrophins, during both target innervation and activity-dependent synaptic rearrangement, which influence the sprouting and retraction of axonal processes (Conner and Varon, 1996). The cell adhesion molecule L1 was reported to express on reinnervating fibers after they make synaptic contacts with other structures (Jucker *et al.*, 1996). Mossy fiber sprouting in the hippocampus is dependent on NMDA receptor activity (Sutula *et al.*, 1996). The process of neural network reorganization involves not only axonal and dendritic sprouting but also the pruning of redundant synaptic connections.

Different hormones exert a variety of modulatory actions at the neuronal level in the brain. Estrogens, leptin, and insulin, can cross the blood brain barrier and influence synaptic transmission, for example, to enhance NMDA receptor-mediated currents, however the mechanisms of their influence differ in detail. All mentioned hormones may act as modulators of synaptic plasticity (Moult and Harvey, 2008). Estrogens have a profound effect upon dendritic morphology, particularly at the synaptic level. The numbers of dendritic spines on the pyramidal cells of the hippocampal CA1 region but not in CA3 or dentate gyrus in adult female animals fluctuate in parallel to the levels of circulating ovarian steroids (Woolley *et al.*, 1990a). In contrast, high levels of corticosterone decreased the complexity and size of the apical dendritic tree of pyramidal neurons in the CA3 but not in CA1 or dentate gyrus (Woolley *et al.*, 1990b).

Less publicized aspects of brain plasticity include angiogenesis, the sprouting of new blood vessels from preexisting vessels, a complex, multicellular phenomenon involving capillary endothelial cell proliferation, migration, and tissue infiltration. Angiogenesis in the postischemic brain is regulated by vascular growth factors (Whitaker *et al.*, 2007). Thus, the adult brain tissue has the potential to recover its function and structure after various injuries. This potential is mainly based on plastic phenomena occurring at the levels of synapses, cells, and neuronal networks being integral parts of tissue remodeling and turnover.

## III. Postischemic Brain Plasticity

In cases when an ischemic insult has no immediate fatal consequence the brain tissue recovers to the extent depending on the severity of brain injury. The recovery involves plastic changes occurring at synaptic, cellular, and tissue levels. Generally, postischemic brain plasticity is based on the modifications of surviving cells and proliferation/differentiation of NSCs.

Large body of experimental evidence suggests that ischemia influences the activity of synapses. Structurally, it is confirmed by a significant increase in the frequency of concave-shaped synapses observed 1 day after ischemia (Kovalenko *et al.*, 2006) presumably reflecting the enhanced rate of SVs exocytosis (Hammond *et al.*, 1994). In addition, ischemia-related potentiation of excitatory transmission similar to LTP has been reported *in vitro* (Jourdain *et al.*, 2002). Selected actin-binding proteins of the PSD get ubiquitinated and degraded after rapid ischemia. Their loss promotes actin reorganization and transient retraction of dendritic spines. It was assumed that this rapid and reversible synaptic remodeling reduces NMDA-mediated currents, rendering the cells refractory to NMDA receptor-mediated toxicity (Meller *et al.*, 2008).

A rapid ischemia-related increase in the number of perforated synapses and multiple synapse boutons was shown both *in vivo* and *in vitro* (Jourdain *et al.*, 2002; Kovalenko *et al.*, 2006). This $Ca^{2+}$ and NMDA receptor-dependent process reflects reactive synaptogenesis and correlates with the growth of filopodia, enlargements of existing spines, and formation of new spines (Jourdain *et al.*, 2002). The proportion of perforated synapses drops almost to the control values 7 days after ischemia in concert with decreasing numbers of concave-shaped synapses (Kovalenko *et al.*, 2006). Later, up to 12 weeks after ischemia, the number of synapses and spines increases to the control level or even exceeds it. In parallel, the number of multiple synapse boutons increases (Ito *et al.*, 2006). Dendrites was reported to disintegrate within minutes of acute ischemia but rapidly reassemble during reperfusion. In the surviving brain tissue they show increased levels of spine turnover for many weeks after stroke (Brown and Murphy, 2008).

Plastic changes in the brain tissue are mediated by second messenger cascades that control protein kinases and the subsequent phosphorylation of various proteins. Calcium/calmodulin-dependent protein kinase II (CaMKII) is the most abundant protein kinase in the brain, and the α-subunit of this kinase is the major protein of synaptic junctions (Kelly, 1991). CaMKII was found to be very susceptible to oxidative modulation (Shetty *et al.*, 2008).

cAMP response element-binding protein (CREB) is a DNA-binding transcription factor constituting a convergence point for many signaling cascades. CREB phosphorylation in neurons is induced by $Ca^{2+}$ influx and subsequent activation of CaMK II–IV (Mabuchi *et al.*, 2001). In ischemia, CaMKIV mediates the early phosphorylation of CREB driving the synthesis of BDNF. This BDNF synthesis, in turn, induces a second peak of CREB phosphorylation to maintain further BDNF synthesis (Blanquet *et al.*, 2006). CREB-mediated transcription of genes remains enhanced in the penumbra but rapidly diminishes in the ischemic core (Tanaka, 2001).

mRNAs of BDNF and TrkB, a component of BDNF receptor, were transiently increased in the dentate gyrus following cerebral ischemia (Merlio *et al.*, 1993). BDNF, together with other neurotrophins, stimulates axonal growth by mediating the polymerization and accumulation of F-actin in growth cones and axon shafts (Lykissas *et al.*, 2007). Recent data indicate that BDNF exerts substantial neuroregenerative effects influencing the postischemic expression of GABA and NMDA receptor subunits. In addition, it increases MAP1B protein expression in the surviving periischemic region (Muller *et al.*, 2009).

At cellular level postischemic plasticity is observed, first of all, in the form of NSCs proliferation. Mild ischemia may stimulate NSCs proliferation in the hippocampal dentate gyrus, SVZ of the temporal horn of the lateral ventricle, and temporal neocortex (Tonchev *et al.*, 2003; Yagita *et al.*, 2001). After ischemic stroke, NPCs in the SVZ proliferate and migrate toward the ischemic boundary region to replenish damaged neurons (Jenny *et al.*, 2009; Morris *et al.*, 2007). BDNF significantly increased neurogenesis in the dentate gyrus and enhanced the migration of NSCs from SVZ to the nearby striatum of the ischemic hemisphere (Schäbitz *et al.*, 2007).

The activity of tissue plasminogen activator (tPA), a serine protease converting inactive plasminogen to the active protease plasmin, produced in the brain tissue primarily by microglia is thought to be involved in neuronal plasticity (Tsirka *et al.*, 1995). t-PA activity is regulated through the activation of serine protease inhibitors (serpins), including neuroserpin released from neurons in response to their depolarization. Following the onset of cerebral ischemia, there is an increase in both tPA activity and neuroserpin expression in the ischemic penumbra, and treatment with neuroserpin results in a significant decrease in the volume of the ischemic area as well as in the number of apoptotic cells (Yepes and Lawrence, 2004).

Protein synthesis in the postischemic brain is generally suppressed, but specific genes (c-fos, c-jun, and zinc finger gene) are activated and particular proteins (heat-shock proteins and amyloid-$\beta$ precursor protein) are expressed. Nerve and fibroblast growth factors are also induced after ischemia and may be related to repair processes with the participation of glial cells (Kogure and Kato, 1993). The level of the 72 kDa heat-shock protein in synapses markedly increased 1–2 days after transient focal ischemia, declining during the following 2 days. During the following 4 days, the rate of local synaptic protein synthesis increased more than twofold (Mariucci *et al.*, 2007).

Immediate early genes represent the first wave of gene expression following ischemia and are induced widely across the ischemic brain. BDNF, neuritin, and activity-regulated cytoskeleton-associated protein (Arc), belong to a subgroup of such genes being implicated in synaptic plasticity (Rickhag *et al.*, 2007). During the first day after ischemia, two waves of gene expression comprising 44 genes involved in cell signaling and plasticity were revealed. The first wave occurred within 0–3 h of reperfusion, and the second—within 9–15 h after ischemia (Rickhag *et al.*, 2006).

One week after focal cerebral ischemia, the levels of gene and protein expression of neurofilament proteins, including MAP1B and MAP2, were increased in the border zone of the infarcted area (Popa-Wagner *et al.*, 1999). The expression of candidate plasticity-related gene 15 encoding a protein that regulates dendritic and axonal arbor growth and synaptic maturation was upregulated in the hippocampus at 1–2 weeks after ischemia (Han *et al.*, 2007).

Functional deficits caused by cerebral ischemia can be compensated by reinnervation of denervated targets by regenerating axons or by collateral branching of undamaged axons, and remodeling of neuronal circuitry. Following cerebral ischemia, pronounced plasticity of central connections is observed in the form of axon sprouting, which could be regarded as regenerative (Carmichael, 2003). In primary visual cortex of kittens, postischemic dendritic sprouting correlates in time with changes in the synaptic transmission and functional reorganization (Zepeda *et al.*, 2004). Ischemic lesions induce axonal sprouting within local and long-distance projections. Such axonal sprouting is induced by a transient pattern of synchronous, low-frequency neuronal activity in cortical areas connected to the infarct. Perinatal ischemia can induce mossy fiber sprouting in the lesioned and contralateral hippocampus (Williams *et al.*, 2004).

*In vitro*, dendritic outgrowth has been observed in surviving neurons months after ischemia. The total dendritic length has reached a peak at 6 h after reoxygenation. This increase was mainly due to the NMDA activity-dependent sprouting (Lei *et al.*, 2006). Sprouting is, in part, ill-targeted and may be a subject for later pruning of redundant connections. Increased dendritic disorientation in the CA1 was observed 48 h after ischemia with many basal dendrites coursed into the territory of apical dendrites and apical dendrites branched into the region of basal dendrites (Ruan *et al.*, 2006).

Most of the functional recovery after stroke takes place during the first 3 months after the insult (Aronen *et al.*, 2007). In adult animals, cortical neurons first adopt wider functional roles as they develop strategies to compensate for loss of specific sensory modalities after stroke (Winship and Murphy, 2008). Data suggest that functional reorganization of remote cortical areas occurs in response to cortical injury and that the greater the damage to reciprocal intracortical pathways, the greater the plasticity in intact areas (Frost *et al.*, 2003).

Many molecular players implicated in the early ischemia-related deterioration of the brain tissue are involved later in the processes of postischemic brain plasticity. MMP-9 is upregulated in periinfarct cortex 7–14 days after focal ischemia, being colocalized with markers of neurovascular remodeling. Treatment with MMP inhibitors suppresses this remodeling, increases ischemic injury, and impairs functional recovery (Zhao *et al.*, 2006). TNF-$\alpha$ and IL-1$\beta$ cytokines probably have dual functions in ischemic injury: in the absence of inducible NOS, they may contribute to neuroprotection and plasticity (Stoll *et al.*, 2000). In addition, IL-6 exerts a neuroprotective effect against ischemia depressing the spread of excitation and evoked glutamate release in the cerebral cortex (D'Arcangelo *et al.*, 2000).

17$\beta$-Estradiol, at physiological concentrations, is thought to have protective effects in acute ischemic stroke by stimulating the release of astrocyte-derived neuroprotective factors (Dhandapani and Brann, 2002). The administration of 17$\beta$-estradiol hours before ischemia was able to protect neurons in the CA1 area from ischemia-related impairments in synaptic transmission (Dai *et al.*, 2007).

Roles in postischemic brain plasticity were also shown for some proteins, including cellular prion protein, producing an antiapoptotic effect on Bax-mediated cell death (Lo *et al.*, 2007), and CRMPs making neurons excitotoxicity-resistant by the downregulation of NR2B subunit of NMDA receptor (Bretin *et al.*, 2006).

Chronic cerebral ischemia can induce collateral circulation with the formation of new microvessels in ischemic brain regions. Postischemic angiogenesis helps to restore blood flow to ischemic tissue. Increased expression of a number of vascular growth factors coincides with increased number of proliferating endothelial cells and vessels in ischemic cortex

(Whitaker *et al.*, 2007). The molecular signals for postischemic angiogenesis begin within hours of initial cerebral ischemia, with vascular growth factors promoting the proliferation of endothelial cells. The overlap in molecular signaling between postischemic angiogenesis, neurogenesis, and axonal sprouting suggests a continuum of vascular and neural reorganization (Carmichael, 2003).

The following conclusions can be made. Despite the well-known fact that the brain tissue has a limited capacity for repair, it reacts against ischemic injury using multiple mechanisms based on plastic phenomena unrolling at the synaptic, cellular, and tissue levels (Fig. 6.2). These mechanisms involve changes in the number and strength of synapses, remodeling of extracellular matrix, neovascularization, proliferation and differentiation of NSCs, axonal and dendritic sprouting, and rewiring of neuronal networks. The inflammatory factors playing destructive roles on the initial stages of ischemic injury later contribute into remodeling processes. The reaction of the brain tissue depends on the type (global or focal) of ischemia and severity of ischemic injury. If the latter is relatively mild, then cell death tends to be delayed and selective for neurons, whereas more severe ischemic insults cause broader and faster destructive changes. Multiple lines of experimental

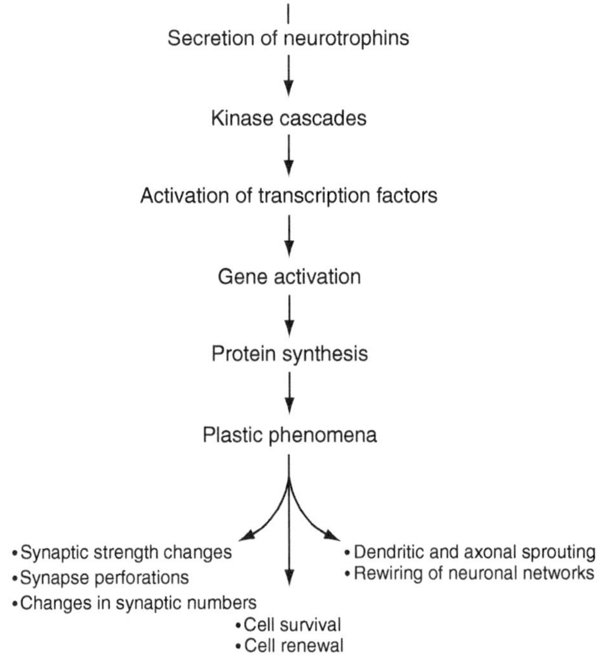


**Figure 6.2** A scheme illustrating sequential molecular factors driving the events of postischemic brain plasticity.

evidence indicate numerous points in which the process of postischemic recovery may be influenced with the aim to restore the full capability of the damaged brain tissue.

## REFERENCES

Amantea, D., Russo, R., Gliozzi, M., Fratto, V., Berliocchi, L., Bagetta, G., Bernardi, G., and Corasaniti, M. T. (2007). Early upregulation of matrix metalloproteinases following reperfusion triggers neuroinflammatory mediators in brain ischemia in rat. *Int. Rev. Neurobiol.* **82,** 149–169.

Aronen, H. J., Laakso, M. P., Moser, M., and Perkio, J. (2007). Diffusion and perfusion-weighted magnetic resonance imaging techniques in stroke recovery. *Eura. Medicophys.* **43,** 271–284.

Avignone, E., Frenguelli, B. G., and Irving, A. J. (2005). Differential responses to NMDA receptor activation in rat hippocampal interneurons and pyramidal cells may underlie enhanced pyramidal cell vulnerability. *Eur. J. NeuroSci.* **22,** 3077–3090.

Bazan, N. G., and Allan, G. (1996). Platelet-activating factor in the modulation of excitatory amino acid neurotransmitter release and of gene expression. *J. Lipid Mediat. Cell. Signal.* **14,** 321–330.

Benveniste, H., Drejer, J., Schousboe, A., and Diemer, N. H. (1984). Elevation of the extracellular concentrations of glutamate and aspartate in rat hippocampus during transient cerebral ischemia monitored by intracerebral microdialysis. *J. Neurochem.* **43,** 1369–1374.

Blanquet, P. R., Mariani, J., and Fournier, B. (2006). Identification of a biphasic signaling pathway involved in ischemic resistance of the hippocampal dentate gyrus. *Exp. Neurol.* **202,** 357–372.

Bonanni, L., Chachar, M., Jover-Mengual, T., Li, H., Jones, A., Yokota, H., Ofengeim, D., Flannery, R. J., Miyawaki, T., Cho, C. H., Polster, B. M., Pypaert, M., *et al.* (2006). Zinc-dependent multi-conductance channel activity in mitochondria isolated from ischemic brain. *J. Neurosci.* **26,** 6851–6862.

Bretin, S., Rogemond, V., Marin, P., Maus, M., Torrens, Y., Honnorat, J., Glowinski, J., Premont, J., and Gauchy, C. (2006). Calpain product of WT-CRMP2 reduces the amount of surface NR2B NMDA receptor subunit. *J. Neurochem.* **98,** 1252–1265.

Brown, C. E., and Murphy, T. H. (2008). Livin' on the edge: Imaging dendritic spine turnover in the peri-infarct zone during ischemic stroke and recovery. *Neuroscientist* **14,** 139–146.

Buddle, M., Eberhardt, E., Ciminello, L. H., Levin, T., Wing, R., DiPasquale, K., and Raley-Susman, K. M. (2003). Microtubule-associated protein 2 (MAP2) associates with the NMDA receptor and is spatially redistributed within rat hippocampal neurons after oxygen–glucose deprivation. *Brain Res.* **978,** 38–50.

Carmichael, S. T. (2003). Gene expression changes after focal stroke, traumatic brain and spinal cord injuries. *Curr. Opin. Neurol.* **16,** 699–704.

Chen, Z., Kontonotas, D., Friedmann, D., Pitts-Kiefer, A., Frederick, J. R., Siman, R., and Neumar, R. W. (2005). Developmental status of neurons selectively vulnerable to rapidly triggered post-ischemic caspase activation. *Neurosci. Lett.* **376,** 166–170.

Choi, D. W., and Rothman, S. M. (1990). The role of glutamate neurotoxicity in hypoxic-ischemic neuronal death. *Annu. Rev. Neurosci.* **13,** 171–182.

Choi, J. S., Kim, H. Y., Chun, M. H., Chung, J. W., and Lee, M. Y. (2006). Differential regulation of cyclooxygenase-2 in the rat hippocampus after cerebral ischemia and ischemic tolerance. *Neurosci. Lett.* **393,** 231–236.

Chu, X., Fu, X., Zou, L., Qi, C., Li, Z., Rao, Y., and Ma, K. (2007). Oncosis, the possible cell death pathway in astrocytes after focal cerebral ischemia. *Brain Res.* **1149,** 157–164.

Cimarosti, H., Lindberg, C., Bomholt, S. F., Ronn, L. C., and Henley, J. M. (2008). Increased protein SUMOylation following focal cerebral ischemia. *Neuropharmacology* **54,** 280–289.

Conner, J. M., and Varon, S. (1996). Maintenance of sympathetic innervation into the hippocampal formation requires a continuous local availability of nerve growth factor. *Neuroscience* **72,** 933–945.

Dai, X., Chen, L., and Sokabe, M. (2007). Neurosteroid estradiol rescues ischemia-induced deficit in the long-term potentiation of rat hippocampal CA1 neurons. *Neuropharmacology* **52,** 1124–1138.

D'Arcangelo, G., Tancredi, V., Onofri, F., D'Antuono, M., Giovedi, S., and Benfenati, F. (2000). Interleukin-6 inhibits neurotransmitter release and the spread of excitation in the rat cerebral cortex. *Eur. J. NeuroSci.* **12,** 1241–1252.

Debanne, D., Daoudal, G., Sourdet, V., and Russier, M. (2003). Brain plasticity and ion channels. *J. Physiol. Paris* **97,** 403–414.

Dhandapani, K. M., and Brann, D. W. (2002). Estrogen–astrocyte interactions: Implications for neuroprotection. *BMC Neurosci.* **3,** 6–9.

Endo, T. (2005). Glycans and glycan-binding proteins in brain: Galectin-1-induced expression of neurotrophic factors in astrocytes. *Curr. Drug Targets* **6,** 427–436.

Frost, S. B., Barbay, S., Friel, K. M., Plautz, E. J., and Nudo, R. J. (2003). Reorganization of remote cortical regions after ischemic brain injury: A potential substrate for stroke recovery. *J. Neurophysiol.* **89,** 3205–3214.

Frotscher, M., Deller, T., Heimrich, B., Förster, E., Haas, C., and Naumann, T. (1996). Survival, regeneration and sprouting of central neurons: The rat septohippocampal projection as a model. *Ann. Anat.* **178,** 311–315.

Gao, J., Duan, B., Wang, D. G., Deng, X. H., Zhang, G. Y., Xu, L., and Xu, T. L. (2005). Coupling between NMDA receptor and acid-sensing ion channel contributes to ischemic neuronal death. *Neuron* **48,** 635–646.

Gould, E., Tanapat, P., McEwen, B. S., Flügge, G., and Fuchs, E. (1998). Proliferation of granule cell precursors in the dentate gyrus of adult monkeys is diminished by stress. *Proc. Natl Acad. Sci. USA* **95,** 3168–3171.

Gubellini, P., Bisso, G. M., Ciofi-Luzzatto, A., Fortuna, S., Lorenzini, P., Michalek, H., and Scarsella, G. (1997). Ubiquitin-mediated stress response in a rat model of brain transient ischemia/hypoxia. *Neurochem. Res.* **22,** 93–100.

Hammond, C., Crépel, V., Gozlan, H., and Ben-Ari, Y. (1994). Anoxic LTP sheds light on the multiple facets of NMDA receptors. *Trends Neurosci.* **17,** 497–503.

Han, Y., Chen, X., Shi, F., Li, S., Huang, J., Xie, M., Hu, L., Hoidal, J. R., and Xu, P. (2007). CPG15, a new factor upregulated after ischemic brain injury, contributes to neuronal network re-establishment after glutamate-induced injury. *J. Neurotrauma* **24,** 722–731.

Hou, S. T., Jiang, S. X., and Smith, R. A. (2008). Permissive and repulsive cues and signalling pathways of axonal outgrowth and regeneration. *Int. Rev. Cell. Mol. Biol.* **267,** 125–181.

Hu, B. R., Martone, M. E., Jones, Y. Z., and Liu, C. L. (2000). Protein aggregation after transient cerebral ischemia. *J. Neurosci.* **20,** 3191–3199.

Hu, N. W., Smith, I. M., Walsh, D. M., and Rowan, M. J. (2008). Soluble amyloid-beta peptides potently disrupt hippocampal synaptic plasticity in the absence of cerebrovascular dysfunction in vivo. *Brain* **131,** 2414–2424.

Ishida, A., Ishiwa, S., Trescher, W. H., Nakajima, W., Lange, M. S., Blue, M. E., and Johnston, M. V. (2001). Delayed increase in neuronal nitric oxide synthase immunoreactivity in thalamus and other brain regions after hypoxic-ischemic injury in neonatal rats. *Exp. Neurol.* **168,** 323–333.

Ishimaru, H., Casamenti, F., Uéda, K., Maruyama, Y., and Pepeu, G. (2001). Changes in presynaptic proteins, SNAP-25 and synaptophysin, in the hippocampal CA1 area in ischemic gerbils. *Brain Res.* **903,** 94–101.

Ito, U., Kuroiwa, T., Nagasao, J., Kawakami, E., and Oyanagi, K. (2006). Temporal profiles of axon terminals, synapses and spines in the ischemic penumbra of the cerebral cortex: Ultrastructure of neuronal remodeling. *Stroke* **37,** 2134–2139.

Iwai, M., Sato, K., Omori, N., Nagano, I., Manabe, Y., Shoji, M., and Abe, K. (2002). Three steps of neural stem cells development in gerbil dentate gyrus after transient ischemia. *J. Cereb. Blood Flow Metab.* **22,** 411–419.

Jenny, B., Kanemitsu, M., Tsupykov, O., Potter, G., Salmon, P., Zgraggen, E., Gascon, E., Skibo, G., Dayer, A. G., and Kiss, J. Z. (2009). Fibroblast growth factor-2 overexpression in transplanted neural progenitors promotes perivascular cluster formation with a neurogenic potential. *Stem Cells* **27,** 1309–1317.

Jourdain, P., Nikonenko, I., Alberi, S., and Muller, D. (2002). Remodeling of hippocampal synaptic networks by a brief anoxia-hypoglycemia. *J. Neurosci.* **22,** 3108–3116.

Jucker, M., D'Amato, F., Mondadori, C., Mohajeri, H., Magyar, J., Bartsch, U., and Schachner, M. (1996). Expression of the neural adhesion molecule L1 in the deafferented dentate gyrus. *Neuroscience* **75,** 703–715.

Kajihara, H., Tsutsumi, E., Kinoshita, A., Nakano, J., Takagi, K., and Takeo, S. (2001). Activated astrocytes with glycogen accumulation in ischemic penumbra during the early stage of brain infarction: Immunohistochemical and electron microscopic studies. *Brain Res.* **909,** 92–101.

Kelly, P. T. (1991). Calmodulin-dependent protein kinase II. Multifunctional roles in neuronal differentiation and synaptic plasticity. *Mol. Neurobiol.* **5,** 153–177.

Kogure, K., and Kato, H. (1993). Altered gene expression in cerebral ischemia. *Stroke* **24,** 2121–2127.

Kovalenko, T., Osadchenko, I., Nikonenko, A., Lushnikova, I., Voronin, K., Nikonenko, I., Muller, D., and Skibo, G. (2006). Ischemia-induced modifications in hippocampal CA1 stratum radiatum excitatory synapses. *Hippocampus* **16,** 814–825.

Kriz, J., and Lalancette-Hebert, M. (2009). Inflammation, plasticity and real-time imaging after cerebral ischemia. *Acta Neuropathol.* **117,** 497–509.

Lei, Z., Ruan, Y., Yang, A. N., and Xu, Z. C. (2006). NMDA receptor mediated dendritic plasticity in cortical cultures after oxygen–glucose deprivation. *Neurosci. Lett.* **407,** 224–229.

Li, Y., Jiang, N., Powers, C., and Chopp, M. (1998). Neuronal damage and plasticity identified by microtubule-associated protein 2, growth-associated protein 43, and cyclin D1 immunoreactivity after focal cerebral ischemia in rats. *Stroke* **29,** 1972–1980.

Liu, B., Liao, M., Mielke, J. G., Ning, K., Chen, Y., Li, L., El Hayek, Y. H., Gomez, E., Zukin, R. S., Fehlings, M. G., and Wan, Q. (2006). Ischemic insults direct glutamate receptor subunit 2-lacking AMPA receptors to synaptic sites. *J. Neurosci.* **26,** 5309–5319.

Lo, E. H. (2008). A new penumbra: Transitioning from injury into repair after stroke. *Nat. Med.* **14,** 497–500.

Lo, R. Y., Shyu, W. C., Lin, S. Z., Wang, H. J., Chen, S. S., and Li, H. (2007). New molecular insights into cellular survival and stress responses: Neuroprotective role of cellular prion protein (PrPC). *Mol. Neurobiol.* **35,** 236–244.

Lykissas, M. G., Batistatou, A. K., Charalabopoulos, K. A., and Beris, A. E. (2007). The role of neurotrophins in axonal growth, guidance, and regeneration. *Curr. Neurovasc. Res.* **4,** 143–151.

Mabuchi, T., Kitagawa, K., Kuwabara, K., Takasawa, K., Ohtsuki, T., Xia, Z., Storm, D., Yanagihara, T., Hori, M., and Matsumoto, M. (2001). Phosphorylation of cAMP response element-binding protein in hippocampal neurons as a protective response after exposure to glutamate in vitro and ischemia in vivo. *J. Neurosci.* **21,** 9204–9213.

Mariucci, G., Tantucci, M., Giuditta, A., and Ambrosini, M. V. (2007). Permanent brain ischemia induces marked increments in hsp72 expression and local protein synthesis in synapses of the ischemic hemisphere. *Neurosci. Lett.* **415,** 77–80.

Martin, L. J., Al-Abdulla, N. A., Brambrink, A. M., Kirsch, J. R., Sieber, F. E., and Portera-Cailliau, C. (1998). Neurodegeneration in excitotoxicity, global cerebral ischemia, and target deprivation: A perspective on the contributions of apoptosis and necrosis. *Brain Res. Bull.* **46,** 281–309.

Martone, M. E., Jones, Y. Z., Young, S. J., Ellisman, M. H., Zivin, J. A., and Hu, B. R. (1999). Modification of postsynaptic densities after transient cerebral ischemia: A quantitative and three-dimensional ultrastructural study. *J. Neurosci.* **19,** 1988–1997.

Matsuda, S., Wen, T. C., Morita, F., Otsuka, H., Igase, K., Yoshimura, H., and Sakanaka, M. (1996). Interleukin-6 prevents ischemia-induced learning disability and neuronal and synaptic loss in gerbils. *Neurosci. Lett.* **204,** 109–112.

McLaughlin, B. (2004). The kinder side of killer proteases: Caspase activation contributes to neuroprotection and CNS remodeling. *Apoptosis* **9,** 111–121.

Meller, R., Thompson, S. J., Lusardi, T. A., Ordonez, A. N., Ashley, M. D., Jessick, V., Wang, W., Torrey, D. J., Henshall, D. C., Gafken, P. R., Saugstad, J. A., Xiong, Z. G., and Simon, R. P. (2008). Ubiquitin proteasome-mediated synaptic reorganization: A novel mechanism underlying rapid ischemic tolerance. *J. Neurosci.* **28,** 50–59.

Merlio, J. P., Ernfors, P., Kokaia, Z., Middlemas, D. S., Bengzon, J., Kokaia, M., Smith, M. L., Siesjö, B. K., Hunter, T., Lindvall, O., and Personn, H. (1993). Increased production of the TrkB protein tyrosine kinase receptor after brain insults. *Neuron* **10,** 151–164.

Miyazaki, S., Katayama, Y., Furuichi, M., Kinoshita, K., Kawamata, T., and Tsubokawa, T. (1993). Impairment of hippocampal long-term potentiation following transient cerebral ischaemia in rat: Effects of bifemelane, a potent inhibitor of ischaemia-induced acetylcholine release. *Neurol. Res.* **15,** 249–252.

Mizushima, H., Zhou, C. J., Dohi, K., Horai, R., Asano, M., Iwakura, Y., Hirabayashi, T., Arata, S., Nakajo, S., Takaki, A., Ohtaki, H., and Shioda, S. (2002). Reduced postischemic apoptosis in the hippocampus of mice deficient in interleukin-1. *J. Comp. Neurol.* **448,** 203–216.

Morris, D. C., Zhang, Z. G., Wang, Y., Zhang, R. L., Gregg, S., Liu, X. S., and Chopp, M. (2007). Wnt expression in the adult rat subventricular zone after stroke. *Neurosci. Lett.* **418,** 170–174.

Moult, P. R., and Harvey, J. (2008). Hormonal regulation of hippocampal dendritic morphology and synaptic plasticity. *Cell. Adh. Migr.* **2,** 269–275.

Muller, H. D., Neder, A., Sommer, C., and Schabitz, W. R. (2009). Different postischemic protein expression of the GABA(A) receptor alpha2 subunit and the plasticity-associated protein MAP1B after treatment with BDNF versus G-CSF in the rat brain. *Restor. Neurol. Neurosci.* **27,** 27–39.

Müller, G. J., Stadelmann, C., Bastholm, L., Elling, F., Lassmann, H., and Johansen, F. F. (2004). Ischemia leads to apoptosis- and necrosis-like neuron death in the ischemic rat hippocampus. *Brain Pathol.* **14,** 415–424.

Naganuma, Y. (1990). Changes of the cerebral microvascular structure and endothelium during the course of permanent ischemia. *Keio J. Med.* **39,** 26–31.

Nakamichi, N., Oikawa, H., Kambe, Y., and Yoneda, Y. (2004). Relevant modulation by ferrous ions of N-methyl-D-aspartate receptors in ischemic brain injuries. *Curr. Neurovasc. Res.* **1,** 429–440.

Neumann-Haefelin, T., Staiger, J. F., Redecker, C., Zilles, K., Fritschy, J. M., Mohler, H., and Witte, O. W. (1998). Immunohistochemical evidence for dysregulation of the GABAergic system ipsilateral to photochemically induced cortical infarcts in rats. *Neuroscience* **87,** 871–879.

Neumar, R. W., Meng, F. H., Mills, A. M., Xu, Y. A., Zhang, C., Welsh, F. A., and Siman, R. (2001). Calpain activity in the rat brain after transient forebrain ischemia. *Exp. Neurol.* **170,** 27–35.

Nicholson, Ch., and Sykova, E. (1998). Extracellular space structure revealed by diffusion analysis. *Trends Neurosci.* **21,** 207–215.

Nikonenko, A. G., and Skibo, G. G. (2004). Technique to quantify local clustering of synaptic vesicles using single section data. *Microsc. Res. Tech.* **65,** 287–291.

Nikonenko, A. G., Radenovic, L., Andjus, P., and Skibo, G. (2009). Structural features of ischemic damage in the hippocampus. *Anat. Rec. (Hoboken)* **292,** 1914–1921.

Paakkari, I., and Lindsberg, P. (1995). Nitric oxide in the central nervous system. *Ann. Med.* **27,** 369–377.

Pérez-Otaño, I., and Ehlers, M. D. (2005). Homeostatic plasticity and NMDA receptor trafficking. *Trends Neurosci.* **28,** 229–238.

Perez-Pinzon, M. A., Dave, K. R., and Raval, A. P. (2005). Role of reactive oxygen species and protein kinase C in ischemic tolerance in the brain. *Antioxid. Redox Signal.* **7,** 1150–1157.

Picconi, B., Tortiglione, A., Barone, I., Centonze, D., Gardoni, F., Gubellini, P., Bonsi, P., Pisani, A., Bernardi, G., Di Luca, M., and Calabresi, P. (2006). NR2B subunit exerts a critical role in postischemic synaptic plasticity. *Stroke* **37,** 1895–1901.

Popa-Wagner, A., Schroder, E., Schmoll, H., Walker, L. C., and Kessler, C. (1999). Upregulation of MAP1B and MAP2 in the rat brain after middle cerebral artery occlusion: Effect of age. *J. Cereb. Blood Flow Metab.* **19,** 425–434.

Rami, A. (2003). Ischemic neuronal death in the rat hippocampus: The calpain-calpastatin-caspase hypothesis. *Neurobiol. Dis.* **13,** 75–88.

Rami, A., and Kögel, D. (2008). Apoptosis meets autophagy-like cell death in the ischemic penumbra: Two sides of the same coin? *Autophagy* **4,** 422–426.

Rickhag, M., Wieloch, T., Gido, G., Elmer, E., Krogh, M., Murray, J., Lohr, S., Bitter, H., Chin, D. J., von Schack, D., Shamloo, M., and Nikolich, K. (2006). Comprehensive regional and temporal gene expression profiling of the rat brain during the first 24 h after experimental stroke identifies dynamic ischemia-induced gene expression patterns, and reveals a biphasic activation of genes in surviving tissue. *J. Neurochem.* **96,** 14–29.

Rickhag, M., Teilum, M., and Wieloch, T. (2007). Rapid and long-term induction of effector immediate early genes (BDNF, Neuritin and Arc) in peri-infarct cortex and dentate gyrus after ischemic injury in rat brain. *Brain Res.* **1151,** 203–210.

Rosenblum, W. I. (1997). Histopathologic clues to the pathways of neuronal death following ischemia/hypoxia. *J. Neurotrauma* **14,** 313–326.

Rossi, D. J., Brady, J. D., and Mohr, C. (2007). Astrocyte metabolism and signaling during brain ischemia. *Nat. Neurosci.* **10,** 1377–1386.

Ruan, Y. W., Zou, B., Fan, Y., Li, Y., Lin, N., Zeng, Y. S., Gao, T. M., Yao, Z., and Xu, Z. C. (2006). Dendritic plasticity of CA1 pyramidal neurons after transient global ischemia. *Neuroscience* **140,** 191–201.

Savonenko, A., Munoz, P., Melnikova, T., Wang, Q., Liang, X., Breyer, R. M., Montine, T. J., Kirkwood, A., and Andreasson, K. (2009). Impaired cognition, sensorimotor gating, and hippocampal long-term depression in mice lacking the prostaglandin E2 EP2 receptor. *Exp. Neurol.* **217,** 63–73.

Schäbitz, W. R., Steigleder, T., Cooper-Kuhn, C. M., Schwab, S., Sommer, C., Schneider, A., and Kuhn, H. G. (2007). Intravenous brain-derived neurotrophic factor enhances poststroke sensorimotor recovery and stimulates neurogenesis. *Stroke* **38,** 2165–2172.

Serou, M. J., DeCoster, M. A., and Bazan, N. G. (1999). Interleukin-1 beta activates expression of cyclooxygenase-2 and inducible nitric oxide synthase in primary hippocampal neuronal culture: Platelet-activating factor as a preferential mediator of cyclooxygenase-2 expression. *J. Neurosci. Res.* **58,** 593–598.

Shackelford, D. A., and Yeh, R. Y. (1998). Dephosphorylation of tau during transient forebrain ischemia in the rat. *Mol. Chem. Neuropathol.* **34,** 103–120.

Shetty, P. K., Huang, F. L., and Huang, K. P. (2008). Ischemia-elicited oxidative modulation of $Ca^{2+}$/calmodulin-dependent protein kinase II. *J. Biol. Chem.* **283,** 5389–5401.

Song, I., and Huganir, R. L. (2002). Regulation of AMPA receptors during synaptic plasticity. *Trends Neurosci.* **25,** 578–588.

Stoll, G., Jander, S., and Schroeter, M. (2000). Cytokines in CNS disorders: Neurotoxicity versus neuroprotection. *J. Neural. Transm. Suppl.* **59,** 81–89.

Sudhof, T. C. (2000). The synaptic vesicle cycle revisited. *Neuron* **28,** 317–320.

Sutula, T., Koch, J., Golarai, G., Watanabe, Y., and McNamara, J. O. (1996). NMDA receptor dependence of kindling and mossy fiber sprouting: Evidence that the NMDA receptor regulates patterning of hippocampal circuits in the adult brain. *J. Neurosci.* **16,** 7398–7406.

Tanaka, K. (2001). Alteration of second messengers during acute cerebral ischemia -adenylate cyclase, cyclic AMP-dependent protein kinase, and cyclic AMP response element binding protein. *Prog. Neurobiol.* **65,** 173–207.

Tanaka, H., Grooms, S. Y., Bennett, M. V., and Zukin, R. S. (2000). The AMPAR subunit GluR2: Still front and center-stage. *Brain Res.* **886,** 190–207.

Tonchev, A. B., Yamashima, T., Zhao, L., Okano, H. J., and Okano, H. (2003). Proliferation of neural and neuronal progenitors after global brain ischemia in young adult macaque monkeys. *Mol. Cell. Neurosci.* **23,** 292–301.

Tsirka, S. E., Gualandris, A., Amaral, D. G., and Strickland, S. (1995). Excitotoxin-induced neuronal degeneration and seizure are mediated by tissue plasminogen activator. *Nature* **377,** 340–344.

Vexler, Z. S., and Yenari, M. A. (2009). Does inflammation after stroke affect the developing brain differently than adult brain? *Dev. Neurosci.* **31,** 378–393.

Wake, H., Moorhouse, A. J., Jinno, S., Kohsaka, S., and Nabekura, J. (2009). Resting microglia directly monitor the functional state of synapses in vivo and determine the fate of ischemic terminals. *J. Neurosci.* **29,** 3974–3980.

Wei, L., Ying, D. J., Cui, L., Langsdorf, J., and Yu, S. P. (2004). Necrosis, apoptosis and hybrid death in the cortex and thalamus after barrel cortex ischemia in rats. *Brain Res.* **1022,** 54–61.

Whitaker, V. R., Cui, L., Miller, S., Yu, S. P., and Wei, L. (2007). Whisker stimulation enhances angiogenesis in the barrel cortex following focal ischemia in mice. *J. Cereb. Blood Flow Metab.* **27,** 57–68.

Williams, P. A., Dou, P., and Dudek, F. E. (2004). Epilepsy and synaptic reorganization in a perinatal rat model of hypoxia-ischemia. *Epilepsia* **45,** 1210–1218.

Winship, I. R., and Murphy, T. H. (2008). In vivo calcium imaging reveals functional rewiring of single somatosensory neurons after stroke. *J. Neurosci.* **28,** 6592–6606.

Witte, O. W., and Stoll, G. (1997). Delayed and remote effects of focal cortical infarctions: Secondary damage and reactive plasticity. *Adv. Neurol.* **73,** 207–227.

Woolley, C. S., Gould, E., Frankfurt, M., and McEwen, B. S. (1990a). Naturally occurring fluctuation in dendritic spine density on adult hippocampal pyramidal neurons. *J. Neurosci.* **10,** 4035–4039.

Woolley, C. S., Gould, E., and McEwen, B. S. (1990b). Exposure to excess glucocorticoids alters dendritic morphology of adult hippocampal pyramidal neurons. *Brain Res.* **531,** 225–231.

Yagita, Y., Kitagawa, K., Ohtsuki, T., Takasawa, K., Miyata, T., Okano, H., Hori, M., and Matsumoto, M. (2001). Neurogenesis by progenitor cells in the ischemic adult rat hippocampus. *Stroke* **32,** 1890–1896.

Yang, H., and Chen, C. (2008). Cyclooxygenase-2 in synaptic signaling. *Curr.Pharm.Des.* **14,** 1443–1451.

Yepes, M., and Lawrence, D. A. (2004). New functions for an old enzyme: Nonhemostatic roles for tissue-type plasminogen activator in the central nervous system. *Exp. Biol. Med. (Maywood)* **229,** 1097–1104.

Yokota, M., Saido, T. C., Kamitani, H., Tabuchi, S., Satokata, I., and Watanabe, T. (2003). Calpain induces proteolysis of neuronal cytoskeleton in ischemic gerbil forebrain. *Brain Res.* **984,** 122–132.

Yu, A. C., Wong, H. K., Yung, H. W., and Lau, L. T. (2001). Ischemia-induced apoptosis in primary cultures of astrocytes. *Glia* **35,** 121–130.

Zepeda, A., Sengpiel, F., Guagnelli, M. A., Vaca, L., and Arias, C. (2004). Functional reorganization of visual cortex maps after ischemic lesions is accompanied by changes in expression of cytoskeletal proteins and NMDA and GABA(A) receptor subunits. *J. Neurosci.* **24,** 1812–1821.

Zhang, F., Liu, C. L., and Hu, B. R. (2006). Irreversible aggregation of protein synthesis machinery after focal brain ischemia. *J. Neurochem.* **98,** 102–112.

Zhao, B. Q., Wang, S., Kim, H. Y., Storrie, H., Rosen, B. R., Mooney, D. J., Wang, X., and Lo, E. H. (2006). Role of matrix metalloproteinases in delayed cortical responses after stroke. *Nat. Med.* **12,** 441–445.

CHAPTER SEVEN

# Hypothalamic Inflammation and Obesity

Eliana P. Araújo,* Márcio A. Torsoni,† and Lício A. Velloso†

## Contents



## Abstract

Obesity is one of the most prevalent diseases in the modern world. It results from the progressive loss of balance between food intake and whole body energy expenditure. Recent studies have shown that consumption of fat-rich diets induces hypothalamic inflammation and dysfunction which is characterized by defective response to anorexygenic and thermogenic hormones, such as leptin and insulin, leading to anomalous neurotransmitter production and favoring body mass gain. In this chapter, we present the main recent advances in this rapidly evolving field, focusing on the role of hypothalamic inflammation on the genesis of obesity. 

## I. Introduction

Currently, obesity is one of the most important health problems in the world. It affects more than 300 million humans and according to the WHO these numbers will continue growing during the next 15 years (Kopelman, 2000). Recent epidemiological data on obesity regarding distinct regions of the world can be found at www.iuns.org/features/obesity/tabfig.htm.

* Department of Nursing, FCM UNICAMP, Campinas, Sao Paulo, Brazil
† Laboratory of Cell Signaling, FCM UNICAMP, Campinas, Sao Paulo, Brazil

*Vitamins and Hormones*, Volume 82
ISSN 0083-6729, DOI: 10.1016/S0083-6729(10)82007-2

Except for some rare types of monogenic defects (Farooqi and O'Rahilly, 2006), obesity occurs because of a complex combination of multiple environmental and genetic factors (Galgani and Ravussin, 2008). The consumption of highly energetic and palatable foods is among the most important epidemiological factors predisposing to this disease (Galgani and Ravussin, 2008). However, not all people exposed to this type of diet develop obesity, and this fact has long intrigued researchers. As currently known, the main reason for the protective phenotype is the intrinsic capacity to maintain the homeostatic control of energy stores in the body (Galgani and Ravussin, 2008). Specialized neurons of the hypothalamus play a central role, connecting the information provided by leptin and insulin, regarding the size of peripheral fat depots, with the mechanisms that regulate hunger and thermogenesis (Velloso *et al.*, 2008). As long as the system is perfectly coupled, changes in energy intake are matched by proportional modifications in energy expenditure. Thus, it is clear that, in order to understand the mechanisms behind most cases of obesity, the phenomena that connect the consumption of highly energetic foods with the loss of energy homeostasis must be deciphered.

A series of recent studies has provided a solid basis to support the hypothesis that in diet-induced obesity, the hypothalamus is targeted by an inflammatory process that leads to a defective regulation of the energy homeostasis (De Souza *et al.*, 2005; Milanski *et al.*, 2009; Wisse and Schwartz, 2009; Yang and Hotamisligil, 2008; Zhang *et al.*, 2008). In this chapter, to discuss this complex mechanism, we start by presenting the physiological role of the hypothalamus in the control of food intake and energy expenditure. Next, we discuss how nutrients can disrupt the correct functioning of this highly specialized organ. Finally, we present the mechanisms involved in diet-induced hypothalamic resistance to adipostatic signals.

## II. Hypothalamic Control of Feeding and Energy Expenditure

Two distinct subpopulations of neurons of the arcuate nucleus of the hypothalamus act as the sensors for the energy stores in the body and coordinate a complex network of neurons that, in due course, control the balance of hunger versus satiety, and pro- versus antithermogenesis (Flier and Maratos-Flier, 1998; Schwartz *et al.*, 2000). These first-order neurons are equipped with receptors and intracellular molecular systems capable of detecting subtle or chronic changes in the levels of hormones and nutrients present in the bloodstream (Schwartz *et al.*, 2000). The response to these changes is based on the modulation of the firing rate and of neurotransmitter production and release by specific neuron bodies (Horvath, 2005).

The subpopulations of neurons of the arcuate nucleus are characterized by the neurotransmitters each one produces. One of the subpopulations expresses the orexygenic peptides, NPY and AgRP, while the other expresses the anorexygenic POMC (α-MSH) and CART (Horvath, 2005; Schwartz *et al.*, 2000) (Fig. 7.1). Both subpopulations project to the lateral (LH) and paraventricular (PVN) nuclei of the hypothalamus where they control the functions of second-order neurons. In the PVN, two distinct subpopulations of neurons produce the anorexigenic and prothermogenic neurotransmitters, TRH and CRH (Cone, 2005), while in the LH, two other subpopulations produce the predominantly orexygenic neurotransmitter orexin and the predominantly antithermogenic MCH (Cone, 2005) (Fig. 7.1). During fasting or when body energy stores are depleted, the expressions of NPY and AgRP are induced, while POMC and CART are inhibited. This coordinated response is dependent on the simultaneous sensing of decreased nutrient availability, reduced levels of the adipostatic hormones leptin and insulin, reduced levels of the gut hormones CCK, GLP-1, and GIP, and increased levels of the gastric hormone ghrelin (Badman and Flier, 2005). Active NPY/AgRPergic neurons send inhibitory

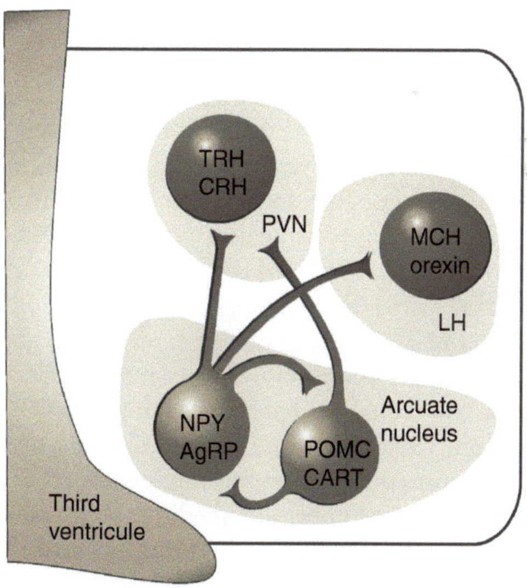


**Figure 7.1** First-order neurons localized in the arcuate nucleus of the hypothalamus provide the signals that modulate food intake and thermogenesis. During fasting, NPY/AgRP neurons are active and control the function of second-order neurons at the paraventricular nucleus (PVN) (inhibiting CRH and TRH neurons) and at the lateral hypothalamus (LH) (stimulating orexin and MCH neurons). After a meal, or following leptin/insulin stimulation, NPY/AgRP neurons are inhibited and POMC/CART neurons become active, providing the opposite signals.

projections to the PVN, reducing the expressions of TRH and CRH, and stimulatory projections to the LH, boosting the activities of the orexin and MCH expressing neurons. Conversely, following a meal, or when body energy stores are replenished, NPY/AgRPergic neurons are inhibited and POMC/CARTergic neurons are active. In this context, nutrient availability and the levels of leptin and insulin as well as the levels of CCK, GLP-1, and GIP are increased. Conversely, the level of ghrelin is reduced. The result is the inhibition of orexin and MCH neurons in the LH and the activation of the CRH and TRH neurons in the PVN (Badman and Flier, 2005; Cone, 2005; Horvath, 2005).

The mechanisms by which second-order neurons effectively control food intake and energy expenditure are under intense investigation. MCH neurons play an important role in the control of energy expenditure (Cone, 2005; Pereira-da-Silva *et al.*, 2003). Increased expression of this neurotransmitter restrains animal motility and reduces the expression of the mitochondrial uncoupling protein, UCP1, in brown adipose tissue, together leading to a reduction in energy output (Cone, 2005; Pereira-da-Silva *et al.*, 2003). Conversely, the knockout of MCH produces a lean phenotype, due to a combined effect on feeding and thermogenesis (Qu *et al.*, 1996; Shimada *et al.*, 1998), while the knockout of the main MCH receptor, MCHR1, produces a lean phenotype, predominantly due to increased energy expenditure (Marsh *et al.*, 2002). Orexin has a predominant role in arousal and the control of feeding (Chemelli *et al.*, 1999; Farr *et al.*, 2005). Injection of this neurotransmitter in the hypothalamus generates a potent orexigenic stimulus, however, little is known about the mechanisms behind this response (Farr *et al.*, 2005). In fact, recent studies shown that orexin signaling through orexin receptor-2 has an anorexigenc instead of orexigenic action and the enhanced expression of this receptor protects against obesity (Funato *et al.*, 2009). In opposition to the neurotransmitters of the LH, the PVN neurotransmitters, TRH and CRH, have rather overlapping roles in the control of hunger and thermogenesis (Appel *et al.*, 1991; Schuhler *et al.*, 2007). TRH biosynthesis and release is controlled by multiple inputs coming from POMC neurons, leptin direct signals, and from other sources such as T3 (Fekete and Lechan, 2007). Although most studies explore the role of TRH upon the control of thyroid function and, consequently, on thermogenesis, there are plenty of data showing its direct action in the control of feeding (Valassi *et al.*, 2008). Similar to TRH, CRH production is modulated by a number of different inputs, such as signals emanating from the arcuate nucleus and direct actions of leptin, GLP-1, and histamine (Nicholson *et al.*, 2004; Sarkar *et al.*, 2003). The reduction of appetite is the most studied effect of CRH, but several studies point to its prothermogenic effects as well (Solinas *et al.*, 2006). However, little is known about the mechanisms controlling these phenomena. Some additional pathways of the central nervous system play modulatory roles in energy balance. The connections of first- and second-order neurons of the hypothalamus with these systems are only beginning to be deciphered (Horvath, 2005). The actions of serotonin and norepinephrine to induce satiety and increase energy expenditure have been known for a long time (Leibowitz and Miller, 1969; Waldbillig *et al.*, 1981). Nevertheless, these neurotransmitters play rather unspecific and minor regulatory roles in this context, as evidenced by the moderate/severe adverse effects produced by drugs acting through the control of these neurotransmitters and by the limited efficiency of all treatment regimens employing such drugs (Mancini and Halpern, 2006).

Recently, a new player came onto the scene. The development of drugs that interact with the receptor for the endogenous cannabinoid system revealed an additional mechanism for the control of food intake and thermogenesis. The first clues about the orexygenic properties of the endogenous cannabinoid system came from the observation that the consumption of exogenous cannabinoids present in marijuana produced a powerful sensation of hunger (Di Marzo and Matias, 2005). The characterization of the main endocannabinoid receptor, CB1, and the development of synthetic antagonists for this receptor created hope for the production of new, safer, and more effective drugs for the treatment of obesity (Di Marzo and Matias, 2005). However, clinical data show that the weight loss produced by this class of drug is only marginally superior to that produced by inhibitors of serotonine reuptake and adverse affects, such as depression and increased rate of suicide, have precluded the widespread use of these compounds for the treatment of obesity (Vinod and Hungund, 2006).

Although, as discussed above, a number of mechanisms play a role in the control of hypothalamic neurons involved in the regulation of energy homeostasis, leptin and insulin are regarded as the most robust peripheral signal providers to the hypothalamus (Schwartz *et al.*, 2000). Leptin is produced predominantly by the adipose tissue in direct proportion to body fat mass (Myers *et al.*, 2008), and although some peripheral actions have been attributed to this hormone, such as the regulation of insulin production by pancreatic $\beta$-cells (Seufert *et al.*, 1999), modulation of insulin action (Barzilai *et al.*, 1997), and control of a number of immune functions (Girasol *et al.*, 2009; Mansour *et al.*, 2006; Matarese and La Cava, 2004), the basomedial hypothalamus, particularly the arcuate nucleus, is the main site of its action (Myers *et al.*, 2008; Schwartz *et al.*, 2000).

Both NPY/AgRPergic and POMC/CARTergic neurons express high levels of the main form of the leptin receptor, the ObRb (Myers *et al.*, 2008; Schwartz *et al.*, 2000) (Fig. 7.2). This is a monomeric transmembrane protein that belongs to the type I cytokine receptor family (Munzberg and Myers, 2005; Myers *et al.*, 2008). Like the other members of this family, the ObRb lacks intrinsic enzymatic activity and depends on at least one associated kinase, JAK2, to transduce its signal (Munzberg and Myers, 2005). Upon leptin binding to the ObRb, a dimmerization of receptor units accompanies JAK2 autophosphorylation and the tyrosine phosphorylation

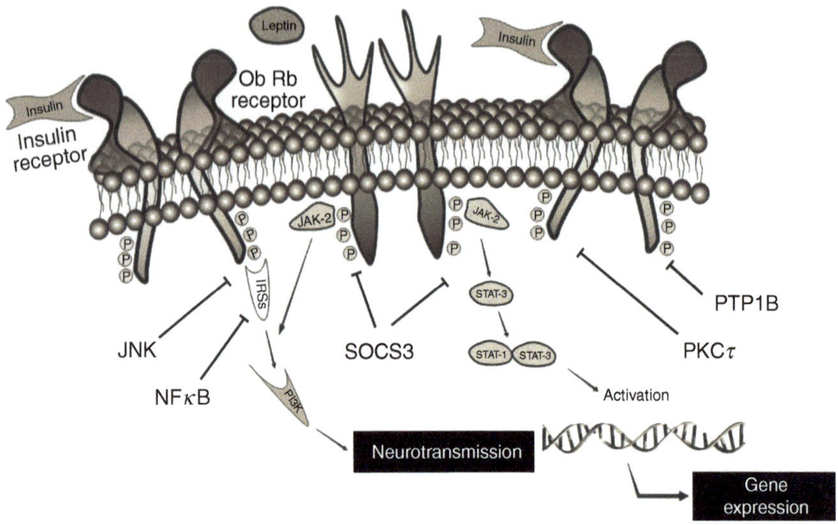


**Figure 7.2** Leptin (ObRb) and insulin receptors transduce the signals generated by their respective ligands though a cascade of signaling molecules of the JAK/STAT and/or IRSs/PI3kinase pathways. So far, five inhibitory mechanisms have been identified that hamper proper signaling through these pathways, SOCS3, PTP1B, JNK, NFκB, and PKCτ.

of two residues (Tyr985 and Tyr1138) in the receptor itself (Munzberg and Myers, 2005). These events generate the possibility of activation of at least three distinct intracellular signaling pathways (Munzberg and Myers, 2005). Tyrosine phosphorylation of Tyr985 recruits the tyrosine phosphatase SHP2, which mediates the activation of the p21ras/ERK signaling pathway (Munzberg and Myers, 2005; Myers *et al.*, 2008). Tyrosine phosphorylation of Tyr1138 recruits STAT3 to the ObRb, leading to STAT3 tyrosine phosphorylation and translocation to the nucleus, providing a fast-track for leptin-induced control of gene expression (Myers *et al.*, 2008). Finally, the autophosphorylation of JAK2 leads to the recruitment and tyrosine phosphorylation of adaptor proteins IRS1/2, which promotes the activation of PI3-kinase and its downstream signaling (Myers *et al.*, 2008). It is possible that several other signaling pathways are also activated through the ObRb, these may include other substrates for JAK2, since a large number of Tyr residues may be phosphorylated following kinase activation, and the engagement of as yet unknown tyrosine kinases, since signal transduction through IRS1/PI3-kinase/Akt can occur even in the absence of the activation of JAK2 (Mansour *et al.*, 2006).

Each of the signaling pathways controlled by leptin plays a role in a specific compartment of the complex response to this hormone. In POMC/CARTergic neurons, the activation of JAK2/STAT3 signaling by leptin leads to increased transcriptional activity, boosting the expression of POMC (Cone, 2005; Schwartz *et al.*, 2000). This effect is enhanced by simultaneous insulin action, but apparently not reproduced by insulin alone (Carvalheira *et al.*, 2001). Conversely, the activation of PI3 kinase activity seems to play a minor role in the control of neurotransmitter expression, but is essential for neurotransmitter release in the synaptic terminals. This effect is achieved through the control of neuronal firing rate. Once activated by either leptin or insulin, PI3 kinase mediates neuronal depolarization by inhibiting ATP-sensitive potassium channels (Horvath, 2005; Niswender *et al.*, 2004). The central role for PI3 kinase in this context is further evidenced by the fact that genetic or pharmacological modulations of phosphatases that control the signaling through PI3 kinase, such as PTEN and 5′-ptase IV, have profound effects on feeding behavior (Bertelli *et al.*, 2006; Plum *et al.*, 2006).

Since not only POMC/CARTergic neurons, but also the orexigenic NPY/AgRPergic neurons harbor the leptin receptor, an important question was raised regarding the mechanisms by which leptin can simultaneously activate the anorexygenic neurons, while inhibiting the orexigenic neurons. The answer to this question came from the demonstration that POMC/CARTergic neurons project inhibitory fibers to NPY/AgRPergic neurons. When leptin levels are high, the activation of POMC/CARTergic neurons leads to a simultaneous inhibition of NPY/AgRPergic neurons, a phenomenon that superimposes the direct signals eventually generated through the ObRb, and also the insulin receptor present in these orexigenic neurons (Horvath, 2005).

Finally, it is important to mention that, besides its predominant actions in the arcuate nucleus, a number of studies have shown that leptin can act through cells present in other regions of the brain. Some of these cells may act only to modulate the main signals delivered by arcuate nucleus neurons, but in some cases specific responses may control other physiological phenomena primarily controlled by leptin (Hayes *et al.*, 2009; Leinninger *et al.*, 2009).

Insulin is the second most important adipostatic signal provider to the hypothalamus. Studies from the late 1960s pioneered the investigation into the roles of insulin in the central nervous system (Margolis and Altszuler, 1967). However, only after the identification of leptin in 1994 were the functions of insulin in the hypothalamus described (Air *et al.*, 2002; Carvalheira *et al.*, 2001). In the arcuate nucleus, both NPY/AgRPergic and POMC/CARTergic neurons express receptors for insulin (Schwartz *et al.*, 2000; Velloso *et al.*, 2008). As in peripheral tissues, in the hypothalamus, insulin activates signal transduction through IRS1 and IRS2, leading to the engagement of the

PI3K/Akt/Foxo1 pathway (Carvalheira *et al.*, 2001; Schwartz *et al.*, 2000; Torsoni *et al.*, 2003). In addition, insulin induces a potent cross-talk with the leptin signaling pathway through the activation of JAK2 (Carvalheira *et al.*, 2001). In fact, one of the most important functions of insulin in the hypothalamus is to enhance leptin's signal (Carvalheira *et al.*, 2001; Schwartz *et al.*, 2000). In the absence of the insulin signal in the hypothalamus, much of the adipostatic function of leptin is lost, as demonstrated using neuron-specific insulin receptor knockout mice (Bruning *et al.*, 2000).

Therefore, under physiological conditions the hypothalamus, acting under the control of peripheral factors, coordinates perfect coupling between food intake and energy expenditure. As long as the system is fully active, body mass is maintained steady.

## III. Nutrient-Induced Dysfunction of the Hypothalamus

Excessive caloric intake, regardless of the type of nutrient consumed, is a primary risk factor for the development of obesity (Cohen, 2008). However, a number of epidemiological studies have shown that populations consuming preferentially diets rich in fat are especially prone to gain body mass (Damiao *et al.*, 2006; Ogden *et al.*, 2007). In addition to its caloric value, fatty acids are known to exert functional modulation of several tissues and cell types. Therefore, we tested the hypothesis that high fat consumption could modulate gene expression in the hypothalamus. Using a macroarray approach, the expressions of more than 1000 hypothlamic genes were simultaneously evaluated. More than 15% of the analyzed genes were somehow modulated by the diet. Clustering the genes by function revealed that immune response related genes were the most affected (De Souza *et al.*, 2005, 2008). Among the fatty acids commonly present in the occidental diet, the long-chain saturated ones are the most harmful. By activating signal transduction through receptors of the toll-like receptor family, especially the TLR4/MyD88, these fatty acids activate an inflammatory response by the microglia cells in the hypothalamus (Kleinridders *et al.*, 2009; Milanski *et al.*, 2009). Signaling through JNK and NF$\kappa$B leads to the induction of cytokine gene transcription and local levels of TNF-$\alpha$, IL-1$\beta$, IL-6, and IFN-$\gamma$ raise and nourish hypothalamic inflammation (Milanski *et al.*, 2009).

Following fatty acid-induced activation of TLR4 signaling, cells turn on an adaptive mechanism that has the biological purpose of adapting protein synthesis to the harms imposed by the inflammation. This mechanism is called endoplasmic reticulum stress and, depending on the magnitude and duration of the harm, can enhance inflammatory signal or induce apoptosis (Milanski *et al.*, 2009; Zhang and Kaufman, 2008).

The ER is the organelle responsible for the synthesis and processing of membrane and secretory proteins (Xu *et al.*, 2005). When the ER homeostasis is disrupted, the accumulation of misfolded and unfolded proteins in the ER lumen ensues (Schroder and Kaufman, 2005; Xu *et al.*, 2005). To deal with this condition, the affected cells activate a complex signaling system known as the unfolded protein response (UPR), aimed at preserving cell integrity, while the harmful condition persists (Schroder and Kaufman, 2005; Xu *et al.*, 2005). One of the outcomes of the activation of UPR is the induction of the expression of cytokines and proteins involved in immune surveillance (Krappmann *et al.*, 2004; Marciniak and Ron, 2006).

If the exposure to saturated fatty acids is prolonged and, depending on some as yet unknown genetic determinants, a proapototic response is induced, leading to the preferential loss of anorexigenic neurons in the arcuate nucleus (Moraes *et al.*, 2009). As time passes, a gradual modification in the relative numbers of orexygenic and anorexygenic neurons takes place and a novel set point for body adiposity is generated. This fact may explain why some obese patients are so resistant to different behavioral and pharmacological approaches employed to treat obesity. Thus, activation of TLR4 signal transduction, followed by the induction of hypothalamic ER-stress, is the main mechanism linking the high consumption of dietary fats to the induction of hypothalamic dysfunction in obesity.

## IV. Hypothalamic Resistance to Anorexygenic Signals

As long as a perfect coupling between caloric intake and energy expenditure is preserved, body adiposity is maintained at a physiological level (Cohen, 2008). One of the mechanisms involved in the breakdown of this equilibrium is the installation of resistance to the anorexigenic and thermogenic effects of leptin and insulin (De Souza *et al.*, 2005). In animal models, the resistance to both these hormones can be quantified by a simple method. In intracerebroventricular cannulated lean rodents, the acute injection of leptin leads to a reduction of up to 60% of spontaneous food intake over 12 h (Carvalheira *et al.*, 2001). Injecting insulin, rather than leptin, produces a reduction of up to 50% in food intake. However, if the same tests are performed in obese animals, the effects of both hormones are blunted by at least 50%. To identify the molecular mechanism responsible for producing the functional resistance to leptin and insulin, several groups have evaluated different models of obesity and the most remarkable findings reveal that, upon diet-induced obesity, the induction of inflammatory activity, specifically in the hypothalamus, leads to the activation of intracellular signaling pathways that promote a negative cross-talk with the leptin and insulin signaling systems, therefore, impairing their physiological anorexygenic activities (Carvalheira *et al.*, 2001; Torsoni *et al.*, 2003).

Currently, five distinct mechanisms are known to play a role in diet- and cytokine-induced resistance to leptin and/or insulin in the hypothalamus of rodents. Both TNF-$\alpha$ and the consumption of fat-rich diets can induce the activation of the serine-kinase JNK in hypothalamic cells. Activated JNK targets IRS1, catalyzing its serine phosphorylation and hampering its capacity to act appropriately as a docking protein for PI3-kinase. This results in reduced insulin-induced activation of Akt and coincides with functional resistance to insulin. Inhibition of JNK by a specific chemical inhibitor or inhibition of TNF-$\alpha$ by a blocking monoclonal antibody reverses the effects of the diet or the cytokine and reestablishes a normal tonus for insulin activity in the hypothalamus (De Souza *et al.*, 2005).

IKK, the serine kinase involved in IkB phosphorylation/degradation, which leads to NF$\kappa$B activation, is another intermediary involved in diet-induced insulin resistance in the hypothalamus (Zhang *et al.*, 2008). In rats fed on a fat-rich diet, activated IKK promotes the serine phosphorylation of IRS1, hampering the signal transduction of insulin. While in peripheral organs of insulin resistant animals, the inhibition of IKK by salicylates was proved to reinstall correct insulin activity (Prattali *et al.*, 2005), no data regarding the pharmacological inhibition of this enzyme in the hypothalamus are available so far, but it is expected to work in a similar manner.

SOCS3 belongs to a family of inducible proteins that respond to stimulus by a number of cytokines, hormones, and growth factors. Once expressed, SOCS proteins provide a negative feedback to the original signal, acting as a regulatory loop to restrain over stimulation. In diet-induced obesity, the expression of SOCS3 is significantly stimulated in the hypothalamus, providing a negative control for leptin and insulin signaling (Bjorbaek *et al.*, 1998; Munzberg *et al.*, 2004). At least two mechanisms are involved in SOCS3 inhibition of leptin and insulin signaling. The first depends on the physical interaction of SOCS3 with either the ObRb or STAT3 (Myers *et al.*, 2008). Under these circumstances, the transduction of the signal is inhibited because the sites for protein–protein interaction are blocked by SOCS3. The second mechanism depends on SOCS3-dependent ubiquitination of IRS proteins. Following ubiquitin tagging, IRS proteins are directed to proteosomic degradation, restraining signal transduction through this pathway (Myers *et al.*, 2008). The important role for SOCS3 in diet-induced obesity is further confirmed by the fact that genetic abrogation of SOCS3 is capable of protecting mice from diet-induced obesity (Myers *et al.*, 2008).

An additional mechanism involved in diet- and cytokine-induced resistance to anorexigenic signaling in the hypothalamus is the induction of expression of the tyrosine phosphatase PTP1B. Once induced, this enzyme catalyzes the dephosphorylation of the IR and IRS proteins, turning off the signals generated by insulin. The knockout of the PTP1B gene or the knockdown of PTP1B by antisense oligonucleotides protects experimental animals from diet- and cytokine-induced insulin resistance in the hypothalamus (Picardi *et al.*, 2008).

Finally, a recent study showed that consumption of saturated fatty acid-rich diet promotes the activation of PKC-$\theta$ in discrete neuronal populations of the arcuate nucleus leading to insulin and leptin resistance in the hypothalamus (Benoit *et al.*, 2009).

## V. Concluding Remarks

Following the identification of leptin, 15 years ago, great advance has been obtained in the understanding the physiological mechanisms of body mass control. This has allowed substantial advance in the characterization of the mechanisms involved in the development of obesity. It is currently believed that, in most cases of obesity, the hypothalamus is the primarily affected organ. Upon high consumption of dietary fat a local inflammatory process is triggered by the activation of TLR4 signaling. Neurons and microglia are affected and ER-stress is induced. Depending on genetic background, specific subpopulations of neurons are lost by apoptosis, therefore enhancing the harmful effects of inflammation. With time, the homeostatic control of body energy stores is lost and obesity emerges. Future studies will focus on the characterization of similar phenomena in human beings and on the evaluation of specific antiinflammatory approaches to treat obesity.

## ACKNOWLEDGMENTS

Studies performed at the Laboratory of Cell Signaling were supported by grants from Fundação de Amparo à Pesquisa do Estado de São Paulo, Conselho Nacional de Desenvolvimento Científico e Tecnológico. The Laboratory of Cell Signaling is a member of the National Institute of Obesity and Diabetes (INCT de obesidade e diabetes—CNPq/MCT). The author thanks Dr. N. Conran for editing the English grammar.

## REFERENCES

Air, E. L., Benoit, S. C., Clegg, D. J., Seeley, R. J., and Woods, S. C. (2002). Insulin and leptin combine additively to reduce food intake and body weight in rats. *Endocrinology* **143,** 2449–2452.

Appel, N. M., Owens, M. J., Culp, S., Zaczek, R., Contrera, J. F., Bissette, G., Nemeroff, C. B., and De Souza, E. B. (1991). Role for brain corticotropin-releasing factor in the weight-reducing effects of chronic fenfluramine treatment in rats. *Endocrinology* **128,** 3237–3246.

Badman, M. K., and Flier, J. S. (2005). The gut and energy balance: Visceral allies in the obesity wars. *Science* **307,** 1909–1914.

Barzilai, N., Wang, J., Massilon, D., Vuguin, P., Hawkins, M., and Rossetti, L. (1997). Leptin selectively decreases visceral adiposity and enhances insulin action. *J. Clin. Invest.* **100,** 3105–3110.

Benoit, S. C., Kemp, C. J., Elias, C. F., Abplanalp, W., Herman, J. P., Migrenne, S., Lefevre, A. L., Cruciani-Guglielmacci, C., Magnan, C., Yu, F., Niswender, K., Irani, B. G., *et al.* (2009). Palmitic acid mediates hypothalamic insulin resistance by altering PKC-theta subcellular localization in rodents. *J. Clin. Invest.* **119,** 2577–2589.

Bertelli, D. F., Araujo, E. P., Cesquini, M., Stoppa, G. R., Gasparotto-Contessotto, M., Toyama, M. H., Felix, J. V., Carvalheira, J. B., Michelini, L. C., Chiavegatto, S., Boschero, A. C., Saad, M. J., *et al.* (2006). Phosphoinositide-specific inositol polyphosphate 5-phosphatase IV inhibits inositide trisphosphate accumulation in hypothalamus and regulates food intake and body weight. *Endocrinology* **147,** 5385–5399.

Bjorbaek, C., Elmquist, J. K., Frantz, J. D., Shoelson, S. E., and Flier, J. S. (1998). Identification of SOCS-3 as a potential mediator of central leptin resistance. *Mol. Cell* **1,** 619–625.

Bruning, J. C., Gautam, D., Burks, D. J., Gillette, J., Schubert, M., Orban, P. C., Klein, R., Krone, W., Muller-Wieland, D., and Kahn, C. R. (2000). Role of brain insulin receptor in control of body weight and reproduction. *Science* **289,** 2122–2125.

Carvalheira, J. B., Siloto, R. M., Ignacchitti, I., Brenelli, S. L., Carvalho, C. R., Leite, A., Velloso, L. A., Gontijo, J. A., and Saad, M. J. (2001). Insulin modulates leptin-induced STAT3 activation in rat hypothalamus. *FEBS Lett.* **500,** 119–124.

Chemelli, R. M., Willie, J. T., Sinton, C. M., Elmquist, J. K., Scammell, T., Lee, C., Richardson, J. A., Williams, S. C., Xiong, Y., Kisanuki, Y., Fitch, T. E., Nakazato, M., *et al.* (1999). Narcolepsy in orexin knockout mice: Molecular genetics of sleep regulation. *Cell* **98,** 437–451.

Cohen, D. A. (2008). Obesity and the built environment: Changes in environmental cues cause energy imbalances. *Int. J. Obes. (Lond.)* **32**(Suppl. 7), S137–S142.

Cone, R. D. (2005). Anatomy and regulation of the central melanocortin system. *Nat. Neurosci.* **8,** 571–578.

Damiao, R., Castro, T. G., Cardoso, M. A., Gimeno, S. G., and Ferreira, S. R. (2006). Dietary intakes associated with metabolic syndrome in a cohort of Japanese ancestry. *Br. J. Nutr.* **96,** 532–538.

De Souza, C. T., Araujo, E. P., Bordin, S., Ashimine, R., Zollner, R. L., Boschero, A. C., Saad, M. J., and Velloso, L. A. (2005). Consumption of a fat-rich diet activates a proinflammatory response and induces insulin resistance in the hypothalamus. *Endocrinology* **146,** 4192–4199.

De Souza, C. T., Pereira-da-Silva, M., Araujo, E. P., Morari, J., Alvarez-Rojas, F., Bordin, S., Moreira-Filho, D. C., Carvalheira, J. B., Saad, M. J., and Velloso, L. A. (2008). Distinct subsets of hypothalamic genes are modulated by two different thermogenesis-inducing stimuli. *Obesity (Silver Spring)* **16,** 1239–1247.

Di Marzo, V., and Matias, I. (2005). Endocannabinoid control of food intake and energy balance. *Nat. Neurosci.* **8,** 585–589.

Farooqi, S., and O'Rahilly, S. (2006). Genetics of obesity in humans. *Endocr. Rev.* **27,** 710–718.

Farr, S. A., Banks, W. A., Kumar, V. B., and Morley, J. E. (2005). Orexin-A-induced feeding is dependent on nitric oxide. *Peptides* **26,** 759–765.

Fekete, C., and Lechan, R. M. (2007). Negative feedback regulation of hypophysiotropic thyrotropin-releasing hormone (TRH) synthesizing neurons: Role of neuronal afferents and type 2 deiodinase. *Front Neuroendocrinol.* **28,** 97–114.

Flier, J. S., and Maratos-Flier, E. (1998). Obesity and the hypothalamus: Novel peptides for new pathways. *Cell* **92,** 437–440.

Funato, H., Tsai, A. L., Willie, J. T., Kisanuki, Y., Williams, S. C., Sakurai, T., and Yanagisawa, M. (2009). Enhanced orexin receptor-2 signaling prevents diet-induced obesity and improves leptin sensitivity. *Cell Metab.* **9,** 64–76.

Galgani, J., and Ravussin, E. (2008). Energy metabolism, fuel selection and body weight regulation. *Int. J. Obes. (Lond.)* **32**(Suppl. 7), S109–S119.

Girasol, A., Albuquerque, G. G., Mansour, E., Araujo, E. P., Degasperi, G., Denis, R. G., Carvalheira, J. B., Saad, M. J., and Velloso, L. A. (2009). Fyn mediates leptin actions in the thymus of rodents. *PLoS ONE* **4,** e7707.

Hayes, M. R., Skibicka, K. P., Bence, K. K., and Grill, H. J. (2009). Dorsal hindbrain AMP-Kinase as an intracellular mediator of energy balance. *Endocrinology* **150,** 2175–2182.

Horvath, T. L. (2005). The hardship of obesity: A soft-wired hypothalamus. *Nat. Neurosci.* **8,** 561–565.

Kleinridders, A., Schenten, D., Konner, A. C., Belgardt, B. F., Mauer, J., Okamura, T., Wunderlich, F. T., Medzhitov, R., and Bruning, J. C. (2009). MyD88 signaling in the CNS is required for development of fatty acid-induced leptin resistance and diet-induced obesity. *Cell Metab.* **10,** 249–259.

Kopelman, P. G. (2000). Obesity as a medical problem. *Nature* **404,** 635–643.

Krappmann, D., Wegener, E., Sunami, Y., Esen, M., Thiel, A., Mordmuller, B., and Scheidereit, C. (2004). The IkappaB kinase complex and NF-kappaB act as master regulators of lipopolysaccharide-induced gene expression and control subordinate activation of AP-1. *Mol. Cell Biol.* **24,** 6488–6500.

Leibowitz, S. F., and Miller, N. E. (1969). Unexpected adrenergic effects of chlorpromazine: Eating elicited by injection into rat hypothalamus. *Science* **165,** 609–611.

Leinninger, G. M., Jo, Y. H., Leshan, R. L., Louis, G. W., Yang, H., Barrera, J. G., Wilson, H., Opland, D. M., Faouzi, M. A., Gong, Y., Jones, J. C., Rhodes, C. J., *et al.* (2009). Leptin acts via leptin receptor-expressing lateral hypothalamic neurons to modulate the mesolimbic dopamine system and suppress feeding. *Cell Metab.* **10,** 89–98.

Mancini, M. C., and Halpern, A. (2006). Pharmacological treatment of obesity. *Arq. Bras. Endocrinol. Metabol.* **50,** 377–389.

Mansour, E., Pereira, F. G., Araujo, E. P., Amaral, M. E., Morari, J., Ferraroni, N. R., Ferreira, D. S., Lorand-Metze, I., and Velloso, L. A. (2006). Leptin inhibits apoptosis in thymus through a janus kinase-2-independent, insulin receptor substrate-1/phosphatidylinositol-3 kinase-dependent pathway. *Endocrinology* **147,** 5470–5479.

Marciniak, S. J., and Ron, D. (2006). Endoplasmic reticulum stress signaling in disease. *Physiol. Rev.* **86,** 1133–1149.

Margolis, R. U., and Altszuler, N. (1967). Insulin in the cerebrospinal fluid. *Nature* **215,** 1375–1376.

Marsh, D. J., Weingarth, D. T., Novi, D. E., Chen, H. Y., Trumbauer, M. E., Chen, A. S., Guan, X. M., Jiang, M. M., Feng, Y., Camacho, R. E., Shen, Z., Frazier, E. G., *et al.* (2002). Melanin-concentrating hormone 1 receptor-deficient mice are lean, hyperactive, and hyperphagic and have altered metabolism. *Proc. Natl. Acad. Sci. USA* **99,** 3240–3245.

Matarese, G., and La Cava, A. (2004). The intricate interface between immune system and metabolism. *Trends Immunol.* **25,** 193–200.

Milanski, M., Degasperi, G., Coope, A., Morari, J., Denis, R., Cintra, D. E., Tsukumo, D. M., Anhe, G., Amaral, M. E., Takahashi, H. K., Curi, R., Oliveira, H. C., *et al.* (2009). Saturated fatty acids produce an inflammatory response predominantly through the activation of TLR4 signaling in hypothalamus: Implications for the pathogenesis of obesity. *J. Neurosci.* **29,** 359–370.

Moraes, J. C., Coope, A., Morari, J., Cintra, D. E., Roman, E. A., Pauli, J. R., Romanatto, T., Carvalheira, J. B., Oliveira, A. L., Saad, M. J., and Velloso, L. A. (2009). High-fat diet induces apoptosis of hypothalamic neurons. *PLoS ONE* **4,** e5045.

Munzberg, H., and Myers, M. G. Jr. (2005). Molecular and anatomical determinants of central leptin resistance. *Nat. Neurosci.* **8,** 566–570.

Munzberg, H., Flier, J. S., and Bjorbaek, C. (2004). Region-specific leptin resistance within the hypothalamus of diet-induced obese mice. *Endocrinology* **145,** 4880–4889.

Myers, M. G., Cowley, M. A., and Munzberg, H. (2008). Mechanisms of leptin action and leptin resistance. *Annu. Rev. Physiol.* **70,** 537–556.

Nicholson, R. C., King, B. R., and Smith, R. (2004). Complex regulatory interactions control CRH gene expression. *Front Biosci.* **9,** 32–39.

Niswender, K. D., Baskin, D. G., and Schwartz, M. W. (2004). Insulin and its evolving partnership with leptin in the hypothalamic control of energy homeostasis. *Trends Endocrinol. Metab.* **15,** 362–369.

Ogden, C. L., Yanovski, S. Z., Carroll, M. D., and Flegal, K. M. (2007). The epidemiology of obesity. *Gastroenterology* **132,** 2087–2102.

Pereira-da-Silva, M., Torsoni, M. A., Nourani, H. V., Augusto, V. D., Souza, C. T., Gasparetti, A. L., Carvalheira, J. B., Ventrucci, G., Marcondes, M. C., Cruz-Neto, A. P., Saad, M. J., Boschero, A. C., *et al.* (2003). Hypothalamic melanin-concentrating hormone is induced by cold exposure and participates in the control of energy expenditure in rats. *Endocrinology* **144,** 4831–4840.

Picardi, P. K., Calegari, V. C., Prada Pde, O., Moraes, J. C., Araujo, E., Marcondes, M. C., Ueno, M., Carvalheira, J. B., Velloso, L. A., and Saad, M. J. (2008). Reduction of hypothalamic protein tyrosine phosphatase improves insulin and leptin resistance in diet-induced obese rats. *Endocrinology* **149,** 3870–3880.

Plum, L., Ma, X., Hampel, B., Balthasar, N., Coppari, R., Munzberg, H., Shanabrough, M., Burdakov, D., Rother, E., Janoschek, R., Alber, J., Belgardt, B. F., *et al.* (2006). Enhanced PIP3 signaling in POMC neurons causes KATP channel activation and leads to diet-sensitive obesity. *J. Clin. Invest.* **116,** 1886–1901.

Prattali, R. R., Barreiro, G. C., Caliseo, C. T., Fugiwara, F. Y., Ueno, M., Prada, P. O., Velloso, L. A., Saad, M. J., and Carvalheira, J. B. (2005). Aspirin inhibits serine phosphorylation of insulin receptor substrate 1 in growth hormone treated animals. *FEBS Lett.* **579,** 3152–3158.

Qu, D., Ludwig, D. S., Gammeltoft, S., Piper, M., Pelleymounter, M. A., Cullen, M. J., Mathes, W. F., Przypek, R., Kanarek, R., and Maratos-Flier, E. (1996). A role for melanin-concentrating hormone in the central regulation of feeding behaviour. *Nature* **380,** 243–247.

Sarkar, S., Fekete, C., Legradi, G., and Lechan, R. M. (2003). Glucagon like peptide-1 (7-36) amide (GLP-1) nerve terminals densely innervate corticotropin-releasing hormone neurons in the hypothalamic paraventricular nucleus. *Brain Res.* **985,** 163–168.

Schroder, M., and Kaufman, R. J. (2005). ER stress and the unfolded protein response. *Mutat. Res.* **569,** 29–63.

Schuhler, S., Warner, A., Finney, N., Bennett, G. W., Ebling, F. J., and Brameld, J. M. (2007). Thyrotrophin-releasing hormone decreases feeding and increases body temperature, activity and oxygen consumption in Siberian hamsters. *J. Neuroendocrinol.* **19,** 239–249.

Schwartz, M. W., Woods, S. C., Porte, D. Jr., Seeley, R. J., and Baskin, D. G. (2000). Central nervous system control of food intake. *Nature* **404,** 661–671.

Seufert, J., Kieffer, T. J., and Habener, J. F. (1999). Leptin inhibits insulin gene transcription and reverses hyperinsulinemia in leptin-deficient ob/ob mice. *Proc. Natl. Acad. Sci. USA* **96,** 674–679.

Shimada, M., Tritos, N. A., Lowell, B. B., Flier, J. S., and Maratos-Flier, E. (1998). Mice lacking melanin-concentrating hormone are hypophagic and lean. *Nature* **396,** 670–674.

Solinas, G., Summermatter, S., Mainieri, D., Gubler, M., Montani, J. P., Seydoux, J., Smith, S. R., and Dulloo, A. G. (2006). Corticotropin-releasing hormone directly stimulates thermogenesis in skeletal muscle possibly through substrate cycling between de novo lipogenesis and lipid oxidation. *Endocrinology* **147,** 31–38.

Torsoni, M. A., Carvalheira, J. B., Pereira-Da-Silva, M., de Carvalho-Filho, M. A., Saad, M. J., and Velloso, L. A. (2003). Molecular and functional resistance to insulin in hypothalamus of rats exposed to cold. *Am. J. Physiol. Endocrinol. Metab.* **285,** E216–E223.

Valassi, E., Scacchi, M., and Cavagnini, F. (2008). Neuroendocrine control of food intake. *Nutr. Metab. Cardiovasc. Dis.* **18,** 158–168.

Velloso, L. A., Araujo, E. P., and de Souza, C. T. (2008). Diet-induced inflammation of the hypothalamus in obesity. *Neuroimmunomodulation* **15,** 189–193.

Vinod, K. Y., and Hungund, B. L. (2006). Role of the endocannabinoid system in depression and suicide. *Trends Pharmacol. Sci.* **27,** 539–545.

Waldbillig, R. J., Bartness, T. J., and Stanley, B. G. (1981). Increased food intake, body weight, and adiposity in rats after regional neurochemical depletion of serotonin. *J. Comp. Physiol. Psychol.* **95,** 391–405.

Wisse, B. E., and Schwartz, M. W. (2009). Does hypothalamic inflammation cause obesity? *Cell Metab.* **10,** 241–242.

Xu, C., Bailly-Maitre, B., and Reed, J. C. (2005). Endoplasmic reticulum stress: Cell life and death decisions. *J. Clin. Invest.* **115,** 2656–2664.

Yang, L., and Hotamisligil, G. S. (2008). Stressing the brain, fattening the body. *Cell* **135,** 20–22.

Zhang, K., and Kaufman, R. J. (2008). From endoplasmic-reticulum stress to the inflammatory response. *Nature* **454,** 455–462.

Zhang, X., Zhang, G., Zhang, H., Karin, M., Bai, H., and Cai, D. (2008). Hypothalamic IKKbeta/NF-kappaB and ER stress link overnutrition to energy imbalance and obesity. *Cell* **135,** 61–73.

CHAPTER EIGHT

# The Role of Functional Postsynaptic NMDA Receptors in the Central Nucleus of the Amygdala in Opioid Dependence

Michael J. Glass

## Contents



## Abstract

Activation of ionotropic *N*-methyl-D-aspartate (NMDA)-type glutamate receptors in limbic system nuclei, such as the central nucleus of the amygdala (CeA), plays an essential role in autonomic, behavioral, and affective processes that are profoundly impacted by exposure to opioids. However, the heterogeneous ultrastructural distribution of the NMDA receptor, its complex pharmacology, and the paucity of genetic models have hampered the development of linkages between functional amygdala NMDA receptors and opioid dependence. To overcome these shortcomings, high-resolution imaging and molecular pharmacology were used to (1) Identify the ultrastructural localization of the essential NMDA-NR1 receptor (NR1) subunit and its relationship to the mu-opioid receptor ($\mu$OR), the major cellular target of abused opioids like morphine, in the CeA and (2) Determine the effect of CeA NR1 deletion on the physical, and particularly,

Department of Neurology and Neuroscience, Weill Cornell Medical College, New York, USA

*Vitamins and Hormones*, Volume 82
ISSN 0083-6729, DOI: 10.1016/S0083-6729(10)82008-4

psychological aspects of opioid dependence. Combined immunogold and immuoperoxidase electron microscopic analysis showed that NR1 was prominently expressed in postsynaptic (i.e., somata, dendrites) locations of CeA neurons, where they were also frequently colocalized with the $\mu$OR. A spatial–temporal deletion of NR1 in postsynaptic sites of CeA neurons was produced by local microinjection of a neurotropic recombinant adeno-associated virus (rAAV), expressing the green fluorescent protein (GFP) reporter and Cre recombinase (rAAV–GFP–Cre), in adult "floxed" NR1 (fNR1) mice. Mice with deletion of NR1 in the CeA showed no obvious impairments in sensory, motor, or nociceptive function. In addition, when administered chronic morphine, these mice also displayed an acute physical withdrawal syndrome precipitated by naloxone. However, opioid-dependent CeA NR1 knockout mice failed to exhibit a conditioned place aversion induced by naloxone-precipitated withdrawal. These results indicate that postsynaptic NMDA receptor activity in central amygdala neurons is required for the expression of a learned affective behavior associated with opioid withdrawal. The neurogenetic dissociation of physical and psychological properties of opioid dependence demonstrates the value of combined ultrastructural analysis and molecular pharmacology in clarifying the neurobiological mechanisms subserving opioid-mediated plasticity. 

## I. Introduction

Opioid use can lead to an interrelated complex of adverse neural and behavioral adaptations that include tolerance, addiction, and dependence (Christie, 2008). Dependence, the focus of this paper, is typically studied within the framework of the withdrawal state produced by cessation of opioid exposure. However, dependence can also have an impact that endures beyond the immediate experience of withdrawal. One persistent effect of dependence is the result of learned associations between the noxious affective properties of withdrawal and environmental cues (O'Brien, 2008), a topic that is an important theme of this chapter. Although the neural mechanisms subserving the complex aspects of dependence are far from clear, there is reason to believe that interactions between the ionotropic *N*-methyl-D-aspartate (NMDA)-type glutamate receptor and the mu-opioid receptor ($\mu$OR), the major target of abused opioids like morphine, play a key role in opioid-adaptive processes. Moreover, the central nucleus of the amygdala (CeA), a critical coordinator of behavioral and emotional processes impacted by opioid exposure (LeDoux, 2000; Phelps and LeDoux, 2005), is a key anatomical substrate of NMDA-dependent forms of neural plasticity that may parallel the development of drug dependence (Hyman *et al.*, 2006). After providing some important background information, this paper describes recent results characterizing the synaptic organization and functional interactions of NMDA and $\mu$ORs in the CeA, particularly with respect to opioid withdrawal. The implications of these results are considered within the context of prior neuropharmacological data on the role of the CeA in psychological and physical aspects of opioid dependence.

## II. Opioids and Dependence

Opioids have been used by humans for millennia (Brownstein, 1993; Scarborough, 1995), and their analgesic properties have made them an important part of current medical care (Inturrisi, 2002). Despite their utility, the complex relationships between opioid pharmacology, use history, genetic factors, and other physiologic and environmental variables present significant problems with respect to their rational clinical use (Kreek, 2008; O'Brien, 2008; Weaver and Schnoll, 2002).

A pernicious complication of opioid consumption, dependence reflects the development of opioid-mediated adaptations that are expressed after termination of drug action by drug withdrawal-induced counter-adaptations, which are typically contrary to the acute effects of the opioid (Christie, 2008; Koob, 2009; Williams *et al.*, 2001). Although it has been reported that only a single opioid exposure may be required (Eisenberg and Sparber, 1979), repetitive use results in a more pronounced degree of dependence (Way *et al.*, 1969). The development of dependence on the prototypic opioid morphine requires functional $\mu$ORs (Matthes *et al.*, 1996), and its severity is significantly related to genetic factors (Kest *et al.*, 2002; Korostynski *et al.*, 2007).

Well characterized in humans, dependence has been studied extensively in rodent models (Martin *et al.*, 1963). Opioid antagonist-precipitated withdrawal is the most common means of studying this phenomenon in animals, and will be the method discussed throughout the remainder of this chapter. Symptoms of opioid withdrawal can be elicited in dependent animals upon peripheral administration of general opioid receptor antagonists, such as naloxone or naltrexone (Hamlin *et al.*, 2001). However, intraventricular (Maldonado *et al.*, 1992) or intracranial (Stinus *et al.*, 1990) microinjection of opioid blockers can reproduce most withdrawal signs, indicating that dependence is critically mediated by diverse neural circuits. A list of some prominent withdrawal symptoms is provided in Table 8.1. As would be expected from the highly unpleasant nature of these symptoms, avoiding or relieving withdrawal is an important factor in continued drug use (Koob, 2009).

In addition to its acute unconditional effects, opioid withdrawal has other consequences that may endure beyond periods of detoxification and abstinence due to learned associations between the withdrawal state and

**Table 8.1** In small rodents, opioid withdrawal symptoms can be categorized into distinct functional groupings, including somatic, autonomic, endocrine, and affective (Buccafusco, 1990; Gonzalvez *et al.*, 1994; Martin *et al.*, 1963; Mucha, 1987)

| | |
|---|---|
| Somatic | Escape-jumping, wet-dog shakes |
| Autonomic | Diarrhea, hypertension |
| Endocrine | ACTH and Cort release |
| Affective | Acute aversion |
| Learning | Conditioned place aversion/avoidence |

Only some of the most prominent examples are listed for each category.
ACTH: adrenocorticotropic hormone; Cort: corticosterone

environmental cues (O'Brien, 2008). Indeed, animals will learn to spend less time in a previously neutral environment paired with aversive periods of naloxone-precipitated withdrawal (Stinus *et al.*, 1990), a phenomenon termed conditioned place aversion (CPA). CPA is a species of classical conditioning whereby a conditional stimulus, such as a particular location with salient cues (i.e., tactile, visual, or other sensory stimuli), acquires secondary negative qualities, due to its pairing with an unconditional stimulus, such as opioid withdrawal (Cunningham *et al.*, 2006). Because the conditioned stimulus can evoke a response similar to the unconditioned stimulus, animals engage in avoidance behavior when subsequently exposed to the aversive conditioning environment.

What is notable about CPA is that it can be produced with doses of naloxone that are so low even physical withdrawal signs are not elicited (Gracy *et al.*, 2001). The latter finding indicates that a negative affective state (Schulteis *et al.*, 1998), rather than physical symptoms *per se*, is the requisite unconditioned stimulus necessary for learning. In addition, place aversion persists for weeks after training (Stinus *et al.*, 2000), and may have a lasting influence on behavior. Learned aversive cues may also impact drug-taking behavior, as demonstrated by evidence that withdrawal-associated stimuli can prime opioid self-administration in dependent animals (Hellemans *et al.*, 2006; Kenny *et al.*, 2006), likely as a means of escaping the noxious experience. Moreover, the role of aversive learning in drug-seeking behavior may also have significant clinical relevance (O'Brien *et al.*, 1975, 1998).

## III. Glutamate Systems and Opioid Dependence

Functional ionotropic glutamate receptors are required for the full development and expression of dependence. In particular, it has been shown that acute or chronic administration of NMDA receptor antagonists attenuates physical withdrawal symptoms. This has proven to be a reliable outcome as NMDA receptor blockade inhibits withdrawal in opioid-dependent mice (McLemore *et al.*, 1997), guinea pigs (Tanganelli *et al.*, 1991), rats (Trujillo and Akil, 1991), and humans (Bisaga *et al.*, 2001). Moreover, in rodent models, NMDA antagonists also suppress the conditioned aversive properties (i.e., CPA) of naloxone-precipitated withdrawal (Higgins *et al.*, 1992).

The NMDA receptor has complex signaling properties and is a potent modulator of cellular adaptability (Dingledine *et al.*, 1999; Kohr, 2006). The NMDA receptor is a tetrameric heteromer composed of the essential NMDA-NR1 subunit (NR1), of which there are eight splice variants. Along with NR1, glutamate-responsive NMDA receptors require some combination of NMDA-NR2 subunits (NR2), of which there are four subtypes (NR2A–D). Critical characteristics of the NMDA receptor include a voltage-dependent $Mg^{2+}$ blockade requiring cellular depolarization for channel activation and high permeability to $Ca^{2+}$. In addition, NMDA receptor activation can modulate numerous intracellular signaling processes including protein kinase activity (Haddad, 2005), protein transport (Shi *et al.*, 1999), transcription factor activity, and gene expression (Yoneda *et al.*, 2001), as well as epigenetic events (Lubin *et al.*, 2008). These properties of NMDA receptor activation may have profound effects on cellular and behavioral plasticity (Tsien, 2000), such as long-term potentiation (LTP) and long-term depression (LTD) (Kullmann *et al.*, 2000), in addition to whole-animal learning and memory (Shapiro and Eichenbaum, 1999; Walker and Davis, 2002), processes that play a role in opioid dependence.

## IV. The Central Nucleus and Dependence

Despite the significant relationship between NMDA receptors and opioid dependence, a brain-map linking sites of functional receptor expression to specific opioid withdrawal behaviors is at best rudimentary. Given that brain motivational systems are a critical substrate of opioid plasticity, NMDA receptors in nuclei such as the CeA are likely to play an important role in dependence. The central amygdala is a critical component of neural motivation pathways that play a role in endocrine (Schulkin *et al.*, 1994) and autonomic function (Brody, 1988), fluid (McKinley *et al.*, 2001) and energy balance (Glass *et al.*, 2000), as well as behavioral (Davis, 1998) and affective (Davis, 1989) processes. Many of these activities are known to be modulated by glutamate and opioids. In order to better appreciate the role of the CeA glutamate system in opioid dependence, a brief overview of its place within the larger context of brain circuitry subserving motivated behavior is illustrated in Fig. 8.1. The basic neurochemical content of the CeA will be outlined below.

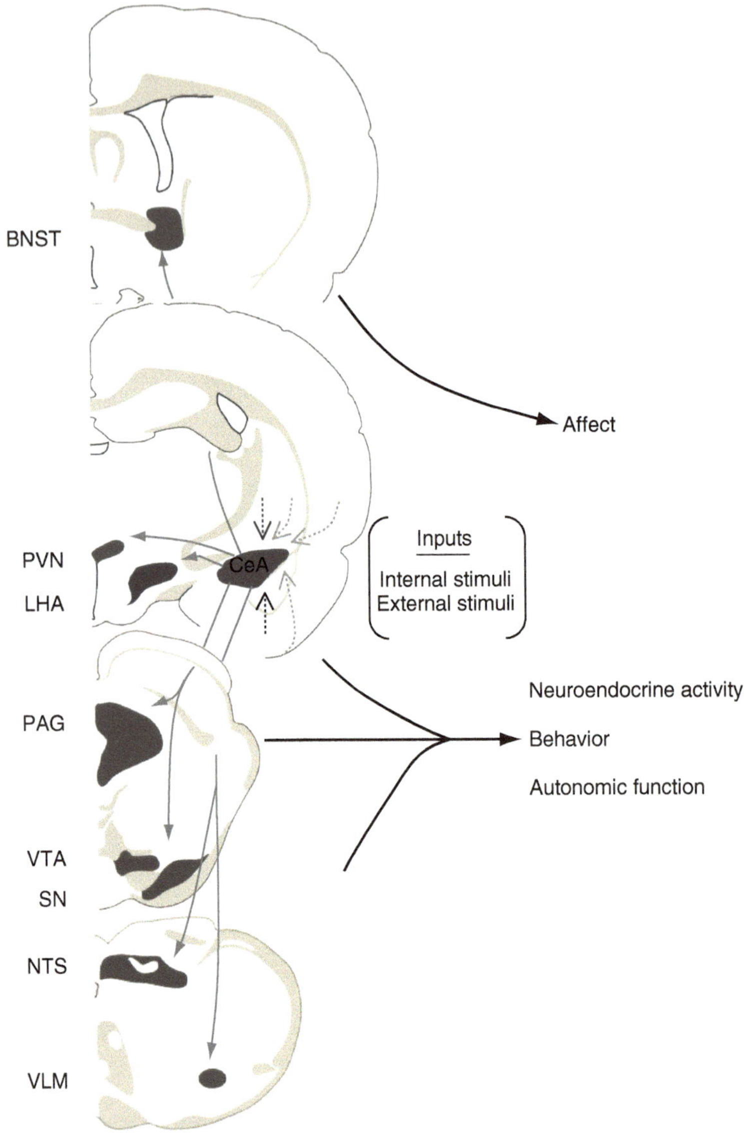


**Figure 8.1** The organization of brain motivational pathways involved in opioid dependence. Brain motivational systems are composed of a highly complex network of extero- and interoceptive processing systems, integrative memory systems, as well as hypothalamic and brainstem endocrine/autonomic and motor activators that play essential roles in maintaining homeostasis (Swanson, 2000). The CeA is a critical coordinator of relevant functional inputs and outputs within this circuitry, and may play a role in coupling homeostatic need with emotional valence and immediate survival behaviors, as well as more slowly developing learned emotional responses to those challenging conditions. A highly schematic representation of neural pathways

The CeA has a diverse compliment of glutamate afferents that arise from areas of brain motivational systems. These include limbic cortices (McDonald, 1998), the thalamus (Turner and Herkenham, 1991), and the basolateral nucleus of the amygdala (BLA) (Pitkanen *et al.*, 1997). The CeA is notable for its abundant expression of glutamate receptors (Petralia and Wenthold, 1992; Petralia *et al.*, 1994; Sato *et al.*, 1993). In particular, neurons in the CeA express the NR1 gene (Sato *et al.*, 1995) and protein (Petralia *et al.*, 1994), as well as NMDA ligand-binding sites (Monaghan and Cotman, 1985).

In addition to glutamate, CeA neurons have a rich array of signaling molecules. Many central amygdala neurons express the enzyme glutamic acid decarboxylase 67 kDa, which is responsible for synthesizing the inhibitory transmitter $\gamma$-aminobutyric acid (GABA) (Carta *et al.*, 2008). In addition to GABA, there are also endogenous CeA neurons and/or axons from extrinsic sources that express a variety of neuroactive peptides as well as receptors for these modulators. These include corticotropin-releasing factor (Beyer *et al.*, 1988; Swanson *et al.*, 1983), components of the renin–angiotensin system (Brown and Gray, 1988; Brownfield *et al.*, 1982; Lavoie *et al.*, 2004), neuropeptide Y (Gray *et al.*, 1986; Heilig *et al.*, 1993), orexin (Baldo *et al.*, 2003), oxytocin (Lee *et al.*, 2005; Roozendaal *et al.*, 1993), and vasopressin (Francis *et al.*, 2002; Roozendaal *et al.*, 1993). In addition, the CeA is also

---

involved in motivational processes is presented as a series of illustrated coronal brain sections. The CeA and its outputs are indicated in gray, while specific targets are highlighted by red fill. Relevant CeA inputs are indicated by differently colored dashed arrows. These include glutamate afferents (green) from areas of limbic cortices (McDonald, 1998), the thalamus (Turner and Herkenham, 1991), and the BLA. The CeA is also innervated by catecholaminergic neurons (black), including dopaminergic neurons from the mesolimbic system and noradrenergic neurons from the medulla (Asan, 1998). Neuropeptides (blue) from these and other brain areas are also expressed by CeA inputs (see text for description). This array of afferents provides the CeA with a complex set of signals about current internal and external stimuli, as well as representations in memory, that are critical in organizing behavioral responses. Both NMDA-type glutamate and $\mu$ORs are expressed within the central amygdala. Critical outputs of the central amygdala include areas of the extended amygdala involved in learned emotional processes such as the bed nucleus of the stria terminalis (BNST; Zahm *et al.*, 1999), hypothalamic systems involved in endocrine (paraventricular nucleus of the hypothalamus [PVN]) and behavioral (lateral hypothalamic area [LHA]) responses (Allen and Cechetto, 1995; Gray *et al.*, 1989; Marcilhac and Siaud 1997), midbrain areas involved in nociception (periaqueductal gray [PAG]), reward (ventral tegmental area [VTA]), and/or motor (substantia nigra [SN]) function (Schmued *et al.*, 1989; Zahm *et al.*, 1999), as well as medullary nuclei (nucleus of the solitary tract [NTS], ventrolateral medulla [VLM]) that mediate cardiovascular, respiratory, and/or gastrointestinal processes (Glass *et al.*, 2002; Wallace *et al.*, 1989). Drawings are adapted from the brain atlas of Swanson (Swanson, 1992).

innervated by catecholamine releasing axonal varicosities (Asan, 1998) and contains neurons that express adrenergic receptors (Glass *et al.*, 2002).

Among the peptide systems active in the central amygdala, the opioids are notably abundant. Enkephalin (Cassell *et al.*, 1986; Petrovich *et al.*, 2000; Wray and Hoffman, 1983), dynorphin (Fallon and Leslie, 1986; Reyes *et al.*, 2008; Zardetto-Smith *et al.*, 1988), and $\beta$-endorphin (Gray *et al.*, 1984) are each present in the CeA. Moreover, opioid receptors, including the $\mu$OR, are expressed in the central amygdala (Mansour *et al.*, 1988, 1994). Deciphering the synaptic organization and functional relationships between NMDA receptors and opioid receptors is a critical issue in understanding of central amygdala function and its relationship to opioid dependence.

## V. The Synaptic Relationship Between NMDA and $\mu$-Opioid Receptors in the CeA

The cellular relationship between NMDA receptors and $\mu$ORs in the CeA has only been inferred by the results of electrophysiological studies, and is a matter of contention (Zhu and Pan, 2004, 2005). Because of its high spatial resolution, immunoelectron microscopy employing gold and peroxidase markers is a valuable tool for examining the fine structural location of functionally interacting proteins in areas of brain motivational pathways (Gracy and Pickel, 1995; Van Bockstaele *et al.*, 2000). Using immunoelectron microscopic analysis, the NR1 subunit was shown to be highly enriched in postsynaptic (i.e., dendrites) sites of central amygdala neurons (Fig. 8.2A and B) (Glass *et al.*, 2009). Labeled dendritic profiles were typically small to intermediate in size (0.5–1 $\mu$m cross-sectional area). NR1 was frequently present in near intracellular endomembranous organelles, reflecting sites of protein storage and trafficking. However, this protein was also present on the plasma membrane, the primary location of functional receptors. Furthermore, many of these dendritic profiles were contacted by axon terminals forming asymmetric excitatory-type synapses, typical of those made by glutamatergic axon terminals (Peters *et al.*, 1991). In summary, NR1 was mainly localized to dendritic (i.e., postsynaptic) structures where it was frequently present in vesicular organelles linked to protein transport, as well as the extrasynaptic and synaptic plasma membrane contacted by excitatory-type axon terminals.

In terms of its subcellular location, the $\mu$OR was found near intracellular membranous organelles and the plasma membrane, a pattern similar to that of NR1. However, relative to the glutamate receptor subunit, the

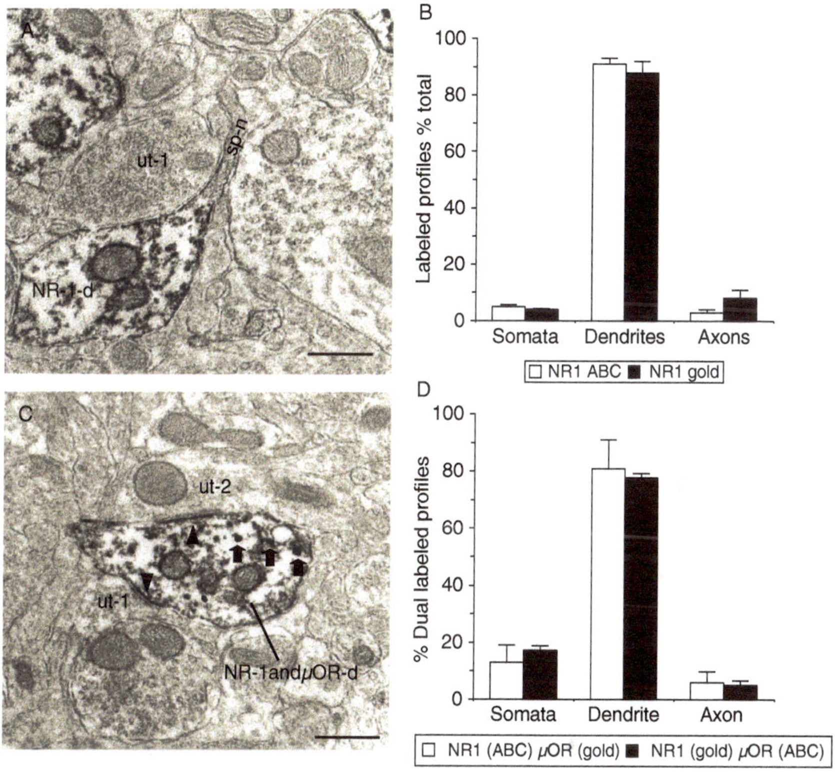


**Figure 8.2** Ultrastructural distribution of the NMDA receptor and its relationship to the $\mu$OR in the central amygdala. (A) A dendritic profile (NR-1-d) expressed diffuse immunoperoxidase reaction product for NR1. This profile exhibited a spine neck (sp-n) and was contacted by an unlabeled axon terminal (ut-1) that did not appear to form a synapse. (B) When labeled by either immunoperoxidase (ABC) or immunogold (Gold) secondary markers, the majority of NR1-labeled processes were dendrites. (C) A dendritic profile (NR-1 and $\mu$OR-d) expressed diffuse immunoperoxidase reaction product for NR1 and immunogold labeling (small arrows) for the $\mu$OR. This dual labeled profile received distinct asymmetric excitatory-type synapses (arrow heads) from two unlabeled axon terminals (ut-1 and ut-2). (D) Of all dual labeled neuronal profiles, the majority were dendrites. This was the case when NR1 was labeled by immunoperoxidase (ABC) and the $\mu$OR by immunogold (Gold) markers, and when secondary antisera were reversed (i.e., NR1 labeled by gold and the $\mu$OR by peroxidase). Scale bars: 0.5 $\mu$m. See Glass *et al.* (2009) for details.

$\mu$OR had a more heterogeneous distribution in neuronal compartments. Although frequently present in somata and dendrites, there were also many instances of $\mu$OR expressing axons and axon terminals (for details see Glass *et al.*, 2009).

There were numerous instances of neuronal profiles in the CeA that expressed labeling for both NR1 and the $\mu$OR. Despite the prominent dendritic distribution of NR1 and the mixed dendritic and axonal localization for the $\mu$OR, dual labeling for these proteins was most commonly found in dendrites (Fig. 8.2C and D). Like the single-labeled NR1 and $\mu$OR containing dendrites, structures that expressed immunoreactivity for both proteins were small to medium in size. Labeling for each protein was present near vesicular organelles characteristic of those involved in protein transport, and was also found on the surface membrane. Synapses formed on these dual labeled dendrites were typically of the asymmetric excitatory kind. In sum, these results indicate that NMDA receptors and $\mu$ORs are strategically positioned for postsynaptic comodulation of glutamate signaling in dendrites of CeA neurons.

## VI. Deletion of Postsynaptic NR1 in Central Amygdala Neurons Attenuates Opioid Withdrawal-Induced Place Aversion

Establishing relationships between functional postsynaptic NMDA receptor expression and opioid dependence is not feasible by traditional pharmacological approaches. Currently available NMDA receptor antagonists cannot discriminate between neuronal dendrites, axon terminals, or other ultrastructural compartments. In addition, interpreting neuropharmacological studies involving NMDA receptor antagonists is problematic given that central amygdala blockade of NMDA receptors can have rewarding or aversive properties, depending on the particular agent used (Watanabe *et al.*, 2002), as well as other effects likely to confound performance of learned behaviors including alterations in motor function (Andrzejewski *et al.*, 2004).

To examine the relationship between opioid dependence and functional postsynaptic central amygdala NMDA receptors a spatial–temporal gene deletion strategy employing Cre-lox technology was used. The NR1 subunit was deleted by local CeA microinjection of a neurotropic replication deficient recombinant rAAV that expressed a fusion protein of the enzyme Cre recombinase (Cre) and a reporter, GFP, termed "rAAV–GFP–Cre" (South *et al.*, 2003). Injections were made in adult male transgenic loxP knockin mice that have strategically placed loxP sites in the NR1 gene (i.e., "floxed NR1" (fNR1) mice) flanking exons that encode for the four membrane domains and the C-terminus (South *et al.*, 2003).

Direct microinjection of rAAV–GFP–Cre into the CeA of adult fNR1 mice resulted in recombination specifically in neurons (Glass *et al.*, 2008). Moreover, there was a significant reduction in NR1 expression in the target area (Fig. 8.3). This reduction occurred in somata and dendritic profiles (i.e., postsynaptic). There were no concomitant reductions of presynaptic NR1 or somatodendritic NR2 immunolabeling, and no effects on local cell number or cellular morphology (Glass *et al.*, 2008).

Bilateral knockout of the NR1 gene in the CeA did not produce obvious basal behavioral deficits. In particular, deletion of CeA NR1 did not impact locomotor activity, body weight, sensory-motor coordination, thermal nociception, or somatosensation (Glass *et al.*, 2008).

In regard to opioid dependence, CeA NR1 knockout mice were chronically exposed to morphine by subcutaneously implanted morphine pellets and then administered naloxone to precipitate withdrawal. The CeA NR1 knockouts did not differ from control animals with respect to somatic signs, such as escape-jumping and wet-dog shakes, or autonomic symptoms, notably diarrhea and weight loss (Fig. 8.4A). Contrary to what was found with physical symptoms, central amygdala NR1 gene deletion did impair naloxone withdrawal-induced place aversion (Fig. 8.4B). Therefore, functional CeA NMDA receptors were not necessary for the induction of many major physical withdrawal symptoms, but were required for the production of a learned withdrawal-induced negative affective state.

The finding that CeA NR1 deletion impaired withdrawal-induced CPA is consistent with prior reports that NMDA receptor activity in the CeA plays an important role in aversive learning and memory (Goosens and Maren, 2003). The precise role of the central amygdala in emotional learning has been controversial. Traditionally, the CeA has been considered less as a sight of learning and more as a modulator of other areas that encode new associations, however, emerging evidence contradicts this view. For example, recent findings have shown that the CeA is required for the normal acquisition and consolidation of conditioned fear (Rabinak and Maren, 2008; Wilensky *et al.*, 2006). These neurological results are supported by neurophysiological data. It has been reported that neurons in the CeA exhibit a form of cellular plasticity (i.e., LTP) that occurs in established neuroanatomical substrates of learning and memory, such as the hippocampus and lateral amygdala (Samson and Paré, 2005). Moreover, CeA LTP is also sensitive to withdrawal from drugs of abuse in an NMDA receptor-dependent manner (Pollandt *et al.*, 2006).

## VII. Does the CeA Selectively Participate in the Conditioned Aversive Properties of Opioid Withdrawal?

The finding that CeA NR1 deletion selectively interferes with CPA has interesting parallels with other reports in the literature. For example, in opioid-dependent animals, withdrawal induced by CeA microinjection of

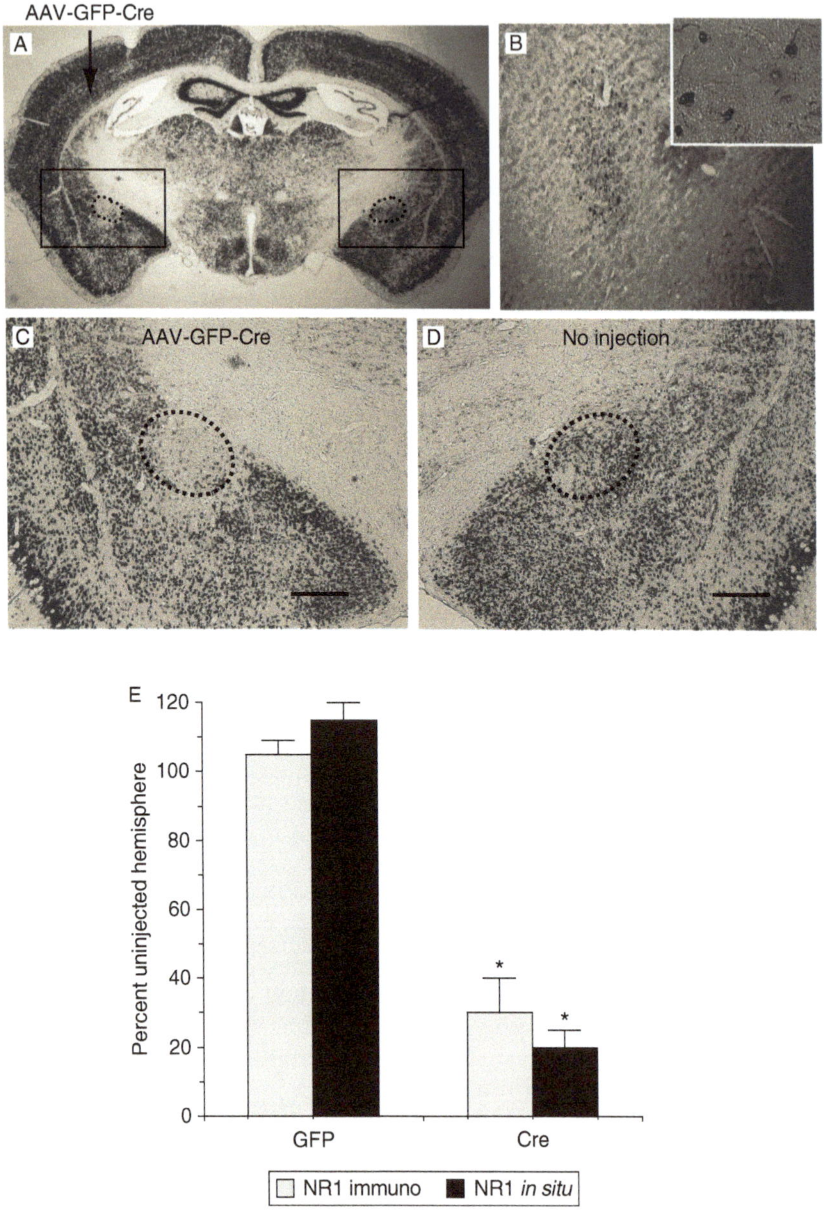


**Figure 8.3** Postsynaptic NR1 deletion in central amygdala neurons. (A) Unilateral microinjection of rAAV–GFP–Cre (indicated by arrow) in the CeA produced a localized knockout of NR1 as shown by *in situ* hybridization. (B) Gene deletion corresponded to expression of the GFP reporter as seen in a serial section. GFP expressing neurons can be seen at a higher magnification in the inset. (C, D) Higher magnification views of the CeA as seen in Fig. 8.3A. Note the significantly diminished NR1

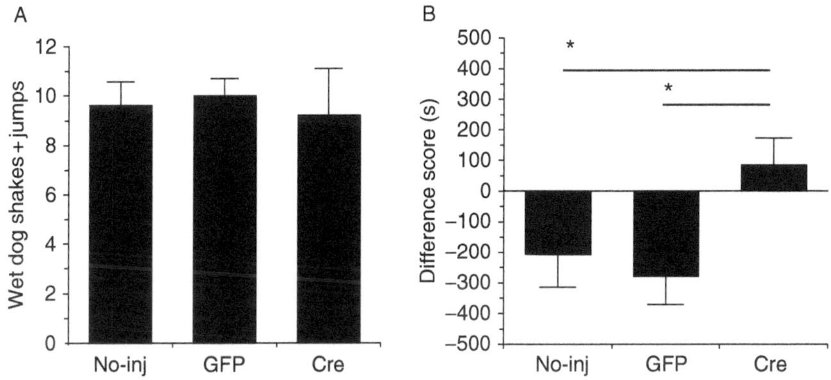


**Figure 8.4** Postsynaptic deletion of NR1 in central amygdala neurons and opioid dependence. (A) Separate groups of fNR1 mice, including those given no intracranial injection (No-inj), or bilaterally microinjected with either rAAV–GFP (GFP) or rAAV–GFP–Cre (Cre), were chronically administered morphine by subcutaneous implantation of a morphine pellet (25 mg), which was replaced by a fresh pellet every fourth day as needed. Dependence was determined by visual observations of physical withdrawal signs including diarrhea, wet-dog shakes, and escape jumping after an acute injection of naloxone (1 mg/kg, i.p.). There were no differences between any of the groups in the number of somatic (wet-dog shakes + jumping) or autonomic (diarrhea; not shown) symptoms. (B). Place aversion training took place in an apparatus consisting of a two-chamber box, each with distinct tactile and visual cues, that was inserted into an automated activity monitor. After measuring baseline selection, during which time animals were allowed to freely explore each chamber, animals began place aversion training. On alternate training days morphine dependent mice were injected with saline (0.9%, i.p.) or naloxone (1 mg/kg, i.p.) and then restricted to the respective chamber for 30 min. On the test day, subjects were allowed to freely explore both chambers for 30 min as during baseline. The difference in time spent in the naloxone-paired chamber during the testing and preconditioning phases served as the measure of place aversion. In distinction to physical withdrawal, there were significant between-group differences in CPA. Unlike animals in both control groups, mice bilaterally microinjected with Cre in the CeA did not spend less time in the withdrawal-paired chamber, indicating that CeA NR1 knockout interfered with a conditioned place aversion in response to opioid withdrawal. Reductions in NR1 mRNA were seen selectively in the animals injected with rAAV–GFP–Cre as previously reported (Glass *et al.*, 2008).

---

expression in the injected hemisphere (C) compared to the uninjected (D) side. (E) Compared to unilateral CeA injection of the control GFP vector, injection of Cre produced significant reductions in NR1 labeling in somatodendritic sites in the injected compared to the uninjected hemisphere as measured by immunoelectron microscopy (NR1 immuno). This corresponded to reductions in NR1 gene expression (NR1 *in situ*) in the CeA (Glass *et al.*, 2008). *$p < 0.05$ compared to GFP. Scale bars: 1 mm.

an opioid antagonist preferentially produces CPA (Stinus *et al.*, 1990). In addition, it has been shown that lesioning the ventral noradrenergic bundle, which provides an important source of norepinephrine containing axons in the CeA and related areas of the extended amygdala, inhibits opioid withdrawal place aversive (Delfs *et al.*, 2000). The latter effect is paralleled by the ability of CeA administered alpha-2-adrenergic agonists to inhibit only affective signs of withdrawal (Taylor *et al.*, 1998). Based on this evidence, one may conclude that the CeA has a privileged role in mediating the conditioned emotional properties of withdrawal.

The notion that the CeA has a special role in affective learning processes related to dependence, however, may be too simplistic. For example, it has been reported that blockade of non-NMDA type glutamate receptors in the CeA of dependent animals reduces somatic signs of opioid withdrawal (Taylor *et al.*, 1998). This result is similar to other findings in the literature showing that blockade of CeA CRF receptors also inhibits physical symptoms of withdrawal from opioids (McNally and Akil, 2002).

It appears that manipulating particular receptor systems in the CeA can produce distinct outcomes with respect to the acute physical and psychological features of opioid withdrawal. One reasonable explanation for these complex outcomes is that particular manifestations of dependence may be coded by distinct neurochemical signals acting within circumscribed central amygdala circuits. One hypothetical scheme detailing how such a system may be organized is illustrated in Fig. 8.5.

## VIII. Conclusion

Both NMDA receptors and $\mu$ORs are strategically positioned for the comodulation of excitatory postsynaptic signaling in CeA neurons. Moreover, genetic deletion of postsynaptic NMDA receptors impairs the conditioned aversive, but not the physical features of opioid withdrawal. Because conditioned cues may promote behaviors to avoid or escape their associated adverse state, NMDA receptor activity in the CeA may be essential in complex neural processes that engage relief-seeking behaviors, like drug taking (Koob, 2009).

The intriguing speculation that the neural substrates of particular behavioral features of dependence may involve differing synaptic organizational arrangements within central amygdala circuitry will require further exploration. This hypothesis can be addressed by, among other techniques, multilabeling immunoelectron microscopic ultrastructural analysis. This can be combined with application of spatial–temporal gene deletion methodology similar to that described in this chapter, as well as a development of this approach that incorporates phenotype-specific promoters to manipulate

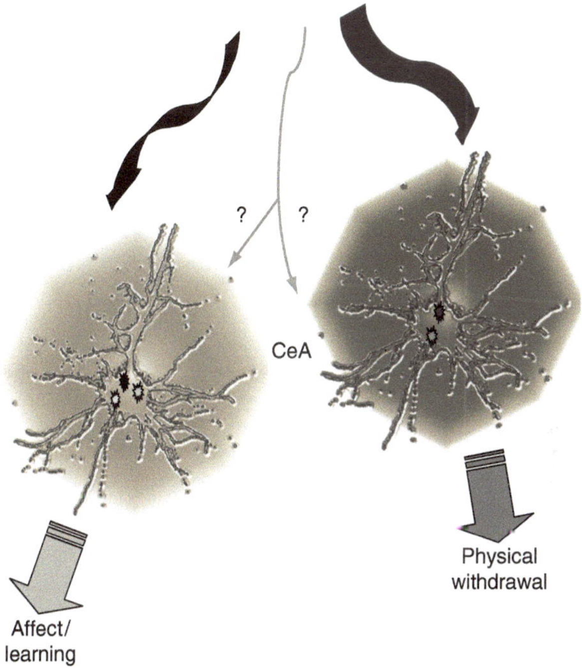


**Figure 8.5** Schematic representation illustrating hypothetical model of central amygdala synaptic coding of physical and psychological signals linked with opioid withdrawal. One group of CeA neurons may express non-NMDA type (i.e., AMPA, kainate), glutamate (dark green stars), and CRF (blue stars) receptors, and receive preferential input from CRF expressing neurons (blue arrow). The physical withdrawal pathway may be engaged by an upstream opioid receptor-dependent pathway, presumably expressing glutamate or CRF. This system would be expected to mediate immediate opioid withdrawal symptoms. The other group (light gray shading) of CeA neurons may express NMDA (pale green stars), mu-opioid (yellow star), and noradrenergic (black star) receptors, and receive input from noradrenergic afferents (black arrow), particularly from areas of the medulla (NTS, VLM), as well as opioid peptides (not shown). This system would be expected to have a slow onset and require consolidation and long-lasting synaptic integration. Glutamate (green arrow) would be expected to activate both systems, however, whether segregation of function is mediated by separate input sources (i.e., source of glutamate), or by distinct processing capacities (i.e., glutamate receptor types) of their targets is uncertain. The role of CRF (Stinus *et al.*, 2005) and non-NMDA (Kawasaki *et al.*, 2005) receptors may not be limited to immediate unconditional responses, and an alternative model would incorporate these signaling systems into both immediate physical and affective/learning processing streams.

gene expression in selected neurochemical populations (Jasnow *et al.*, 2009; Oh *et al.*, 2009). Such efforts should elucidate the synaptic and molecular bases of the many manifestations of dependence, knowledge that may be critical in developing the next generation of neurobiological tools to manage the long-term consequences of opioid abuse. Moreover, this approach also has implications for other psychiatric disorders involving aversive learning, such as anxiety disorders.

## REFERENCES


Allen, G. V., and Cechetto, D. F. (1995). Neurotensin in the lateral hypothalamic area: Origin and function. *Neuroscience* **69,** 533–544.

Andrzejewski, M. E., Sadeghian, K., and Kelley, A. E. (2004). Central amygdalar and dorsal striatal NMDA receptor involvement in instrumental learning and spontaneous behavior. *Behav. Neurosci.* **118,** 715–729.

Asan, E. (1998). The catecholaminergic innervation of the rat amygdala. *Adv. Anat. Embryol. Cell Biol.* **142,** 1–118.

Baldo, B. A., Daniel, R. A., Berridge, C. W., and Kelley, A. E. (2003). Overlapping distributions of orexin/hypocretin- and dopamine-beta-hydroxylase immunoreactive fibers in rat brain regions mediating arousal, motivation, and stress. *J. Comp. Neurol.* **464,** 220–237.

Beyer, H. S., Matta, S. G., and Sharp, B. M. (1988). Regulation of the messenger ribonucleic acid for corticotropin-releasing factor in the paraventricular nucleus and other brain sites of the rat. *Endocrinology* **123,** 2117–2123.

Bisaga, A., Comer, S. D., Ward, A. S., Popik, P., Kleber, H. D., and Fischman, M. W. (2001). The NMDA antagonist memantine attenuates the expression of opioid physical dependence in humans. *Psychopharmacology* **157,** 1–10.

Brody, M. J. (1988). Central nervous system and mechanisms of hypertension. *Clin. Physiol. Biochem.* **6,** 230–239.

Brown, M. R., and Gray, T. S. (1988). Peptide injections into the amygdala of conscious rats: Effects on blood pressure, heart rate, and plasma catecholamines. *Reg. Peptides* **21,** 95–106.

Brownfield, M. S., Reid, I. A., Ganten, D., and Ganong, W. F. (1982). Differential distribution of immunoreactive angiotensin and angiotensin-converting enzyme in rat brain. *Neuroscience* **7,** 1759–1769.

Brownstein, M. J. (1993). A brief history of opiates, opioid peptides, and opioid receptors. *Proc. Natl. Acad. Sci.* **90,** 5391–5393.

Buccafusco, J. J. (1990). Participation of different brain regions in the anti-narcotic withdrawal action of clonidine in the dependent rat. *Brain Res.* **513,** 8–14.

Carta, A. R., Moreno, C. C., Cadoni, C., Tronci, E., and Di Chiara, G. (2008). Long-term increase in GAD67 mRNA expression in the central amygdala of rats sensitized by drugs and stress. *Eur. J. Neurosci.* **27,** 1220–1230.

Cassell, M. D., Gray, T. S., and Kiss, J. Z. (1986). Neuronal architecture in the rat central nucleus of the amygdala: A cytological, hodological, and immunocytochemical study. *J. Comp. Neurol.* **246,** 478–499.

Christie, M. J. (2008). Cellular neuroadaptations to chronic opioids: Tolerance, withdrawal and addiction. *Br. J. Pharmacol.* **154,** 384–396.

Cunningham, C. L., Gremel, C. M., and Groblewski, P. A. (2006). Drug-induced conditioned place preference and aversion in mice. *Nat. Protocols* **1,** 1662–1670.

Davis, M. (1989). Neural systems involved in fear-potentiated startle. *Ann. N.Y. Acad. Sci.* **563,** 165–183.

Davis, M. (1998). Are different parts of the extended amygdala involved in fear versus anxiety? *Biol. Psychiatry* **44,** 1239–1247.

Delfs, J. M., Zhu, Y., Druhan, J. P., and Aston-Jones, G. (2000). Noradrenaline in the ventral forebrain is critical for opiate withdrawal-induced aversion. *Nature* **403,** 430–434.

Dingledine, R., Borges, K., Bowie, D., and Traynelis, S. F. (1999). The glutamate receptor ion channels. *Pharmacol. Rev.* **51,** 7–61.

Eisenberg, R. M., and Sparber, S. B. (1979). Changes in plasma corticosterone levels as a measure of acute dependence upon levorphanol in rats. *J. Pharmacol. Exp. Ther.* **211,** 364–369.

Fallon, J. H., and Leslie, F. M. (1986). Distribution of dynorphin and enkephalin peptides in the rat brain. *J. Comp. Neurol.* **249,** 293–336.

Francis, D. D., Young, L. J., Meaney, M. J., and Insel, T. R. (2002). Naturally occurring differences in maternal care are associated with the expression of oxytocin and vasopressin (V1a) receptors: Gender differences. *J. Neuroendocrinol.* **14,** 349–353.

Glass, M. J., Billington, C. J., and Levine, A. S. (2000). Naltrexone administered to central nucleus of amygdala or PVN: Neural dissociation of diet and energy. *Am. J. Physiol.* **279,** R86–R92.

Glass, M. J., Colago, E. E., and Pickel, V. M. (2002). Alpha-2A-adrenergic receptors are present on neurons in the central nucleus of the amygdala that project to the dorsal vagal complex in the rat. *Synapse* **46,** 258–268.

Glass, M. J., Hegarty, D. M., Oselkin, M., Quimson, L., South, S. M., Xu, Q., Pickel, V. M., and Inturrisi, C. E. (2008). Conditional deletion of the NMDA-NR1 receptor subunit gene in the central nucleus of the amygdala inhibits naloxone-induced conditioned place aversion in morphine dependent mice. *Exp. Neurol.* **213,** 57–70.

Glass, M. J., Vanyo, L., Quimson, L., and Pickel, V. M. (2009). Ultrastructural relationship between NMDA-NR1 and mu-opioid receptor in the mouse central nucleus of the amygdala. *Neuroscience* **163,** 857–867.

Gonzalvez, M. L., Milanes, M. V., Martinez-Pinero, M. G., Marin, M. T., and Vargas, M. L. (1994). Effects of intracerebroventricular clonidine on the hypothalamic noradrenaline and plasma corticosterone levels of opiate naive rats and after naloxone-induced withdrawal. *Brain Res.* **647,** 199–203.

Goosens, K. A., and Maren, S. (2003). Pretraining NMDA receptor blockade in the basolateral complex, but not the central nucleus, of the amygdala prevents savings of conditional fear. *Behav. Neurosci.* **117,** 738–750.

Gracy, K. N., and Pickel, V. M. (1995). Comparative ultrastructural localization of the NMDAR1 glutamate receptor in the rat basolateral amygdala and bed nucleus of the stria terminalis. *J. Comp. Neurol.* **362,** 71–85.

Gracy, K. N., Dankiewicz, L. A., and Koob, G. F. (2001). Opiate withdrawal-induced FOS immunoreactivity in the rat extended amygdala parallels the development of conditioned place aversion. *Neuropsychopharmacology* **24,** 152–160.

Gray, T. S., Cassell, M. D., and Kiss, J. Z. (1984). Distribution of pro-opiomelanocortin-derived peptides and enkephalins in the rat central nucleus of the amygdala. *Brain Res.* **306,** 354–358.

Gray, T. S., O'Donohue, T. L., and Magnuson, D. J. (1986). Neuropeptide Y innervation of amygdaloid and hypothalamic neurons that project to the dorsal vagal complex in rat. *Peptides* **7,** 341–349.

Gray, T. S., Carney, M. E., and Magnuson, D. J. (1989). Direct projections from the central amygdaloid nucleus to the hypothalamic paraventricular nucleus: Possible role in stress-induced adrenocorticotropin release. *Neuroendocrinology* **50,** 433–446.

Haddad, J. J. (2005). N-methyl-D-aspartate (NMDA) and the regulation of mitogen-activated protein kinase (MAPK) signaling pathways: A revolving neurochemical axis for therapeutic intervention? *Prog. Neurobiol.* **77,** 252–282.

Hamlin, A., Buller, K. M., Day, T. A., and Osborne, P. B. (2001). Peripheral withdrawal recruits distinct central nuclei in morphine-dependent rats. *Neuropharmacology* **41,** 574–581.

Heilig, M., McLeod, S., Brot, M., Heinrichs, S. C., Menzaghi, F., Koob, G. F., and Britton, K. T. (1993). Anxiolytic-like action of neuropeptide Y: Mediation by Y1 receptors in amygdala, and dissociation from food intake effects. *Neuropsychopharmacology* **8,** 357–363.

Hellemans, K. G., Dickinson, A., and Everitt, B. J. (2006). Motivational control of heroin seeking by conditioned stimuli associated with withdrawal and heroin taking by rats. *Behav. Neurosci.* **120,** 103–114.

Higgins, G. A., Nguyen, P., and Sellers, E. M. (1992). The NMDA antagonist dizocilpine (MK801) attenuates motivational as well as somatic aspects of naloxone precipitated opioid withdrawal. *Life Sci.* **50,** PL167–PL172.

Hyman, S. E., Malenka, R. C., and Nestler, E. J. (2006). Neural mechanisms of addiction: The role of reward-related learning and memory. *Annu. Rev. Neurosci.* **29,** 565–598.

Inturrisi, C. E. (2002). Clinical pharmacology of opioids for pain. *Clin. J. Pain* **18**(4 Suppl.), S3–S13.

Jasnow, A. M., Rainnie, D. G., Maguschak, K. A., Chhatwal, J. P., and Ressler, K. J. (2009). Construction of cell-type specific promoter lentiviruses for optically guiding electrophysiological recordings and for targeted gene delivery. *Methods Mol. Biol.* **515,** 199–213.

Kawasaki, Y., Jin, C., Suemaru, K., Kawasaki, H., Shibata, K., Choshi, T., Hibino, S., Gomita, Y., and Araki, H. (2005). Effect of glutamate receptor antagonists on place aversion induced by naloxone in single-dose morphine-treated rats. *Br. J. Pharmacol.* **145,** 751–757.

Kenny, P. J., Chen, S. A., Kitamura, O., Markou, A., and Koob, G. F. (2006). Conditioned withdrawal drives heroin consumption and decreases reward sensitivity. *J. Neurosci.* **26,** 5894–5900.

Kest, B., Palmese, C. A., Hopkins, E., Adler, M., Juni, A., and Mogil, J. S. (2002). Naloxone-precipitated withdrawal jumping in 11 inbred mouse strains: Evidence for common genetic mechanisms in acute and chronic morphine physical dependence. *Neuroscience* **115,** 463–469.

Kohr, G. (2006). NMDA receptor function: Subunit composition versus spatial distribution. *Cell Tissue Res.* **326,** 439–446.

Koob, G. F. (2009). Neurobiological substrates for the dark side of compulsivity in addiction. *Neuropharmacology* **56**(Suppl. 1), 18–31.

Korostynski, M., Piechota, M., Kaminska, D., Solecki, W., and Przewlocki, R. (2007). Morphine effects on striatal transcriptome in mice. *Genome Biol.* **8,** R128.

Kreek, M. J. (2008). Role of a functional human gene polymorphism in stress responsivity and addictions. *Clin. Pharmacol. Ther.* **83,** 615–618.

Kullmann, D. M., Asztely, F., and Walker, M. C. (2000). The role of mammalian ionotropic receptors in synaptic plasticity: LTP, LTD, and epilepsy. *Cell. Mol. Life Sci.* **57,** 1551–1561.

Lavoie, J. L., Cassell, M. D., Gross, K. W., and Sigmund, C. D. (2004). Localization of renin expressing cells in the brain, by use of a REN-eGFP transgenic model. *Physiol. Genomics* **16,** 240–246.

LeDoux, J. E. (2000). Emotion circuits in the brain. *Ann. Rev. Neurosci.* **23,** 155–184.

Lee, P. R., Brady, D. L., Shapiro, R. A., Dorsa, D. M., and Koenig, J. I. (2005). Social interaction deficits caused by chronic phencyclidine administration are reversed by oxytocin. *Neuropsychopharmacology* **30,** 1883–1894.

Lubin, F. D., Roth, T. L., and Sweatt, J. D. (2008). Epigenetic regulation of BDNF gene transcription in the consolidation of fear memory. *J. Neurosci.* **28,** 10576–10586.

Maldonado, R., Stinus, L., Gold, L. H., and Koob, G. F. (1992). Role of different brain structures in the expression of the physical morphine withdrawal syndrome. *J. Pharmacol. Exp. Ther.* **261,** 669–677.

Mansour, A., Khachaturian, H., Lewis, M. E., Akil, H., and Watson, S. J. (1988). Anatomy of CNS opioid receptors. *Trends Neurosci.* **11,** 308–314.

Mansour, A., Fox, C. A., Thompson, R. C., Akil, H., and Watson, S. J. (1994). mu-Opioid receptor mRNA expression in the rat CNS: Comparison to mu-receptor binding. *Brain Res.* **643,** 245–265.

Marcilhac, A., and Siaud, P. (1997). Identification of projections from the central nucleus of the amygdala to the paraventricular nucleus of the hypothalamus which are immunoreactive for corticotrophin-releasing hormone in the rat. *Exp. Physiol.* **82,** 273–281.

Martin, W. R., Wikler, A., Eades, C. G., and Pescor, F. T. (1963). Tolerance to and physical dependence on morphine in rats. *Psychopharmacologia* **4,** 247–260.

Matthes, H. W. D., Maldonado, R., Simonin, F., Valverde, O., Slowe, S., Kitchen, I., Befort, K., Dierich, A., Le Meur, M., Dolle, P., Tzavara, E., Hanoune, J., *et al.* (1996). Loss of morphine-induced analgesia, reward effect and withdrawal symptoms in mice lacking the mu-opioid-receptor gene. *Nature* **383,** 819–823.

McDonald, A. J. (1998). Cortical pathways to the mammalian amygdala. *Prog. Neurobiol.* **55,** 257–332.

McKinley, M. J., Allen, A. M., Mathai, M. L., May, C., McAllen, R. M., Oldfield, B. J., and Weisinger, R. S. (2001). Brain angiotensin and body fluid homeostasis. *Jpn. J. Physiol.* **51,** 281–289.

McLemore, G. L., Kest, B., and Inturrisi, C. E. (1997). The effects of LY293558, an AMPA receptor antagonist, on acute and chronic morphine dependence. *Brain Res.* **778,** 120–126.

McNally, G. P., and Akil, H. (2002). Role of corticotropin-releasing hormone in the amygdala and bed nucleus of the stria terminalis in the behavioral, pain modulatory, and endocrine consequences of opiate withdrawal. *Neuroscience* **112,** 605–617.

Monaghan, D. T., and Cotman, C. W. (1985). Distribuiton of N-methyl-D-aspartate-sensitive L-[3H]glutamate-binding sites in rat brain. *J. Neurosci.* **5,** 2909–2919.

Mucha, R. F. (1987). Is the motivational effect of opiate withdrawal reflected by common somatic indices of precipitated withdrawal? A place conditioning study in the rat. *Brain Res.* **418,** 214–220.

O'Brien, C. P. (2008). Evidence-based treatments of addiction. *Philos. Trans. R. Soc. Lond. B Biol. Sci.* **363,** 3277–3286.

O'Brien, C. P., O'Brien, T. J., Mintz, J., and Brady, J. P. (1975). Conditioning of narcotic abstinence symptoms in human subjects. *Drug Alcohol Depend.* **1,** 115–123.

O'Brien, C. P., Childress, A. R., Ehrman, R., and Robbins, S. J. (1998). Conditioning factors in drug abuse: Can they explain compulsion? *J. Psychopharmacol.* **12,** 15–22.

Oh, M. S., Hong, S. J., Huh, Y., and Kim, K. S. (2009). Expression of transgenes in midbrain dopamine neurons using the tyrosine hydroxylase promoter. *Gene Ther.* **16,** 437–440.

Peters, A., Palay, S. L., and Webster, H. (1991). The Fine Structure of the Nervous System. Oxford University Press, New York.

Petralia, R. S., and Wenthold, R. J. (1992). Light and electron immunocytochemical localization of AMPA-selective glutamate receptors in the rat brain. *J. Comp. Neurol.* **318,** 329–354.

Petralia, R. S., Yokotani, N., and Wenthold, R. J. (1994). Light and electron microscope distribution of the NMDA receptor subunit NMDAR1 in the rat nervous system using a selective anti-peptide antibody. *J. Neurosci.* **14,** 667–696.

Petrovich, G. D., Scicli, A. P., Thompson, R. F., and Swanson, L. W. (2000). Associative fear conditioning of enkephalin mRNA levels in central amygdalar neurons. *Behav. Neurosci.* **114,** 681–686.

Phelps, E. A., and LeDoux, J. E. (2005). Contributions of the amygdala to emotion processing: From animal models to human behavior. *Neuron* **48,** 175–187.

Pitkanen, A., Savander, V., and LeDoux, J. E. (1997). Organization of intra-amygdaloid circuitries in the rat: An emerging framework for understanding functions of the amygdala. *Trends Neurosci.* **20,** 517–523.

Pollandt, S., Liu, J., Orozco-Cabal, L., Grigoriadis, D. E., Vale, W. W., Gallagher, J. P., and Shinnick-Gallagher, P. (2006). Cocaine withdrawal enhances long-term potentiation induced by corticotropin-releasing factor at central amygdala glutamatergic synapses via CRF, NMDA receptors and PKA. *Eur. J. Neurosci.* **24,** 1733–1743.

Rabinak, C. A., and Maren, S. (2008). Associative structure of fear memory after basolateral amygdala lesions in rats. *Behav. Neurosci.* **122,** 1284–1294.

Reyes, B. A., Drolet, G., and Van Bockstaele, E. J. (2008). Dynorphin and stress-related peptides in rat locus coeruleus: Contribution of amygdalar efferents. *J. Comp. Neurol.* **508,** 663–675.

Roozendaal, B., Schoorlemmer, G. H., Koolhaas, J. M., and Bohus, B. (1993). Cardiac, neuroendocrine, and behavioral effects of central amygdaloid vasopressinergic and oxytocinergic mechanisms under stress-free conditions in rats. *Brain Res. Bull.* **32,** 573–579.

Samson, R. D., and Paré, D. (2005). Activity-dependent synaptic plasticity in the central nucleus of the amygdala. *J. Neurosci.* **25,** 1847–1855.

Sato, K., Kiyama, H., and Tohyama, M. (1993). The differential expression patterns of messenger RNAs encoding non-N-methyl-D-aspartate glutamate receptor subunits (GluR1-4) in the rat brain. *Neuroscience* **52,** 515–539.

Sato, K., Mick, G., Kiyama, H., and Tohyama, M. (1995). Expression patterns of a glutamate-binding protein in the rat central nervous system: Comparison with N-methyl-D-aspartate receptor subunit 1 in rat. *Neuroscience* **64,** 459–475.

Scarborough, J. (1995). The opium poppy in hellenistic and roman medicine. *In* "Drugs and narcotics in history" (R. Porter and M. Teich, Eds.). Cambridge University Press, New York, 4–23.

Schmued, L., Phermsangngam, P., Lee, H., Thio, S., Chen, E., Truong, P., Colton, E., and Fallon, J. (1989). Collateralization and GAD immunoreactivity of descending pallidal efferents. *Brain Res.* **487,** 131–142.

Schulkin, J., McEwen, B. S., and Gold, P. W. (1994). Allostasis, amygdala, and anticipatory angst. *Neurosci. Biobehav. Rev.* **18,** 385–396.

Schulteis, G., Yackey, M., Risbrough, V., and Koob, G. F. (1998). Anxiogenic-like effects of spontaneous and naloxone-precipitated opiate withdrawal in the elevated plus-maze. *Pharmacol. Biochem. Behav.* **60,** 727–731.

Shapiro, M. L., and Eichenbaum, H. (1999). Hippocampus as a memory map: Synaptic plasticity and memory encoding by hippocampal neurons. *Hippocampus* **9,** 365–384.

Shi, S. H., Hayashi, Y., Petralia, R. S., Zaman, S. H., Wenthold, R. J., Svoboda, K., and Malinow, R. (1999). Rapid spine delivery and redistribution of AMPA receptors after synaptic NMDA receptor activation. *Science* **284,** 1811–1816.

South, S. M., Kohno, T., Kaspar, B. K., Hegarty, D., Vissel, B., Drake, C. T., Ohata, M., Jenab, S., Sailer, A. W., Malkmus, S., Masuyama, T., Horner, P., *et al.* (2003). A conditional deletion of the NR1 subunit of the NMDA receptor in adult spinal cord dorsal horn reduces NMDA currents and injury-induced pain. *J. Neurosci.* **23,** 5031–5040.

Stinus, L., Le Moal, M., and Koob, G. F. (1990). Nucleus accumbens and amygdala are possible substrates for the aversive stimulus effects of opiate withdrawal. *Neuroscience* **37,** 767–773.

Stinus, L., Caille, S., and Koob, G. F. (2000). Opiate withdrawal-induced place aversion lasts for up to 16 weeks. *Psychopharmacology* **149,** 115–120.

Stinus, L., Cador, M., Zorrilla, E. P., and Koob, G. F. (2005). Buprenorphine and a CRF1 antagonist block the acquisition of opiate withdrawal-induced conditioned place aversion in rats. *Neuropsychopharmacology* **30,** 90–98.

Swanson, L. W. (1992). Brain Maps: Structure of the Rat Brain. Elsevier, Amsterdam.

Swanson, L. W. (2000). Cerebral hemisphere regulation of motivated behavior. *Brain Res.* **886,** 113–164.

Swanson, L. W., Sawchenko, P. E., Rivier, J., and Vale, W. W. (1983). Organization of ovine corticotropin-releasing factor immunoreactive cells and fibers in the rat brain: An immunohistochemical study. *Neuroendocrinology* **36,** 165–186.

Tanganelli, S., Antonelli, T., Morari, M., Bianchi, C., and Beani, L. (1991). Glutamate antagonists prevent morphine withdrawal in mice and guinea pigs. *Neurosci. Lett.* **122,** 270–272.

Taylor, J. R., Punch, L. R., and Elsworth, J. D. (1998). A comparisopn of the effects of clonidine and CNQX infusion into the locus coeruleus and the amygdala on naloxone-precipitated opiate withdrawal in the rat. *Psychopharmacology* **138,** 133–142.

Trujillo, K. A., and Akil, H. (1991). Inhibition of morphine tolerance and dependence by the NMDA receptor antagonist MK-801. *Science* **251,** 85–87.

Tsien, J. Z. (2000). Linking Hebb's coincidence-detection to memory formation. *Curr. Opin. Neurobiol.* **10,** 266–273.

Turner, B. H., and Herkenham, M. (1991). Thalamoamygdaloid projections in the rat: A test of the amygdala's role in sensory processing. *J. Comp. Neurol.* **313,** 295–325.

Van Bockstaele, E. J., Saunders, A., Commons, K. G., Liu, X. B., and Peoples, J. (2000). Evidence for coexistence of enkephalin and glutamate in axon terminals and cellular sites for functional interactions of their receptors in the rat locus coeruleus. *J. Comp. Neurol.* **417,** 103–114.

Walker, D. L., and Davis, M. (2002). The role of amygdala glutamate receptors in fear learning, fear-potentiated startle, and extinction. *Pharmacol. Biochem. Behav.* **71,** 379–392.

Wallace, D. M., Magnuson, D. J., and Gray, T. S. (1989). The amygdalo-brainstem pathway: Selective innervation of dopaminergic, noradrenergic and adrenergic cells in the rat. *Neurosci. Lett.* **97,** 252–258.

Watanabe, T., Nakagawa, T., Yamamoto, R., Maeda, A., Minami, M., and Satoh, M. (2002). Involvement of glutamate receptors within the central nucleus of the amygdala in naloxone-precipitated withdrawal-induced conditioned place aversion in rats. *Jpn. J. Pharmacol.* **88,** 399–406.

Way, E. L., Loh, H. H., and Shen, F. H. (1969). Simultaneous quantitative assessment of morphine tolerance and physical dependence. *J. Pharmacol. Exp. Ther.* **167,** 1–8.

Weaver, M., and Schnoll, S. (2002). Abuse liability in opioid therapy for pain treatment in patients with an addiction history. *Clin. J. Pain* **18**(4 Suppl.), S61–S69.

Wilensky, A. E., Schafe, G. E., Kristensen, M. P., and LeDoux, J. E. (2006). Rethinking the fear circuit: The central nucleus of the amygdala is required for the acquisition, consolidation, and expression of Pavlovian fear conditioning. *J. Neurosci.* **26,** 12387–12396.

Williams, J. T., Christie, M. J., and Manzoni, O. (2001). Cellular and synaptic adaptations mediating opioid dependence. *Physiol. Rev.* **81,** 299–343.

Wray, S., and Hoffman, G. E. (1983). Organization and interrelationship of neuropeptides in the central amygdaloid nucleus of the rat. *Peptides* **4,** 525–541.

Yoneda, Y., Kuramoto, N., Kitayama, T., and Hinoi, E. (2001). Consolidation of transient ionotropic glutamate signals through nuclear transcription factors in the brain. *Prog. Neurobiol.* **63,** 697–719.

Zahm, D. S., Jensen, S. L., Williams, E. S., and Martin, J. R. (1999). Direct comparison of projections from the central amygdaloid region and nucleus accumbens shell. *Eur. J. Neurosci.* **11,** 1119–1126.

Zardetto-Smith, A. M., Moga, M. M., Magnuson, D. J., and Gray, T. S. (1988). Lateral hypothalamic dynorphinergic efferents to the amygdala and brainstem in the rat. *Peptides* **9,** 1121–1127.

Zhu, W., and Pan, Z. Z. (2004). Synaptic properties and postsynaptic opioid effects in rat central amygdala neurons. *Neuroscience* **127,** 871–879.

Zhu, W., and Pan, Z. Z. (2005). $\mu$-opioid-mediated inhibition of glutamate synaptic transmission in rat central amygdala neurons. *Neuroscience* **133,** 97–103.

CHAPTER NINE

# Hippocampal Kainate Receptors

Erik B. Bloss* *and* Richard G. Hunter†

## Contents



## Abstract

Glutamate is the major fast excitatory amino acid transmitter in the CNS, and exerts its action through receptors that function as ion channels such as NMDA receptors (NMDARs), AMPA receptors (AMPARs), and kainate receptors (KARs), and also through signaling cascades via metabotropic receptors. Of the ionotropic receptors, NMDARs and AMPARs have been extensively studied for decades, while relatively fewer studies have focused on the role of the KARs

* Fishberg Department of Neuroscience and Kastor Neurobiology of Aging Laboratory, Mount Sinai School of Medicine, New York, New York, USA
† Laboratory of Neuroendocrinology, Laboratory of Neurobiology and Behavior, Rockefeller University, New York, New York, USA

*Vitamins and Hormones*, Volume 82
ISSN 0083-6729, DOI: 10.1016/S0083-6729(10)82009-6

in the glutamatergic synapse. Despite this, there is considerable experimental data that suggest a major role for KARs in modulating synaptic transmission and plasticity, particularly in the hippocampal formation, as well as an involvement in disease states. KARs mediate most aspects of kainate-induced seizures and excitotoxic cell death, and thus, are a rational drug target for antiepileptic drug discovery. Recent data from human studies have also highlighted a role for KARs in certain psychiatric diseases, such as schizophrenia and major depression, and a recent association of KAR gene variants with response to antidepressants has brought considerable interest in developing a clearer understanding of KAR action in the brain. We have recently found that exposure to stress and stress hormone administration can produce contrasting changes in KAR subunit expression in the rat hippocampus, suggesting that a modification of hippocampal KARs by stress may be a mechanism for predisposing individuals to stress-related psychiatric diseases. Here, we review the anatomical and functional characteristics of hippocampal KARs, their role in synaptic plasticity, their regulation by certain hormones, and briefly review what is known about their involvement in disease states such as epilepsy and depression. 

# I. Introduction

Glutamate is the major fast excitatory amino acid transmitter in the CNS, and exerts its actions on target cells via synaptic glutamate receptors (GluRs). Receptors for glutamate are found throughout the brain and spinal cord, and mediate responses to synaptic glutamate release through receptors that function as both ion channels (ionotropic GluRs, iGluRs) and through G-protein signaling (metabotropic GluRs, mGluRs). iGluRs have been well studied, as these proteins generally provide rapid postsynaptic depolarizations in response to synaptically released glutamate, and have a major role in neural transmission, synaptic plasticity, and learning and memory.

The ionotropic GluRs consist of three distinct families originally named for their sensitivity to pharmacological agents: *N*-methyl-D-aspartate sensitive receptors (NMDARs), α-amino-3-hydroxyl-5-methyl-4-isoxazole-propionate sensitive receptors (AMPARs), and kainate receptors (KARs). An initial technical difficulty in recognizing KARs as a distinct family of iGluRs is that both KARs and AMPARs can bind kainic acid, though KARs have roughly a 1000-fold higher affinity (Herb *et al.*, 1992).

NMDARs and AMPARs have been extensively studied for decades; in contrast, fewer studies have focused on the role of the KAR family in the glutamatergic synapse. KARs have been more difficult than NMDARs or AMPARs to study for several reasons, including lack of antibodies for specific KAR subunits, inadequate subunit-specific agonists and antagonists, action at both pre- and postsynaptic sites, and conflicting results using pharmacological methods versus transgenic mouse knockout models. While these technical problems have slowed the progress in understanding KAR composition and function in the CNS, there remains sufficient evidence to suggest a role for KARs in neural transmission and synaptic plasticity, as well as an involvement in disease states where KAR function may be impaired or excessive. Low doses of kainic acid in rodent models cause seizure activity reminiscent of epilepsy (Ben-Ari, 1985) and kainate seizure models are among the most widely used in preclinical research; furthermore, the current evidence supports KARs in mediating a majority of these effects (Vincent and Mulle, 2009). Because other agents that target other iGluRs (e.g., NMDARs) have failed as antiepileptic therapeutics, KARs represent a novel and rational drug target. In addition, human studies have found connections between KAR subunit genes and psychiatric diseases such as depression (Schiffer and Heinemann, 2007a), schizophrenia (Beneyto *et al.*, 2007; Pickard *et al.*, 2006), and bipolar disorder (Blackwood *et al.*, 2007; Shaltiel *et al.*, 2008); thus, there is renewed interest in resolving the KAR contribution to glutamatergic signaling and synaptic transmission. Our laboratory and other groups have found that glucocorticoid hormones involved in the stress response can regulate the expression of certain KAR subunits, with a surprise finding that psychological stress and stress hormone administration can produce contrasting effects on KAR gene expression (Hunter *et al.*, 2009). The focus of this chapter is to explore how KARs function in the brain, particularly in the hippocampal formation, how changes to KARs may play a role in human diseases, such as epilepsy and depression, and the role of glucocorticoid hormones in regulating hippocampal KAR expression.

# II. Structure of Kainate Receptors

## A. Molecular structure of KAR subunits

The KAR family of GluRs consists of five distinct genes: GluR5, GluR6, GluR7, KA1, and KA2. GluR5–7 form one distinct subfamily, sharing 75–80% amino acid sequence homology, while KA1 and KA2 subunits form a second receptor subfamily, sharing about 70% homology. These two KAR subfamilies share only about 40% homology to each other, but are more closely related to each other than either subfamily is to the AMPAR GluR1–4 and NMDAR NR1–NR3 subunits. All KAR subunits display a signal sequence on the extracellular N-terminus, four transmembrane domains (TMDs), and an elongated intracellular stretch of amino acids between the third and fourth TMD. Posttranslational modifications of

KARs include alternative RNA splicing (GluR5), RNA editing from a glutamine to arginine in the second TMD (GluR5 and GluR6 only), several N-linked glycosylation sites prior to the first TMD, and several possible phosphorylation sites between TMD 3 and 4. For a detailed review on the structural aspects of KARs and their similarities and differences to other iGluRs, see Hollmann and Heinemann (1994).

## B. Composition of native KARs

Despite the cloning of the individual KAR subunits over a decade ago, it remains debated as to what subunit(s) comprise native KARs, if and how these homomeric or heteromeric receptors differ in regional or circuit manner, and how the different admixtures might relate to the multiple functional roles of KARs at glutamatergic synapses. Importantly, these KAR gene subfamilies differ in terms of affinity for glutamate: KA1 and 2 bind glutamate with an approximate 10-fold higher affinity than GluR5–7 (i.e., 5 nM compared to 50 nM, respectively), which suggests a mechanism by which KARs of diverse subunit compositions may differentially contribute to synaptic responses depending on the level of released glutamate. Recent studies *in vitro* demonstrated an important aspect of KARs; that, like AMPA receptors, individual subunits have their own agonist binding domains with unique properties. However, in contrast to AMPARs where most subunits have similar gating and current characteristics, the individual GluR5–7 subunits have drastically different profiles. This property of heteromeric KARs was demonstrated in HEK293 cells by expressing GluR5, GluR6, and KA2, alone or in combination, and exposing these cells to the compound dysiherbaine, which caused differential inward currents and receptor desensitization depending on the subunit composition (Swanson *et al.*, 2002).

Early *in vitro* expression studies in *Xenopus laevis* oocytes and cell culture models provided evidence that functional KAR receptors do not associate with NMDA or AMPA-type iGluRs (Bettler and Mulle, 1995), and that functional KARs can be comprised of subunit homomers or heteromers. For example, the selective expression of GluR5 (Sommer *et al.*, 1992) or GluR6 (Egebjerg *et al.*, 1991) alone in oocytes or transfected cell lines can yield responses to agonists such as domoic acid, kainic acid, and glutamate, but most studies report no responses with the expression of GluR7 alone. Though there is some evidence for the existence of a GluR7 splice variant homomer with unusually low sensitivity to glutamate (Schiffer *et al.*, 1997), it remains unknown if these receptors exist *in vivo*. Most other data have suggested that GluR7 likely forms heteromeric receptor complexes with additional subunits, including GluR6 and/or KA2, a hypothesis that has been strengthened by studies in hippocampal slice culture. In contrast to GluR5–7, expression studies suggest that neither KA1 nor KA2 subunits form homomers that act as functional ionotropic receptors (Herb *et al.*, 1992; Werner *et al.*, 1991); however, both can form functional heteromeric receptors when coexpressed with either GluR5 or GluR6 alone (but apparently not with GluR7 alone) (Herb *et al.*, 1992).

Thus, the composition of the native KARs *in vivo* is still a matter of some debate, a question likely complicated by differential expression of these subunits across different brain regions (e.g., cerebellum vs. hippocampus), circuits within a region (e.g., perforant path vs. mossy fiber terminals (MFTs)), and cell types (e.g., granule cells, pyramidal cells, and interneurons). The molecular understanding of KAR composition has been extended *in vivo* using coimmunoprecipitation, immunolabeling, and transgenic knockout approaches. Despite some concern over antisera specificity, coimmunoprecipitation and immunolabeling approaches have thus far largely agreed with the body of *in vitro* studies; for example, Darstein *et al.* found KA1 and KA2 subunits coimmunoprecipitated with GluR6 but not GluR7 at mossy fiber synapses in mouse tissue (Darstein *et al.*, 2003). A subsequent study by Pinhiero *et al.* demonstrated that GluR6 coimmunoprecipitated with GluR7 to form functional KARs, thus strengthening the notion of a functionally relevant GluR6/7 heteromeric receptor on MFTs (Pinheiro *et al.*, 2007). Thus, the emerging *in vivo* data suggest that KA1 and KA2 form high-affinity KARs with GluR6 subunits, while GluR6 also assembles with GluR7 to form lower affinity KARs.

## III. Anatomical Localization of KARs in the Hippocampal Formation

Further clues to the functional compositions of the KARs have been found in studies mapping subunit mRNA, protein expression, and $^{3}$H kainate binding in the distinct hippocampal subfields. *In situ* hybridization and radioligand binding studies have been particularly successful in mapping distinct anatomical changes in KAR expression; $^{3}$H KAR binding is highest in the CA3 stratum lucidum (Monaghan and Cotman, 1982), where DG mossy fibers synapse onto dendritic spines and thorny excrescences of CA3 pyramidal cells. Colchicine-induced elimination of hippocampal DG cells drastically reduces high-affinity $^{3}$H KAR binding in the stratum lucidum, suggesting that a large portion of KARs is present at presynaptic sites at MFTs rather than CA3 dendritic spines (Represa *et al.*, 1987). Despite these approaches to KAR study, a major impediment to determining specific expression patterns at the synapse has been the absence of subunit-specific antibodies, particularly for GluRs 5–7.

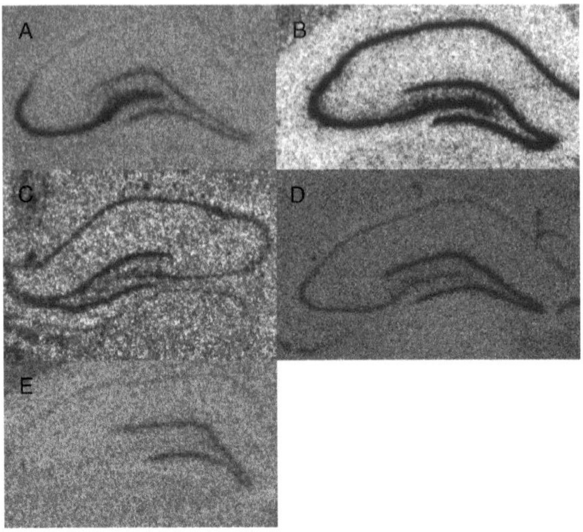


**Figure 9.1** Autoradiograms of hippocampal KAR gene expression in rat brain. (A) KA1 mRNA is present in CA3 neurons and in DG granule cells; whereas KA2 (B) is expressed highly throughout the granule and pyramidal cell layers of the hippocampal formation. GluR5 (C) is expressed at low levels throughout the hippocampal formation and (D) GluR6 mRNA is present in CA3, DG, and CA1 hippocampal neurons; (E) GluR7 is predominantly expressed in DG granule cells.

## A. GluR5

*In situ* hybridization studies have provided evidence that GluR5 is expressed in low levels in all hippocampal subfields (Bahn *et al.*, 1994), and its expression in principal neurons remains in question (see Fig. 9.1C). A more recent study in rat hippocampus utilized nonradioactive *in situ* hybridization double labeling of GluR5 and glutamic acid decarboxylase-65 (GAD-65) mRNA in an attempt to identify the cell type of GluR5 expressing cells (Paternain *et al.*, 2000). The data demonstrated that GluR5 cells are found scattered across all hippocampal subfields (stratum oriens, SO; stratum pyramidal, SP; and stratum radiatum, SR). Quantitative analysis determined that >80% of GluR5 mRNA positive cells in rat CA1 and CA3 also express GAD-65, suggesting that a large majority of GluR5-containing cells are GABAergic interneurons. However, not all GAD-65 positive cells express GluR5 mRNA, thus, the GluR5-positive interneurons are likely a subpopulation of hippocampal GABAergic interneurons. Further studies on the cellular localization of GluR5 have been confounded by the low levels of GluR5 expression in the hippocampus, the complete lack of GluR5-specific antibodies, and by the use of antibodies that recognize an epitope common to GluR5/6/7 (Siegel *et al.*, 1995). Therefore, while it is not inconceivable that GluR5 is expressed in

DG granule cells or CA3 pyramidal cells at levels below the threshold of detection, the current consensus is that a majority of GluR5s actions are likely confined to hippocampal GABAergic interneurons in CA1 and CA3. The presence of these receptors in GABA neurons is currently under exploration as a potential drug target to develop novel antiepileptic agents (Khalilov *et al.*, 2002) (see section below).

## B. GluR6

In contrast to the limited expression of GluR5, GluR6 mRNA (shown in Fig. 9.1D) is expressed in both DG granule cells and CA3 pyramidal cells (Bahn *et al.*, 1994). Paternain's study also demonstrated GluR6 expression in hippocampal interneurons across the SP, SO, and SR, and could, therefore, conceivably contribute to KARs action at multiple sites, including presynaptic MFTs, postsynaptic dendritic spines on CA3 pyramidal cells, and in GABA cells (Paternain *et al.*, 2000). GluR6 mRNA was expressed in a large majority of principal cells in CA1, CA3, and DG, and surprisingly in ~90% of GAD-65 expressing cells in the pyramidal layers; thus, it is likely there are populations of GluR6-containing KARs in GABAergic interneurons. However, this analysis also noted that because there are far more pyramidal cells compared to interneurons in the SP, very few GluR6 cells coexpressed GAD-65 in the CA1 and CA3 SP. At the protein level, the study of GluR6 expression has been somewhat confounded by the use of antibodies that label common epitopes to both GluR6 and GluR7 protein. To circumvent this problem, many studies have utilized this antibody in GluR7 knockout mice in hopes to specifically label GluR6 protein. In GluR7 knockout mice, staining using the antibody that recognizes GluR6/7 prominently labels the CA3 stratum lucidum and lightly labels CA3 pyramidal cell bodies (Darstein *et al.*, 2003). This staining pattern is greatly reduced in GluR6 knockout mice, but only marginally reduced in GluR7 knockout mice, thus, the authors have suggested that most labeling in a wild-type mouse hippocampus is indeed GluR6-specific immunoreactivity.

In addition, a set of coimmunoprecipitation experiments have provided evidence that demonstrates GluR6 coassembles with KA1 protein, which was predominantly localized to presynaptic vesicular, extrasynaptic, and synaptic sites in MFTs, though postsynaptic KA1 localization at axospinous synapses was also observed (Darstein *et al.*, 2003). However, the same study also demonstrated that GluR6 coassembles with KA2 protein, which was predominantly localized to postsynaptic axospinous sites. It can be said that GluR6 cannot be definitively localized to either synaptic site using these approaches, and likely exists at both sides of the MFT–CA3 synapse; unfortunately, we are not aware of any study that utilizes an ultrastructural approach to characterize the qualitative and quantitative features of subunit-specific labeling of GluR6 at MFT synapses.

## C. GluR7

GluR7 mRNA (shown in Fig. 9.1E) is largely confined to the granule cells of the DG (Bahn *et al.*, 1994), suggesting that GluR7 may be acting at a presynaptic site at MFTs. Not surprisingly, GluR7 protein is enriched in synaptic preparations, and immunoprecipitates with GluR6 (Pinheiro *et al.*, 2007) and, as with GluR6, a lack of adequate antibodies to specifically label GluR7 subunits remains the largest impediment to characterizing further the ultrastructural sites of GluR7 action; as a result, most of the aforementioned results have been obtained by using the hippocampi of GluR6 knockout mice. Future microscopy studies are obviously needed to confirm the functional data that largely suggest presynaptic sites at MFTs for GluR7-containing heteromeric receptors.

## D. KA1

mRNA expression studies show high levels of KA1 expression (see Fig. 9.1A) in DG and CA3 cells in the rat hippocampus (Bahn *et al.*, 1994). In contrast to GluR5–7, specific antibodies have been made against a synthetic C-terminal fragment of KA1 (Darstein *et al.*, 2003; Fogarty *et al.*, 2000), and immunolabeling approaches have demonstrated KA1 immunoreactivity in the polymorphic layer of the DG, the CA3 stratum lucidum, and some somatic labeling in CA1 pyramidal cells. However, there are notable differences between the described immunoreactivity in these two studies, at both the light level and with the electron microscope. While Fogarty *et al.* (2000) describe intense reaction product in DG, CA3, and CA1 cell bodies, Darstein *et al.* (2003) describe only lightly labeled somatic staining, with the majority of labeling in the CA3 lucidum. In addition, at the electron microscopic (EM) level, these studies find opposite locations for KA1 at MFT synapses. Qualitative preembed ultrastructural observations of KA1 immunoreactivity in rat hippocampus largely demonstrated postsynaptic KA1 labeling (Fogarty *et al.*, 2000); in contrast, analysis using quantitative postembed EM immunolabeling has revealed a predominant presynaptic localization for KA1 receptors in granule mossy fiber synapses, though labeling is occasionally found at CA3 dendritic spines (Darstein *et al.*, 2003). The discrepancy may be the result of technical reasons, such as immunoperoxidase labeling versus nanogold secondaries, inherent differences in the antibodies chosen, and pre- versus postembed labeling, or differences in KA1 expression between rat and mouse. Future immunolabeling studies could clarify this issue by using an identical antisera and technical approach in both species.

### E. KA2

KA2 mRNA is widely expressed throughout the brain, including the hippocampus (Fig. 9.1B), cerebellum, and cerebral cortex (Bahn *et al.*, 1994). Several groups have produced antisera to different portions of the C-terminus of the KA2 protein (Darstein *et al.*, 2003; Petralia *et al.*, 1994), and both studies demonstrate light microscopic KA2 immunolabeling in CA3 cells and in the stratum lucidum of the CA3. Ultrastructural data, from both pre- and postembedding studies, suggest a predominant postsynaptic localization for KA2 receptors on CA3 dendritic spines, though both studies also note rare presynaptic labeling in MFTs.

## IV. Electrophysiological Functions of KARs in Hippocampal Synapses

While most studies have focused on AMPAR and NMDAR-dependent postsynaptic mechanisms of synaptic plasticity on CA3–CA1 synapses, synaptic plasticity such as long-term potentiation (LTP) at DG-CA3 MFTs may involve NMDAR-independent, KAR-dependent *pre*synaptic modulation of glutamate release. In hippocampal synapses, KARs play a role in glutamatergic transmission through multiple mechanisms, including bidirectionally modulating MFT glutamate release (Schmitz *et al.*, 2001a), increasing excitability on commissural/perforant path inputs to CA3 neurons (Contractor *et al.*, 2000), and exerting direct actions on postsynaptic neurons via KARs on CA3 neurons (Castillo *et al.*, 1997). In addition, KARs are postulated to have indirect effects on CA1 neurons through the modulation of GABA release via KAR-containing CA1 interneurons, a mechanism that is presumed to play a role in KA-induced epileptogenesis (Min *et al.*, 1999).

### A. Presynaptic actions of KARs in MFT transmission and synaptic plasticity

The initial observations by Monaghan and Cotman (1982), of a remarkably high amount of high-affinity KAR binding in the CA3 stratum lucidum, prompted speculation that KARs might have a prominent role in MFT–CA3 synaptic plasticity, especially considering the low abundance of NMDARs found in the MFT termination zone (Watanabe *et al.*, 1998). Very low applications of kainate (50 nM) enhance both AMPAR and NMDAR-mediated excitatory postsynaptic currents, probably through an increase in transmitter release via presynaptic KARs (Schmitz *et al.*, 2001b). In contrast, higher doses (3 $\mu$M or greater in most studies) of kainate can depolarize MFTs and profoundly depress MFT transmission, an effect that is reported to be mediated by presynaptic KARs (Contractor *et al.*, 2000; Kamiya and Ozawa, 2000; Vignes *et al.*, 1998). These actions on MFT transmission can be elicited by endogenous KAR activation by synaptically released glutamate (Schmitz *et al.*, 2000). At the present vantage, the bulk of the evidence supports the idea that KARs can both facilitate and depress mossy fiber transmission through presynaptic autoreceptors.

Research into the subunit(s) responsible for the presynaptic KAR actions have produced conflicting data; pharmacological studies using compounds that selectively activate and inhibit (ATPA and LY382884, respectively) GluR5-containing receptors have implicated GluR5 as mediating the KAR-dependent actions in MFT transmission (Bortolotto *et al.*, 1999), though studies from Heinemann and colleagues using GluR5 and GluR6 transgenic mouse models have found that GluR6 knockout mice fail to exhibit many actions of kainate-induced MFT plasticity (Contractor *et al.*, 2000, 2001), suggesting that it is the GluR6 subunit, and not the GluR5, that plays a presynaptic autoreceptor role in facilitating MFT plasticity. A more recent study has also evaluated the role of presynaptic GluR7-containing receptors using transgenic mice, and reported that disruption of the GluR7 gene reduced paired-pulse facilitation, low-frequency facilitation, and MFT LTP (see below); thus, GluR7-containing presynaptic autoreceptors also facilitate plasticity at the MFT–CA3 synapse (Pinheiro *et al.*, 2007). Considering this study also demonstrated a physical interaction between GluR6 and GluR7 proteins, reasonable hypothesis is that a GluR6/7 KAR complex in MFTs may be critical in these plasticity modes. Generation of KA2 knockout mice has demonstrated altered heterosynaptic facilitation in the presynaptic MFT, likely as a result of a decreased affinity for endogenously released glutamate; furthermore, these mice also demonstrate altered postsynaptic excitatory postsynaptic currents (Contractor *et al.*, 2003).

A similar discrepancy exists as to which KARs contribute to mossy fiber LTP. Although initial observations suggested MFT LTP was not blocked by AMPAR/KAR antagonists (Ito and Sugiyama, 1991), subsequent studies have demonstrated that the induction of MFT LTP can be impaired by blocking KARs (Bortolotto *et al.*, 2003). Collingridge and colleagues have provided substantial evidence that KARs, particularly those containing the GluR5 subunit, serve as a trigger for the induction of LTP at MFT–CA3 synapses. Ten micromolar application of LY382884, which does not affect the basal mossy fiber synaptic transmission, blocks MFT LTP induction in a number of electrophysiological protocols, but not preexisting LTP or LTP induced by Protein Kinase A (e.g., bath forskolin treatment) (Bortolotto *et al.*, 1999). Transgenic mouse studies using GluR5 and GluR6 knockout mice, however, have found that MFT LTP is impaired by GluR6 but not GluR5 knockout (Contractor *et al.*, 2001); furthermore, as noted earlier, MFT LTP is also impaired in GluR7 knockout mice (though not completely abolished) (Pinheiro *et al.*, 2007). As mentioned earlier, GluR6 and GluR7 mRNA are abundantly expressed in granule cells, whereas GluR5 mRNA is minimal, thus, the anatomical evidence would support a role for GluR6-containing receptors rather than GluR5-containing KARs. A more recent paper by Lauri *et al.* (2003) demonstrates that GluR5-containing receptors may utilize internal $Ca^{2+}$ stores to trigger MFT LTP; changing the divalent cation balance in the perfusing medium could inhibit the ability to block MFT MTP with a selective GluR5 antagonist. While these results may reconcile the discrepancies concerning the role of KAR subunits in mossy fiber plasticity, future studies will be needed to clarify these observations.

## V. KARs in Disease

Kainate has a long and successful history as a model of seizure disorders, most notably, temporal-lobe epilepsy. More recent work has revealed that KARs may have a role in other brain disorders as well.

### A. KAR-mediated models of epilepsy

Historically, KA has been widely used as an agent to induce seizures and excitotoxic cell death in animal models (Ben-Ari, 1985; Nadler *et al.*, 1981; Vincent and Mulle, 2009). Much progress has been made at the cellular and molecular levels in understanding the actions of KARs in mediating these actions of kainate, and these models have been a fruitful avenue of research for antiepileptic drugs (Rogawski, 2006).

Mechanistic studies that focused on determining the neuronal circuits that mediate kainate's seizure-like activity have provided evidence that kainate-induced seizures are primarily generated by KARs in the MFT–CA3 synapse; however, there is also evidence for secondary KAR action in CA1 interneurons. Experimental lesions of DG cells reduce the epileptogenic actions of kainate (Gaiarsa *et al.*, 1994), and disconnecting CA3 and CA1 pyramidal cells reduces the kainate-induced activity in the CA1 (Rodriguez-Moreno *et al.*, 1997). Because higher doses of kainate are required for seizure activity and KA neurotoxicity in GluR6 knockout mice (Mulle *et al.*, 1998) and overexpression of GluR6 produces seizures (Telfeian *et al.*, 2000), a plausible hypothesis is that GluR6-containing KARs on mossy fiber synapses may mediate the KAR-related epileptogenic actions of kainate. Kainate also induces postsynaptic currents in CA3 neurons that can induce spontaneous action potentials (Robinson and Deadwyler, 1981), which may also be dependent on GluR6-containing subunits (Mulle *et al.*, 1998). As yet, due in part to a lack of selective tools, the role of GluR6 in epileptogenesis cannot be completely defined.

As noted earlier, there is some evidence that hippocampal KAR-containing interneurons may play a role in modulating GABAergic transmission in the hippocampus, though the data are somewhat controversial regarding the exact effects of kainate on GABA transmission. An intuitive hypothesis is that kainate inhibits GABAergic transmission, and indeed, there is evidence that kainate administration can suppress inhibitory postsynaptic currents and hyperpolarize cells through GluR5-containing KARs (Clarke *et al.*, 1997; Rodriguez-Moreno *et al.*, 1997). In fact, drugs which antagonize GluR5, such as topiramate (Gryder and Rogawski, 2003), are anticonvulsant (Kaminski *et al.*, 2004; Smolders *et al.*, 2002). However, it has also been reported that a majority of hippocampal interneurons depolarize and "overinhibit" pyramidal neurons in response to kainate, possibly also through GluR5-containing receptors (Cossart *et al.*, 1998). How this response might work to limit or contribute to seizure activity is unclear, though, paradoxically, given the effects of GluR5 antagonists, a subsequent study has actually demonstrated the antiepileptic activity of the GluR5 agonist ATPA (Khalilov *et al.*, 2002). GluR7 may also play a role in epilepsy, as the GluR7 selective drug LU97175 is anticonvulsant in at least one model (Loscher *et al.*, 1999).

## B. Hippocampal neurotoxicity

With regard to neuronal loss in both seizure and ischemia models, KAR selective drugs, particularly GluR5 selective antagonists, have been shown to be robustly neuroprotective (Jane *et al.*, 2009; Vincent and Mulle, 2009). Indeed, the GluR5 antagonist LY377770 provided better protection than a battery of NMDAR and AMPAR antagonists and was able to do so up to an hour after ischemic injury occurred (O'Neill *et al.*, 2000). Similarly, antisense knockdown of GluR6 and KA2 mRNA also reduces neuronal loss after ischemia (Jiang *et al.*, 2007) and GluR6 knockout mice are resistant to KA neurotoxicity (Mulle *et al.*, 1998).

Interestingly, given the role corticosteroids play in hippocampal neurotoxicity (Sapolsky, 1996), a number of groups, including our own, have shown that these hormones are capable of regulating KAR subunit levels in the hippocampal formation (Hunter *et al.*, 2009; Joels *et al.*, 1996; Strutz-Seebohm *et al.*, 2005; Watanabe *et al.*, 1995). Our study showed that the KA1, GluR6, and GluR7 subunits were upregulated by adrenalectomy and that this effect was reversed by corticosterone in the case of KA1 and GluR7 and by aldosterone in the case of GluR6. Of particular interest is the finding that chronic stress elevates the levels of the KA1 receptor in the dentate gyrus, demonstrating a mechanism by which stress may alter susceptibility to seizure in both humans and animal models (Hunter *et al.*, 2009). Stress is also known to produce structural and synaptic plasticity in the hippocampus (McEwen, 2007), and it seems quite possible that KARs may significantly affect the hippocampal structure and function in response to stress, as do the other classes of ionotropic GluRs.

### C. Kainate receptors in mental disorders

While the KA model of epilepsy has a long history, it has only recently become apparent that KARs may play a role in other brain disorders. The aforementioned work showing that the expression of some KAR subunits is regulated by corticosteroid levels and stress points to the possibility that mental disorders, such as depression and schizophrenia, which are known to correlate with aberrant corticosteroid secretion, might be linked in part to KARs. In point of fact, a number of recent studies have demonstrated genetic linkages between KAR subunits and several distinct mental disorders. Postmortem brains taken from patients diagnosed with schizophrenia show reduced levels of GluR5 and GluR6 expression (Benes *et al.*, 2001; Beneyto *et al.*, 2007; Woo *et al.*, 2007), and gene association studies have shown a relation between KA1 and schizophrenia and bipolar disorder (Blackwood *et al.*, 2007; Djurovic *et al.*, 2009), as well as associating the KA1 gene with suicidal ideation and response to antidepressant treatment (Laje *et al.*, 2007; Paddock *et al.*, 2007). Other recent studies have shown an association between polymorphisms in the human GluR7 gene and schizophrenia (Ahmad *et al.*, 2009; Djurovic *et al.*, 2009), as well as major depression (Schiffer and Heinemann, 2007b). Given the social deficits present in schizophrenia, it is perhaps unsurprising that two studies of autistic populations have also associated polymorphisms in the GluR6 gene with the presence of the disorder (Jamain *et al.*, 2002; Shuang *et al.*, 2004). Although all of these studies point to a prominent role in human brain disorders for kainate receptors, it should also be apparent that very little is understood mechanistically about how KARs might influence the development of diseases such as schizophrenia or depression. It is to be hoped that future advances in the study of KARs will correct this situation.

## VI. Conclusion

Though KARs remain the least studied of the ionotropic GluRs, they show great promise as an avenue to deepen our understanding of the function of the hippocampal formation and of glutamatergic circuits within it.

While a long-established body of literature has shown the importance of hippocampal kainate receptors in epileptogenesis, more recent work has established that they, like their better-studied cousins, the AMPA and NMDA receptors, have a role to play in a variety of aspects of normal brain function and disease. Presently, however, our potential understanding of the total context of KAR activity is hindered by the lack of tools with which to distinguish KARs from AMPARs and different KAR subunits from one another. A recent review of KAR pharmacology and function (Jane *et al.*, 2009) lists dozens of KAR agonists and antagonists, none of which is selective for GluR6 or GluR7, and a similar situation exists with regard to antibodies, which renders much anatomical and physiological work with KARs much more difficult than work with AMPA or NMDA channels. It is to be hoped that the recent work reviewed here will attract more interest in developing the tools to understand KARs and their role in the nervous system.

## REFERENCES

Ahmad, Y., Bhatia, M. S., Mediratta, P. K., Sharma, K. K., Negi, H., Chosdol, K., and Sinha, S. (2009). Association between the ionotropic glutamate receptor kainate3 (GRIK3) Ser310Ala polymorphism and schizophrenia in the Indian population. *World J. Biol. Psychiatry*. 1–4.

Bahn, S., Volk, B., and Wisden, W. (1994). Kainate receptor gene expression in the developing rat brain. *J. Neurosci.* **14,** 5525–5547.

Ben-Ari, Y. (1985). Limbic seizure and brain damage produced by kainic acid: Mechanisms and relevance to human temporal lobe epilepsy. *Neuroscience* **14,** 375–403.

Benes, F. M., Todtenkopf, M. S., and Kostoulakos, P. (2001). GluR5, 6, 7 subunit immunoreactivity on apical pyramidal cell dendrites in hippocampus of schizophrenics and manic depressives. *Hippocampus* **11,** 482–491.

Beneyto, M., Kristiansen, L. V., Oni-Orisan, A., McCullumsmith, R. E., and Meador-Woodruff, J. H. (2007). Abnormal glutamate receptor expression in the medial temporal lobe in schizophrenia and mood disorders. *Neuropsychopharmacology* **32,** 1888–1902.

Bettler, B., and Mulle, C. (1995). Review: Neurotransmitter receptors. II. AMPA and kainate receptors. *Neuropharmacology* **34,** 123–139.

Blackwood, D. H., Pickard, B. J., Thomson, P. A., Evans, K. L., Porteous, D. J., and Muir, W. J. (2007). Are some genetic risk factors common to schizophrenia, bipolar disorder and depression? Evidence from DISC1, GRIK4 and NRG1. *Neurotox. Res.* **11,** 73–83.

Bortolotto, Z. A., Clarke, V. R., Delany, C. M., Parry, M. C., Smolders, I., Vignes, M., Ho, K. H., Miu, P., Brinton, B. T., Fantaske, R., Ogden, A., Gates, M., *et al.* (1999). Kainate receptors are involved in synaptic plasticity. *Nature* **402,** 297–301.

Bortolotto, Z. A., Lauri, S., Isaac, J. T., and Collingridge, G. L. (2003). Kainate receptors and the induction of mossy fibre long-term potentiation. *Philos. Trans. R Soc. Lond. B Biol. Sci.* **358,** 657–666.

Castillo, P. E., Malenka, R. C., and Nicoll, R. A. (1997). Kainate receptors mediate a slow postsynaptic current in hippocampal CA3 neurons. *Nature* **388,** 182–186.

Clarke, V. R., Ballyk, B. A., Hoo, K. H., Mandelzys, A., Pellizzari, A., Bath, C. P., Thomas, J., Sharpe, E. F., Davies, C. H., Ornstein, P. L., Schoepp, D. D.,

Kamboj, R. K., *et al.* (1997). A hippocampal GluR5 kainate receptor regulating inhibitory synaptic transmission. *Nature* **389,** 599–603.

Contractor, A., Sailer, A. W., Darstein, M., Maron, C., Xu, J., Swanson, G. T., and Heinemann, S. F. (2003). Loss of kainate receptor-mediated heterosynaptic facilitation of mossy-fiber synapses in KA2−/− mice. *J. Neurosci.* **23,** 422–429.

Contractor, A., Swanson, G., and Heinemann, S. F. (2001). Kainate receptors are involved in short- and long-term plasticity at mossy fiber synapses in the hippocampus. *Neuron* **29,** 209–216.

Contractor, A., Swanson, G. T., Sailer, A., O'Gorman, S., and Heinemann, S. F. (2000). Identification of the kainate receptor subunits underlying modulation of excitatory synaptic transmission in the CA3 region of the hippocampus. *J. Neurosci.* **20,** 8269–8278.

Cossart, R., Esclapez, M., Hirsch, J. C., Bernard, C., and Ben-Ari, Y. (1998). GluR5 kainate receptor activation in interneurons increases tonic inhibition of pyramidal cells. *Nat. Neurosci.* **1,** 470–478.

Darstein, M., Petralia, R. S., Swanson, G. T., Wenthold, R. J., and Heinemann, S. F. (2003). Distribution of kainate receptor subunits at hippocampal mossy fiber synapses. *J. Neurosci.* **23,** 8013–8019.

Djurovic, S., Kahler, A. K., Kulle, B., Jonsson, E. G., Agartz, I., Le Hellard, S., Hall, H., Jakobsen, K. D., Hansen, T., Melle, I., Werge, T., Steen, V. M., *et al.* (2009). A possible association between schizophrenia and GRIK3 polymorphisms in a multicenter sample of Scandinavian origin (SCOPE). *Schizophr. Res.* **107,** 242–248.

Egebjerg, J., Bettler, B., Hermans-Borgmeyer, I., and Heinemann, S. (1991). Cloning of a cDNA for a glutamate receptor subunit activated by kainate but not AMPA. *Nature* **351,** 745–748.

Fogarty, D. J., Perez-Cerda, F., and Matute, C. (2000). KA1-like kainate receptor subunit immunoreactivity in neurons and glia using a novel anti-peptide antibody. *Brain Res. Mol. Brain Res.* **81,** 164–176.

Gaiarsa, J. L., Zagrean, L., and Ben-Ari, Y. (1994). Neonatal irradiation prevents the formation of hippocampal mossy fibers and the epileptic action of kainate on rat CA3 pyramidal neurons. *J. Neurophysiol.* **71,** 204–215.

Gryder, D. S., and Rogawski, M. A. (2003). Selective antagonism of GluR5 kainate-receptor-mediated synaptic currents by topiramate in rat basolateral amygdala neurons. *J. Neurosci.* **23,** 7069–7074.

Herb, A., Burnashev, N., Werner, P., Sakmann, B., Wisden, W., and Seeburg, P. H. (1992). The KA-2 subunit of excitatory amino acid receptors shows widespread expression in brain and forms ion channels with distantly related subunits. *Neuron* **8,** 775–785.

Hollmann, M., and Heinemann, S. (1994). Cloned glutamate receptors. *Annu. Rev. Neurosci.* **17,** 31–108.

Hunter, R. G., Bellani, R., Bloss, E., Costa, A., McCarthy, K., and McEwen, B. S. (2009). Regulation of kainate receptor subunit mRNA by stress and corticosteroids in the rat hippocampus. *PLoS ONE* **4,** e4328.

Ito, I., and Sugiyama, H. (1991). Roles of glutamate receptors in long-term potentiation at hippocampal mossy fiber synapses. *Neuroreport* **2,** 333–336.

Jamain, S., Betancur, C., Quach, H., Philippe, A., Fellous, M., Giros, B., Gillberg, C., Leboyer, M., and Bourgeron, T. (2002). Linkage and association of the glutamate receptor 6 gene with autism. *Mol. Psychiatry.* **7,** 302–310.

Jane, D. E., Lodge, D., and Collingridge, G. L. (2009). Kainate receptors: Pharmacology, function and therapeutic potential. *Neuropharmacology* **56,** 90–113.

Jiang, H. X., Guan, Q. H., Pei, D. S., and Zhang, G. Y. (2007). Functional cooperation between KA2 and GluR6 subunits is involved in the ischemic brain injury. *J. Neurosci. Res.* **85,** 2960–2970.

Joels, M., Bosma, A., Hendriksen, H., Diegenbach, P., and Kamphuis, W. (1996). Corticosteroid actions on the expression of kainate receptor subunit mRNAs in rat hippocampus. *Brain Res. Mol. Brain Res.* **37,** 15–20.
Kaminski, R. M., Banerjee, M., and Rogawski, M. A. (2004). Topiramate selectively protects against seizures induced by ATPA, a GluR5 kainate receptor agonist. *Neuropharmacology* **46,** 1097–1104.
Kamiya, H., and Ozawa, S. (2000). Kainate receptor-mediated presynaptic inhibition at the mouse hippocampal mossy fibre synapse. *J. Physiol.* **523**(Pt 3), 653–665.
Khalilov, I., Hirsch, J., Cossart, R., and Ben-Ari, Y. (2002). Paradoxical anti-epileptic effects of a GluR5 agonist of kainate receptors. *J. Neurophysiol.* **88,** 523–527.
Laje, G., Paddock, S., Manji, H., Rush, A. J., Wilson, A. F., Charney, D., and McMahon, F. J. (2007). Genetic markers of suicidal ideation emerging during citalopram treatment of major depression. *Am. J. Psychiatry.* **164,** 1530–1538.
Lauri, S. E., Bortolotto, Z. A., Nistico, R., Bleakman, D., Ornstein, P. L., Lodge, D., Isaac, J. T., and Collingridge, G. L. (2003). A role for Ca2+ stores in kainate receptor-dependent synaptic facilitation and LTP at mossy fiber synapses in the hippocampus. *Neuron* **39,** 327–341.
Loscher, W., Lehmann, H., Behl, B., Seemann, D., Teschendorf, H. J., Hofmann, H. P., Lubisch, W., Hoger, T., Lemaire, H. G., and Gross, G. (1999). A new pyrrolyl-quinoxalinedione series of non-NMDA glutamate receptor antagonists: Pharmacological characterization and comparison with NBQX and valproate in the kindling model of epilepsy. *Eur. J. Neurosci.* **11,** 250–262.
McEwen, B. S. (2007). Physiology and neurobiology of stress and adaptation: Central role of the brain. *Physiol. Rev.* **87,** 873–904.
Min, M. Y., Melyan, Z., and Kullmann, D. M. (1999). Synaptically released glutamate reduces gamma-aminobutyric acid (GABA)ergic inhibition in the hippocampus via kainate receptors. *Proc. Natl. Acad. Sci. USA* **96,** 9932–9937.
Monaghan, D. T., and Cotman, C. W. (1982). The distribution of [3H]kainic acid binding sites in rat CNS as determined by autoradiography. *Brain Res.* **252,** 91–100.
Mulle, C., Sailer, A., Perez-Otano, I., Dickinson-Anson, H., Castillo, P. E., Bureau, I., Maron, C., Gage, F. H., Mann, J. R., Bettler, B., and Heinemann, S. F. (1998). Altered synaptic physiology and reduced susceptibility to kainate-induced seizures in GluR6-deficient mice. *Nature* **392,** 601–605.
Nadler, J. V., Evenson, D. A., and Smith, E. M. (1981). Evidence from lesion studies for epileptogenic and non-epileptogenic neurotoxic interactions between kainic acid and excitatory innervation. *Brain Res.* **205,** 405–410.
O'Neill, M. J., Bogaert, L., Hicks, C. A., Bond, A., Ward, M. A., Ebinger, G., Ornstein, P. L., Michotte, Y., and Lodge, D. (2000). LY377770, a novel iGlu5 kainate receptor antagonist with neuroprotective effects in global and focal cerebral ischaemia. *Neuropharmacology* **39,** 1575–1588.
Paddock, S., Laje, G., Charney, D., Rush, A. J., Wilson, A. F., Sorant, A. J., Lipsky, R., Wisniewski, S. R., Manji, H., and McMahon, F. J. (2007). Association of GRIK4 with outcome of antidepressant treatment in the STAR*D cohort. *Am. J. Psychiatry* **164,** 1181–1188.
Paternain, A. V., Herrera, M. T., Nieto, M. A., and Lerma, J. (2000). GluR5 and GluR6 kainate receptor subunits coexist in hippocampal neurons and coassemble to form functional receptors. *J. Neurosci.* **20,** 196–205.
Petralia, R. S., Wang, Y. X., and Wenthold, R. J. (1994). Histological and ultrastructural localization of the kainate receptor subunits, KA2 and GluR6/7, in the rat nervous system using selective antipeptide antibodies. *J. Comp. Neurol.* **349,** 85–110.
Pickard, B. S., Malloy, M. P., Christoforou, A., Thomson, P. A., Evans, K. L., Morris, S. W., Hampson, M., Porteous, D. J., Blackwood, D. H., and Muir, W. J.

(2006). Cytogenetic and genetic evidence supports a role for the kainate-type glutamate receptor gene, GRIK4, in schizophrenia and bipolar disorder. *Mol. Psychiatry* **11,** 847–857.

Pinheiro, P. S., Perrais, D., Coussen, F., Barhanin, J., Bettler, B., Mann, J. R., Malva, J. O., Heinemann, S. F., and Mulle, C. (2007). GluR7 is an essential subunit of presynaptic kainate autoreceptors at hippocampal mossy fiber synapses. *Proc. Natl. Acad. Sci. USA* **104,** 12181–12186.

Represa, A., Tremblay, E., and Ben-Ari, Y. (1987). Kainate binding sites in the hippocampal mossy fibers: Localization and plasticity. *Neuroscience* **20,** 739–748.

Robinson, J. H., and Deadwyler, S. A. (1981). Kainic acid produces depolarization of CA3 pyramidal cells in the vitro hippocampal slice. *Brain Res.* **221,** 117–127.

Rodriguez-Moreno, A., Herreras, O., and Lerma, J. (1997). Kainate receptors presynaptically downregulate GABAergic inhibition in the rat hippocampus. *Neuron* **19,** 893–901.

Rogawski, M. A. (2006). Molecular targets versus models for new antiepileptic drug discovery. *Epilepsy Res.* **68,** 22–28.

Sapolsky, R. M. (1996). Stress, glucocorticoids, and damage to the nervous system: The current state of confusion. *Stress* **1,** 1–19.

Schiffer, H. H., and Heinemann, S. F. (2007a). Association of the human kainate receptor GluR7 gene (GRIK3) with recurrent major depressive disorder. *Am. J. Med. Genet. B Neuropsychiatr. Genet.* **144B,** 20–26.

Schiffer, H. H., and Heinemann, S. F. (2007b). Association of the human kainate receptor GluR7 gene (GRIK3) with recurrent major depressive disorder. *Am. J. Med. Genet. B Neuropsychiatr. Genet.* **144,** 20–26.

Schiffer, H. H., Swanson, G. T., and Heinemann, S. F. (1997). Rat GluR7 and a carboxy-terminal splice variant, GluR7b, are functional kainate receptor subunits with a low sensitivity to glutamate. *Neuron* **19,** 1141–1146.

Schmitz, D., Frerking, M., and Nicoll, R. A. (2000). Synaptic activation of presynaptic kainate receptors on hippocampal mossy fiber synapses. *Neuron* **27,** 327–338.

Schmitz, D., Mellor, J., Frerking, M., and Nicoll, R. A. (2001a). Presynaptic kainate receptors at hippocampal mossy fiber synapses. *Proc. Natl. Acad. Sci. USA* **98,** 11003–11008.

Schmitz, D., Mellor, J., and Nicoll, R. A. (2001b). Presynaptic kainate receptor mediation of frequency facilitation at hippocampal mossy fiber synapses. *Science* **291,** 1972–1976.

Shaltiel, G., Maeng, S., Malkesman, O., Pearson, B., Schloesser, R. J., Tragon, T., Rogawski, M., Gasior, M., Luckenbaugh, D., Chen, G., and Manji, H. K. (2008). Evidence for the involvement of the kainate receptor subunit GluR6 (GRIK2) in mediating behavioral displays related to behavioral symptoms of mania. *Mol. Psychiatry* **13,** 858–872.

Shuang, M., Liu, J., Jia, M. X., Yang, J. Z., Wu, S. P., Gong, X. H., Ling, Y. S., Ruan, Y., Yang, X. L., and Zhang, D. (2004). Family-based association study between autism and glutamate receptor 6 gene in Chinese Han trios. *Am. J. Med. Genet. B Neuropsychiatr. Genet.* **131B,** 48–50.

Siegel, S. J., Janssen, W. G., Tullai, J. W., Rogers, S. W., Moran, T., Heinemann, S. F., and Morrison, J. H. (1995). Distribution of the excitatory amino acid receptor subunits GluR2(4) in monkey hippocampus and colocalization with subunits GluR5–7 and NMDAR1. *J. Neurosci.* **15,** 2707–2719.

Smolders, I., Bortolotto, Z. A., Clarke, V. R., Warre, R., Khan, G. M., O'Neill, M. J., Ornstein, P. L., Bleakman, D., Ogden, A., Weiss, B., Stables, J. P., Ho, K. H., *et al.* (2002). Antagonists of GLU(K5)-containing kainate receptors prevent pilocarpine-induced limbic seizures. *Nat. Neurosci.* **5,** 796–804.

Sommer, B., Burnashev, N., Verdoorn, T. A., Keinanen, K., Sakmann, B., and Seeburg, P. H. (1992). A glutamate receptor channel with high affinity for domoate and kainate. *EMBO J.* **11,** 1651–1656.

Strutz-Seebohm, N., Seebohm, G., Shumilina, E., Mack, A. F., Wagner, H. J., Lampert, A., Grahammer, F., Henke, G., Just, L., Skutella, T., Hollmann, M., and Lang, F. (2005). Glucocorticoid adrenal steroids and glucocorticoid-inducible kinase isoforms in the regulation of GluR6 expression. *J. Physiol.* **565,** 391–401.

Swanson, G. T., Green, T., Sakai, R., Contractor, A., Che, W., Kamiya, H., and Heinemann, S. F. (2002). Differential activation of individual subunits in heteromeric kainate receptors. *Neuron* **34,** 589–598.

Telfeian, A. E., Federoff, H. J., Leone, P., During, M. J., and Williamson, A. (2000). Overexpression of GluR6 in rat hippocampus produces seizures and spontaneous non-synaptic bursting in vitro. *Neurobiol. Dis.* **7,** 362–374.

Vignes, M., Clarke, V. R., Parry, M. J., Bleakman, D., Lodge, D., Ornstein, P. L., and Collingridge, G. L. (1998). The GluR5 subtype of kainate receptor regulates excitatory synaptic transmission in areas CA1 and CA3 of the rat hippocampus. *Neuropharmacology* **37,** 1269–1277.

Vincent, P., and Mulle, C. (2009). Kainate receptors in epilepsy and excitotoxicity. *Neuroscience* **158,** 309–323.

Watanabe, M., Fukaya, M., Sakimura, K., Manabe, T., Mishina, M., and Inoue, Y. (1998). Selective scarcity of NMDA receptor channel subunits in the stratum lucidum (mossy fibre-recipient layer) of the mouse hippocampal CA3 subfield. *Eur. J. Neurosci.* **10,** 478–487.

Watanabe, Y., Weiland, N. G., and McEwen, B. S. (1995). Effects of adrenal steroid manipulations and repeated restraint stress on dynorphin mRNA levels and excitatory amino acid receptor binding in hippocampus. *Brain Res.* **680,** 217–225.

Werner, P., Voigt, M., Keinanen, K., Wisden, W., and Seeburg, P. H. (1991). Cloning of a putative high-affinity kainate receptor expressed predominantly in hippocampal CA3 cells. *Nature* **351,** 742–744.

Woo, T. U., Shrestha, K., Amstrong, C., Minns, M. M., Walsh, J. P., and Benes, F. M. (2007). Differential alterations of kainate receptor subunits in inhibitory interneurons in the anterior cingulate cortex in schizophrenia and bipolar disorder. *Schizophr. Res.* **96,** 46–61.

CHAPTER TEN

# Role of Neurotrophic Factors in Behavioral Processes: Implications for the Treatment of Psychiatric and Neurodegenerative Disorders

Marie-Christine Pardon

## Contents



## Abstract

Neurotrophins are important regulators of neuronal function in the developing and adult brain and thus play a critical role in sustaining normal behavioral function. Brain-derived neurotrophic factor (BDNF) has been the most widely studied neurotrophin because of its important role as modulator of synaptic plasticity, which is essential to the regulation of experience-dependent behavior. Extensive work implicates BDNF in hippocampus-dependent forms of learning and memory, although it also regulates other cognitive processes. A role for BDNF in anxiety-related disorders and aggressive behavior can also be suspected. More importantly, BDNF signaling has recently emerged as a key player in the development of drug addiction and is well known to be involved in adaptation to stress and stress-related disorders. NGF in the other hand is thought to be involved in aggression and alcohol dependence. Finally, BDNF

School of Biomedical Sciences, Institute of Neuroscience, Queen's Medical Centre, University of Nottingham Medical School, Nottingham, United Kingdom

*Vitamins and Hormones*, Volume 82
ISSN 0083-6729, DOI: 10.1016/S0083-6729(10)82010-2

**appears to participate in the therapeutic effects of drugs and interventions capable of reversing or attenuating behavioral disturbances relevant to psychiatric and neurodegenerative disorders. Compounds mimicking BDNF signaling, however, are unlikely to be used in a clinical context, given their adverse side effects and pharmacokinetic limitations.** 

## I. Introduction

Neurotrophic factors are small proteins known to control many aspects of development and adult neuronal function in both the peripheral and central nervous system (for review see Reichardt, 2006). In the developing brain, they play a major role in controlling the number of surviving neurons to ensure an optimal density of neurons in a given target, and in cell differentiation. In addition to these survival effects, neurotrophic factors play a significant role in the maintenance of adult neuronal function by regulating the growth of neurons, synaptic function, and plasticity as well as associated metabolic functions such as protein synthesis and neurotransmission. Thus, neurotrophic factors play a key role in sustaining normal behavioral processes and are therapeutic targets for a number of brain disorders, both psychiatric and neurodegenerative, as discussed below.

There are two families of neurotrophic factors: the neurotrophins and the glial cell line-derived neurotrophic factor (GDNF). The biology of neurotrophic factors has been the topic of a number of reviews (for review see Huang and Reichardt, 2003; Reichardt, 2006; Skaper, 2008) and is only summarized below. In mammals, the four members of the neurotrophin family consist of nerve growth factor (NGF), brain-derived neurotrophic factor (BDNF), neurotrophin-3 (NT-3), and neurotrophin-4/5 (NT 4/5). They have similar structure and biochemical characteristics and bind to two types of receptors: tyrosine kinase receptors TrkA, TrkB, or TrkC, or a common low-affinity neurotrophin receptor p75 (p75NTR) that has no tyrosine kinase domain. NGF binds specifically to TrkA, BDNF and NT-4/5 to TrkB, and NT-3 to TrkC with high efficiency and to the other Trk receptors with less efficiency. In addition to these specific receptors, all neurotrophins bind to p75NTR. Four members of the GDNF family have been identified so far: GDNF, neurturin, artemin, and persephin, also acting via activation of receptor tyrosine kinases. Neurotrophins are most widely studied for their involvement in behavioral function and brain disorders. They are therefore the focus of this chapter.

By maintaining neuronal function, neurotrophic factors have widespread actions on the brain, and thus mediate or modulate a wide range of behavioral processes. BDNF has been the most widely studied neurotrophin because of its important role as modulator of synaptic plasticity, which is essential to the regulation of experience-dependent behavior (Cohen and Greenberg, 2008). The cognition-enhancing properties of BDNF are its best characterized behavioral effects. In recent years, BDNF was shown to be involved in aggressive behavior, although the involvement of NGF has been prior established. There is also evidence to suggest a role for neurotrophins in anxiety-related behavior, reward, and adaptation to stress. As a consequence, neurotrophins have become targets of interest for the treatment of brain disorders, although in practice, the development of synthetic neurotrophins for clinical use suffers several limits, as discussed below.

## II. Cognition-Enhancing Effects of Neurotrophins

### A. BDNF

Preliminary evidence supporting a role for BDNF in learning and memory processes focused on hippocampus-dependent tasks and showed that memory acquisition and consolidation are associated with an increase in BDNF mRNA expression and the activation of its receptor TrkB in this area (Yamada and Nabeshima, 2003). Studies using genetic and pharmacological alterations of brain BDNF levels further indicate that performance in multiple cognitive domains can be improved by raising the availability of the neurotrophin and impaired by its depletion, although the effects are both region- and dose-dependent. Indeed, in heterozygous BDNF knockout mice, endogenous BDNF levels are reduced by about 50% in the brain, leading to cognitive deficits in contextual, but not cue, fear memory (Liu *et al.*, 2004) and spatial learning impairments in the Morris water maze task (Linnarsson *et al.*, 1997) but unaltered emotional behavior (Chourbaji *et al.*, 2004). Altogether, this suggests a preferential effect on hippocampus-dependent cognitive function. Forebrain-restricted BDNF mutant mice exhibit a similar behavioral profile, with the absence of BDNF localized to the dorsal cortex, hippocampus, and parts of the ventral cortex and amygdala (Gorski *et al.*, 2003). By contrast, heterozygous transgenic mice overexpressing BDNF showed that double levels of the neurotrophin in the brain result in improved spatial learning and memory performance (Nakajo *et al.*, 2008).

Consistent with genetic manipulation studies, pharmacological elevation of hippocampal BDNF levels, via local injection of recombinant BDNF in the hippocampus, was also shown to facilitate spatial learning in the Morris water maze (Pietropaolo *et al.*, 2007) and contextual fear memory formation (Alonso *et al.*, 2002). Consolidation of contextual fear, however, was improved when the neurotrophin was injected in the parietal cortex (Alonso *et al.*, 2005). This indicates that BDNF can also modulate nonhippocampus-dependent cognitive processes. Conversely, hippocampus-specific deletion of the BDNF gene or local injection of anti-BDNF antibodies, to specifically block hippocampal-dependent functions, further confirmed its critical role in learning and memory processes. In adult mice, targeted deletion of the BDNF gene in the hippocampus selectively impaired declarative memory in the novel object recognition test, spatial learning, and extinction of contextual fear, but acquisition and retention of aversive memory, locomotion, and anxiety-related behaviors were not altered (Heldt *et al.*, 2007). In addition, hippocampal BDNF levels increased during acquisition of declarative memory in the object recognition task, whereas posttraining intrahippocampal infusion of anti-BDNF antibody impairs the long-term retention of the task (Furini *et al.*, 2009), suggesting that the rise in neurotrophin levels during learning contributes to subsequent consolidation of object recognition memory.

Thus, the work reported above suggests that BDNF is a potential target for the future treatment of cognitive disorders. In support of this, recent evidence indicates that the cognition-enhancing effects of exercise after traumatic brain injury, or of neural stem cells implantation in a mouse model of Alzheimer's disease, are dependent upon BDNF activation, as they are antagonized by treatment with a BDNF inhibitor (Blurton-Jones *et al.*, 2009; Griesbach *et al.*, 2009). Indirect evidence comes from a range of studies showing that the cognitive benefits of a number of treatments, such as caffeine (Costa *et al.*, 2008), allosteric modulation of the glutamatergic AMPA receptor (Woolley *et al.*, 2009), or voluntary exercise (Khabour *et al.*, 2009), are associated with an increase in BDNF mRNA or protein levels in the hippocampus. Other brain areas are often not investigated. There is, however, some indication that excess of BDNF will have opposite or no effects on cognitive performance, thus limiting its potential therapeutic use. Indeed, when using recombinant adeno-virus-associated viral vectors to induced localized BDNF over expression, Pietropaolo *et al.* (2007) observed bidirectional effects on cognitive performance as a function of the levels of hippocampal BDNF levels: when those are moderately increased, spatial memory is improved, but high levels of the neurotrophin disrupt cognitive performance. Thus, a major challenge for the future development of pharmacological treatments is to develop methods to increase BDNF levels in effective doses.

As mentioned before, the cognitive benefits of BDNF may not be limited to hippocampus-dependent behaviors, but this has been little studied. There is evidence to suggest that olfactory discrimination learning and memory in rats are, at least in part, mediated by an increased synthesis of BDNF and NT-3 in the frontal cortex (Naimark *et al.*, 2007). In addition, the formation of social recognition memory was found associated with increased BDNF mRNA levels in the inferior part of the temporal cortex, subfield CA1 of the hippocampus, the diagonal band, basolateral amygdala, and the anterior cingulate, medial frontal, entorhinal, and pyriform cortices (Broad *et al.*, 2002). Furthermore, infusion of BDNF in the rat insular cortex was found to enhance retention of conditioned taste aversion (Castillo *et al.*, 2006) and to reverse the memory deficits induced in this task by inhibition of protein synthesis (Moguel-Gonzalez *et al.*, 2008). This further confirms that BDNF may have widespread cognition-enhancing properties.

### B. Other neurotrophins

NGF is another potential cognition enhancer. Together with BDNF, NGF plays important roles in the maintenance of cholinergic neurons of the basal forebrain, which are critically involved in cognitive processes (Mufson *et al.*, 1999). Studies in young naïve rodents did not reveal any improvement in spatial learning and memory performance following NGF treatment, although positive effects were usually observed in cognitively impaired, aged, lesioned, or ethanol-treated rats (Frick *et al.*, 1997; Frielingsdorf *et al.*, 2006; Lukoyanov *et al.*, 2003; Markowska *et al.*, 1994, 1996; Pizzo and Thal, 2004). Intrahippocampal injections of function-blocking NGF antibodies, however, were found to impair spatial memory (Conner *et al.*, 2009), suggesting an involvement of this neurotrophin in memory formation.

NT-3 and NT-4/5 have hardly been studied in cognition. A role for the former has been found in olfactory discrimination learning and memory (Naimark *et al.*, 2007) while the latter appears essential for hippocampus- and amygdala-dependent long-term contextual fear memory (Xie *et al.*, 2000).

## III. Neurotrophins Mediate Aggressive and Defensive Behavior

NGF has long been associated with aggressive behavior, and more recently, a role for BDNF has been suspected, but it remains inconclusive so far. Heterozygous BDNF knockout or conditional knockout mice showing either prenatal or postnatal depletion of the neurotrophin in the brain all display enhanced aggressiveness (Chan *et al.*, 2006; Lyons *et al.*, 1999), although a recent study failed to replicate differences in aggressive behavior in heterozygous BDNF knockout mice (Ibarguen-Vargas *et al.*, 2009). By contrast, the comparison between two inbred mouse strains contrasted in this behavioral trait revealed elevated BDNF protein levels in the hippocampus, cortex, and striatum of the highly aggressive strain together with elevated hippocampal NGF levels, suggesting that the neurotrophins in these areas are critical to the expression of aggression (Lang *et al.*, 2009). A history of repeated winning experiences, however, was not found associated with changes in BDNF mRNA levels in the ventral tegmental area, measured 24 h after the last aggressive interaction (Bondar *et al.*, 2009). The fact that these levels were only measured in one brain area is not sufficient to rule out the involvement of BDNF levels in aggression, but experiments looking at regional BDNF levels after acute aggressive episodes are currently lacking. Our work also provides indirect evidence for a role of BDNF in aggressive behavior. We have devised a social threat procedure whereby experimental mice are confronted with repeated sensory contact with aggressive mice in an unfamiliar cage bisected longitudinally by a perforated steel partition, allowing the animals to see, hear, and smell each other but preventing them from direct physical contact (Pardon *et al.*, 2004). We have shown that the repeated exposure to such a procedure was associated with increased hippocampal BDNF protein levels in NMRI mice, although these were unaltered by an acute social threat session (Pardon *et al.*, 2005). NMRI mice are prone to aggression and we subsequently showed that the rise in hippocampal levels induced by repeated indirect confrontations was associated with increased aggressiveness (Pardon *et al.* unpublished data), providing further indirect evidence for a role of BDNF in aggression. In addition, BDNF mRNA or protein levels in several brain areas were found reduced following acute (Pizarro *et al.*, 2004) and repeated social defeat (Haenisch *et al.*, 2009), suggesting that submissive behavior may be in part caused by a reduction in neurotrophin levels. However, in aged mice, the condition of subordination enhances the level of NGF in the subventricular zone and hippocampus, whereas dominance elevates BDNF (Fiore *et al.*, 2003, 2005). Thus, there seem to be a link between impaired BDNF signaling and aggression, but it needs to be further characterized.

NGF appears as the major neurotrophin regulating aggressive behavior. A number of studies in mice have shown that fighting induces NGF release from the salivary glands into the bloodstream, the levels of which correlate with the number of fights and the resulting social status (for review see Alleva and Francia, 2009). In the brain, NGF mRNA and protein levels are induced in the hypothalamus after successful aggressive experience (Aloe *et al.*, 1990; Spillantini *et al.*, 1989), although in aggressive mice basal hippocampal NGF levels increase only in an age-dependent manner (Lang *et al.*, 2009). Unlike males, females usually do not display aggressive behavior outside the lactating period, and circulating NGF levels were also found related to maternal defense behavior (for review see Alleva and Francia, 2009).

## IV. Involvement of Neurotrophins in Anxiety-Like Behavior

The work performed genetically on animals suggests that BDNF does not directly influence anxiety-related behaviors, since the performance in a number of conflict tests is unaltered by the lack or the excess of BDNF proteins (Chourbaji *et al.*, 2004; Gorski *et al.*, 2003; Koizumi *et al.*, 2006). BDNF is thought to contribute indirectly to the expression of anxiety *via* regulation of the activity of neuronal networks involved in this emotional state (Martinowich *et al.*, 2007). Infusion of BDNF directly into the hippocampus, however, was found to have an anxiogenic-like profile in the elevated-plus maze and open-field tests (Deltheil *et al.*, 2009), although anxiolytic-like effects were also observed in another rodent study (Cirulli *et al.*, 2004). Furthermore, deletion of TrkB in newborn neuron results in enhanced anxiety-like behavior as seen in the open-field and elevated-plus maze tests in 4- to 6-week-old mutant mice (Bergami *et al.*, 2009), consistent with the attenuated anxiety-like behavior seen in mice overexpressing this receptor (Koponen *et al.*, 2004). Overall, this work is rather in favor of anxiolytic-like properties of BDNF. Whether or not basal levels of neurotrophins determine susceptibility to anxiety-related behavior is still a matter of debate (Cirulli and Alleva, 2009), but in inbred C57BL6 mice, the magnitude of anxiogenic-like reactions in the elevated-plus maze was found positively related to dorsal hippocampal BDNF protein levels, and negatively related to NGF protein levels in dorsal hippocampus and in the amygdala (Yee *et al.*, 2007). In human, NGF serum concentrations rise after successful cognitive-behavioral therapy of generalized anxiety disorder (Jockers-Scherubl *et al.*, 2007b), suggesting that peripheral levels of neurotrophin could serve as a biomarker for this affective disorder. Further work is needed in this area to better understand the role played by neurotrophins in emotional responses and anxiety-related disorders.

## V. Role of Neurotrophins in Rewarding and Addictive Behavior

BDNF is expressed in the dopaminergic mesocorticolimbic pathway originating in the ventral tegmental area and nucleus accumbens, known as the reward pathway. Recent findings implicate the neurotrophin as a key mediator of reward processes and drug addiction. For example, psychoactive drugs such as cocaine, amphetamine, or ethanol, either injected or self-administered, increase BDNF mRNA and protein levels in the paraventricular hypothalamus and in several mesocorticolimbic areas to include the prefrontal cortex, the basolateral amygdala, the piriform and cingulate cortices, and the striatum (Fumagalli *et al.*, 2007; Grimm *et al.*, 2003; Kerns *et al.*, 2005; Le Foll *et al.*, 2005; Logrip *et al.*, 2009; Meredith *et al.*, 2002). In contrast, reduced cortical BDNF levels were seen following long-term ethanol self-administration (Logrip *et al.*, 2009). Furthermore, infusion or overexpression of BDNF in the ventral tegmental area increases intake (Graham *et al.*, 2007) and response to psychostimulant drugs, as reflected by enhanced locomotor response and conditioned place preference (Bahi *et al.*, 2008).

Increased BDNF levels are also thought to facilitate the transition from the acute reward response to a state of dependence characteristic of addiction. Indeed, infusions or overexpression of BDNF within the reward pathway sensitizes the dopaminergic response to drugs (Vargas-Perez *et al.*, 2009) and facilitates drug-seeking behavior (Lu *et al.*, 2004) and relapse (Graham *et al.*, 2007), whereas cocaine seeking can be suppressed by BDNF when infused in the medial prefrontal cortex (Berglind *et al.*, 2007). This indicates that the development of drug addiction is in part mediated by alterations in BDNF signaling. Consistent with this statement, human findings have reported decreased serum BDNF levels in heroin-dependent patients whereas NGF levels were decreased in both heroin- and cocaine-dependent users (Angelucci *et al.*, 2007). The implication of NGF in addiction, however, has been better characterized in alcoholism. Plasma NGF levels are increased in alcohol-dependent patients and decreased during withdrawal (Heberlein *et al.*, 2008; Jockers-Scherubl *et al.*, 2007a). Another study suggested that plasma NGF levels may represent a trait marker for the development of alcohol dependence: these were found decreased in dependent patients but increased in those patients with a positive history of alcoholism (Yoon *et al.*, 2006). Finally, there are only two studies so far that has looked at serum neurotrophin levels in cannabis-dependent users free from any associated psychiatric conditions. Reduced NGF levels but no changes in BDNF levels were observed (Alleva *et al.*, 1996; Angelucci *et al.*, 2008).

NT-3 is also thought to contribute to the rewarding properties of drugs of abuse. Its effects are yet to be fully characterized but NT-3 appears to mediate processes other than BDNF. Indeed, pharmacological blockade studies suggest that NT-3 prevents the occurrence of a behavioral response to cocaine in the nucleus accumbens and inhibits the development of behavioral sensitization to repeated cocaine injections in the ventral tegmental area (Freeman and Pierce, 2002) although earlier work from the same group indicated, on the contrary, that infusion of NT-3 directly into the ventral tegmental area contributes to the sensitization to cocaine (Pierce *et al.*, 1999). Thus, the involvement of neurotrophins in drug addiction is a recent but promising area of research. To date, they appear to facilitate the development of drug addiction but in a neurotrophin-dependent and psychostimulant-dependent manner.

## VI. Neurotrophins Facilitate Adaptation to Stress

Recent evidence shows that neurotrophins are induced in response to stress and may promote successful adaptation to environmental challenges. Indeed, we found enhanced hippocampal BDNF protein levels after a 1-h exposure to a novel environment (Pardon *et al.*, 2005), a valid inducer of mild stress (Pardon *et al.*, 2004). In addition, immobilization stress, a more severe procedure, rapidly induces BDNF mRNA and protein levels in the hypothalamus and hippocampus of rats, but these effects seen up to 1 h after the onset of stress turns into decreased BDNF mRNA levels when the duration of the adverse experience increases (Marmigere *et al.*, 2003; Rage *et al.*, 2002). This suggests that BDNF induction is part of a normal stress response, possibly as a mechanism designed to prevent harmful outcomes. In support of this hypothesis, in the learned helplessness paradigm, heterozygous BDNF knockout mice were found slower to escape footshocks after training than wild-type mice (MacQueen *et al.*, 2001), indicating that BDNF facilitates the occurrence of an adequate coping response. Furthermore, deficient BDNF levels in the brain are usually associated with failure to adapt to stress, and thus, seen after chronic inescapable stress procedures, although this is not a consistent observation. Indeed, repeated restraint stress decreases hippocampal BDNF mRNA levels (Murakami *et al.*, 2005). And, using the chronic mild stress model, where animals are intermittently exposed to a variety of mild stressors, some authors have reported reduced mRNA and protein levels in the hippocampus (Gronli *et al.*, 2006; Li *et al.*, 2007a), while others found that the procedure effectively induced behavioral disturbances without altering neurotrophin levels (Allaman *et al.*, 2008; Lucca *et al.*, 2008). By contrast, positive effects were also found in another study where exposure to multiple stressors enhanced BDNF mRNA and protein levels in the hippocampus, and interestingly this was associated with increased learning and memory performance (Li *et al.*, 2007b). Furthermore, we did not find reduced hippocampal BDNF levels following repeated novelty stress (Pardon *et al.*, 2005), although the procedure induces associative learning deficits (Pardon *et al.*, 2004). Thus, these data suggest that the behavioral outcomes of chronic stress procedure are not caused by the reduction of BDNF levels, despite the fact that antidepressant treatments effectively reverse both (Schmidt and Duman, 2007).

Given the involvement of BDNF in the stress response, another possibility is that it mediates vulnerability to stress, as suggested by some of the work below. Reduction in BDNF levels in knockout mice did not exacerbate adverse behavioral outcomes of chronic mild stress, although it attenuated the ability of antidepressants to reverse the behavioral outcomes (Ibarguen-Vargas *et al.*, 2009). This suggests that variations in BDNF levels may help predict responsiveness to antidepressants. Work using the social defeat model has linked BDNF signaling within the reward pathway with the occurrence of adverse effects. Mice experiencing adverse outcomes of the procedure showed increased BDNF levels in the ventral tegmental area, whereas no changes were observed in resistant mice (Krishnan *et al.*, 2007). In addition, deletion of BDNF in the ventral tegmental area promotes resistance to social defeat (Berton *et al.*, 2006). More recently, another group found that upregulation of BDNF in the hippocampus was associated with resilience to chronic mild stress (Bergstrom *et al.*, 2008).

## VII. Neurotrophins: A Therapeutic Target for the Treatment of Brain Disorders?

In summary, neurotrophins, and in particular BDNF, have widespread effects on behavioral function, suggesting that they could be targeted for the treatment of brain disorders. In favor of this, regulation of neurotrophin signaling appears to be a mechanism mediating the therapeutic-like effects of exercise on behavioral disturbances relevant to stress-related and neurodegenerative disorders (Ang and Gomez-Pinilla, 2007; Strohle, 2009), and the efficacy of antidepressant treatments (Schmidt and Duman, 2007). Nevertheless, there are a number of barriers to the development of compounds that mimic neurotrophin signaling. First, the behavioral effects of neurotrophins can be both positive and negative depending on the dose and the brain area, as indicated above. Thus, in theory, systemic administration of BDNF could enhance cognitive performance or reverse cognitive deficits by increasing hippocampal levels, together with creating vulnerability to stress or to addiction by increasing levels in the ventral tegmental area. Other limits come from the fact that neurotrophins also act outside the brain, on nonneuronal cells, and are involved in the proliferation of cancer cells: increasing their levels could thus lead to severe side effects, such as abnormal nerve sprouting, vulnerability to cardiovascular diseases and cancer, as well as altered immune response (for review see Price *et al.*, 2007). Other limits to the development of synthetic neurotrophins for clinical trials are their relatively large size, which requires administration via injections, and short half-lives, requiring several doses per day (for review see Price *et al.*, 2007). Therefore, a more appropriate strategy relies on the development of molecules potentiating the effect of endogenous neurotrophins, as this may avoid the potential complications of systemic neurotrophin stimulation (Webster and Pirrung, 2008).

## *REFERENCES*

Allaman, I., Papp, M., Kraftsik, R., Fiumelli, H., Magistretti, P. J., and Martin, J. L. (2008). Expression of brain-derived neurotrophic factor is not modulated by chronic mild stress in the rat hippocampus and amygdala. *Pharmacol. Rep.* **60,** 1001–1007.

Alleva, E., and Francia, N. (2009). Psychiatric vulnerability: Suggestions from animal models and role of neurotrophins. *Neurosci. Biobehav. Rev.* **33,** 525–536.

Alleva, L., Aloe, F., Cirulli, D., Seta, Della, and Tirassa, P. (1996). Serum NGF levels increase during lactation and following maternal aggression in mice. *Physiol. Behav.* **59,** 461–466.

Aloe, L., Alleva, E., and De Simone, R. (1990). Changes of NGF level in mouse hypothalamus following intermale aggressive behaviour: Biological and immunohistochemical evidence. *Behav. Brain Res.* **39,** 53–61.

Alonso, M., Bekinschtein, P., Cammarota, M., Vianna, M. R., Izquierdo, I., and Medina, J. H. (2005). Endogenous BDNF is required for long-term memory formation in the rat parietal cortex. *Learn Mem.* **12,** 504–510.

Ang, E. T., and Gomez-Pinilla, F. (2007). Potential therapeutic effects of exercise to the brain. *Curr. Med. Chem.* **14,** 2564–2571.

Angelucci, F., Ricci, V., Pomponi, M., Conte, G., Mathe, A. A., Attilio Tonali, P., and Bria, P. (2007). Chronic heroin and cocaine abuse is associated with decreased serum concentrations of the nerve growth factor and brain-derived neurotrophic factor. *J. Psychopharmacol.* **21,** 820–825.

Angelucci, F., Ricci, V., Spalletta, G., Pomponi, M., Tonioni, F., Caltagirone, C., and Bria, P. (2008). Reduced serum concentrations of nerve growth factor, but not brain-derived neurotrophic factor, in chronic cannabis abusers. *Eur. Neuropsychopharmacol.* **18,** 882–887.

Bahi, A., Boyer, F., Chandrasekar, V., and Dreyer, J. L. (2008). Role of accumbens BDNF and TrkB in cocaine-induced psychomotor sensitization, conditioned-place preference, and reinstatement in rats. *Psychopharmacology (Berl.)* **199,** 169–182.

Bergami, M., Berninger, B., and Canossa, M. (2009). Conditional deletion of TrkB alters adult hippocampal neurogenesis and anxiety-related behavior. *Commun. Integr. Biol.* **2,** 14–16.

Berglind, W. J., See, R. E., Fuchs, R. A., Ghee, S. M., Whitfield, T. W. Jr., Miller, S. W., and McGinty, J. F. (2007). A BDNF infusion into the medial prefrontal cortex suppresses cocaine seeking in rats. *Eur. J. Neurosci.* **26,** 757–766.

Bergstrom, A., Jayatissa, M. N., Mork, A., and Wiborg, O. (2008). Stress sensitivity and resilience in the chronic mild stress rat model of depression; an in situ hybridization study. *Brain Res.* **1196,** 41–52.

Berton, O., McClung, C. A., Dileone, R. J., Krishnan, V., Renthal, W., Russo, S. J., Graham, D., Tsankova, N. M., Bolanos, C. A., Rios, M., Monteggia, L. M., Self, D. W., *et al.* (2006). Essential role of BDNF in the mesolimbic dopamine pathway in social defeat stress. *Science* **311,** 864–868.

Blurton-Jones, M., Kitazawa, M., Martinez-Coria, H., Castello, N. A., Muller, F. J., Loring, J. F., Yamasaki, T. R., Poon, W. W., Green, K. N., and LaFerla, F. M. (2009). Neural stem cells improve cognition via BDNF in a transgenic model of Alzheimer disease. *Proc. Natl. Acad. Sci. USA* **106,** 13594–13599.

Bondar, N. P., Boyarskikh, U. A., Kovalenko, I. L., Filipenko, M. L., and Kudryavtseva, N. N. (2009). Molecular implications of repeated aggression: Th, Dat1 Snca and Bdnf gene expression in the VTA of victorious male mice. *PLoS ONE* **4,** e4190.

Broad, K. D., Mimmack, M. L., Keverne, E. B., and Kendrick, K. M. (2002). Increased BDNF and trk-B mRNA expression in cortical and limbic regions following formation of a social recognition memory. *Eur. J. Neurosci.* **16,** 2166–2174.

Castillo, D. V., Figueroa-Guzman, Y., and Escobar, M. L. (2006). Brain-derived neurotrophic factor enhances conditioned taste aversion retention. *Brain Res.* **1067,** 250–255.

Chan, J. P., Unger, T. J., Byrnes, J., and Rios, M. (2006). Examination of behavioral deficits triggered by targeting Bdnf in fetal or postnatal brains of mice. *Neuroscience* **142,** 49–58.

Chourbaji, S., Hellweg, R., Brandis, D., Zorner, B., Zacher, C., Lang, U. E., Henn, F. A., Hortnagl, H., and Gass, P. (2004). Mice with reduced brain-derived neurotrophic factor expression show decreased choline acetyltransferase activity, but regular brain monoamine levels and unaltered emotional behavior. *Brain Res. Mol. Brain Res.* **121,** 28–36.

Cirulli, F., and Alleva, E. (2009). The NGF saga: From animal models of psychosocial stress to stress-related psychopathology. *Front. Neuroendocrinol.* **30,** 379–395.

Cirulli, F., Berry, A., Chiarotti, F., and Alleva, E. (2004). Intrahippocampal administration of BDNF in adult rats affects short-term behavioral plasticity in the Morris water maze and performance in the elevated plus-maze. *Hippocampus* **14,** 802–807.

Cohen, S., and Greenberg, M. E. (2008). Communication between the synapse and the nucleus in neuronal development, plasticity, and disease. *Annu. Rev. Cell Dev. Biol.* **24,** 183–209.

Conner, J. M., Franks, K. M., Titterness, A. K., Russell, K., Merrill, D. A., Christie, B. R., Sejnowski, T. J., and Tuszynski, M. H. (2009). NGF is essential for hippocampal plasticity and learning. *J. Neurosci.* **29,** 10883–10889.

Costa, M. S., Botton, P. H., Mioranzza, S., Ardais, A. P., Moreira, J. D., Souza, D. O., and Porciuncula, L. O. (2008). Caffeine improves adult mice performance in the object recognition task and increases BDNF and TrkB independent on phospho-CREB immunocontent in the hippocampus. *Neurochem. Int.* **53,** 89–94.

Deltheil, T., Tanaka, K., Reperant, C., Hen, R., David, D. J., and Gardier, A. M. (2009). Synergistic neurochemical and behavioural effects of acute intrahippocampal injection of brain-derived neurotrophic factor and antidepressants in adult mice. *Int. J. Neuropsychopharmacol.* 1–11.

Fiore, M., Amendola, T., Triaca, V., Tirassa, P., Alleva, E., and Aloe, L. (2003). Agonistic encounters in aged male mouse potentiate the expression of endogenous brain NGF and BDNF: Possible implication for brain progenitor cells' activation. *Eur. J. Neurosci.* **17,** 1455–1464.

Fiore, M., Amendola, T., Triaca, V., Alleva, E., and Aloe, L. (2005). Fighting in the aged male mouse increases the expression of TrkA and TrkB in the subventricular zone and in the hippocampus. *Behav. Brain Res.* **157,** 351–362.

Freeman, A. Y., and Pierce, R. C. (2002). Neutralization of neutrophin-3 in the ventral tegmental area or nucleus accumbens differentially modulates cocaine-induced behavioral plasticity in rats. *Synapse* **46,** 57–65.

Frick, K. M., Price, D. L., Koliatsos, V. E., and Markowska, A. L. (1997). The effects of nerve growth factor on spatial recent memory in aged rats persist after discontinuation of treatment. *J. Neurosci.* **17,** 2543–2550.

Frielingsdorf, H., Thal, L. J., and Pizzo, D. P. (2006). The septohippocampal cholinergic system and spatial working memory in the Morris water maze. *Behav. Brain Res.* **168,** 37–46.

Fumagalli, F., Di Pasquale, L., Caffino, L., Racagni, G., and Riva, M. A. (2007). Repeated exposure to cocaine differently modulates BDNF mRNA and protein levels in rat striatum and prefrontal cortex. *Eur. J. Neurosci.* **26,** 2756–2763.

Furini, C. R., Rossato, J. I., Bitencourt, L. L., Medina, J. H., Izquierdo, I., and Cammarota, M. (2009). beta-Adrenergic receptors link NO/sGC/PKG signaling to BDNF expression during the consolidation of object recognition long-term memory. *Hippocampus.*

Gorski, J. A., Balogh, S. A., Wehner, J. M., and Jones, K. R. (2003). Learning deficits in forebrain-restricted brain-derived neurotrophic factor mutant mice. *Neuroscience* **121,** 341–354.

Graham, D. L., Edwards, S., Bachtell, R. K., DiLeone, R. J., Rios, M., and Self, D. W. (2007). Dynamic BDNF activity in nucleus accumbens with cocaine use increases self-administration and relapse. *Nat. Neurosci.* **10,** 1029–1037.

Griesbach, G. S., Hovda, D. A., and Gomez-Pinilla, F. (2009). Exercise-induced improvement in cognitive performance after traumatic brain injury in rats is dependent on BDNF activation. *Brain Res.* **1288,** 105–115.

Grimm, J. W., Lu, L., Hayashi, T., Hope, B. T., Su, T. P., and Shaham, Y. (2003). Time-dependent increases in brain-derived neurotrophic factor protein levels within the mesolimbic dopamine system after withdrawal from cocaine: Implications for incubation of cocaine craving. *J. Neurosci.* **23,** 742–747.

Gronli, J., Bramham, C., Murison, R., Kanhema, T., Fiske, E., Bjorvatn, B., Ursin, R., and Portas, C. M. (2006). Chronic mild stress inhibits BDNF protein expression and CREB activation in the dentate gyrus but not in the hippocampus proper. *Pharmacol. Biochem. Behav.* **85,** 842–849.

Haenisch, B., Bilkei-Gorzo, A., Caron, M. G., and Bonisch, H. (2009). Knockout of the norepinephrine transporter and pharmacologically diverse antidepressants prevent behavioral and brain neurotrophin alterations in two chronic stress models of depression. *J. Neurochem.* **111,** 403–416.

Heberlein, A., Bleich, S., Bayerlein, K., Frieling, H., Groschl, M., Kornhuber, J., and Hillemacher, T. (2008). NGF plasma levels increase due to alcohol intoxication and decrease during withdrawal. *Psychoneuroendocrinology* **33,** 999–1003.

Heldt, S. A., Stanek, L., Chhatwal, J. P., and Ressler, K. J. (2007). Hippocampus-specific deletion of BDNF in adult mice impairs spatial memory and extinction of aversive memories. *Mol. Psychiatry* **12,** 656–670.

Huang, E. J., and Reichardt, L. F. (2003). Trk receptors: Roles in neuronal signal transduction. *Annu. Rev. Biochem.* **72,** 609–642.

Ibarguen-Vargas, Y., Surget, A., Vourc'h, P., Leman, S., Andres, C. R., Gardier, A. M., and Belzung, C. (2009). Deficit in BDNF does not increase vulnerability to stress but dampens antidepressant-like effects in the unpredictable chronic mild stress. *Behav. Brain Res.* **202,** 245–251.

Jockers-Scherubl, M. C., Bauer, A., Kuhn, S., Reischies, F., Danker-Hopfe, H., Schmidt, L. G., Rentzsch, J., and Hellweg, R. (2007a). Nerve growth factor in serum is a marker of the stage of alcohol disease. *Neurosci. Lett.* **419,** 78–82.

Jockers-Scherubl, M. C., Zubraegel, D., Baer, T., Linden, M., Danker-Hopfe, H., Schulte-Herbruggen, O., Neu, P., and Hellweg, R. (2007b). Nerve growth factor serum concentrations rise after successful cognitive-behavioural therapy of generalized anxiety disorder. *Prog. Neuropsychopharmacol. Biol. Psychiatry* **31,** 200–204.

Kerns, R. T., Ravindranathan, A., Hassan, S., Cage, M. P., York, T., Sikela, J. M., Williams, R. W., and Miles, M. F. (2005). Ethanol-responsive brain region expression networks: Implications for behavioral responses to acute ethanol in DBA/2J versus C57BL/6J mice. *J. Neurosci.* **25,** 2255–2266.

Khabour, O. F., Alzoubi, K. H., Alomari, M. A., and Alzubi, M. A. (2009). Changes in spatial memory and BDNF expression to concurrent dietary restriction and voluntary exercise. *Hippocampus.*

Koizumi, H., Hashimoto, K., and Iyo, M. (2006). Dietary restriction changes behaviours in brain-derived neurotrophic factor heterozygous mice: Role of serotonergic system. *Eur. J. Neurosci.* **24,** 2335–2344.

Koponen, E., Voikar, V., Riekki, R., Saarelainen, T., Rauramaa, T., Rauvala, H., Taira, T., and Castren, E. (2004). Transgenic mice overexpressing the full-length neurotrophin receptor trkB exhibit increased activation of the trkB-PLCgamma pathway, reduced anxiety, and facilitated learning. *Mol. Cell Neurosci.* **26,** 166–181.

Krishnan, V., Han, M. H., Graham, D. L., Berton, O., Renthal, W., Russo, S. J., Laplant, Q., Graham, A., Lutter, M., Lagace, D. C., Ghose, S., Reister, R., *et al.* (2007). Molecular adaptations underlying susceptibility and resistance to social defeat in brain reward regions. *Cell* **131,** 391–404.

Lang, U. E., Gunther, L., Scheuch, K., Klein, J., Eckhart, S., Hellweg, R., Danker-Hopfe, H., and Oehler, J. (2009). Higher BDNF concentrations in the hippocampus and cortex of an aggressive mouse strain. *Behav. Brain Res.* **197,** 246–249.

Le Foll, B., Diaz, J., and Sokoloff, P. (2005). A single cocaine exposure increases BDNF and D3 receptor expression: Implications for drug-conditioning. *Neuroreport* **16,** 175–178.

Li, S., Wang, C., Wang, M., Li, W., Matsumoto, K., and Tang, Y. (2007a). Antidepressant like effects of piperine in chronic mild stress treated mice and its possible mechanisms. *Life Sci.* **80,** 1373–1381.

Li, X. H., Liu, N. B., Zhang, M. H., Zhou, Y. L., Liao, J. W., Liu, X. Q., and Chen, H. W. (2007b). Effects of chronic multiple stress on learning and memory and the expression of Fyn, BDNF, TrkB in the hippocampus of rats. *Chin. Med. J. (Engl.)* **120,** 669–674.

Linnarsson, S., Bjorklund, A., and Ernfors, P. (1997). Learning deficit in BDNF mutant mice. *Eur. J. Neurosci.* **9,** 2581–2587.

Liu, I. Y., Lyons, W. E., Mamounas, L. A., and Thompson, R. F. (2004). Brain-derived neurotrophic factor plays a critical role in contextual fear conditioning. *J. Neurosci.* **24,** 7958–7963.

Logrip, M. L., Janak, P. H., and Ron, D. (2009). Escalating ethanol intake is associated with altered corticostriatal BDNF expression. *J. Neurochem.* **109,** 1459–1468.

Lu, L., Dempsey, J., Liu, S. Y., Bossert, J. M., and Shaham, Y. (2004). A single infusion of brain-derived neurotrophic factor into the ventral tegmental area induces long-lasting potentiation of cocaine seeking after withdrawal. *J. Neurosci.* **24,** 1604–1611.

Lucca, G., Comim, C. M., Valvassori, S. S., Pereira, J. G., Stertz, L., Gavioli, E. C., Kapczinski, F., and Quevedo, J. (2008). Chronic mild stress paradigm reduces sweet food intake in rats without affecting brain derived neurotrophic factor protein levels. *Curr. Neurovasc. Res.* **5,** 207–213.

Lukoyanov, N. V., Pereira, P. A., Paula-Barbosa, M. M., and Cadete-Leite, A. (2003). Nerve growth factor improves spatial learning and restores hippocampal cholinergic fibers in rats withdrawn from chronic treatment with ethanol. *Exp. Brain Res.* **148,** 88–94.

Lyons, W. E., Mamounas, L. A., Ricaurte, G. A., Coppola, V., Reid, S. W., Bora, S. H., Wihler, C., Koliatsos, V. E., and Tessarollo, L. (1999). Brain-derived neurotrophic factor-deficient mice develop aggressiveness and hyperphagia in conjunction with brain serotonergic abnormalities. *Proc. Natl. Acad. Sci. USA* **96,** 15239–15244.

MacQueen, G. M., Ramakrishnan, K., Croll, S. D., Siuciak, J. A., Yu, G., Young, L. T., and Fahnestock, M. (2001). Performance of heterozygous brain-derived neurotrophic factor knockout mice on behavioral analogues of anxiety, nociception, and depression. *Behav. Neurosci.* **115,** 1145–1153.

Markowska, A. L., Koliatsos, V. E., Breckler, S. J., Price, D. L., and Olton, D. S. (1994). Human nerve growth factor improves spatial memory in aged but not in young rats. *J. Neurosci.* **14,** 4815–4824.

Markowska, A. L., Price, D., and Koliatsos, V. E. (1996). Selective effects of nerve growth factor on spatial recent memory as assessed by a delayed nonmatching-to-position task in the water maze. *J. Neurosci.* **16,** 3541–3548.

Marmigere, F., Givalois, L., Rage, F., Arancibia, S., and Tapia-Arancibia, L. (2003). Rapid induction of BDNF expression in the hippocampus during immobilization stress challenge in adult rats. *Hippocampus* **13,** 646–655.

Martinowich, K., Manji, H., and Lu, B. (2007). New insights into BDNF function in depression and anxiety. *Nat. Neurosci.* **10,** 1089–1093.

Meredith, G. E., Callen, S., and Scheuer, D. A. (2002). Brain-derived neurotrophic factor expression is increased in the rat amygdala, piriform cortex and hypothalamus following repeated amphetamine administration. *Brain Res.* **949,** 218–227.

Moguel-Gonzalez, M., Gomez-Palacio-Schjetnan, A., and Escobar, M. L. (2008). BDNF reverses the CTA memory deficits produced by inhibition of protein synthesis. *Neurobiol. Learn. Mem.* **90,** 584–587.

Mufson, E. J., Kroin, J. S., Sendera, T. J., and Sobreviela, T. (1999). Distribution and retrograde transport of trophic factors in the central nervous system: Functional implications for the treatment of neurodegenerative diseases. *Prog. Neurobiol.* **57,** 451–484.

Murakami, S., Imbe, H., Morikawa, Y., Kubo, C., and Senba, E. (2005). Chronic stress, as well as acute stress, reduces BDNF mRNA expression in the rat hippocampus but less robustly. *Neurosci. Res.* **53,** 129–139.

Naimark, A., Barkai, E., Matar, M. A., Kaplan, Z., Kozlovsky, N., and Cohen, H. (2007). Upregulation of neurotrophic factors selectively in frontal cortex in response to olfactory discrimination learning. *Neural. Plast.* **2007,** 13427.

Nakajo, Y., Miyamoto, S., Nakano, Y., Xue, J. H., Hori, T., and Yanamoto, H. (2008). Genetic increase in brain-derived neurotrophic factor levels enhances learning and memory. *Brain Res.* **1241,** 103–109.

Pardon, M. C., Kendall, D. A., Perez-Diaz, F., Duxon, M. S., and Marsden, C. A. (2004). Repeated sensory contact with aggressive mice rapidly leads to an anticipatory increase in core body temperature and physical activity that precedes the onset of aversive responding. *Eur. J. Neurosci.* **20,** 1033–1050.

Pardon, M. C., Roberts, R. E., Marsden, C. A., Bianchi, M., Latif, M. L., Duxon, M. S., and Kendall, D. A. (2005). Social threat and novel cage stress-induced sustained extracellular-regulated kinase1/2 (ERK1/2) phosphorylation but differential modulation of brain-derived neurotrophic factor (BDNF) expression in the hippocampus of NMRI mice. *Neuroscience* **132,** 561–574.

Pierce, R. C., Pierce-Bancroft, A. F., and Prasad, B. M. (1999). Neurotrophin-3 contributes to the initiation of behavioral sensitization to cocaine by activating the Ras/Mitogen-activated protein kinase signal transduction cascade. *J. Neurosci.* **19,** 8685–8695.

Pietropaolo, S., Paterna, J. C., Bueler, H., Feldon, J., and Yee, B. K. (2007). Bidirectional changes in water-maze learning following recombinant adenovirus-associated viral vector (rAAV)-mediated brain-derived neurotrophic factor expression in the rat hippocampus. *Behav. Pharmacol.* **18,** 533–547.

Pizarro, J. M., Lumley, L. A., Medina, W., Robison, C. L., Chang, W. E., Alagappan, A., Bah, M. J., Dawood, M. Y., Shah, J. D., Mark, B., Kendall, N., Smith, M. A., *et al.* (2004). Acute social defeat reduces neurotrophin expression in brain cortical and subcortical areas in mice. *Brain Res.* **1025,** 10–20.

Pizzo, D. P., and Thal, L. J. (2004). Intraparenchymal nerve growth factor improves behavioral deficits while minimizing the adverse effects of intracerebroventricular delivery. *Neuroscience* **124,** 743–755.

Price, R. D., Milne, S. A., Sharkey, J., and Matsuoka, N. (2007). Advances in small molecules promoting neurotrophic function. *Pharmacol. Ther.* **115,** 292–306.

Rage, F., Givalois, L., Marmigere, F., Tapia Arancibia, L., and Arancibia, S. (2002). Immobilization stress rapidly modulates BDNF mRNA expression in the hypothalamus of adult male rats. *Neuroscience* **112,** 309–318.

Reichardt, L. F. (2006). Neurotrophin-regulated signalling pathways. *Philos. Trans. R. Soc. Lond. B Biol. Sci.* **361,** 1545–1564.

Schmidt, H. D., and Duman, R. S. (2007). The role of neurotrophic factors in adult hippocampal neurogenesis, antidepressant treatments and animal models of depressive-like behavior. *Behav. Pharmacol.* **18,** 391–418.

Skaper, S. D. (2008). The biology of neurotrophins, signalling pathways, and functional peptide mimetics of neurotrophins and their receptors. *CNS Neurol. Disord. Drug. Targets* **7,** 46–62.

Spillantini, M. G., Aloe, L., Alleva, E., De Simone, R., Goedert, M., and Levi-Montalcini, R. (1989). Nerve growth factor mRNA and protein increase in hypothalamus in a mouse model of aggression. *Proc. Natl. Acad. Sci. USA* **86,** 8555–8559.

Strohle, A. (2009). Physical activity, exercise, depression and anxiety disorders. *J. Neural. Transm.* **116,** 777–784.

Vargas-Perez, H., Kee, R. T., Walton, C. H., Hansen, D. M., Razavi, R., Clarke, L., Bufalino, M. R., Allison, D. W., Steffensen, S. C., and van der Kooy, D. (2009). Ventral tegmental area BDNF induces an opiate-dependent-like reward state in naive rats. *Science* **324,** 1732–1734.

Webster, N. J., and Pirrung, M. C. (2008). Small molecule activators of the Trk receptors for neuroprotection. *BMC Neurosci.* **9**(Suppl. 2), S1.

Woolley, M. L., Waters, K. A., Gartlon, J. E., Lacroix, L. P., Jennings, C., Shaughnessy, F., Ong, A., Pemberton, D. J., Harries, M. H., Southam, E., Jones, D. N., and Dawson, L. A. (2009). Evaluation of the pro-cognitive effects of the AMPA receptor positive modulator, 5-(1-piperidinylcarbonyl)-2,1,3-benzoxadiazole (CX691), in the rat. *Psychopharmacology (Berl.)* **202,** 343–354.

Xie, C. W., Sayah, D., Chen, Q. S., Wei, W. Z., Smith, D., and Liu, X. (2000). Deficient long-term memory and long-lasting long-term potentiation in mice with a targeted deletion of neurotrophin-4 gene. *Proc. Natl. Acad. Sci. USA* **97,** 8116–8121.

Yamada, K., and Nabeshima, T. (2003). Brain-derived neurotrophic factor/TrkB signaling in memory processes. *J. Pharmacol. Sci.* **91,** 267–270.

Yee, B. K., Zhu, S. W., Mohammed, A. H., and Feldon, J. (2007). Levels of neurotrophic factors in the hippocampus and amygdala correlate with anxiety- and fear-related behaviour in C57BL6 mice. *J. Neural. Transm.* **114,** 431–444.

Yoon, S. J., Roh, S., Lee, H., Lee, J. Y., Lee, B. H., Kim, Y. K., and Kim, D. J. (2006). Possible role of nerve growth factor in the pathogenesis of alcohol dependence. *Alcohol Clin. Exp. Res.* **30,** 1060–1065.

CHAPTER ELEVEN

# Postnatal Development of Hypothalamic Leptin Receptors

Elizabeth C. Cottrell,*[,1] Julian G. Mercer,[†] *and* Susan E. Ozanne*

## Contents



## Abstract

The hormone, leptin, plays a key role in the regulation of energy balance and neuroendocrine function, as well as modulating a range of other physiological systems from immunity to cognition. In the adult brain, leptin regulates food intake and energy expenditure primarily via the hypothalamus. In addition to these well-defined actions in adult life, there is increasing evidence for a role of leptin during development. Leptin receptors are widely expressed in the developing brain from an early stage, and leptin is known to have profound effects on

* Institute of Metabolic Science, Metabolic Research Laboratories, University of Cambridge, Addenbrooke's Hospital, Cambridge, United Kingdom

† Division of Obesity and Metabolic Health, Rowett Institute of Nutrition and Health, University of Aberdeen, Bucksburn, Aberdeen, Scotland, United Kingdom

1 Current address: Centre for Cardiovascular Science, Queen's Medical Research Institute, University of Edinburgh, Edinburgh, Scotland, United Kingdom

*Vitamins and Hormones,* Volume 82
ISSN 0083-6729, DOI: 10.1016/S0083-6729(10)82011-4

the proliferation, maintenance, and differentiation of neuronal and glial cells. During the early postnatal period, in both rats and mice, there is a surge in circulating leptin concentrations. Despite this elevation in leptin, neonates maintain a high level of food intake, and both feeding behavior and metabolic responses to exogenous leptin administration are absent until around the time of weaning. However, it is during this period that direct neurotrophic actions of leptin have been demonstrated, with leptin promoting neurite outgrowth and the establishment of hypothalamic circuitry. Exactly how leptin exerts these effects remains unknown, but changes in the distribution of hypothalamic leptin receptors during this period may, at least in part, underlie these age-specific effects of leptin. 

# I. Introduction

Leptin, a hormone derived predominantly from adipocytes, is critically involved in the regulation of energy homeostasis. Discovered in 1994 (Zhang *et al.*, 1994), leptin was initially thought to act only as a satiety factor, signaling to the brain the repletion of body fat stores. Since this time, however, leptin receptors (ObRs) have been identified in essentially every tissue, and leptin has been shown to play a role in a diverse range of physiological systems, including reproduction, immunity, cardiovascular function, and cognition (Ahima and Flier, 2000; Harvey *et al.*, 2005; La Cava and Matarese, 2004; Tune and Considine, 2007). In terms of body weight regulation, leptin promotes negative energy balance by inhibiting feeding and stimulating energy expenditure, acting primarily through central nervous system (CNS) ObRs. Thus, the most apparent phenotype in those individuals lacking either leptin or functional ObR is severe obesity, driven by extreme hyperphagia and unabated adipose tissue accumulation. Numerous other metabolic abnormalities result from a lack of leptin signaling, as well as impairments in brain development, bone formation, and immune function.

In addition, a developmental role for leptin has been increasingly recognized. In line with these findings, it has been established that there are restricted periods in early life in which leptin replacement (in leptin-deficient *ob/ob* mice) or supplementation (in leptin-replete animals) can exert long-term effects on tissue structure and organization, as well as influencing the metabolic phenotype of the adult. These age-specific effects of leptin may be related to developmental changes in receptor expression, as recent evidence has shown that there are dynamic changes in the distribution of ObR and leptin actions during the early postnatal period in the rodent.

## II. The Leptin System

### A. Sources and regulation of leptin

Leptin is the protein product of the *ob* gene, which encodes a 4.5-kb mRNA and is translated to produce a circulating protein of 16 kDa (Zhang *et al.*, 1994). Leptin is mainly expressed in adipose tissue and circulating concentrations are, in the fed adult, highly correlated with the degree of adiposity (Maffei *et al.*, 1995). Additional tissues have been shown to produce leptin, including the stomach, placenta, mammary epithelium, pituitary, and hypothalamus (Bado *et al.*, 1998; Hoggard *et al.*, 2001; Morash *et al.*, 1999; Smith-Kirwin *et al.*, 1998). Over any relatively extended period of time, leptin concentrations are reflective of total body fat mass; however, leptin concentrations are acutely modulated by nutritional state. In both humans and adult rats, leptin concentrations are decreased in response to fasting, correlating with a reduction in blood glucose and insulin levels (Ahren *et al.*, 1997a,b). On fasting, the fall in leptin concentrations mediates a number of other physiological adaptations to this disruption in energy homeostasis (reducing thyroid hormone concentrations and altering gonadal hormones), with administration of leptin during fasting preventing the changes in these axes (Ahima *et al.*, 1996). These adaptations presumably function to promote feeding behavior, in order that body fat does not become overly depleted, and to minimize energy expended and delay reproduction until such time that body energy stores are replaced. *In vitro* studies investigating the molecular regulation of leptin have shown that incubation of isolated rat adipocytes with glucose and insulin increases both the production and secretion of leptin, with the rate of secretion being proportional to the degree of insulin-stimulated glucose uptake (Barr *et al.*, 1997; Mueller *et al.*, 1998). In addition, a number of other factors are also involved in the regulation of leptin expression, including glucocorticoids, catecholamines, and cytokines.

In the obese adult *ob/ob* mouse, there is no detectable leptin in the circulation, as the mutation responsible for this genetic obesity causes the production of a truncated protein that is not secreted (Zhang *et al.*, 1994). In accordance with this defective secretion, *ob/ob* mice have markedly elevated adipose tissue leptin mRNA, approximately 20-fold that of wild-type mice. Administration of leptin to *ob/ob* animals results in a normalization of body weight, due to a curbing of the extreme hyperphagia exhibited by these animals and a concurrent stimulation of metabolic rate (Campfield *et al.*, 1995; Halaas *et al.*, 1995; Pelleymounter *et al.*, 1995).

## B. Leptin receptors

The ObRs are members of the class I cytokine receptor family, and to date there are at least six alternatively spliced isoforms (ObRa–f; reviewed in Fruhbeck, 2006). Each of these isoforms share common extracellular and transmembrane domains, but differ in their intracellular sequence. ObRa, c, d, and f possess relatively short cytoplasmic domains whereas ObRb, the so-called "long" isoform, has an extended C-terminal region and is the only variant capable of complete intracellular signal transduction (Baumann *et al.*, 1996). However, a degree of intracellular signaling has been demonstrated for the ObRa isoform (Bjorbaek *et al.*, 1997). ObRe is a truncated form of the receptor, which is secreted into the circulation and thought to play a role as a binding protein, regulating leptin bioavailability. In addition, recent studies have implicated ObRe in the regulation of leptin transport at the blood–brain barrier (BBB) by antagonizing the uptake of leptin and inhibiting transport into the CNS (Tu *et al.*, 2008).

Binding of leptin to ObRb results in autophosphorylation of JAK2 and subsequent activation of the janus kinase/signal transducer and activator of transcription (JAK/STAT-3) pathway (Banks *et al.*, 2000). Parallel signaling pathways are also activated, including the phosphoinositide 3-kinase (PI3K) and mitogen-activated protein kinase/extracellular signal-regulated kinase (MAPK/ERK) pathways. Suppressor of cytokine signaling-3 (SOCS3) is robustly induced in response to leptin (Bjorbaek *et al.*, 1998), and acts as a negative regulator of this signaling pathway. As such, SOCS3 has been used extensively as a marker of ObR activation by leptin (Cottrell *et al.*, 2009; Proulx *et al.*, 2002). Furthermore, SOCS3 is also implicated as a key molecule in the development of central leptin resistance and progression to obesity in mice fed a high-fat diet (HFD; Enriori *et al.*, 2007). Mice with either a complete lack of SOCS3 in the brain (Mori *et al.*, 2004) or whole body haploinsufficiency of SOCS3 (Howard *et al.*, 2004) have both increased leptin sensitivity and are relatively resistant to the obesity-inducing effects of high-fat feeding.

## C. Central regulation of energy balance by leptin

Leptin was initially thought to act solely as a regulator of energy balance through hypothalamic sites. However, as mentioned previously, leptin has wide-ranging effects in the body and acts on many tissues. For the purposes of this review, we focus on the central actions of leptin and its role in development.

Within the CNS, leptin acts to inhibit feeding and stimulate energy expenditure. The arcuate nucleus (ARC) of the hypothalamus is considered to be the primary site of leptin action (Satoh *et al.*, 1997). Recent evidence has also shown that a population of ARC/ObR-expressing neurons extend processes into the median eminence and are, therefore, in direct contact with the circulation (Faouzi *et al.*, 2007). Leptin receptors are widely expressed within this brain region, and are colocalized within orexigenic neurons expressing neuropeptide Y (NPY) and agouti-related protein (AgRP) as well as within the anorexigenic proopiomelanocortin (POMC) and cocaine- and amphetamine-regulated transcript (CART)-expressing neurons (Schwartz *et al.*, 2000). These neurons are responsive to changes in circulating leptin and insulin (indicative of body energy stores) and send projections to downstream nuclei, including the paraventricular nucleus (PVN) where integration of inputs from hypothalamic and other sites, including brainstem and higher cortical centers, occurs. Disruption of projections from the ARC to these downstream nuclei is associated with an inability to regulate energy balance appropriately (Bell *et al.*, 2000; Bouret *et al.*, 2004a; Dawson *et al.*, 1997), indicating a key role for this region in energy homeostasis.

Within the ARC, leptin inhibits the activity of NPY/AgRP neurons, while activating POMC/CART containing neurons (Schwartz *et al.*, 2000). Thus, in states of positive energy balance (where adipose stores are replete), raised leptin concentrations inhibit further feeding (reducing orexigenic and increasing anorexigenic signaling) and concurrently stimulate metabolic activity through increased sympathetic nervous system activation to peripheral tissues. Direct effects of leptin in ObR-expressing peripheral tissues have also been shown, including the mobilization of stored triglycerides in adipose tissue (Siegrist-Kaiser *et al.*, 1997) and stimulation of skeletal muscle thermogenesis (Dulloo *et al.*, 2002; Maroni *et al.*, 2003; Minokoshi *et al.*, 2002). Finally, leptin also plays a key role in preventing the lipotoxic effects of ectopically stored lipid in tissues such as pancreas and cardiomyocytes. Excess lipid, deposited in these tissues, can impair normal cellular physiology and contribute to the development of type 2 diabetes and cardiovascular disease (Unger, 2003). Thus, the effects of leptin to regulate whole body energy status are widespread.

## III. Developmental Roles of Leptin

### A. Leptin and brain development

In addition to the metabolic effects of leptin in the adult, leptin plays a key role in brain development during early life. Several decades ago, *ob/ob* mice were found to have perturbed CNS development, including impaired myelination, structural abnormalities within the hypothalamus and altered expression of neuronal and glial cell markers (Bereiter and Jeanrenaud, 1979, 1980; Sena *et al.*, 1985). More recently, studies have shown that the brain abnormalities in *ob/ob* mice are present even at embryonic ages.

In the developing rodent, ObRs are widely expressed throughout the CNS from early in mid-gestation (Hoggard *et al.*, 1997; Matsuda *et al.*, 1999; Udagawa *et al.*, 2000) and leptin protein is detected in the circulation of fetal rodents (Udagawa *et al.*, 2006). There is thus evidence for the presence of both ligand and the relevant receptor within the developing brain. Further direct evidence that leptin is involved during fetal life in brain growth and development was obtained recently, with the observation that *ob/ob* fetuses at E16 exhibit reductions in neuroepithelial cell number within the cortex and reduced proliferative capacity compared with wild-type animals (Udagawa *et al.*, 2006). Administration of leptin to E14 embryos was able to increase cell number at E16, and there is some *in vitro* evidence that leptin can directly increase BrdU incorporation (a marker of proliferation) in cultured neurospheres (Udagawa *et al.*, 2006).

Postnatally, the administration of leptin to *ob/ob* animals has been shown to increase brain size (Ahima *et al.*, 1999; Steppan and Swick, 1999). However, restoration of brain weight to wild-type levels required leptin to be administered for 6 weeks, from 4 weeks of age. Administration of leptin for 2 weeks from 8 to 10 weeks of age was ineffective in the rescue of brain weight or protein content (Ahima *et al.*, 1999), suggesting perhaps a critical period for leptin administration in the rescue of brain development during periods of greater plasticity at younger ages. Further evidence for restricted periods of leptin action in brain development is discussed in more detail below.

## B. Postnatal leptin surge

In both rats and mice, there is a transient increase in circulating leptin concentrations during the first 2 postnatal weeks (Ahima *et al.*, 1998; Devaskar *et al.*, 1997; Morash *et al.*, 2001, 2003; Rayner *et al.*, 1997; Yura *et al.*, 2005). This rise in leptin is independent of body fat mass, and the regulation of this "surge" is at present unknown. Several studies have indicated that the source of postnatal leptin is likely to be the adipose tissue of neonatal rodents (Ahima *et al.*, 1998; Devaskar *et al.*, 1997), although alternative sites of production have been demonstrated. In one study, pituitary and cerebral cortex leptin expression was shown to correlate with the elevated circulating leptin concentrations, and it was suggested that this rise in leptin concentrations locally with the brain–pituitary axis may play a role in the development of these systems (Morash *et al.*, 2001). If the leptin surge is generated by the developing adipose tissue, then the regulation of its production and secretion is clearly quite different to that in the adult. The rise in leptin appears to occur independently of circulating glucose or insulin (Cottrell *et al.*, 2009; Srinivasan *et al.*, 2008), key factors in the stimulation of leptin secretion in the adult (as discussed above). However, nutritional manipulations are able to alter leptin concentrations in the neonatal rodent. Exposure to a reduced plane of nutrition during the postnatal period, through restriction of maternal food intake, was shown to lead to a significant reduction in circulating leptin concentrations in the offspring (Delahaye *et al.*, 2008). Interestingly, the feeding of a high-carbohydrate diet during the first 2 postnatal weeks to rats, which leads to significant elevation of glucose and insulin concentrations, was also found to result in markedly reduced levels of circulating leptin in these animals (Srinivasan *et al.*, 2008). This further suggests that glucose and insulin are not the key stimulators of leptin production in the early postnatal period.

As well as endogenous production, maternal breast milk contains leptin both in the human (Miralles *et al.*, 2006) and rodent (Stocker *et al.*, 2004). In rat pups, ingested leptin has been shown to be taken up by the immature gastrointestinal tract and to enter the neonate's circulation (Casabiell *et al.*, 1997). It is to be expected that the circulation of a neonatal *ob/ob* mouse would have no detectable leptin. However, given the evidence for uptake of maternally derived leptin by the neonatal stomach, it is theoretically possible that *ob/ob* offspring may have a low level of circulating leptin, if hormone is transferred from the maternal milk. We recently investigated whether this was the case by measuring leptin concentrations in the serum of *ob/ob* mice and their wild-type littermates across a range of postnatal ages between birth and weaning. We also included measurement of heterozygous animals (*ob/+*) from these litters, to determine whether there was a gene–dosage effect in the production of a postnatal leptin surge. As shown in Fig. 11.1 (unpublished observations), wild-type animals exhibited the expected postnatal leptin surge, peaking around postnatal days 9–11 (P9–11). However, at no age did the serum of *ob/ob* offspring contain detectable leptin. Interestingly, *ob/+* animals (those with a single functional copy of the leptin gene) exhibited lower leptin concentrations across the lactation period, and there also appeared to be a lack of evidence for a "surge," as occurs in the wild-type animals. Whether or not this absence of a leptin surge might have long-term effects on adult phenotype in the *ob/+* mice is not clear; however, it has been reported that although these animals are phenotypically normal, they are more susceptible to diet-induced obesity when fed a high-calorie diet, compared with wild-type animals (Begriche *et al.*, 2008).

## C. Leptin insensitivity in the early postnatal period

Despite the marked elevations in circulating leptin concentrations in the developing rodent, there is a lack of effect of these raised hormone levels on food intake in the early postnatal period (Mistry *et al.*, 1999). This lack of sensitivity to leptin presumably allows neonates to maximize their food intake during a period of rapid growth. The acute administration of exogenous leptin to P10 rats was found to increase POMC and decrease NPY

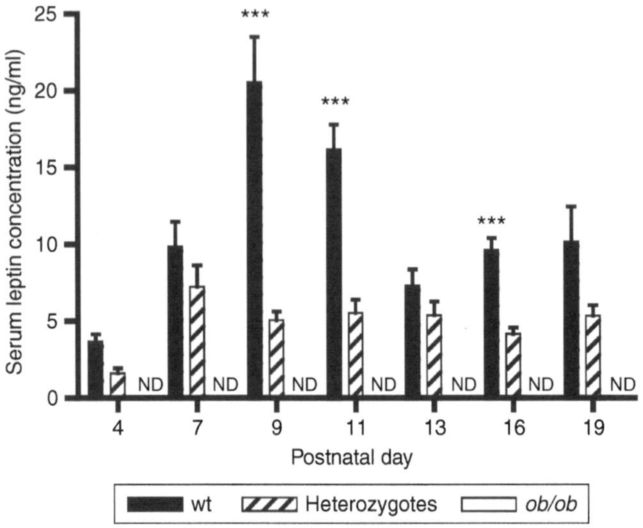


**Figure 11.1** Leptin concentrations in the serum of postnatal mice. Data are shown as mean ± SEM and were examined by two-way ANOVA followed by Fisher's LSD *post hoc* tests. *P* values indicate significant differences in heterozygous (*ob*/+) offspring compared with wild-type littermates at each age. ***$P < 0.001$. $n = 7$–15 animals per group. Leptin was not detectable at any age in the serum of *ob*/*ob* offspring (ND).

in the ARC, as would occur in the adult; however, there was no effect on food intake or body weight in these pups (Proulx *et al.*, 2002). This possibly reflects a lack of functional circuitry at such a young age at least in terms of feeding circuitry. However, it was found that leptin could reduce fat pad weight at this young age. Similarly, it was reported that in both wild-type and *ob*/*ob* mice, leptin was unable to reduce food intake or stimulate oxygen consumption in the first 2 postnatal weeks, but by P17 leptin administration was able to increase energy expenditure (Mistry *et al.*, 1999). By 28 days of age, intracerebroventricular administration of leptin was able to inhibit food intake as well as stimulate oxygen consumption. Together, these findings indicate that there are different developmental trajectories for the circuits regulating feeding versus energy expenditure. In agreement with this, relatively recent studies indicated that indeed, downstream of ARC leptin actions, there is a divergence in melanocortin signaling pathways to regulate the feeding and expenditure arms of energy balance independently (Balthasar *et al.*, 2005).

In humans, the body weights of leptin-deficient infants do not diverge until around the time of weaning (S. Farooqi, personal communication), suggesting that, as in the rodent, leptin does not act to regulate energy balance during the early postnatal period. Importantly, this divergence in body weight occurs in both breast- and bottle-fed infants, negating a role for breastmilk-derived leptin in the regulation of energy balance in early life.

## D. Neurotrophic actions of leptin

Although the physiological function of the leptin surge remains incompletely understood, one of the roles for leptin in early life that has emerged in recent years is the establishment of hypothalamic neuroendocrine systems (Bouret and Simerly, 2007). Some of the most compelling evidence to date for a neurotrophic action of leptin during restricted periods of development is derived from the work of Bouret and colleagues. Using DiI tracing of neuronal projections in wild-type and *ob/ob* mice, they showed that in the absence of leptin there is a failure to form projections from the ARC to the downstream PVN (Bouret *et al.*, 2004a). *ob/ob* mice have a reduction in fiber density within the PVN at P12, and this reduced innervation persists in the adult animal. However, administration of leptin during the period corresponding to the postnatal surge in wild-type mice was able to increase the density of ARC projections to the PVN at P12 in *ob/ob* mice, whereas leptin treatment in adult life had no effect. Further to these tracing studies, a direct action of leptin on neurite outgrowth was shown in ARC explants taken from neonates at P6 (Bouret *et al.*, 2004b). Thus, the current hypothesis is that leptin may play an active role in the establishment of the hypothalamic energy balance circuits it will later regulate (Bouret and Simerly, 2007).

In addition to these actions in the hypothalamus, neurotrophic effects of leptin have also been demonstrated in extrahypothalamic sites. Leptin administration induces ERK1/2 signaling in the cortex of both neonatal and adult mice, and in cultured embryonic cortical neurons affects growth cone size and spreading (Valerio *et al.*, 2006). Taken together, these actions of leptin indicate a widespread role for this hormone in neuronal development and circuit formation.

## E. Developmental changes in leptin receptor

Although there is plentiful evidence that ObRs are expressed throughout embryonic development and early postnatal life, whether or not there might be changes in the distribution of these receptors over this period has only recently been addressed. In the rat, there is an increase in both ObRb mRNA expression and protein binding in the hypothalamus between E18 and early postnatal life (Carlo *et al.*, 2007). We recently showed that, in the rat, there is a marked change in ObR distribution between birth and weaning (Cottrell *et al.*, 2009). Specifically, there was dense ObR mRNA expression within the third ventricle (3V) of the hypothalamus, which progressively decreased and was completely absent by P19, and also was not present in the adult brain. Conversely, ObR expression in the ARC and VMH increased over the period of suckling. Leptin administration at P4 was found to stimulate a robust induction of SOCS3 within the 3V, a response that was essentially absent by P14. By this later age, leptin-induced SOCS3 was prominent within both the ARC and VMH, reflecting a developmental change in leptin responses over a period in which endogenous leptin levels are raised, and during which time hypothalamic circuits are still developing. It is also interesting to note that the location of 3V leptin receptors may be relevant for the reported neurotrophic actions of leptin in the hypothalamic explants used by Bouret and colleagues (discussed above). A similar localization of leptin-induced signaling has been shown in mice, where between P5 and P13, leptin-induced P-STAT3 was detected in an unidentified population of cells within the subependymal region of the 3V (Frontini *et al.*, 2008).

In terms of the leptin insensitivity in the postnatal mouse, we recently studied the ontogeny of hypothalamic ObRs in wild-type and *ob/ob* mice over the postnatal period. Given that (1) in the adult *ob/ob* mouse, ARC ObR mRNA is markedly upregulated in the absence of leptin and downregulated following exogenous leptin administration (Mercer *et al.*, 1997); and (2) there is no significant effect of leptin administration during the early postnatal period on body weights of either wild-type or *ob/ob* neonates (Yura *et al.*, 2008), we wanted to determine when hypothalamic leptin receptors become responsive to circulating endogenous leptin. We hypothesized that *ob/ob* and wild-type animals would have similar levels of ObR in the ARC up until the point at which responsiveness to circulating leptin occurs. Similar to previous observations in the neonatal rat, we determined that ObR increased progressively from P4 into adult life in both wild-type and *ob/ob* offspring (Fig. 11.2, unpublished observations). However, there were clear differences between wild-type and *ob/ob* animals, such that from P9 onward, *ob/ob* offspring exhibited an upregulation of ARC ObR compared with wild-type mice. Whether this divergence represents an onset of sensitivity to circulating leptin will require the administration of a leptin challenge to neonates across this period, to determine the timing of onset of activation of leptin signaling pathways *in vivo* in the hypothalamus.

## IV. Developmental Programming: Role of Altered Neonatal Leptin Signaling

The field of development programming—encompassing the concept that an altered environment during critical periods of development can induce permanent changes in an individual's physiology—continues to grow. There is now a wealth of evidence indicating that neonatal feeding

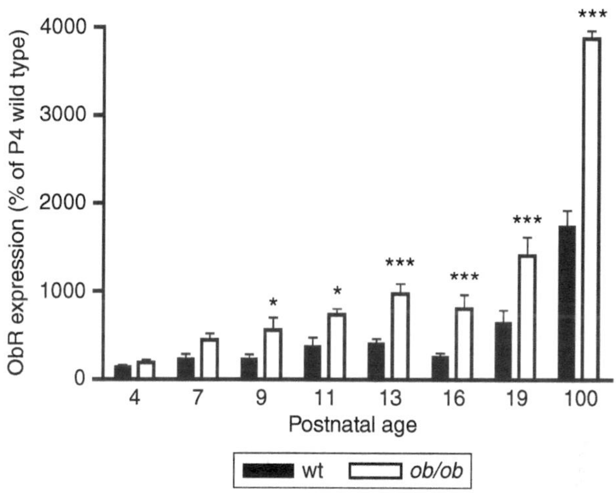


**Figure 11.2** Leptin receptor expression in the postnatal hypothalamus. Leptin receptor expression in the ARC of wild-type and *ob/ob* mice, from neonatal to adult life. Data are shown as mean ± SEM and were examined by two-way ANOVA followed by Fisher's LSD *post hoc* tests. *P* values indicate significant differences in *ob/ob* offspring compared with wild-type littermates at each age. $^*P < 0.05$, $^{***}P < 0.001$. $n = 7$–12 animals per group.

and growth trajectories can modulate predisposition to obesity, and as such, strategies that might be able to prevent such increased risks of adult disease are vital, as obesity rates continue to rise worldwide.

Many models used to study the long-term effects of neonatal nutrition have paid much attention to leptin and to the leptin surge itself in recent years. It has become apparent that not only a complete absence of leptin but also exposure to altered concentrations of this hormone during development can have long-term effects on adult physiology, and in particular can affect subsequent obesity susceptibility and a host of other metabolic complications. In the mouse, intrauterine growth restriction (IUGR) induced by maternal undernutrition (UN) leads to reduced birth weight and increased susceptibility to HFD-induced obesity in the UN offspring compared with normally nourished animals (Yura *et al.*, 2005). In the postnatal period, these offspring exhibited a premature leptin surge, which was of greater amplitude than in control animals. By administering exogenous leptin to mimic this altered surge, it was shown that leptin-treated animals gained more weight on a HFD, implicating excess leptin exposure in early life as causal in the development of obesity susceptibility. Similarly, a recent study in rats demonstrated that maternal obesity during pregnancy is associated with an amplified and prolonged leptin surge in the offspring (Kirk *et al.*, 2009). This increased leptin exposure was associated with subsequent leptin insensitivity, thought to drive the hyperphagia and increased adiposity in these animals. In contrast, the blocking of leptin action in neonatal rats, through administration of a ObR antagonist, was shown to predispose to later leptin resistance and increased body weight gain on a high-energy diet in adult life (Attig *et al.*, 2008).

In a model of relatively severe maternal UN in the rat, there is again an adverse metabolic phenotype in the adult offspring (Vickers *et al.*, 2000). However, in apparent contrast to the situation in mice, leptin treatment in early postnatal life was able to ameliorate these adverse programming effects in female offspring (Stocker *et al.*, 2004; Vickers *et al.*, 2005), but not in males (Vickers *et al.*, 2008). Although female rats as adults have greater ObR expression relative to males, there is apparently no difference in hypothalamic receptor levels before puberty (Smith and Waddell, 2003), when differences in the effects of leptin administration between sexes have been reported. Thus, the biological basis of these observed sex differences remains to be determined.

Similar to the beneficial effects of leptin in the UN rat model, leptin administration during gestation and lactation to rat dams fed a low protein diet also prevents the adverse metabolic programming associated with this dietary manipulation (Stocker *et al.*, 2004). Furthermore, this effect of leptin dosing to the mother protected offspring of normally nourished rat dams from the obesity-inducing effects of high-fat/high-energy feeding (Stocker *et al.*, 2007). In another study looking at the effects of neonatal leptin supplementation, oral dosing with leptin within a physiological range during the lactation period was able to attenuate weight gain on a HFD postweaning (Pico *et al.*, 2007). Although it is not yet clear how leptin supplementation during pregnancy and/or early life might impart long-term metabolic benefits, current data do suggest that early life leptin exposure can have lasting effects on energy homeostasis. However, the idea that leptin might be able to be used therapeutically in the prevention of obesity is still far from a reality.

## V. Conclusions and Future Directions

In contrast to the actions of leptin in adult life, numerous studies have shown that appetite and energy expenditure pathways in the rodent are relatively insensitive to leptin during the first postnatal weeks, a period of rapid growth. Leptin receptors are, however, present and functional in early life, and leptin has a clear role in brain development. It is established that there is an increase in ObR expression postnatally in both rats and mice, and more recently that there are clear differences in the sites of leptin action over the early postnatal period in the rodent. The precise function of the transiently expressed ObRs in the 3V of the hypothalamus is not known, but the location and timing of this receptor expression may imply that they are involved in the early life establishment of hypothalamic circuitry. Changes in the actions of leptin with age might suggest distinct roles for this hormone at different stages of development (summarized in Fig. 11.3).

Clearly the interplay between ObR expression and dynamic changes in leptin production will be of importance in determining the ultimate signal that is transmitted to the developing system in question. The role of leptin in terms of metabolic programming is not yet clear, although there is growing evidence that perturbed leptin signaling or an altered leptin profile in the postnatal period can have long-lasting effects on adult physiology. The challenge now is to determine whether there may be realistic interventions to alter neonatal nutrition or hormone profiles, with the aim of improving adult metabolic health and reducing disease.

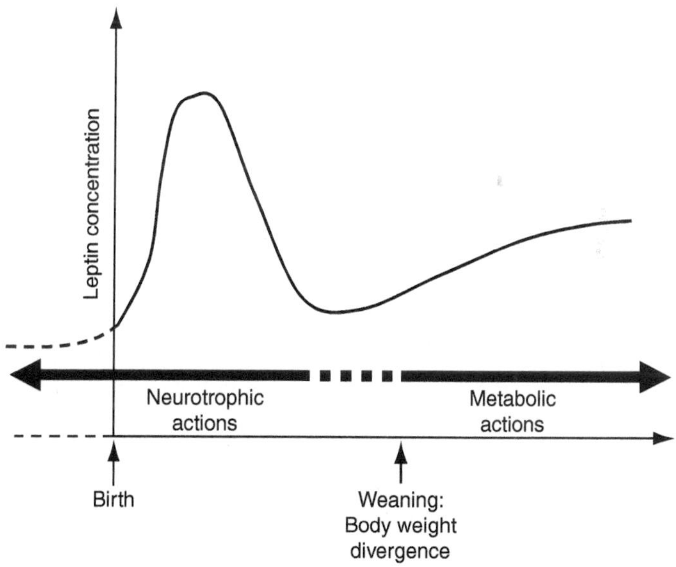


**Figure 11.3** Summary for the developmentally regulated actions of leptin during early life. Recent evidence has accumulated that leptin regulates the development of neuronal and glial cells during fetal life, and that during the early postnatal period leptin signaling plays a role in the establishment of neuronal circuitry. At around the time of weaning, when independent feeding is initiated, leptin begins to exert metabolic effects, presumably reflecting the maturation of circuitry that regulates feeding and energy expenditure. The precise timing of this switch in the actions of leptin is unclear but likely involves changes in the expression and distribution of functional leptin receptors during development.

## ACKNOWLEDGMENTS

We thank Adrian Wayman, Delia Hawkes, and Jacqueline Duncan for their excellent technical assistance in the experimental work described in this chapter.

## REFERENCES

Ahima, R. S., and Flier, J. S. (2000). Leptin. *Annu. Rev. Physiol.* **62,** 413–437.

Ahima, R. S., *et al.* (1996). Role of leptin in the neuroendocrine response to fasting. *Nature* **382,** 250–252.

Ahima, R. S., *et al.* (1998). Postnatal leptin surge and regulation of circadian rhythm of leptin by feeding. Implications for energy homeostasis and neuroendocrine function. *J. Clin. Invest.* **101,** 1020–1027.

Ahima, R. S., *et al.* (1999). Regulation of neuronal and glial proteins by leptin: Implications for brain development. *Endocrinology* **140,** 2755–2762.

Ahren, B., *et al.* (1997a). Regulation of circulating leptin in humans. *Endocrinology* **7,** 1–8.

Ahren, B., *et al.* (1997b). Regulation of plasma leptin in mice: Influence of age, high-fat diet, and fasting. *Am. J. Physiol.* **273,** R113–R120.

Attig, L., *et al.* (2008). Early postnatal leptin blockage leads to a long-term leptin resistance and susceptibility to diet-induced obesity in rats. *Int. J. Obes. (Lond.)* **32,** 1153–1160.

Bado, A., *et al.* (1998). The stomach is a source of leptin. *Nature* **394,** 790–793.

Balthasar, N., *et al.* (2005). Divergence of melanocortin pathways in the control of food intake and energy expenditure. *Cell* **123,** 493–505.

Banks, A. S., *et al.* (2000). Activation of downstream signals by the long form of the leptin receptor. *J. Biol. Chem.* **275,** 14563–14572.

Barr, V. A., *et al.* (1997). Insulin stimulates both leptin secretion and production by rat white adipose tissue. *Endocrinology* **138,** 4463–4472.

Baumann, H., *et al.* (1996). The full-length leptin receptor has signaling capabilities of interleukin 6-type cytokine receptors. *Proc. Natl. Acad. Sci. USA* **93,** 8374–8378.

Begriche, K., *et al.* (2008). Partial leptin deficiency favors diet-induced obesity and related metabolic disorders in mice. *Am. J. Physiol. Endocrinol. Metab.* **294,** E939–E951.

Bell, M. E., *et al.* (2000). Disruption of arcuate/paraventricular nucleus connections changes body energy balance and response to acute stress. *J. Neurosci.* **20,** 6707–6713.

Bereiter, D. A., and Jeanrenaud, B. (1979). Altered neuroanatomical organization in the central nervous system of the genetically obese (*ob/ob*) mouse. *Brain Res.* **165,** 249–260.

Bereiter, D. A., and Jeanrenaud, B. (1980). Altered dendritic orientation of hypothalamic neurons from genetically obese (*ob/ob*) mice. *Brain Res.* **202,** 201–206.

Bjorbaek, C., *et al.* (1997). Divergent signaling capacities of the long and short isoforms of the leptin receptor. *J. Biol. Chem.* **272,** 32686–32695.

Bjorbaek, C., *et al.* (1998). Identification of SOCS-3 as a potential mediator of central leptin resistance. *Mol. Cell.* **1,** 619–625.

Bouret, S. G., and Simerly, R. B. (2007). Development of leptin-sensitive circuits. *J. Neuroendocrinol.* **19,** 575–582.

Bouret, S. G., *et al.* (2004a). Formation of projection pathways from the arcuate nucleus of the hypothalamus to hypothalamic regions implicated in the neural control of feeding behavior in mice. *J. Neurosci.* **24,** 2797–2805.

Bouret, S. G., *et al.* (2004b). Trophic action of leptin on hypothalamic neurons that regulate feeding. *Science* **304,** 108–110.

Campfield, L. A., *et al.* (1995). Recombinant mouse OB protein: Evidence for a peripheral signal linking adiposity and central neural networks. *Science* **269,** 546–549.

Carlo, A. S., *et al.* (2007). Early developmental expression of leptin receptor gene and [$^{125}$I] leptin binding in the rat forebrain. *J. Chem. Neuroanat.* **33,** 155–163.
Casabiell, X., *et al.* (1997). Presence of leptin in colostrum and/or breast milk from lactating mothers: A potential role in the regulation of neonatal food intake. *J. Clin. Endocrinol. Metab.* **82,** 4270–4273.
Cottrell, E. C., *et al.* (2009). Developmental changes in hypothalamic leptin receptor: Relationship with the postnatal leptin surge and energy balance neuropeptides in the postnatal rat. *Am. J. Physiol. Regul. Integr. Comp. Physiol.* **296,** R631–R639.
Dawson, R., *et al.* (1997). Attenuation of leptin-mediated effects by monosodium glutamate-induced arcuate nucleus damage. *Am. J. Physiol.* **273,** E202–E206.
Delahaye, F., *et al.* (2008). Maternal perinatal undernutrition drastically reduces postnatal leptin surge and affects the development of arcuate nucleus proopiomelanocortin neurons in neonatal male rat pups. *Endocrinology* **149,** 470–475.
Devaskar, S. U., *et al.* (1997). Developmental changes in ob gene expression and circulating leptin peptide concentrations. *Biochem. Biophys. Res. Commun.* **238,** 44–47.
Dulloo, A. G., *et al.* (2002). Leptin directly stimulates thermogenesis in skeletal muscle. *FEBS Lett.* **515,** 109–113.
Enriori, P. J., *et al.* (2007). Diet-induced obesity causes severe but reversible leptin resistance in arcuate melanocortin neurons. *Cell Metab.* **5,** 181–194.
Faouzi, M., *et al.* (2007). Differential accessibility of circulating leptin to individual hypothalamic sites. *Endocrinology* **148,** 5414–5423.
Frontini, A., *et al.* (2008). Leptin-dependent STAT3 phosphorylation in postnatal mouse hypothalamus. *Brain Res.* **1215,** 105–115.
Fruhbeck, G. (2006). Intracellular signalling pathways activated by leptin. *Biochem. J.* **393,** 7–20.
Halaas, J. L., *et al.* (1995). Weight-reducing effects of the plasma protein encoded by the obese gene. *Science* **269,** 543–546.
Harvey, J., *et al.* (2005). Leptin: A potential cognitive enhancer? *Biochem. Soc. Trans.* **33,** 1029–1032.
Hoggard, N., *et al.* (1997). Leptin and leptin receptor mRNA and protein expression in the murine fetus and placenta. *Proc. Natl. Acad. Sci. USA* **94,** 11073–11078.
Hoggard, N., *et al.* (2001). Leptin expression in placental and fetal tissues: Does leptin have a functional role? *Biochem. Soc. Trans.* **29,** 57–63.
Howard, J. K., *et al.* (2004). Enhanced leptin sensitivity and attenuation of diet-induced obesity in mice with haploinsufficiency of Socs3. *Nat. Med.* **10,** 734–738.
Kirk, S. L., *et al.* (2009). Maternal obesity induced by diet in rats permanently influences central processes regulating food intake in offspring. *PLoS ONE* **4,** e5870.
La Cava, A., and Matarese, G. (2004). The weight of leptin in immunity. *Nat. Rev. Immunol.* **4,** 371–379.
Maffei, M., *et al.* (1995). Leptin levels in human and rodent: Measurement of plasma leptin and ob RNA in obese and weight-reduced subjects. *Nat. Med.* **1,** 1155–1161.
Maroni, P., *et al.* (2003). Early intracellular events induced by in vivo leptin treatment in mouse skeletal muscle. *Mol. Cell. Endocrinol.* **201,** 109–121.
Matsuda, J., *et al.* (1999). Development changes in long-form leptin receptor expression and localization in rat brain. *Endocrinology* **140,** 5233–5238.
Mercer, J. G., *et al.* (1997). Regulation of leptin receptor and NPY gene expression in hypothalamus of leptin-treated obese (*ob/ob*) and cold-exposed lean mice. *FEBS Lett.* **402,** 185–188.
Minokoshi, Y., *et al.* (2002). Leptin stimulates fatty-acid oxidation by activating AMP-activated protein kinase. *Nature* **415,** 339–343.
Miralles, O., *et al.* (2006). A physiological role of breast milk leptin in body weight control in developing infants. *Obesity (Silver Spring)* **14,** 1371–1377.

Mistry, A. M., *et al.* (1999). Leptin alters metabolic rates before acquisition of its anorectic effect in developing neonatal mice. *Am. J. Physiol.* **277,** R742–R747.
Morash, B., *et al.* (1999). Leptin gene expression in the brain and pituitary gland. *Endocrinology* **140,** 5995–5998.
Morash, B., *et al.* (2001). Developmental regulation of leptin gene expression in rat brain and pituitary. *Mol. Cell. Endocrinol.* **185,** 151–159.
Morash, B. A., *et al.* (2003). Leptin receptors are developmentally regulated in rat pituitary and hypothalamus. *Mol. Cell. Endocrinol.* **210,** 1–8.
Mori, H., *et al.* (2004). Socs3 deficiency in the brain elevates leptin sensitivity and confers resistance to diet-induced obesity. *Nat. Med.* **10,** 739–743.
Mueller, W. M., *et al.* (1998). Evidence that glucose metabolism regulates leptin secretion from cultured rat adipocytes. *Endocrinology* **139,** 551–558.
Pelleymounter, M. A., *et al.* (1995). Effects of the obese gene product on body weight regulation in *ob/ob* mice. *Science* **269,** 540–543.
Pico, C., *et al.* (2007). The intake of physiological doses of leptin during lactation in rats prevents obesity in later life. *Int. J. Obes. (Lond.)* **31,** 1199–1209.
Proulx, K., *et al.* (2002). Leptin regulates appetite-related neuropeptides in the hypothalamus of developing rats without affecting food intake. *Endocrinology* **143,** 4683–4692.
Rayner, D. V., *et al.* (1997). Postnatal development of the ob gene system: Elevated leptin levels in suckling fa/fa rats. *Am. J. Physiol.* **273,** R446–R450.
Satoh, N., *et al.* (1997). The arcuate nucleus as a primary site of satiety effect of leptin in rats. *Neurosci. Lett.* **224,** 149–152.
Schwartz, M. W., *et al.* (2000). Central nervous system control of food intake. *Nature* **404,** 661–671.
Sena, A., *et al.* (1985). Brain myelin of genetically obese mice. *J. Neurol. Sci.* **68,** 233–243.
Siegrist-Kaiser, C. A., *et al.* (1997). Direct effects of leptin on brown and white adipose tissue. *J. Clin. Invest.* **100,** 2858–2864.
Smith, J. T., and Waddell, B. J. (2003). Developmental changes in plasma leptin and hypothalamic leptin receptor expression in the rat: Peripubertal changes and the emergence of sex differences. *J. Endocrinol.* **176,** 313–319.
Smith-Kirwin, S. M., *et al.* (1998). Leptin expression in human mammary epithelial cells and breast milk. *J. Clin. Endocrinol. Metab.* **83,** 1810–1813.
Srinivasan, M., *et al.* (2008). A high-carbohydrate diet in the immediate postnatal life of rats induces adaptations predisposing to adult-onset obesity. *J. Endocrinol.* **197,** 565–574.
Steppan, C. M., and Swick, A. G. (1999). A role for leptin in brain development. *Biochem. Biophys. Res. Commun.* **256,** 600–602.
Stocker, C., *et al.* (2004). Modulation of susceptibility to weight gain and insulin resistance in low birthweight rats by treatment of their mothers with leptin during pregnancy and lactation. *Int. J. Obes. Relat. Metab. Disord.* **28,** 129–136.
Stocker, C. J., *et al.* (2007). Prevention of diet-induced obesity and impaired glucose tolerance in rats following administration of leptin to their mothers. *Am. J. Physiol. Regul. Integr. Comp. Physiol.* **292,** R1810–R1818.
Tu, H., *et al.* (2008). Soluble receptor inhibits leptin transport. *J. Cell. Physiol.* **214,** 301–305.
Tune, J. D., and Considine, R. V. (2007). Effects of leptin on cardiovascular physiology. *J. Am. Soc. Hypertens.* **1,** 231–241.
Udagawa, J., *et al.* (2000). Expression of the long form of leptin receptor (Ob-Rb) mRNA in the brain of mouse embryos and newborn mice. *Brain Res.* **868,** 251–258.
Udagawa, J., *et al.* (2006). The role of leptin in the development of the cerebral cortex in mouse embryos. *Endocrinology* **147,** 647–658.
Unger, R. H. (2003). The physiology of cellular liporegulation. *Annu. Rev. Physiol.* **65,** 333–347.

Valerio, A., *et al.* (2006). Leptin increases axonal growth cone size in developing mouse cortical neurons by convergent signals inactivating glycogen synthase kinase-3beta. *J. Biol. Chem.* **281,** 12950–12958.
Vickers, M. H., *et al.* (2000). Fetal origins of hyperphagia, obesity, and hypertension and postnatal amplification by hypercaloric nutrition. *Am. J. Physiol. Endocrinol. Metab.* **279,** E83–E87.
Vickers, M. H., *et al.* (2005). Neonatal leptin treatment reverses developmental programming. *Endocrinology* **146,** 4211–4216.
Vickers, M. H., *et al.* (2008). The effect of neonatal leptin treatment on postnatal weight gain in male rats is dependent on maternal nutritional status during pregnancy. *Endocrinology* **149**, 1906–1913.
Yura, S., *et al.* (2005). Role of premature leptin surge in obesity resulting from intrauterine undernutrition. *Cell Metab.* **1,** 371–378.
Yura, S., *et al.* (2008). Neonatal exposure to leptin augments diet-induced obesity in leptin-deficient *ob/ob* mice. *Obesity (Silver Spring)* **16**, 1289–1295.
Zhang, Y., *et al.* (1994). Positional cloning of the mouse obese gene and its human homologue. *Nature* **372,** 425–432.

CHAPTER TWELVE

# Regulation of Hippocampal Synaptic Plasticity by Estrogen and Progesterone

Michael R. Foy,* Michel Baudry,† Garnik K. Akopian,† *and* Richard F. Thompson†

## Contents



## Abstract

Accumulating evidence indicates that the ovarian steroid hormones estrogen and progesterone regulate a wide variety of nonreproductive functions in the central nervous system by interacting with several molecular and cellular processes. A growing literature reporting results obtained in rodent models suggests that 17$\beta$-estradiol, the most potent of the biologically relevant estrogens, facilitates some forms of learning and memory, and in particular, those involving hippocampus-dependent tasks. Hippocampal long-term potentiation and long-term depression of synaptic transmission are types of synaptic plasticity that have been extensively studied, as they are considered as cellular models of memory formation in the brain. In this chapter, we review the literature that analyzes and compares the effects of estrogen and progesterone on synaptic transmission and synaptic plasticity in rodents. Understanding the nonreproductive functions of estrogen and progesterone in the

* Department of Psychology, Loyola Marymount University, Los Angeles, California, USA
† Neuroscience Program, University of Southern California, Los Angeles, California, USA

*Vitamins and Hormones*, Volume 82
ISSN 0083-6729, DOI: 10.1016/S0083-6729(10)82012-6

hippocampus has far-reaching implications not only for our basic understanding of neuroendocrinology and neurobiology, but also for developing better treatment of age-related diseases such as Alzheimer's disease. 

# I. Introduction

Steroid hormones play an essential role in a variety of key functions, including reproduction, sexual differentiation, brain development, cognition, memory, and behavior. The nervous system is a major target of steroid hormone action and contains specific receptors for circulating hormones secreted from peripheral organs such as the adrenal cortex, testis, and ovary. Gonadal steroids, such as the estrogens, androgens, and progestins, function not only at the genomic level through classic receptors that belong to the superfamily of nuclear receptors (Guerriero, 2009), but also nongenomically through G protein-coupled steroid receptors and membrane-localized steroid receptors (Hammes and Levin, 2007). While many of the effects of estrogen and other hormones that have a prolonged latency and duration of action in the brain can be readily explained by the genomic mechanism of action (McEwen and Alves, 1999), steroid hormones have also been found to produce rapid, short-term effects on the electrophysiological properties of neurons with latencies and durations on the scale of milliseconds to minutes (Foy, 2001; Teyler *et al.*, 1980). Since the activation of transcriptional and translational mechanisms by intracellular steroid hormone receptors requires a longer latency, physiological responses occurring within extremely short latencies have been presumed to involve nongenomic, specific membrane receptors. This chapter will focus on the ovarian hormones, 17$\beta$-estradiol (E2) and progesterone (P4), as they are the primary hormones that we, and others, have shown to regulate hippocampal synaptic plasticity.

Synaptic plasticity is a term used to describe several forms of long-lasting, activity-dependent changes in synaptic strength; it also refers to the ability of neuronal circuits to undergo functional or organizational changes due to previous activity, and is strongly associated with forebrain structures, including the hippocampal formation. In the late 1940s, Hebb postulated that synaptic plasticity was an essential way for the brain to be modified by experience and practice, a concept that has since been associated with the notion of hebbian synapses (Berlucchi and Buchtel, 2009). A considerable body of work has now demonstrated that hippocampal long-term potentiation (LTP) and long-term depression (LTD) of synaptic transmission were two forms of synaptic plasticity that could be considered as cellular models of memory trace formation in the brain (Baudry *et al.*, 2000; Bear and Malenka, 1994; Bliss and Collingridge, 1993). While the molecular and synaptic mechanisms underlying LTP have been studied extensively, there is a relative paucity of studies demonstrating the critical role of LTP in behavioral learning and memory (Shors and Matzel, 1997; but see Lee and Silva, 2009). Nonetheless, whether LTP (and/or LTD) is or is not the substrate of synaptic modifications that occur during learning in the forebrain structures of vertebrates, the studies of its mechanisms have revealed the existence of a number of processes that undoubtedly play critical roles in memory formation (Bi and Poo, 2001; Fanselow and Poulos, 2005).

In area CA1 of the hippocampus, LTP, the most widely studied form of synaptic plasticity, represents an activity-dependent modification of synaptic efficacy that results in a long-lasting increase in synaptic transmission following a brief burst of high-frequency stimulation of the Schaffer collateral input to pyramidal neurons, and may act as a potential mechanism for learning and memory function. Hippocampal LTP requires *N*-methyl-D-aspartate (NMDA) receptor activation for its induction, and an increase in α-amino-3-hydroxy-5-methyl-4-isoxazoleproprionate (AMPA) receptor function for its expression and maintenance (Lisman, 2003; Malenka and Nicoll, 1999; Stanton, 1996). Conversely, hippocampal LTD is a form of activity-dependent synaptic plasticity where the repeated activation of specific hippocampal inputs results in a decrease in the responsiveness of activated neurons, requires the activation of NMDA and metabotropic glutamate receptors, and is due to the internalization of AMPA receptors (Malenka and Bear, 2004; Stanton, 1996). In this chapter, we will discuss the relationship between steroid hormone action and hippocampal synaptic plasticity by analyzing the effects of E2 and P4 on hippocampal LTP and LTD.

## II. Estrogen and Hippocampus

Over more than 40 years ago, estrogen was found to promote changes in synaptic plasticity within the nervous system. In one pioneering study, decreased hippocampal seizure thresholds were found in animals primed with estrogen and also during proestrus, the time of the estrous cycle when estrogen levels are at their highest levels (Terasawa and Timiras, 1968). In humans, changes in the electrical activity of nervous system tissue correlate with hormonal factors that appear to play a role in catamenial epilepsy, a form of epilepsy in which the likelihood of seizures varies during the menstrual cycle. Many women with catamenial epilepsy experience a sharp increase in seizure frequency immediately before menstruation, when estrogen concentrations relative to those of progesterone are at their highest levels (Backstrom, 1976; Reddy, 2009). Changes in hippocampal responsiveness correlate with estrogen activity, as LTP is maximal in female rats during the afternoon of proestrus, when endogenous estrogen concentrations are highest (Warren *et al.*, 1995). Furthermore, the induction of hippocampal LTP is facilitated in ovariectomized rats treated with E2 as compared to untreated ovariectomized rats (Cordoba Montoya and Carrer, 1997).

The development of *in vitro* models to study the mechanisms of synaptic plasticity has provided researchers better tools to investigate how estrogen regulates synaptic excitability in the nervous system, and, in particular, in the hippocampus (Teyler, 1980). It should be noted, however, that the binding of $^3$H-estradiol in the hippocampus is less than that seen in hypothalamus and related diencephalic structures (McEwen and Alves, 1999; McEwen *et al.*, 1975). Nonetheless, studies by Teyler and colleagues using *in vitro* hippocampal slice preparations showed that gonadal steroids dramatically affected neuronal excitability in specific pathways of the rodent hippocampus (Teyler *et al.*, 1980; Vardaris and Teyler, 1980). In the initial series of experiments, extracellular monosynaptic population field responses recorded from area CA1 of hippocampal slices from male and female rats were monitored before and after the addition of E2 (100 pM) to the slice incubation medium (artificial cerebrospinal fluid; aCSF). In male rats, E2 produced a rapid (<10 min) enhancement of population field responses evoked by the stimulation of the afferents to CA1 pyramidal cells (Fig. 12.1). This was the first published report demonstrating that picomolar concentrations of the gonadal steroid E2 directly enhanced what is now known to be glutamatergic synaptic transmission in the hippocampus (Teyler *et al.*, 1980).

## III. Estrogen, NMDA, and AMPA Receptor Regulation

Although the mechanism of action of gonadal steroids in the hippocampus is not entirely understood, it is likely to be receptor-mediated via the DNA-binding domains of estrogen receptor $\alpha$ and $\beta$ (ER$\alpha$, ER$\beta$), and plasma membrane estrogen receptors (mER). Electrophysiological experiments conducted in the hippocampus found that there was no facilitation of field responses when the inactive estrogen, 17$\alpha$-estradiol, was added to the hippocampal slice medium (Foy and Teyler, 1983), and the further addition of E2 to slices already treated with 17$\alpha$-estradiol no longer resulted in an increased response, as observed in the presence of E2 alone (Foy and Teyler, 1983; Wong and Moss, 1991, 1992). Similar results were found when the estrogen receptor antagonist tamoxifen was applied to hippocampal slices before the addition of E2 (Foy, 1983). The ability of 17$\alpha$-estradiol and tamoxifen to block the effects of E2 on hippocampal excitability provides

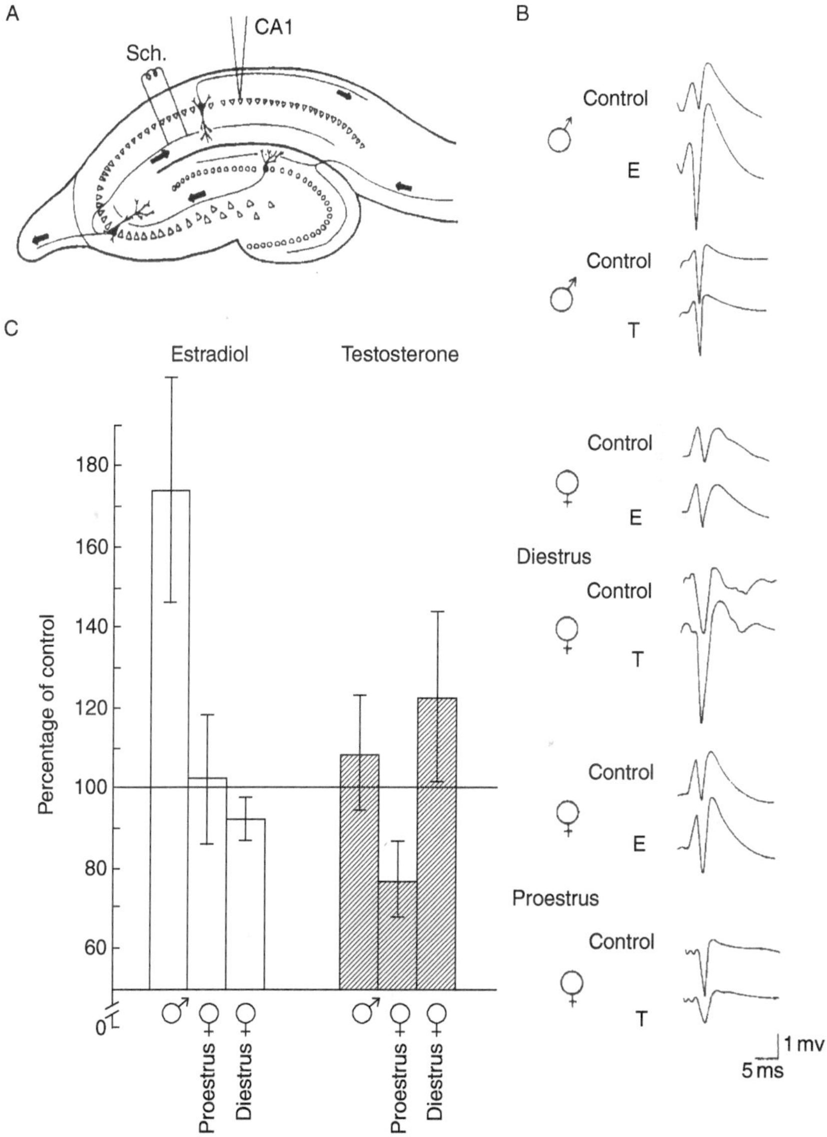


**Figure 12.1** (A) *Diagram of a transverse hippocampal slice.* Stimulating electrodes were located in the afferent pathway, which contains the Schaffer (Sch.) collaterals. Recording micropipettes were situated in the pyramidal cell body layer in CA1. Cells of this subfield receive monosynaptic input from the CA3 pyramids via the Schaffer collateral system. (B) Representative field potentials from slice preparations in the various experimental conditions. Extracellular population spike responses to a given stimulus intensity are shown from the control period (before steroid administration) and after the administration of $10^{-10}$ M 17$\beta$-estradiol (E) or $10^{-10}$ M testosterone (T). Potentials from slices obtained from males and from proestrus and diestrus females are shown for

strong evidence that the rapid physiological modulation of gonadal hormones is most likely due to the activation of mER.

*In vitro* intracellular recordings of CA1 neurons from adult ovariectomized female rats have shown that the addition of E2 increases synaptic excitability in part by enhancing the magnitude of AMPA receptor-mediated synaptic responses (Wong and Moss, 1992). The rapid onset of the increased excitability, and its blockade by 6-cyano-7-nitroquinaxaline (CNQX, an AMPA receptor antagonist) but not by D-2-amino-5-phosphonovalerate (D-APV a competitive NMDA receptor antagonist), supported a postsynaptic membrane site of action resulting in enhanced nonNMDA glutamate receptor function. Later studies using whole cell recordings found that acute E2 application also potentiated kainate-induced currents in a subpopulation of CA1 cells (Gu and Moss, 1996), although a direct interaction between estrogen and the receptor channel was not indicated (Wong and Moss, 1994).

A large body of evidence demonstrates that E2-mediated regulation of synapse formation is dependent on NMDA receptor activation. Morphological studies during the course of neuronal development conducted in cultured neurons prepared from embryonic day 18 rat fetuses have shown that estrogenic steroids exert a growth-promoting, neurotrophic effect on hippocampal and cortical neurons via a mechanism that requires NMDA receptor activation (Brinton *et al.*, 1997a,b). *In vivo* studies using adult ovariectomized female rats have also revealed a proliferation of dendritic spines in hippocampal CA1 pyramidal cells after E2 treatment that could be prevented by the blockade of NMDA receptors, but not of AMPA or muscarinic receptors (Woolley and McEwen, 1994). Other reports using adult ovariectomized female rats provided evidence that chronic E2 treatment increases the number of NMDA receptor-binding sites and NMDA receptor-mediated responses (Gazzaley *et al.*, 1996; Woolley *et al.*, 1997). Collectively, these studies indicate that estrogen and NMDA receptors are heavily involved in synapse formation.

Because of the voltage-dependent blockade of the NMDA receptor channel by $Mg^{2+}$ and the slow kinetics of the channel opening relative to that of the AMPA receptor, there is only a minor NMDA receptor-mediated component of the excitatory postsynaptic potential (EPSP) evoked by the low-frequency stimulation of glutamatergic afferents. This NMDA receptor component can be enhanced with low $Mg^{2+}$ concentrations or the high-frequency stimulation patterns used to induce LTP, which eliminate

the purposes of comparison. All potentials are single sweeps recorded at the same voltage and time scales. (C) Bar graph summarizing the major experimental outcomes. Values on the ordinate are mean percentages of spike amplitudes after steroid administration. Data for each condition are from 6 to 10 animals, each contributing one slice. Cursors representing magnitude of variability (standard error of the mean) are shown for each bar. (Reprinted with permission from Teyler *et al.* (1980). Copyright 2009 American Association for the Advancement of Science).

the $Mg^{2+}$ blockade due to membrane depolarization resulting from the summation of overlapping EPSPs (Xie *et al.*, 1992). In experiments using low $Mg^{2+}$ concentrations and in the presence of the AMPA receptor antagonist 6,7-dinitroquinoxaline-2,3-dione (DNQX), an acute application of E2 in adult male rat hippocampal slices resulted in a rapid increase in the amplitude of NMDA receptor-mediated EPSPs evoked by the stimulation of the Schaffer collaterals (Foy *et al.*, 1999). The effect of E2 on pharmacologically isolated NMDA receptor-mediated synaptic responses was such that concentrations of E2 greater than 10 nM induced seizure activity in hippocampal neurons, and lower concentrations (1 nM) markedly increased the amplitude of NMDA receptor-mediated EPSPs.

## IV. Estrogen and Hippocampal LTP

To investigate the effect of estrogen on synaptic plasticity, hippocampal slices from adult male rats were treated with E2 before applying high-frequency stimulation to induce LTP at the Schaffer collateral inputs to CA1. When LTP magnitude was assessed after high-frequency stimulation, field EPSP (fEPSP) values were increased significantly in E2-treated slices as compared to vehicle-treated slices (Fig. 12.2). fEPSP mean increases in

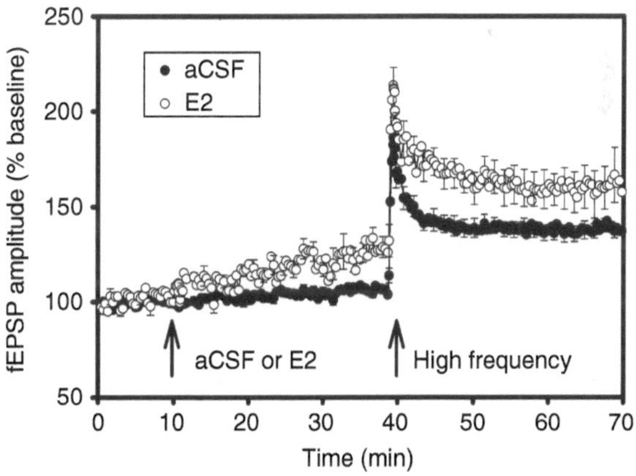


**Figure 12.2** *Field EPSP (fEPSP) recordings in area CA1.* All hippocampal slices were perfused with aCSF for 10 min to obtain fEPSP slope and amplitude percentage baseline data. After 10 min of baseline recording, experimental slices were perfused with 100 pM 17$\beta$-estradiol (E2). Control slices continued to be perfused with aCSF. After 30 min of either E2 or aCSF perfusion, all slices received high-frequency stimulation, designed to induce long-term potentiation. Data points represent averaged fEPSP slope $\pm$ SEM (taken at each 20 s sweep) for experimental (E2-treated) and control (aCSF) hippocampal slices. (Reprinted with permission from Foy *et al.* (1999). Copyright 2009 American Physiology Society).

slope was 192% (experimental) versus 154% (control). Thus, hippocampal slices from adult male rats treated with E2 exhibited a pronounced, persisting, and significant increase in LTP as measured by both population fEPSP slope and fEPSP amplitude (Foy *et al.*, 1999, 2008b).

To further evaluate the effects of E2 on the magnitude of hippocampal LTP, the intensity of afferent stimulation to Schaffer collaterals in slices perfused with E2 was decreased in order to reduce baseline values to preE2 levels immediately before the delivery of the high-frequency stimulation train used to elicit LTP (Bi *et al.*, 2000). Under these conditions, E2 still produced an increase in the magnitude of LTP in adult male rat hippocampal slices as compared to that obtained in vehicle-treated slices (Fig. 12.3). These findings indicate that E2-induced enhancement of hippocampal LTP is not due to simply a change in the basal EPSP level, but is

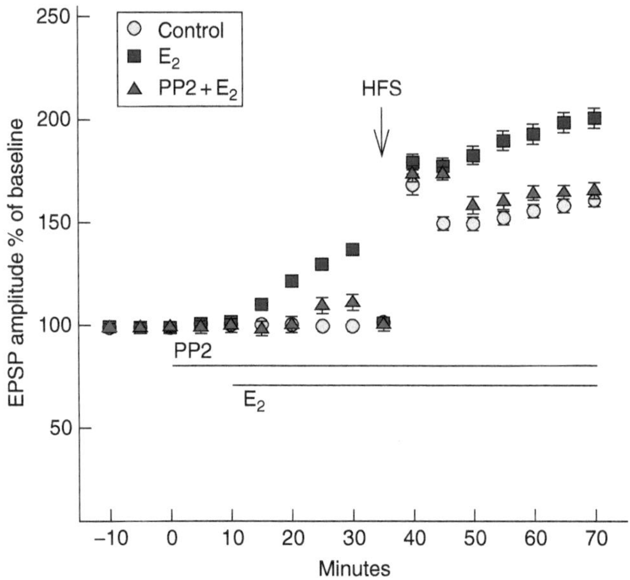


**Figure 12.3** Effects of an src inhibitor (PP2) on 17$\beta$-estradiol (E2)-mediated enhancement of EPSP amplitude and degree of LTP in acute hippocampal slices. A stimulating electrode was located in CA3, and a recording electrode was located in the stratum radiatum of CA1. Extracellular EPSPs were evoked by stimulation every 30 s, and the amplitude of the EPSP was measured. After a stable baseline was recorded, PP2 (10 $\mu$M) was added at the indicated time. Likewise, E2 (1 nM) was added at the indicated time. After resetting the stimulation intensity to obtain an EPSP of the same amplitude as before treatment with PP2 or E2, high-frequency stimulation was delivered and low-frequency stimulation was resumed. PP2 blocked the estrogen enhancement of LTP, but had no effect on LTP itself. Results are expressed as percentages of predrug values and are means ± SEM of 6–10 experiments. HFS, high-frequency stimulation. (Reprinted with permission from Bi *et al.* (2000). Copyright 2009 National Academy of Sciences, USA.)

more likely due to the biochemical activation of an intracellular cascade, presumably mediated by the activation of a src tyrosine pathway that enhances NMDA receptor function. More recent studies have provided evidence that the effects of E2 on LTP induction are due to the activation of an intracellular cascade, leading to increased actin polymerization and increased numbers of AMPA receptors (Kramar *et al.*, 2009; Zadran *et al.*, 2009).

In another series of animal studies, hormonal changes during the estrous cycle were correlated with changes in synaptic plasticity in rats. Hippocampal slices were prepared from cycling female rats in diestrus (low estrogen concentration) or proestrus (high estrogen concentration), and LTP was elicited by high-frequency stimulation. The difference in LTP magnitude between these groups following high-frequency stimulation was dramatic: slices from rats in proestrus exhibited LTP representing about a 50% increase over baseline, whereas slices from rats in diestrus had LTP values representing about a 25% increase over baseline (Bi *et al.*, 2001) (Fig. 12.4). These findings support the original observations of Teyler *et al.* (1980) who identified changes in baseline synaptic transmission that were correlated with the phase of the estrous cycle in female rats at the time of hippocampal slice preparation.

The electrophysiological work discussed earlier has shown that female rats in proestrus exhibited an increased magnitude of hippocampal LTP compared to females in diestrus. A more recent study (Foy *et al.*, 2004) reexamined the effects of E2 on hippocampal LTP during these two critical time periods in the rat estrous cycle, proestrus, and diestrus. Estrous cycles of adult (3–5 months) Sprague–Dawley rats were monitored for 10 days prior to any physiological experiments, and hippocampal slices were prepared from rats that were either in proestrus or diestrus. Recording and stimulating electrodes were positioned in the dendrites of area CA1 and Schaffer collaterals, respectively. Baseline stimulation was adjusted to elicit 50% of the maximum fEPSP amplitude. After 10 min of stable baseline stimulation, aCSF or E2 at a concentration of 100 pM (experimental group) was perfused for 30 min, and LTP was induced by a brief period of high-frequency stimulation. Subsequent synaptic responses were monitored for 30 min postLTP induction. The magnitude of LTP induced in area CA1 was larger in vehicle-treated slices from proestrus rats, compared to slices from diestrus rats, as previously reported (Bi *et al.*, 2001). Surprisingly, addition of E2 increased LTP in slices from diestrus rats, while it decreased LTP in slices from proestrus rats (Fig. 12.5). These observations indicate that E2 alters hippocampal LTP in female rats, depending on the state of their estrous cycle (i.e., on the levels of circulating estrogen). In cycling female rats, when endogenous circulating levels of estrogen are at their highest levels (i.e., during proestrus), LTP magnitude is high, and exogenously applied estrogen during proestrus decreases LTP magnitude, possibly through the activation of an inhibitory or ceiling effect. When endogenous

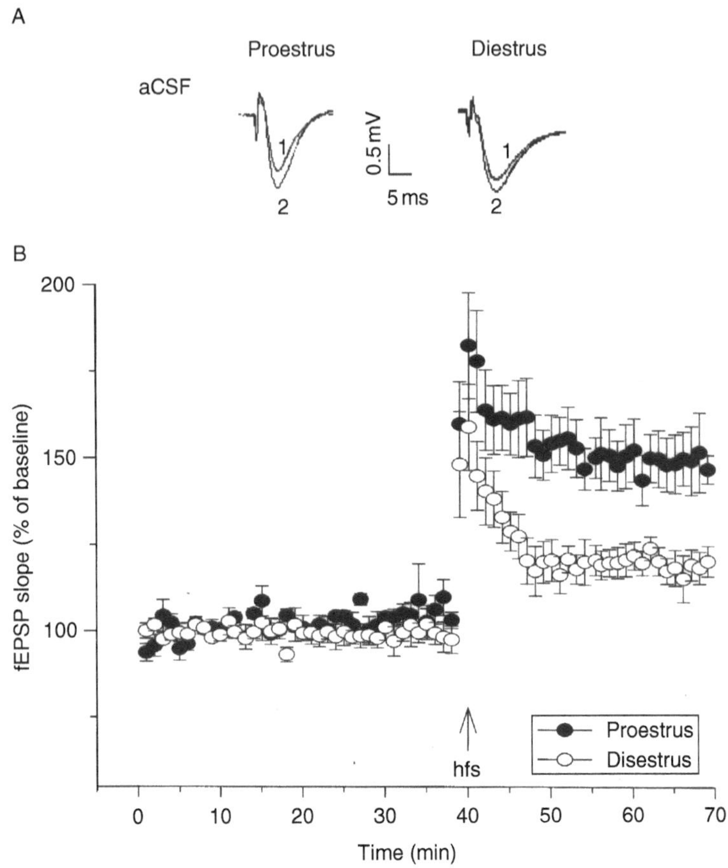


**Figure 12.4** Changes in LTP in field CA1 of hippocampal slices from female rats in proestrus and diestrus. Hippocampal slices from female rats in proestrus or diestrus were prepared and fEPSP amplitude and slope values were obtained for each slice and averaged across slices to produce one average before and after the train of high-frequency stimulation (HFS). fEPSP amplitudes and slopes were normalized for the 10 min preHFS period for each slice. Separate ANOVAs and planned two-tailed *t* tests for the pre- and postHFS periods were used to evaluate the effects of estrous cycle on fEPSP slope and amplitude. (A) Representative waveforms from female rats in proestrus and diestrus for preHFS (1) and postHFS (2) periods. (B) Means ± SEM of fEPSP slopes recorded in slices from female rats in proestrus (filled circles; $n = 6$) and diestrus (open circles; $n = 5$). (Reprinted with permission from Bi *et al.* (2001). Copyright 2009 National Academy of Sciences, USA.)

circulating levels of estrogen are at their lowest levels (i.e., during diestrus), the situation is completely reversed from that observed in the proestrus state. Here, LTP magnitude is low, and exogenously applied estrogen increases LTP magnitude under this condition.

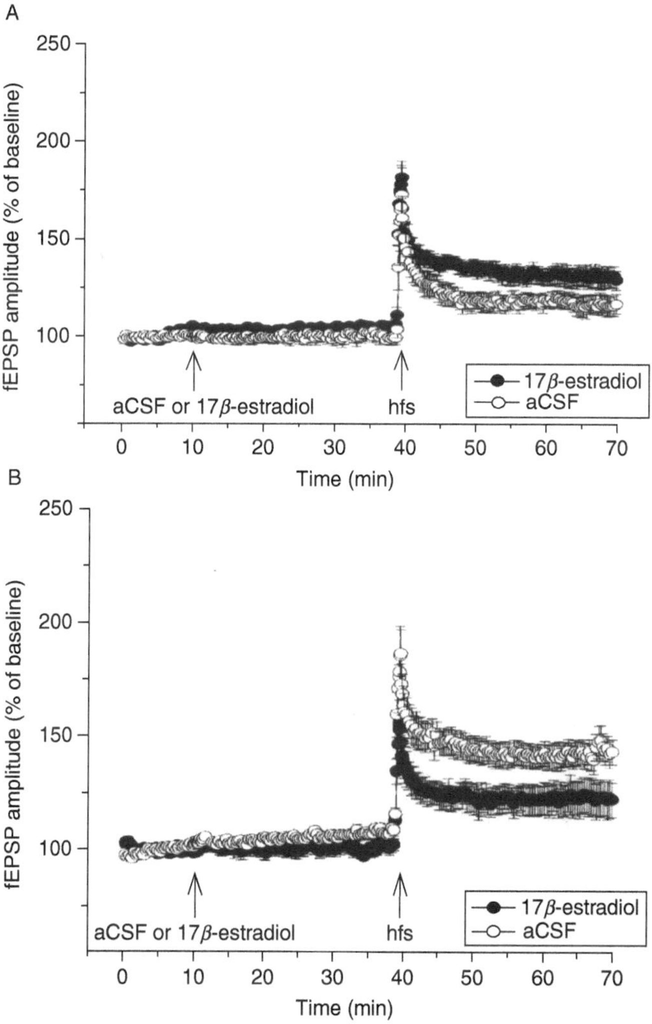


**Figure 12.5** Changes in LTP in field CA1 of hippocampal slices from female rats in proestrus and diestrus, and with 17$\beta$-estradiol (experimental) or aCSF (control) treatment. Hippocampal slices from female rats in proestrus or diestrus were prepared as described. (A) Means $\pm$ SEM of fEPSP amplitudes recorded following tetanus in slices from female rats in diestrus. 17$\beta$-estradiol (filled circles) enhanced LTP relative to control aCSF (open circles). (B) Means $\pm$ SEM of fEPSP amplitudes recorded following tetanus in slices from female rats in proestrus. 17$\beta$-estradiol (filled circles) impaired LTP relative to control aCSF (open circles). (Reprinted with permission from (Foy *et al.*, 2004). Copyright 2009 Cambridge University Press.)

These results indicate that cyclic changes in estrogen levels occurring during the estrous cycle in female rats are associated with changes in the magnitude of LTP recorded from hippocampal CA1 neurons. They also corroborate several results mentioned earlier indicating the facilitation of LTP induction by estrogen in ovariectomized female rats (Cordoba Montoya and Carrer, 1997) and the increased LTP in the afternoon of proestrus of female rats (Warren *et al.*, 1995), and are in good agreement with the results of a study showing improved memory performance with high estrogen levels in female rats (Leuner *et al.*, 2004).

## V. Estrogen and Hippocampal LTD

Several studies have shown that during aging, when memory function declines, the processes of synaptic plasticity in the hippocampus are altered; more specifically, while LTP is impaired, LTD is enhanced (Barnes, 1979, 1994; Barnes *et al.*, 1992; Foster, 1999; Foster and Norris, 1997; Geinisman *et al.*, 1995; Landfield and Lynch, 1977; Landfield *et al.*, 1986; Norris *et al.*, 1996, 1998). The effect of aging on LTD was replicated in a group of aged male rats, and a profound action of estrogen on this process in aged male rats was discovered (Foy *et al.*, 2008b; Vouimba *et al.*, 2000). LTD was induced using standard conditions in CA1 region of hippocampal slices prepared from adult (3–5 months) and aged (18–24 months) rats. In agreement with earlier studies (Foster, 1999; Foster and Norris, 1997), the standard protocol for inducing LTD (1 Hz for 15 min) resulted in little or no LTD in slices from adult animals, but in marked LTD in slices from aged animals (Fig. 12.6A) (Foy *et al.*, 2008b). Infusion of E2 in slices caused a slight increase in synaptic transmission (baseline), as in previous studies. It had little effect on LTD in slices from adult animals, but markedly attenuated LTD in slices from aged animals (Fig. 12.6B). Thus, the prevention by E2 of age-related LTD enhancement may account, in part, for the protective effects of estrogen on memory functions in aged organisms.

## VI. Progesterone and Hippocampus

In contrast to the extensive studies examining the physiological role of estrogen in the hippocampus, the effects of P4 on hippocampal synaptic transmission and plasticity have yet to be fully investigated. One study reported that P4 ($10^{-5}$ M) had no effect on LTP recorded *in vitro* from hippocampal (CA1) slices in adult rats, but no nondrug control was provided to compare the experimental (P4) condition, and the experimental subjects were a combined group of gonadally intact male and female rats (Ito

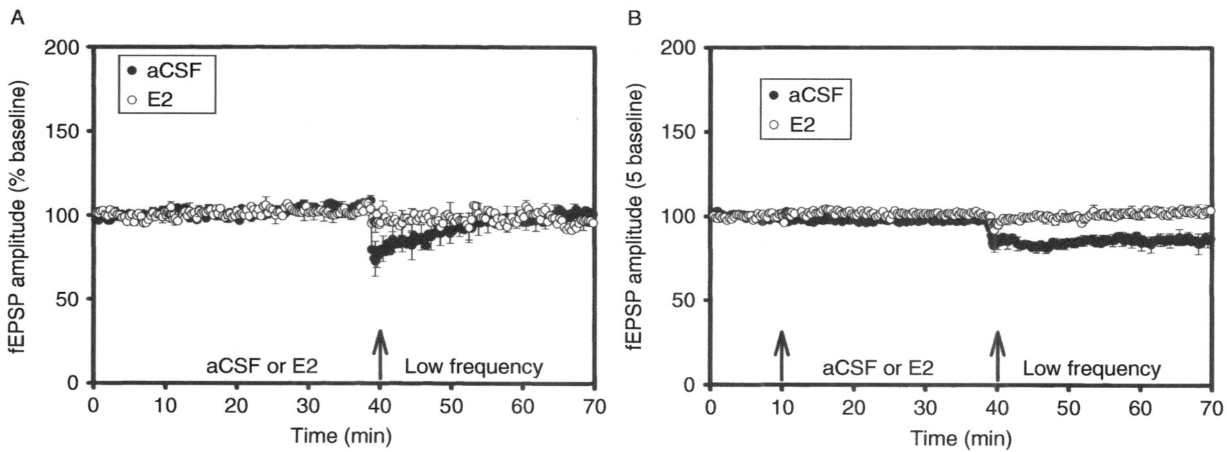


**Figure 12.6** *Long-term depression.* (A) LTD. Adult aCSF versus E2. In this figure, following baseline and drug/aCSF periods, slices received low-frequency stimulation (low frequency) to elicit long-term depression. LTD that was initially induced in the aCSF adult males quickly diminished. The arrows indicate the time at which aCSF/E2 was applied, and when low-frequency stimulation (900 pulses at 1 Hz) was delivered. (B) LTD. Aged aCSF versus E2. LTD was examined in slices from aged rats, with fEPSP comparisons between aCSF versus E2. LFS delivered to aged rat slices perfused with 17$\beta$-estradiol failed to induce robust LTD. (Reprinted with permission from Foy *et al.* (2008b). Copyright 2009 American Psychological Association).

*et al.*, 1999). Another study reported that P4 ($10^{-8}$ M) significantly enhanced synaptic transmission in CA1, but following seizure-inducing tetanus, P4 decreased both field potential and population spike responses, and decreased the duration of after-discharges (Edwards *et al.*, 2000). In a study using whole cell patch clamp of pyramidal neurons from the slices of prelimbic cortex, it was reported that P4 (100 $\mu$M) had no effect on the frequency of EPSCs, but inhibited dopamine-induced increases in EPSCs (Feng *et al.*, 2004). P4 dose–response functions were not performed in any of these studies, and no previous work had investigated the effect of P4 on hippocampal LTD. In summary, the existing data support the hypothesis that progestagens and hormone therapy modify certain properties of hippocampal physiology and function, as discussed subsequently in more detail.

## VII. Progesterone and Progesterone Receptors

The classical nuclear progesterone receptor (cPR) was characterized in the 1970s and has been localized to many regions of the CNS, including the hippocampus, cortex, hypothalamus, and cerebellum (Camacho-Arroyo *et al.*, 1998; Guerra-Araiza *et al.*, 2003; Hagihara *et al.*, 1992). P4 exerts its effects by binding and activating specific cellular receptors through its cognate receptors (PR), classically defined as ligand-activated transcription factors (Brinton *et al.*, 2008). Two major isoforms of cPR are known to exist, the full-length B isoform (PRB) and the N-terminal-truncated A isoform (PRA) (Conneely *et al.*, 1987). While many of P4's effects can be explained by the classical nuclear DNA-binding model of steroid action, P4 can also exert rapid effects on diverse signaling pathways that are independent of transcriptional or genomic regulation (Losel and Wehling, 2003). The discovery of plasma membrane progestagen receptors occurred in 2003, and included the progesterone receptor membrane component 1 (PGRMCI) and a family of membrane progesterone receptors (mPR), for example, mPR$\alpha$, $\beta$, and $\gamma$ (Gellersen *et al.*, 2008, 2009). It is through the activation of these membrane receptors (PGRMCI and mPR) that P4 is suspected of rapidly regulating basic cellular metabolism and homeostasis.

## VIII. Progesterone and Hippocampal LTP and LTD

A recent study examined the effects of several concentrations of P4 ($10^{-9}$, $10^{-8}$, $10^{-7}$, and $10^{-6}$ M) on hippocampal basal synaptic transmission, LTP and LTD (Foy *et al.*, 2008a). P4 resulted in a rapid ($\sim$30 min) decrease in baseline synaptic transmission only at the $10^{-6}$ M concentration (Fig. 12.7A). In addition, P4 concentrations of $10^{-7}$ and $10^{-6}$ M markedly decreased LTP in CA1 neurons from adult, ovariectomized rats (Fig. 12.7B). No changes in synaptic responses were found following low-frequency stimulation to induce hippocampal LTD at any of the P4 concentrations tested (Fig. 12.8A). Thus, in contrast to P4's effect on LTP, P4 did not affect LTD, suggesting that the high-frequency stimulation used to induce hippocampal LTP might increase the inhibitory action of P4 on excitatory synaptic transmission in the hippocampus.

The attenuation of baseline synaptic transmission and decreased magnitude of LTP in hippocampal slices resulting from the application of $10^{-6}$ M P4 could be due to $GABA_A$ receptor activation (Mitchell *et al.*, 2008), either by a direct action of P4 or the action of one of its metabolites. Intracellular recordings were made from individual pyramidal neurons in the presence of NMDA and AMPA receptor antagonists, APV (50 $\mu$M) and CNQX (10 $\mu$M), respectively (Foy *et al.*, 2008a). The isolated currents were proven to be $GABA_A$ receptor-mediated since they were completely blocked by the $GABA_A$ receptor antagonist picrotoxin (100 $\mu$M). Under these conditions, in two of five cells that were recorded, P4 at $10^{-6}$ M increased $GABA_A$ receptor-mediated current (Fig. 12.8B).

## IX. Conclusions

This chapter summarized several fundamental characteristics of the effects of the steroid hormones, estrogen and progesterone, on synaptic plasticity in the mammalian nervous system. Estrogen rapidly enhances both NMDA and AMPA receptor/channel responses elicited by glutamate released from excitatory presynaptic terminals via presumed plasma membrane mechanisms triggering several intracellular signaling cascades. Studies have yet to confirm whether the E2 effects are due to an action directly on the receptors, or indirectly via second messenger processes that in turn influence NMDA and AMPA receptor/channel function.

Estrogen also markedly enhances hippocampal LTP. The LTP enhancement after acute E2 application is apparently due to an increase in NMDA and AMPA receptor function, as well as to the activation of cascades, leading to cytoskeleton regulation. Both possibilities (e.g., NMDA and AMPA activation) are consistent with the effects of E2 on complex signaling intracellular pathways. Changes in estrogen levels in cycling female rats have also been correlated with changes in synaptic plasticity, as measured by LTP magnitude.

Similar to estrogen, progesterone also acts rapidly in the hippocampus, but instead of increasing basal synaptic transmission and LTP, P4 decreases hippocampal basal synaptic transmission and LTP. Intracellular recordings so far suggest that this effect is mediated, at least in part, by the activation of inhibitory $GABA_A$ receptor-mediated activity. This interpretation is

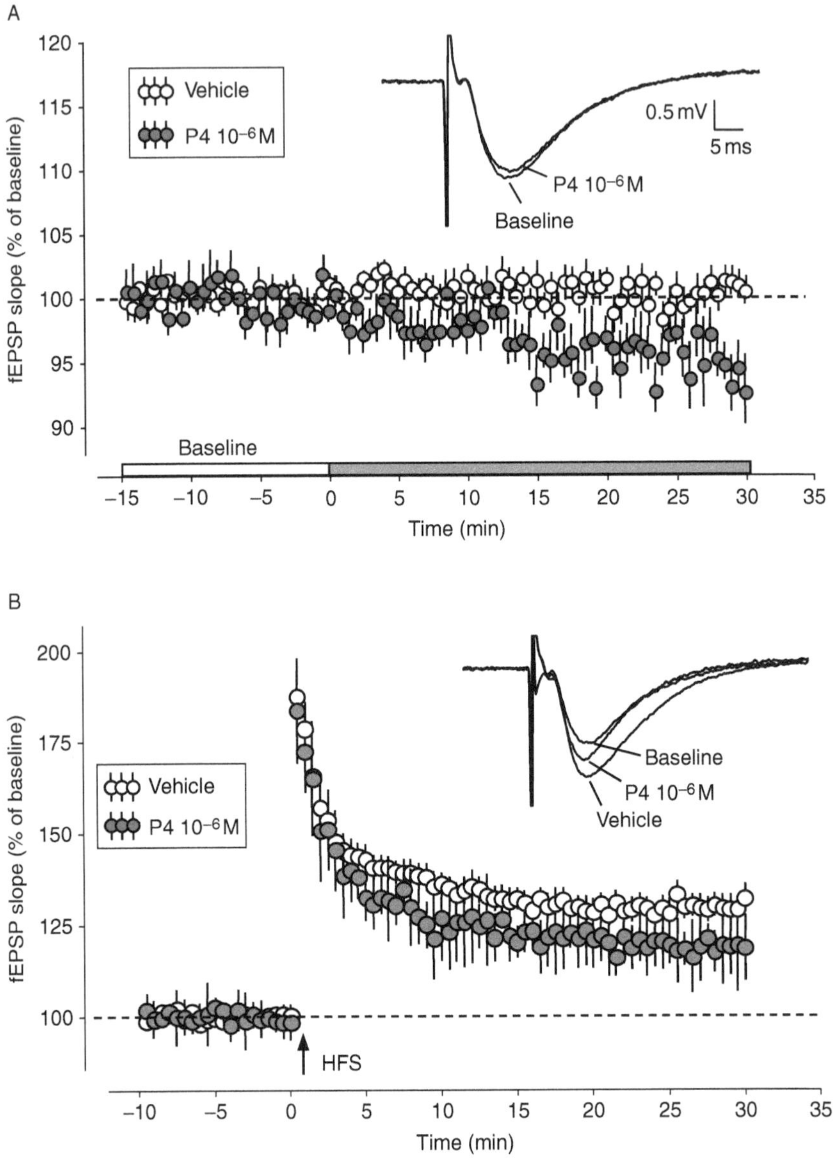


**Figure 12.7** fEPSP baseline synaptic transmission and LTP with progesterone (A). fEPSP baseline slope values (vehicle) were measured for 15 min prior to the infusion of P4 ($10^{-6}$ M). P4 began to elicit a significant decrease in fEPSP during the next 30 min. (B) fEPSP slope values (vehicle and P4) were measured prior to and following high-frequency stimulation (HFS) designed to induce long-term potentiation (LTP). Following HFS, slices perfused with P4 showed a significant decrease in fEPSP slope values compared to vehicle, resulting in decreased LTP. (Reprinted with permission from Foy *et al.* (2008a). Copyright 2009 Cold Spring Harbor Laboratory Press).

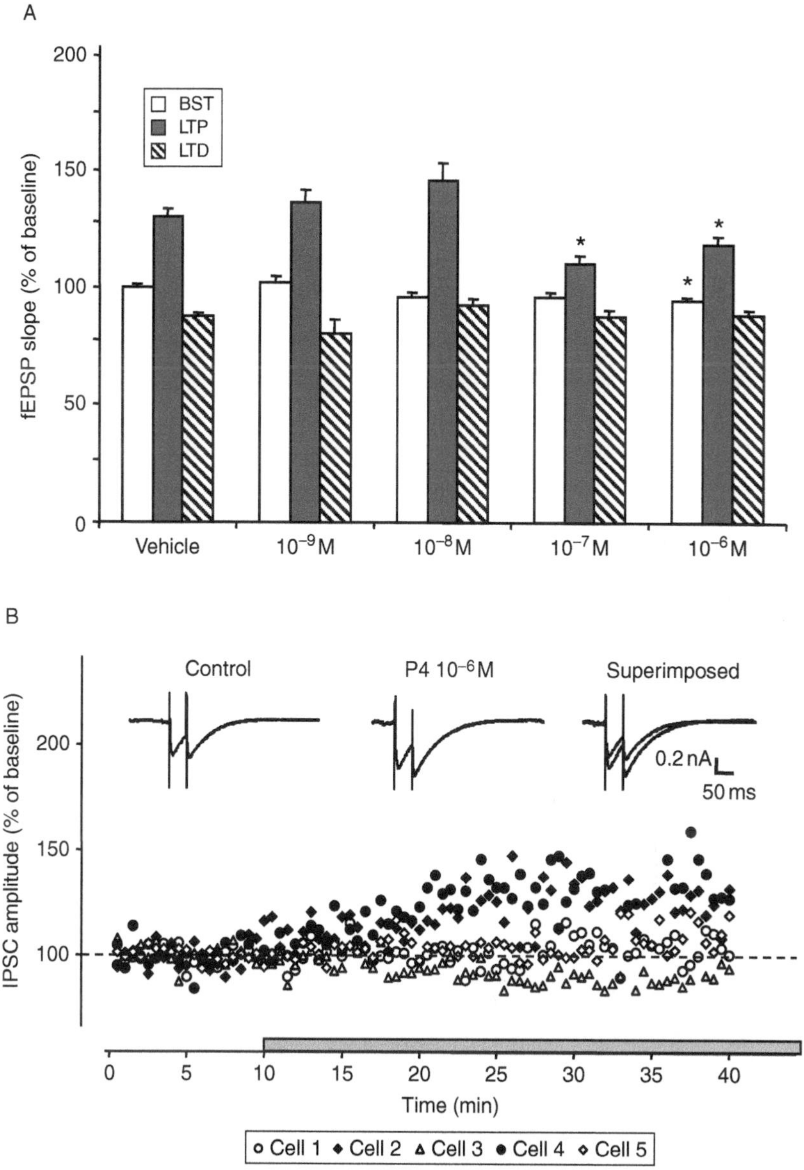


**Figure 12.8** Progesterone concentration, LTP, LTD, and IPSC recordings (A). Vehicle and P4 at different concentrations ($10^{-9}$, $10^{-8}$, $10^{-7}$, and $10^{-6}$ M) was compared with fEPSP values during the baseline synaptic transmission (BST), LTP, and LTD periods. At the two highest concentrations tested, P4 attenuated BST and LTP, but had no effect on LTD. (B) Using intracellular recordings of inhibitory postsynaptic currents (IPSCs) from individual CA1 pyramidal neurons, P4 at $10^{-6}$ M increased $GABA_A$ receptor-mediated current in two of the five recorded cells. The currents were recorded with sharp electrodes in discontinuous single electrode voltage clamp mode. (Representative current traces recorded before and during P4 application at $10^{-6}$ M. Reprinted with permission from Foy *et al.* (2008a). Copyright 2009 Cold Spring Harbor Laboratory Press).

consistent with results from other studies linking $GABA_A$ with P4 or its metabolites (Herd *et al.*, 2007, 2008).

The precise function of the limbic system as a steroid hormone-sensitive target in the CNS is not entirely clear, even though limbic structures in general have long been implicated in the feedback control of the hypothalamo–pituitary–gonadal axis. The hippocampus, purported to be involved in a myriad of memory, mood-modulating, and attentional functions, is regulated by steroid hormones via mechanisms involving structural and functional plasticity. Perhaps, it is through these mechanisms, steroid hormones act in concert with the hippocampus to regulate behavior.

This chapter has examined studies that suggest the existence of mechanisms by which naturally fluctuating endogenous steroid hormone levels can impact cellular process associated with learning and memory function in mammalian CNS. To the extent that LTP is a mechanism involved in the processes of coding and storing information, that is, in memory formation, E2 appears to enhance these processes, while P4 appears to decrease them. Indeed, the E2-induced enhancement of LTP suggests a possible mechanism by which E2 can exert facilitatory effects on memory processes. Clinical evidence indicates that estrogenic steroids can enhance cognitive function in humans, particularly in postmenopausal women (Henderson, 1997, 2000; Kawas *et al.*, 1997); however, prospective observational studies have yet to find a protective effect of estrogen on either age-related decline in cognition or the incidence of dementia (Barrett-Conner and Kritz-Silverstein, 1993; Matthews *et al.*, 1999). A detailed understanding of the mechanisms of action of estrogen and progesterone on hippocampal plasticity, and the roles these steroid hormones play in nonreproductive functions, may have far-reaching implications for hormone therapy to maintain neurological health and function throughout the human life span.

## ACKNOWLEDGMENTS

This work received institutional support from Loyola Marymount University (MRF) and a grant from the National Institute of Aging P01 AG026572 (Roberta Diaz Brinton), Project 2 (RFT).

## REFERENCES

Backstrom, T. (1976). Epileptic seizures in women related to plasma estrogen and progesterone during the menstrual cycle. *Acta Neurol. Scand.* **54,** 321–347.

Barnes, C. A. (1979). Memory deficits associated with senescence: A neurophysiological and behavioral study in the rat. *J. Comp. Physiol. Psychol.* **93,** 74–104.

Barnes, C. A. (1994). Normal aging: Regionally specific changes in hippocampal synaptic transmission. *Trends Neurosci.* **17,** 13–18.

Barnes, C. A., *et al.* (1992). Region-specific age effects on AMPA sensitivity: Electrophysiological evidence for loss of synaptic contacts in hippocampal field CA1. *Hippocampus* **2,** 457–468.

Barrett-Conner, E., and Kritz-Silverstein, D. (1993). Estrogen replacement therapy and cognitive function in older women. *J. Am. Med. Assoc.* **269,** 2637–2641.

Baudry, M., *et al.* (2000). Advances in Synaptic Plasticity. MIT Press, Cambridge, MA.

Bear, M. F., and Malenka, R. C. (1994). Synaptic plasticity: LTP and LTD. *Curr. Opin. Neurobiol.* **4,** 389–399.

Berlucchi, G., and Buchtel, H. A. (2009). Neuronal plasticity: Historical roots and evolution of meaning. *Exp. Brain Res.* **192,** 307–319.

Bi, G., and Poo, M. (2001). Synaptic modification by correlated activity: Hebb's postulate revisited. *Annu. Rev. Neurosci.* **24,** 139–166.

Bi, R., *et al.* (2000). The tyrosine kinase and mitogen-activated protein kinase pathways mediate multiple effects of estrogen in hippocampus. *Proc. Natl. Acad. Sci. USA* **97,** 3602–3607.

Bi, R., *et al.* (2001). Cyclic changes in estradiol regulate synaptic plasticity through the MAP kinase pathway. *Proc. Natl. Acad. Sci. USA* **98,** 13391–13395.

Bliss, T. V., and Collingridge, G. L. (1993). A synaptic model of memory: Long-term potentiation in the hippocampus. *Nature* **361,** 31–39.

Brinton, R. D., *et al.* (1997a). Equilin, a principal component of the estrogen replacement therapy, premarin, increases the growth of cortical neurons via an NMDA receptor dependent mechanism. *Exp. Neurol.* **147,** 211–220.

Brinton, R. D., *et al.* (1997b). 17$\beta$-estradiol increases the growth and survival of cultured cortical neurons. *Neurochem. Res.* **22,** 1339–1351.

Brinton, R. D., *et al.* (2008). Progesterone receptors: Form and function in brain. *Front. Neuroendocrinol.* **29,** 313–339.

Camacho-Arroyo, I., *et al.* (1998). Progesterone receptor isoforms are differentially regulated by sex steroids in the rat forebrain. *Neuroreport* **9,** 3993–3996.

Conneely, O. M., *et al.* (1987). The A and B forms of the chicken progesterone receptor arise by alternate initiation of translation of a unique mRNA. *Biochem. Biophys. Res. Commun.* **149,** 493–501.

Cordoba Montoya, D. A., and Carrer, H. F. (1997). Estrogen facilitates induction of long term potentiation in the hippocampus of awake rats. *Brain Res.* **778,** 430–438.

Edwards, H. E., *et al.* (2000). Progestin receptors mediate progesterone suppression of epileptiform activity in tetanized hippocampal slices in vitro. *Neuroscience* **101,** 895–906.

Fanselow, M. S., and Poulos, A. M. (2005). The neuroscience of mammalian associative learning. *Annu. Rev. Psychol.* **56,** 207–234.

Feng, X. Q., *et al.* (2004). Progesterone inhibition of dopamine-induced increase in frequency of spontaneous excitatory postsynaptic currents in rat prelimbic cortical neurons. *Neuropharmacology* **46,** 211–222.

Foster, T. C. (1999). Involvement of hippocampal synaptic plasticity in age-related memory decline. *Brain Res. Rev.* **30,** 236–249.

Foster, T. C., and Norris, C. M. (1997). Age-associated changes in $Ca^{2+}$-dependent processes: Relation to hippocampal synaptic plasticity. *Hippocampus* **7,** 602–612.

Foy, M. R. (1983). Neuromodulation: Effects of estradiol and THC on brain excitability. Neurobiology Program. Kent State University, Kent.

Foy, M. R. (2001). 17beta-estradiol: Effect on CA1 hippocampal synaptic plasticity. *Neurobiol. Learn. Mem.* **76,** 239–252.

Foy, M. R., and Teyler, T. J. (1983). 17-alpha-estradiol and 17-beta-estradiol in hippocampus. *Brain Res. Bull.* **10,** 735–739.

Foy, M. R., *et al.* (1999). 17beta-estradiol enhances NMDA receptor-mediated EPSPs and long-term potentiation. *J. Neurophysiol.* **81,** 925–929.

Foy, M., *et al.* (2004). Estrogen and hippocampal synaptic plasticity. *Neuron Glia Biol.* **1,** 327–338.

Foy, M. R., *et al.* (2008a). Progesterone regulation of synaptic transmission and plasticity in rodent hippocampus. *Learn Mem.* **15,** 820–822.

Foy, M. R., *et al.* (2008b). 17beta-estradiol modifies stress-induced and age-related changes in hippocampal synaptic plasticity. *Behav. Neurosci.* **122,** 301–309.

Gazzaley, A. H., *et al.* (1996). Differential regulation of NMDAR1 mRNA and protein by estradiol in the rat hippocampus. *J. Neurosci.* **16,** 6830–6838.

Geinisman, Y., *et al.* (1995). Hippocampal markers of age-related memory dysfunction: Behavioral, electrophysiological and morphological perspectives. *Prog. Neurobiol.* **45,** 223–252.

Gellersen, B., *et al.* (2009). Non-genomic progesterone actions in female reproduction. *Hum. Reprod. Update* **15,** 119–138.

Gu, Q., and Moss, R. L. (1996). 17B-estradiol potentiates kainite-induced currents via activation of the camp cascade. *J. Neurosci.* **16,** 3620–3629.

Guennoun, R., *et al.* (2008). The membrane-associated progesterone-binding protein 25-Dx: Expression, cellular localization and up-regulation after brain and spinal cord injuries. *Brain Res. Rev.* **57,** 493–505.

Guerra-Araiza, C., *et al.* (2003). Changes in progesterone receptor isoforms content in the rat brain during the oestrous cycle and after oestradiol and progesterone treatments. *J. Neuroendocrinol.* **15,** 984–990.

Guerriero, G. (2009). Vertebrate sex steroid receptors: Evolution, ligands, and neurodistribution. *Ann. NY Acad. Sci.* **1163,** 154–168.

Hagihara, K., *et al.* (1992). Distribution of cells containing progesterone receptor mRNA in the female rat di- and telencephalon: An in situ hybridization study. *Brain Res. Mol. Brain Res.* **14,** 239–249.

Hammes, S. R., and Levin, E. R. (2007). Extranuclear steroid receptors: Nature and actions. *Endocr. Rev.* **28,** 726–741.

Henderson, V. W. (1997). The epidemiology of estrogen replacement therapy and Alzheimer's disease. *Neurology* **48,** S27–S35.

Henderson, V. W. (2000). Hormone Therapy and the Brain: A Clinical Perspective on the Role of Estrogen. Parthenon Publishing, New York.

Herd, M. B., *et al.* (2007). Neurosteroid modulation of synaptic and extrasynaptic GABA(A) receptors. *Pharmacol. Ther.* **116,** 20–34.

Herd, M. B., *et al.* (2008). The expression of GABAA beta subunit isoforms in synaptic and extrasynaptic receptor populations of mouse dentate gyrus granule cells. *J. Physiol.* **586,** 989–1004.

Ito, K., *et al.* (1999). Age-dependent, steroid-specific effects of oestrogen on long-term potentiation in rat hippocampal slices. *J. Physiol.* **515**(Pt 1), 209–220.

Kawas, C., *et al.* (1997). A prospective study of estrogen replacement therapy and the risk of developing Alzheimer's disease: The Baltimore longitudinal study of aging. *Neurology* **48,** 1517–1521.

Kramar, E. A., *et al.* (2009). Cytoskeletal changes underlie estrogen's acute effects on synaptic transmission and plasticity. *J. Neurosci.* **29,** 12982–12993.

Landfield, P. W., and Lynch, G. (1977). Impaired monosynaptic potentiation in in vitro hippocampal slices from age, memory-deficient rats. *J. Gerontol.* **32,** 523–533.

Landfield, P. W., *et al.* (1986). The effects of high $Mg^{2+}$-$Ca^{2+}$ ratios on frequency potentiation in hippocampal slices of young and aged rats. *J. Neurophysiol.* **56,** 797–811.

Lee, Y. S., and Silva, A. J. (2009). The molecular and cellular biology of enhanced cognition. *Nat. Rev. Neurosci.* **10,** 126–140.

Leuner, B., *et al.* (2004). High levels of estrogen enhance associative memory formation in ovariectomized females. *Psychoneuroendocrinology* **29,** 883–890.

Lisman, J. (2003). Long-term potentiation: Outstanding questions and attempted synthesis. *Philos. Trans. R. Soc. Lond. B Biol. Sci.* **358,** 829–842.

Losel, R., and Wehling, M. (2003). Nongenomic actions of steroid hormones. *Nat. Rev. Mol. Cell Biol.* **4,** 46–56.

Malenka, R. C., and Bear, M. F. (2004). LTP and LTD: An embarrassment of riches. *Neuron* **44,** 5–21.
Malenka, R. C., and Nicoll, R. A. (1999). Long-term potentiation–A decade of progress? *Science* **285,** 1870–1874.
Matthews, K., *et al.* (1999). Estrogen replacement therapy and cognitive decline in older community women. *J. Am. Geriatr. Soc.* **47,** 518–523.
McEwen, B. S., and Alves, S. E. (1999). Estrogen actions in the central nervous system. *Endocr. Rev.* **20,** 279–307.
McEwen, B. S., *et al.* (1975). Putative glucocorticoid receptors in hippocampus and other regions of the rat brain. *In* "The Hippocampus, Vol 2: Neurophysiology and Behavior" (R. Isaacson and K. Pribram, Eds.). Plenum Press, New York.
Mitchell, E. A., *et al.* (2008). Neurosteroid modulation of GABAA receptors: Molecular determinants and significance in health and disease. *Neurochem. Int.* **52,** 588–595.
Norris, C. M., *et al.* (1996). Increased susceptibility to induction of long-term depression and long-term potentiation reversal during aging. *J. Neurosci.* **16,** 5382–5392.
Norris, C. M., *et al.* (1998). Reversal of age-related alterations in synaptic plasticity by blockage of L-type $Ca^{2+}$ channels. *J. Neurosci.* **18,** 3171–3179.
Reddy, D. S. (2009). The role of neurosteroids in the pathophysiology and treatment of catamenial epilepsy. *Epilepsy Res.* **85,** 1–30.
Shors, T. J., and Matzel, L. D. (1997). Long-term potentiation: What's learning got to do with it? *Behav. Brain Sci.* **20,** 597–614, discussion 614–655.
Stanton, P. K. (1996). LTD, LTP, and the sliding threshold for long-term synaptic plasticity. *Hippocampus* **6,** 35–42.
Terasawa, E., and Timiras, P. S. (1968). Electrical activity during the estrous cycle of the rat: Cyclic changes in limbic structures. *Endocrinology* **83,** 207–216.
Teyler, T. J. (1980). Brain slice preparation: Hippocampus. *Brain Res. Bull.* **5,** 391–403.
Teyler, T. J., *et al.* (1980). Gonadal steroids: Effects on excitability of hippocampal pyramidal cells. *Science* **209,** 1017–1018.
Vardaris, R. M., and Teyler, T. J. (1980). Sex differences in the response of hippocampal CA1 pyramids to gonadal steroids: Effects of testosterone and estradiol on the *in vitro* slice preparation. *Soc. Neurosci. Abstr.* A153–A159.
Vouimba, R. M., *et al.* (2000). 17B-estradiol suppresses expression of long-term depression in aged rats. *Brain Res. Bull.* **53,** 783–787.
Warren, S. G., *et al.* (1995). LTP varies across the estrous cycle: Enhanced synaptic plasticity in proestrus rats. *Brain Res.* **703,** 26–30.
Wong, M., and Moss, R. L. (1991). Electrophysiological evidence for a rapid membrane action of the gonadal steroid, 17 beta-estradiol, on CA1 pyramidal neurons of the rat hippocampus. *Brain Res.* **543,** 148–152.
Wong, M., and Moss, R. L. (1992). Long-term and short-term electrophysiological effects of estrogen on the synaptic properties of hippocampal CA1 neurons. *J. Neurosci.* **12,** 3217–3225.
Wong, M., and Moss, R. L. (1994). Patch-clamp analysis of direct steroidal modulation of glutamate receptor-channels. *J. Neuroendocrinol.* **6,** 347–355.
Woolley, C. S., and McEwen, B. S. (1994). Estradiol regulates hippocampal dendritic spine density via an N-methyl-D-aspartate receptor-dependent mechanism. *J. Neurosci.* **14,** 7680–7687.
Woolley, C. S., *et al.* (1997). Estradiol increases the sensitivity of hippocampal CA1 pyramidal cells to NMDA receptor-mediated synaptic input: Correlation with dendritic spine density. *J. Neurosci.* **17,** 1848–1859.
Xie, X., *et al.* (1992). Isolated NMDA receptor-mediated synaptic responses express both LTP and LTD. *J. Neurophysiol.* **67,** 1009–1013.
Zadran, S., *et al.* (2009). 17-Beta-estradiol increases neuronal excitability through MAP kinase-induced calpain activation. *Proc. Natl. Acad. Sci. USA* **106,** 21936–21941.

CHAPTER THIRTEEN

# Hormones and Sexual Reward

Raúl G. Paredes

## Contents



## Abstract

There are different physiological processes that influence behavior. One of this processes that produces approach behavior to a stimuli that induces a positive affective (PA) state, commonly known as reward, plays an important role in modulating behavior. There is an extensive literature in which the rewarding effects of drugs have been investigated. Less research has been devoted to the study of naturally occurring behaviors that produce a PA or reward state. Hormones modulate different behaviors, including sex. However, little attention has been devoted to study the possible role of hormones in reward states. One of the methods most frequently used to study reward or PA states is the conditioned place preference (CPP) paradigm. Hopefully this review will encourage researchers to directly address the effects of hormones on reward, research that is much needed. © 2010 Elsevier Inc.

Instituto de Neurobiología, Universidad Nacional Autónoma de México, Querétaro, Querétaro, Mexico

*Vitamins and Hormones*, Volume 82
ISSN 0083-6729, DOI: 10.1016/S0083-6729(10)82013-8

# I. Introduction

The study of sexual reward in animals has attracted a renewed attention probably due to the fact that finer methods are now available to determine under what conditions sex behavior can induce a positive affective (PA) state. While in humans the evaluation of sex as a reward experience can be obtained directly from the subjects involved in the experience or trough questioners and descriptions of the physiological states associated with sex, in animals we have to make inferences from the behavior displayed by the animals. Sexual behavior in males has been believed to induce a PA state in several species. This is probably due to the fact that a sexually experienced male will attempt to copulate whenever presented with a female in estrous, and in many occasion he will even attempt to mate when anestrous females constantly reject the male. Moreover, early studies showed that males will try to overcome obstacles like crossing an electric grid to get from a starting cage to a goal cage, which held a sexually receptive female. Although females in estrous were also capable of crossing an electric field to be in contact with a sexually vigorous male, it was assumed that sex was not rewarding for females. This was concluded from the observations where female rats showed several rejection behaviors after receiving repeated intromissions from a male. It was postulated that sexual behavior in females has appetitive and aversive consequences and that in many instances these aversive consequences prevented the females from experiencing a PA state associated with sex. It was not until we combined some time ago, the conditioned place preference (CPP) paradigm and paced mating that became evident that sex for the female can induce a PA or reward state. In paced mating, the females controls or paces the sexual interaction and the stimulation received by the males enhances the appetitive components of mating.

In this chapter, I give a brief definition of reward and explain how this PA state can be measured. Most of the studies associated with reward and sex have been done in rats. The few studies done in other species will be explicitly mentioned in the text. Since sexual behavior is highly dependent on hormones, both in males and females, I describe how hormones can modify CPP directly and how can they modulate drug-induced reward. In the final part of the chapter, I describe how hormones contribute to sex to induce a PA state.

# II. Reward

Since in animals the reward or PA state can be inferred only from the behavior displayed by the subject, it is not surprising that the operational definition of reward refers to the ability of eliciting an approach behavior toward a particular stimulus (Young, 1959; Young and Christensen, 1962). In this way, the stimuli can be divided in three categories (Agmo 1999, 2007): (1) stimuli that activate approach behavior are called positive incentives (Di Chiara and Bassareo, 2007), for example, food or a conspecific of the opposite sex; (2) stimuli that activate withdrawal are negative incentives, for example, an electric shock; (3) the third category includes stimuli that are neutral. In this review, I will focus on approach behaviors in response to a positive incentive. The positive incentive induces an approach behavior associated with a PA state. PA states in humans can be evaluated through self-reports or questionnaires that ask subjects about their level of happiness or other feelings, whereas evaluation in animals relies largely on the observation of approach and consummatory behaviors (Burgdorf and Panksepp, 2006). In this chapter, the use of the term "reward" is not intended to describe the psychological processes associated with the development of an approach behavior, this will be anthropomorphic, but rather it simply reflects that a positive physiological state in the subject produces an approach behavior.

Many authors use reward and reinforcement as synonymous because reward induces changes in observable behavior and serves as a positive reinforcer, increasing the frequency of the behavior (Schultz, 2006). Reinforcement means strengthening a stimulus–response habit (as defined by Hull, 1943) or increasing the rate or probability of a response (as defined by Skinner, 1938). It measures the strength or rate of a behavioral response (Berridge and Robinson, 1998). This increase in the response can be associated with the presentation of a positive stimulus or with removal of an aversive one, making evident that something that is reinforcing is not necessarily rewarding (see White, 1989; for a classical discussion). For example, pressing a lever may serve to measure the strength or rate of a behavioral response but does not necessarily reflect an approach behavior associated with a reward state (Agmo, 2007). In this chapter, I emphasize the appetitive aspects of hormones and behavior that induce a positive affect.

Several methods have been used in trying to evaluate reward or PA states. For example, rats can be trained to press a lever to obtain food, water, or the company of a conspecific from the opposite sex (operant behavior). In addition, subjects can be given the choice between two different stimuli. They can be placed in a partner preference box, in a Y- or a T-maze. In the case of evaluating different aspects of sexual reward, in any of these tests, a sexually experience male could get to choose between approaching a sexually receptive female in one side or a nonreceptive female in the other side. When the subject reaches the goal box he can interact with the stimulus animal he chooses. In some tests, physical interaction between the subject and the stimulus animals is prevented by a wire mesh. In this case the subjects can hear, see, and smell the stimulus animals but no physical interaction is possible, in this way the sexual incentive motivation for either stimulus animal is evaluated. Many of these procedure that evaluate approach behaviors depend on measures of response speed or rate, making them very sensitive to manipulations that affect motor functions, or they involve different kinds of learned responses, complicating interpretation of the results. A detailed description and discussion of the different methods that can be used to evaluate sexual reward can be found elsewhere (Paredes, 2009).

## III. Conditioned Place Preference

In this chapter, I focus on a method that evaluates PA states which reduces many of the limitations that other methods have. The first study that described the use of CPP to evaluate if rats spend more time in a compartment where they received a subcutaneous injection of morphine was done by Beach (1957). Latter Rossi and Reid (1976) coin the term positive affect when they demonstrated that morphine injections produced a positive affect and rats preferred to spend more time where they experienced this affective state. This simple method that does not require sophisticated equipment has been extensively used to study the positive affect induced by drugs. It was latter used to study PA states induced by natural stimuli like water (Agmo *et al.*, 1993) or food consumption (Bechara and van der Kooy, 1992), odors from conspecifics (Pankevich *et al.*, 2006; Pierman *et al.*, 2006) or from their home cage (Fitchett *et al.*, 2006), exercising or the use of the running wheel (Lett *et al.*, 2000, 2001). The PA states induced by different social behaviors have also been evaluated by CPP. These behaviors include sexual behavior (which will be described in detail below), maternal behavior and exposure to pups (Mattson *et al.*, 2001, 2003), social play (Calcagnetti and Schechter, 1992; Crowder and Hutto, 1992), and "tickling stimulation" (Burgdorf *et al.*, 2007; Knutson *et al.*, 2002; Panksepp, 2007; Panksepp *et al.*, 2007). A discussion of the natural stimuli or the social behaviors that induce CPP is beyond the scope of this chapter, and the interested reader should look at the original references.

There are different variations in the methods use to determine whether a stimulus induces CPP. The original design used by Beach consisted of a Y-choice discrimination box that had interconnected different goal boxes (Beach, 1957), latter Rossi and Reid (1976) used an alley divided in three different compartments. In both cases, the goal boxes or lateral compartments were different in color and texture. Nowadays, there are differences in conditioning methods and in the number of sessions required to test the subjects (Carboni and Vacca, 2003; Schechter and Calcagnetti, 1993; Tzschentke, 2007). The method that we continuously use to evaluate the reward state induced by sex consists of a three-compartment box in which the lateral compartments are different in odor, color, and texture. In our case, one lateral compartment is painted white and has sawdust on the floor. The contra lateral compartment is painted black and is moistened with a 2% acetic acid solution. Both lateral compartments are interconnected through sliding doors by a middle gray-painted compartment. As already mentioned, the number of sessions can vary, but in our case, we do a pretest, a test, and between them six alternate conditioning sessions, three reinforced and three nonreinforced. In the pretest, we placed the subject in the middle compartment for 1 min before opening the doors to the lateral compartments for a 10-min test, during which we record the time spent in each of the lateral compartments and note the animal's initial preference. If our interest is to determine the positive affect of a drug (e.g., 1 mg/kg of morphine induces a clear CPP), the subjects will be injected with saline and placed in the original preferred compartment for 30 min. In the following day, subjects are injected with morphine and placed in the originally nonpreferred (reinforced) compartment. After three alternating reinforced and nonreinforced sessions, we test the animals in exactly the same way as in the pretest. The basic idea behind the procedure is that the potential affective state induced by the drug is associated with the cues of the originally nonpreferred compartment. Therefore, through conditioning we modified the original preference and the animals will spent more time in the compartment associated with the injection of morphine. This change in preference reflects a CPP. The injection of morphine can be replaced by a natural stimuli or behavior to evaluate other PA states, in the case of the present review sexual behavior.

Several measures can be obtained after a CPP, including the time in the reinforced compartment, the time in the nonreinforced compartment, the preference score [time in reinforced compartment/(time in reinforced compartment + time in nonreinforced compartment)], and the difference between compartments [time in the nonreinforced − time in the reinforced compartment]. They are compared before and after conditioning. However, the most important measure is the time in the reinforced compartment. As already explained, the basic idea of the CPP paradigm is that the potential affective state induced by the stimulus is associated with the cues of the originally nonpreferred compartment, increasing the time spent in it and modifying the original preference. The use of the other variables could only reflect a reduction in the time spent in the original preferred compartment without any significant increase in the time in the originally nonpreferred compartment, the one actually rewarded during conditioning. The time in the reinforced compartment is the most important measure to define a CPP and an induction of a PA state (Bardo *et al.*, 1995; Carboni and Vacca, 2003; Schechter and Calcagnetti, 1993). Without this increase there is no conclusive evidence of a CPP or induction of a reward state.

The use of the CPP procedure has several advantages, including the possibility to observe rewarding and aversive effects after conditioning, relatively low doses of drugs are sufficient to induce CPP, drugs that produce CPP are also rewarding in other behavioral paradigms, only one pairing of the drug or stimuli could be require to induce CPP and during the test the animals is drug-free. Some disadvantages have been associated with this method. For example, novelty-seeking behavior can confound the evaluation of preference. It has also been argued that it is not clear what the procedure measures, and when a drug is used, motor, sedative, or anxiolitic actions can confound the results. Some of these limitations, which are important to consider when testing the rewarding properties of drugs are eliminated or greatly reduced in evaluations of the PA of naturally occurring behaviors. An additional advantage in the way in which we evaluate sex and CPP is the fact that mating takes place in a mating cage and afterward animals are transferred to the conditioning cage. In this way, what we evaluate is the physiological state induced by mating, not the effect of the behavior itself on conditioning.

## IV. Hormones and CPP

The idea that hormones can have rewarding properties is not new; in fact, it has been around for some time (see, e.g., Piazza and Le Moal, 1997). However, the assumption that the release of hormones associated with stimuli that might be potentially rewarding or with naturally occurring behaviors that could induce a PA state is not conclusive evidence that indeed that hormone is involved in reward or in a PA state. It is also obvious that hormonal levels can modify or influence different behaviors. For example, stress, and the hormones involved in this response, can modify the effects of drugs. A discussion of these hormonal effects and the sex differences in drug induce reward is beyond the scope of this manuscript and there are many excellent reviews that address these topics. I will briefly describe how some hormones directly induce a PA state evaluated by CPP and how some of them modify drug induce CPP. It is surprising that only few papers have directly evaluated the effects of hormones on CPP, compared to those of drugs or neurotransmitters (Tzschentke, 2007). Another point that will not be covered in this chapter is the literature associated with hormone self-administration. For example, extensive evidence has demonstrated anabolic steroids self-administration. As already described, self-administration is associated with reinforcement and not necessarily with reward (see above). Moreover, there are recent reviews addressing the topic of anabolic steroids self-administration (Clark and Henderson, 2003; Sato *et al.*, 2008; Wood, 2008). This section is intended to give the reader a general overview of how different hormones modify PA states evaluated through CPP. The hormones will be grouped to facilitate the presentation and discussion and not necessarily because they are related in function.

## A. Oxytocin, cholecystokinin, and ghrelin

Intracerebroventricular administration of different doses of oxytocin (OT) inhibited the acquisition and facilitated the extinction of CPP induced by methamphetamine in mice. The effect of OT was blocked by a selective antagonist to this hormone, suggesting that OT inhibits the reward state induced by methamphetamine (Qi *et al.*, 2009). Cholecystokinin (CCK) has been involved in modulating rewarding effects of opioids. For example, it has been shown that the coadministration of a CCK antagonist with a subthreshold dose of morphine induces CPP (Valverde *et al.*, 1996). There are other studies indicating that a CCKA receptor antagonist attenuates CPP but a CCKB receptor antagonist does not modify morphine-induced CPP, indicating a differential role of CCK receptor subtypes on morphine (Higgins *et al.*, 1992) and cocaine (Lu *et al.*, 2002) induced CPP. Pretreatment with the gut peptide ghrelin enhances the rewarding properties of cocaine. While small doses of cocaine (0.312 or 0.625 mg/kg) does not produce CPP, when rats are pretreated with ghrelin, but not with saline, these doses induce CPP (Davis *et al.*, 2007).

## B. Melatonin and substance P

The administration of melatonin (1–40 mg/kg i.p.) alone does not induce CPP but the combination of melatonin (5–20 mg/kg) and a subeffective dose of morphine (0.5 mg/kg) produced a reliable PA state, as evaluated by CPP (Yahyavi-Firouz-Abadi *et al.*, 2007). Mice lacking the melanin-concentrating receptor (MCH1 KO) as well as wild-type mice developed CPP to cocaine and amphetamine (Tyhon *et al.*, 2008). The administration of substance P (0.1, 1.0, or 10.0 ng) into the shell of the nucleus accumbens failed to induce CPP (Schildein *et al.*, 1998).

## C. Corticosteroids

The role of corticosteroids on reward is not clear, for example, the administration of different doses of corticosterone (0, 2.5, or 10 mg/kg; i.p.) failed to produce either CPP or conditioned place aversion in male rats (Dietz *et al.*, 2007). In addition, the manipulation of corticosterone levels by the injection of a synthesis inhibitor, aminoglutethimide, did not alter the acquisition of place preference in mice induced by ethanol (Chester and Cunningham, 1998). On the other hand, it has been shown that systemic administration or intrahippocampal infusion of the glucocorticoid receptors (GR) antagonist RU38486 blocked or impaired the formation of CPP in a dose-dependent manner. When the compound was administered in the nucleus accumbens, it prevented the formation of CPP. These results suggest that GRs could mediate opiate reward (Dong *et al.*, 2006). There is evidence indicating that in rats, corticosteroids participate in the reinstatement of CPP induced by morphine (Der-Avakian *et al.*, 2005; Wang *et al.*, 2006).

## D. Luteinizing hormone releasing hormone

In gonadectomized males and females that received silastic implants, containing testosterone (T) and estradiol (E), respectively the injection of luteinizing hormone releasing hormone (LHRH) in a dose of 5 $\mu$g/kg i.p. 15 min before conditioning induced CPP in males, but not in females, suggesting that LHRH is rewarding for male but not for female rats (de Beun *et al.*, 2004). The rewarding effect of LHRH is hormone dependent because gonadectomized males with silastic implants of T or E developed CPP while those males with empty implants did not, suggesting that sufficient levels of circulating sex steroids are required to develop CPP after LHRH treatment (de Beun *et al.*, 1991a,b). In a follow-up study, the authors demonstrated that the injection of LHRH had to be done between 15 and 45 min before the conditioning trial to develop CPP. Intervals higher or lower between injection of LHRH and the conditioning session failed to develop CPP (De Beun *et al.*, 1992).

This brief description of the effects of different hormones on CPP reveals that no systematic efforts have been done to evaluate the effects of hormones on reward, and therefore, no conclusive evidence can be presented except for the obvious argument that more studies are needed that directly evaluate the rewarding aspects of hormones. There is more data regarding the rewarding aspects of estradiol (E), progesterone (P), and testosterone (T) and they are discussed in the following sections.

## E. Estradiol and progesterone

The administration of E (s.c.) during conditioning days increased the time spent in the reinforced compartment during the test, indicating a clear CPP. When E was administered with sesame oil, a dose of 10 $\mu$g was sufficient to induce CPP, but when the hormone was administered with propylene glycol, a dose of 1 mg of E was required to induce conditioning (Frye and Rhodes, 2006), suggesting that the effects of E on CPP might vary depending on the route of administration. The CPP induced by E is modified by E receptors antagonist. For example, the subcutaneous administration of tamoxifen alone increased the time spent in the nonpreferred compartment but attenuated the CPP induced by E. The administration of another antagonist, ICI 182-780, into the nucleus accumbens did not produce CCP when administered alone but attenuated the effects of E on CPP.

Moreover, the administration of antisense oligonucleotides for E receptors in the nucleus accumbens significantly decreased CPP induced by E. Based on these results, the authors suggest that the rewarding effects of E might be mediated by E receptors in the nucleus accumbens (Walf *et al.*, 2007). There is one study reporting that E did not induce CPP (Galea *et al.*, 2001). However, in that study a radial-arm maze was adapted as a CPP cage, this might explain the different results compared to other groups. It should also be mentioned that there is data indicating that low doses of E (0–250 $\mu$g/kg, s.c.) induced conditioned taste aversion in a two-bottle choice test and place aversion in a CPP test, that is, subjects avoided the compartment paired with E treatment. The aversive effects were dose-dependent and males were more susceptible to the aversive effects of E than females (de Beun *et al.*, 1991a,b).

There is also data suggesting that E modulates CPP induced by drugs. It has been shown that there are no sex difference in the sensitivity to cocaine (5 mg/kg) induced CPP and gonadectomy in male rats has no effect on conditioning. But in females, the magnitude of cocaine-induced CPP is attenuated after gonadectomy. When ovariectomized (OVX) females are pretreated with P, cocaine CPP is inhibited (Russo *et al.*, 2003). Progesterone treatment during both the acquisition and the expression phases of cocaine conditioning blocked cocaine-induced CPP (Russo *et al.*, 2008). However, the combined treatment of E plus P potentiated the magnitude of CPP, suggesting that ovarian hormones may influence cocaine reward (Russo *et al.*, 2003). There are also studies that have evaluated the effect of E and P on methamphetamine-induced CPP. When gonadectomized mice were treated with P, no effect on methamphetamine-induced CPP in males or females was observed. However, E treatment before and during conditioning facilitated methamphetamine-induced CPP in female but not in male mice, suggesting that the facilitating effects of E in this type of conditioning are sex-dependent (Chen *et al.*, 2003). From the above described evidence, it appears that E can indeed modulate reward.

The rewarding effects of neurosteroids have also been evaluated. A low dose of the neuroactive steroid 3-alpha-hydroxy-5-alpha-pregnan-20-one (allopregnanolone) failed to modify the initial preference in mice, but higher doses increased the time spent in the compartment paired with the drug, indicative of a reward state (Finn *et al.*, 1997). When allopregnanolone was administered intracerebroventricularlly at different doses to male rats place aversion or no effect was observed (Beauchamp *et al.*, 2000). We evaluated the effects of different progestins on CPP. The compounds were administered intravenously (i.v.) through an indwelling jugular catheter in doses that induce the full reproductive pattern, including pacing behavior. None of the compounds tested induced CPP (Gonzalez-Flores *et al.*, 2004). It appears that the effects of neurosteroids on reward might vary depending on the route of administration and the species studied.

## F. Testosterone

Early studies reported contradictory results in the ability of T to induce CPP. For example, a high dose (1 mg) of T produced an increase in the amount of time spent in the compartment paired with T suggestive of a CPP. However, lower doses of T (10 and 100 $\mu$g) did not induce CPP. This led the authors to suggest that it was unlikely that T could produce PA states (Caldarone *et al.*, 1996). However, the group of Packard and colleagues reported that rats displayed a preference for the compartment paired with T (800 and 1200 $\mu$g/kg) as opposed to an environment paired with saline administration, indicating that T has rewarding affective properties (Alexander *et al.*, 1994). This groups has since then evaluated systematically the possible rewarding effects of T. In follow-up studies, they showed that the bilateral administration of T into the nucleus accumbens at different doses (0.25 or 0.5 $\mu$g/0.5 $\mu$l) significantly increase the time spent in the compartment paired with T, indicating a clear CPP and suggesting that intra-accumbens injections of testosterone are sufficient to produce reward (Packard *et al.*, 1997). When T was administered into the medial preoptic area (MPOA), in a dose of 0.1 $\mu$g, CPP was induced. However, a higher dose (0.2 $\mu$g) produced a conditioned place aversion. The authors go on to suggest that the rewarding effects of intra-MPOA testosterone may in part mediate the facilitatory effects of testosterone on motivational aspects of sexual behavior (King *et al.*, 1999). The interpretation of these results might be partially incorrect because in the study by King *et al.*, injections of T were done in the MPOA, using unilateral cannulae and there is extensive evidence that in order to eliminate male sexual behavior and reduce sexual motivation, in theory the effect mediated by T, bilateral lesions of the MPOA are required (Paredes, 2003; Paredes and Baum, 1997).

In another series of experiments, Packard and colleagues evaluated the effects of different DA antagonist on T-induced CPP. In the first study, the peripheral (20 min prior; 0.2, 0.3 mg/kg) or intranucleus accumbens injection (2 min prior, 5.0 $\mu$g) of the dopamine receptor antagonist flupenthixol on the test day, blocked the expression of T CPP (Packard *et al.*, 1998). In a follow-up study, the DA antagonists were administered during conditioning days when subjects were exposed to T and later placed in the reward compartment. Injections of the mixed D1/D2 receptor antagonist $\alpha$-flupenthixol (0.3 mg/kg), the selective D1 antagonist SCH23390 (0.1 mg/kg), or the selective D2 antagonist sulpiride (20 mg/kg), each blocked acquisition of the testosterone CPP (Schroeder and Packard, 2000). The authors interpreted these results as suggesting that the rewarding affective properties of testosterone are mediated, at least in part, via an interaction with the mesolimbic dopamineric system (Packard *et al.*, 1998) and that dopamine D1 and D2 receptor subtypes participate in T-induced CPP (Schroeder and Packard, 2000). In most of the studies of Packard and colleagues, a change of preference is defined when the subjects spend significantly a longer time in the paired versus the unpaired side during the test. Although this could certainly be a way to evaluate preference, the reduction might simply reflect that the animal spent more time in the middle compartment. That is why, as described above, a crucial measure to consider a significant change of preference is an increase in the time in the reinforced compartment during the test versus the pretest.

Not only is T capable of inducing CPP, it has also been shown that the administration of the T metabolite and neurosteroid 3α-androstanediol (3α-diol; an) in a dose of 1.0 mg daily for 6 days, 30 min prior to exposure to the nonpreferred compartment of the CPP chamber significantly increased the time spent in this compartment during testing, indicating the induction of a reward state (Frye *et al.*, 2001).

Contrary to what has been described in rats, the role of T in inducing CPP in mice is not as clear. It has been shown that male mice that received T do not develop CPP (Arnedo *et al.*, 2000; Minerly *et al.*, 2008). In addition, chronic T administration did not alter cocaine-induced CPP, and concurrent administration of T and cocaine failed to induce CPP (Minerly *et al.*, 2008). The only evidence indicating that T induces CPP in mice is when T was paired with the black compartment, suggesting that in mice, CPP to T depend on environmental cues used as conditioned stimuli (Arnedo *et al.*, 2000). However, this interpretation is rather weak and no easy explanation can be put forward to explain why T induces CPP only if paired with a black compartment (Arnedo *et al.*, 2002). Taken together, the above described evidence suggests that T modulates reward and drug-induced reward in rats, while in mice, most evidence indicate that T is not important in reward.

## V. Reward and Sexual Behavior

In this section, I briefly describe the evidence supporting that sex, for both the male and the female, induces a PA or reward state. Since this topic has been the subject of a recent review (Paredes, 2009), I emphasize the hormonal aspects of sex reward, when they have been evaluated. A more detailed description of the behavioral factors affecting sex reward can be found elsewhere (Paredes, 2009).

### A. Males

There is extensive evidence indicating that sexual behavior in males induces CPP. This has been clearly established in male rats (Agmo and Berenfeld, 1990; Harding and McGinnis, 2004; Hughes *et al.*, 1990; Kippin and van der Kooy, 2003; Mehrara and Baum, 1990; Miller and Baum, 1987) and mice (Kudwa *et al.*, 2005; Popik *et al.*, 2003). Sexual behavior induces CPP in intact (Agmo and Berenfeld 1990; Harding and McGinnis 2004; Hughes *et al.* 1990; Kippin and van der Kooy 2003; Mehrara and Baum 1990; Miller and Baum 1987) and recently castrated rats (Mehrara and Baum, 1990). However, in some cases, castration attenuates CPP before males showed reductions in their copulatory performance (Hughes *et al.*, 1990). Administration of the opioid receptor antagonist naloxone blocked the expression but not the acquisition of CPP in both gonadally intact and castrated males (Mehrara and Baum, 1990). If naloxone was injected 7 or 14 days after conditioning and surgery, naloxone-injected males spent significantly less time than controls in the initially nonpreferred compartment (Miller and Baum, 1987). Naloxone also blocked the CCP induced by ejaculation in intact males (Agmo and Berenfeld, 1990). These and other evidence indicate that endogenous opioids modulate the reward state induced by sexual behavior and ejaculation (see Paredes and Fernández-Guasti, 2008; for a discussion).

There have also been studies trying to elucidate which brain area might be involved in the rewarding aspects of male sexual behavior. For example, sexually experienced male rats were castrated and then received silastic capsules filled with testosterone (T) plus intracranial (IC) implants filled with the antiandrogen hydroxyflutamide (OHF) to selectively block androgen receptors in either the MPOA or ventromedial nucleus (VMN) of the hypothalamus. CPP associated with sexual behavior was blocked when OHF was implanted in the VMN but not in the MPOA, suggesting that androgen receptors in the VMN are important for copulation and sexual motivation (Harding and McGinnis, 2004). Another brain area that has been implicated in the reward state induced by copulation is the tegmental pedunculopontine nucleus (TPP) of the brainstem, which mediates food reward in food-satiated animals and opiate reward in drug-naive animals (see Kippin and van der Kooy, 2003; and references therein). Excitotoxic lesions of the TPP produced aversion for the compartment paired with copulation in sexually experienced males, suggesting that this region mediates the rewarding consequences of sexual behavior (Kippin and van der Kooy, 2003). Not strangely, the most important area in the control of male sexual behavior, the MPOA (Paredes, 2003; Paredes and Baum, 1997), has also been involved in the rewarding aspects of male sexual behavior. As described earlier, the systemic injection of naloxone blocks the reward state induced by ejaculation (Agmo and Berenfeld, 1990). The bilateral infusion of this opioid antagonist, into the MPOA, before mating during conditioning sessions blocked the reward state induced by ejaculation, suggesting that opioids in the MPOA are involved in sexual reward (Agmo and Gomez, 1993). This contention is further supported by the observation that the injection of an opioid agonist into the MPOA also induced CPP (Agmo and Gomez, 1991). Interestingly, when naloxone was infused in the nucleus accumbens, males developed a clear reward state after mating, suggesting that this brain region, important in DA-mediated reward, is not involved in sexual reward (Agmo and Gomez, 1993). Moreover, the systemic administration of the DA antagonist pimozide did not block CPP induced by ejaculation, further supporting the lack of a role of DA in sexual reward (Agmo and Berenfeld, 1990) and sexual behavior in male rats (Paredes and Agmo, 2004).

There is a study indicating that DA might indeed participate in the rewarding aspects of male mice sexual behavior. Copulatory activity and CPP were evaluated in D5 knockout (D5KO) mice. Wild-type males displayed CPP for intromissions and ejaculation, but D5KO mice showed CPP only for ejaculation, indicating that D5 receptors mediate rewarding aspects of mating in male mice (Kudwa *et al.*, 2005). As observed in mice, male rats can develop CPP for intromissions only. Coolen and coworkers found that only sexually naïve males, and not sexually experienced, developed CPP for intromissions (Tenk *et al.*, 2009). However, we found that males with one previous sexual experience developed CPP when allowed to display 15 intromissions without ejaculation (Camacho *et al.*, 2007). We have suggested that male rats require a minimum number of intromissions (around 15) to reach a threshold in order for sex to induce a reward state as evaluated by CPP (Camacho *et al.*, 2009a). This is further supported by the observation that male rats injected with 8-OHDPAT, a compound that makes male rats ejaculate after few intromissions, do not developed CPP (Camacho *et al.*, 2007).

Most of the studies that have evaluated the rewarding properties of mating in males have used the classic paradigm in which the male controls the sexual interaction. However, when females control or pace the sexual interaction, a different picture emerges. As will be described below, pacing the sexual interaction has important physiological and behavioral consequences for the female. In order for the female to pace the rate of sexual interaction, we made sufficiently large a hole in the middle of the mating cage to allow the female, but not the male, to move back and forth from the compartment in which the male was confined (Paredes and Vazquez, 1999). When the animals mate without the barrier, the male controls the sexual interaction. In a series of studies, it has been shown that only when the males copulated without the barrier they developed a clear CPP. Males that mated with the barrier (females controlling the sexual interaction) did not develop CPP (Martinez and Paredes, 2001). In a follow-up study, males that mated once a week, for 10 weeks, in a situation where the female (through the barrier) paced the sexual interaction, did not developed CPP, indicating that sexual behavior and estrous females are very powerful incentives for the male, even in situations in which the rewarding value of sex is reduced (Camacho *et al.*, 2004).

## B. Females

One important characteristic of female rat sexual behavior is the ability that females have to control or pace the sexual interaction. A detailed description of the behavioral and physiological changes that favor reproduction associated with paced mating is beyond the scope of this chapter and excellent reviews are available (Blaustein and Erskine, 2002; Erskine, 1989; Paredes and Vazquez, 1999). Sufficient is to say that studies in natural and naturalistic settings have shown that even though males and females have a different temporal pattern of copulation, both sexes control or pace the sexual interaction and obtain the same amount of copulation (McClintock and Adler, 1978; McClintock and Anisko, 1982; McClintock *et al.*, 1982; Robitaille and Bouvet, 1976). This assures that the rewarding state induced by mating outlasts the execution of the behavior, making sex rewarding for both sexes (Agmo, 1999; Martinez and Paredes, 2001). Sexual interaction in female Syrian hamsters also induces CPP (Kohlert and Olexa, 2005).

Most of the studies that have evaluated whether sex can induce a PA state in females have used OVX rats treated with EB and P to induce the full reproductive pattern. We have repeatedly shown that females that paced their sexual interaction develop CPP, while those that mate in a situation in which they do not control the sexual interaction do not change their initial preference, suggesting that they do not find sexual behavior rewarding (Martinez and Paredes, 2001; Paredes and Alonso, 1997; Paredes and Vazquez, 1999). There are studies suggesting that pacing behavior is not necessary for females to develop a reward state after mating. Females allowed to mate with only one male in three different conditioning sessions showed an increase in the preference score and a reduction in the differences between compartments, which the authors interpret as a change of preference, and hence as a reward state (Meerts and Clark, 2007). However, the authors do not report an increase in the preference score after conditioning, which as described above is the most important factor to consider a CPP and hence a reward state. Moreover, in our laboratory, it was directly tested whether females that mated with the same male along the conditioning sessions developed CPP and no changes were observed, that is, no CPP was induced (Camacho *et al.*, 2009). We have also shown that females that received 10 paced intromissions developed CPP, suggesting the existence of a threshold of stimulation, as in the case of the male, for sex to induce a reward state (Paredes and Vazquez, 1999).

Females that were allowed to pace the sexual interaction and treated only with E did not develop CPP (Gonzalez-Flores *et al.*, 2004; Paredes and Vazquez, 1999). But those treated with E and one of different progestins (megestrol acetate, 5$\alpha$-pregnan-20-dione, or 5$\beta$-pregnan-3$\alpha$-ol-20-one) developed CPP after paced mating (Gonzalez-Flores *et al.*, 2004). It has also been shown that females treated with another neuroesteroid 3a, 5a-THP, before paced mating develop CPP without significantly modifying pacing behavior (Frye *et al.*, 1998). These observations suggest that E and P or one of its metabolites are required for sex to induce a reward state after paced mating. We recently showed that intact females can also developed CPP after mating. Regularly cycling females were divided into different groups; two were allowed to pace the sexual interaction and receive one or three ejaculations from the males. The other two groups were not allowed to pace the sexual interaction and received one or three ejaculations. Only the group of females that received three ejaculations and were allowed to pace the sexual interaction developed a reward state as evaluated by CPP, suggesting that a reward state is induced in naturally cycling females only when they paced the rate of sexual interaction (Camacho *et al.*, 2009b).

Different studies clearly indicate that there are several factors that contribute to the reward state induced by sexual behavior in females. For example, females that received vaginal lavage after conditioning did not modify the initial preference but those females that receive the lavage immediately before the conditioning session developed CPP (Walker *et al.*, 2002). In addition, artificial vaginocervical stimulation induces CPP (Meerts and Clark, 2009a). It has also been shown that clitoral stimulation (CLS) induces CPP. Female rats received either distributed (1 stimulation every 5 s for 1 min) or continuous (1 stimulation/s for 1 min) CLS before being placed in one compartment of the conditioning box. Only the females that received distributed CLS stimulation developed CPP (Parada *et al.*, 2009). The reward state induced by mating is not mediated by the pelvic nerve. Females that received bilateral pelvic nerve transections were able to develop CPP after mating, although contact return latencies in paced mating tests were shorten. The authors conclude that the reinforcing effect of mating stimulation does not depend on the integrity of the pelvic nerve (Meerts and Clark, 2009b).

As with male sexual reward, opioids and the MPOA play an important role in the PA state induced by sexual behavior in females. The systemic administration of naloxone (Paredes and Martinez, 2001) but not of DA antagonists (Garcia Horsman and Paredes, 2004) blocked the reward state induced by paced mating, suggesting that, as in the case of male rats, opioids modulate the reward state induced by mating in females (Paredes and Martinez, 2001). Infusion of naloxone into the MPOA, VMN, and the medial amygdala blocked the reward state induced by paced mating, that is, no CPP was developed, suggesting that these areas are important for the rewarding aspect of paced mating. The group of females infused with naloxone into the nucleus accumbens as well as the control group developed a clear CPP after paced mating. We postulated that the medial amygdale and VMN are important for the transmission of sensory information produced by mating, while the MPOA is important for the reward state (Garcia-Horsman *et al.*, 2008). Moreover, in the aforementioned study by

Pfaus and colleagues in which CLS induced CPP, they also observed that distributed CLS induced more Fos in the MPOA than did continuous CLS or no stimulation. On the other hand, continuous CLS induced more Fos in the posteroventral medial amygdala compared to no stimulation (Parada *et al.*, 2009). Further support for the contention that the MPOA is involved in the rewarding aspects of mating come from the observations that bilateral lesions of the MPOA produced by ibotenic acid blocked the CPP induced by vaginocervial stimulation (Meerts and Clark, 2009c).

## VI. Conclusion

It is clear that sexual behavior induces a PA or reward state that can be evaluated by CPP. This method has several advantages, including the possibility of evaluating approach behavior to a positive stimulus. In addition, it can be used to measure the physiological state induced by sex, or any other natural occurring behavior, independently of the execution of that particular behavior. The studies that have evaluated the possible reward state induced by hormones are few and some of them have evaluated the effects of hormones on drug-induced CPP. Undoubtedly, more research is needed to elucidate the possible rewarding effects of hormones.

## *Acknowledgments*

Research supported by CONACyT projects 53547 and DGAPA project IN213609.

## *References*

Agmo, A. (1999). Sexual motivation—An inquiry into events determining the occurrence of sexual behavior. *Behav. Brain Res.* **105**(1), 129–150.

Agmo, A. (2007). Learning and Sex: Sexual Activity as Reinforcement and Reward. Functional and Dysfunctional Sexual Behavior, pp. 257–291. Academic Press, London.

Agmo, A., and Berenfeld, R. (1990). Reinforcing properties of ejaculation in the male rat: Role of opioids and dopamine. *Behav. Neurosci.* **104**(1), 177–182.

Agmo, A., and Gomez, M. (1991). Conditioned place preference produced by infusion of Met-enkephalin into the medial preoptic area. *Brain Res.* **550**(2), 343–346.

Agmo, A., and Gomez, M. (1993). Sexual reinforcement is blocked by infusion of naloxone into the medial preoptic area. *Behav. Neurosci.* **107**(5), 812–818.

Agmo, A., Federman, I., Navarro, V., Padua, M., and Velazquez, G. (1993). Reward and reinforcement produced by drinking water: Role of opioids and dopamine receptor subtypes. *Pharmacol. Biochem. Behav.* **46,** 183–194.

Alexander, G. M., Packard, M. G., and Hines, M. (1994). Testosterone has rewarding affective properties in male rats: Implications for the biological basis of sexual motivation. *Behav. Neurosci.* **108**(2), 424–428.

Arnedo, M. T., Salvador, A., Martinez-Sanchis, S., and Gonzalez-Bono, E. (2000). Rewarding properties of testosterone in intact male mice: A pilot study. *Pharmacol. Biochem. Behav.* **65**(2), 327–332.

Arnedo, M. T., Salvador, A., Martinez-Sanchis, S., and Pellicer, O. (2002). Similar rewarding effects of testosterone in mice rated as short and long attack latency individuals. *Addict. Biol.* **7**(4), 373–379.

Bardo, M. T., Rowlett, J. K., and Harris, M. J. (1995). Conditioned place preference using opiate and stimulant drugs: A meta-analysis. *Neurosci. Biobehav. Rev.* **19**(1), 39–51.

Beach, H. D. (1957). Morphine addiction in rats. *Can. J. Psychol.* **11**(2), 104–112.

Beauchamp, M. H., Ormerod, B. K., Jhamandas, K., Boegman, R. J., and Beninger, R. J. (2000). Neurosteroids and reward: Allopregnanolone produces a conditioned place aversion in rats. *Pharmacol. Biochem. Behav.* **67**(1), 29–35.

Bechara, A., and van der Kooy, D. (1992). A single brain stem substrate mediates the motivational effects of both opiates and food in nondeprived rats but not in deprived rats. *Behav. Neurosci.* **106**(2), 351–363.

Berridge, K. C., and Robinson, T. E. (1998). What is the role of dopamine in reward: Hedonic impact, reward learning, or incentive salience? *Brain Res. Rev.* **28**(3), 309–369.

Blaustein, J. D., and Erskine, M. S. (2002).Feminine Sexual Behavior: Cellular Integration of Hormonal and Afferent Information in the Rodent Brain.Vol. 1. Academic Press, New York.

Burgdorf, J., and Panksepp, J. (2006). The neurobiology of positive emotions. *Neurosci. Biobehav. Rev.* **30**(2), 173–187.

Burgdorf, J., Wood, P. L., Kroes, R. A., Moskal, J. R., and Panksepp, J. (2007). Neurobiology of 50-kHz ultrasonic vocalizations in rats: Electrode mapping, lesion, and pharmacology studies. *Behav. Brain Res.* **182**(2), 274–283.

Calcagnetti, D. J., and Schechter, M. D. (1992). Place conditioning reveals the rewarding aspect of social interaction in juvenile rats. *Physiol. Behav.* **51**(4), 667–672.

Caldarone, B. J., Stock, H. S., Abrahamsen, G. C., Boechler, M. L., Svare, B. B., and Rosellini, R. A. (1996). Nonassociative processes and place preferences conditioned by testosterone. *Psychol. Rec.* **46**(2), 373–390.

Camacho, F., Sandoval, C., and Paredes, R. G. (2004). Sexual experience and conditioned place preference in male rats. *Pharmacol. Biochem. Behav.* **78**(3), 419–425.

Camacho, F. J., Castro, M., Hernandez, V., and Paredes, R. G. (2007). Facilitation of ejaculation induced by 8-OH-DPAT does not produce conditioned place preference in male rats. *Behav. Neurosci.* **121**(3), 579–585.

Camacho, F. J., Portillo, W., Quintero-Enríquez, O., and Paredes, R. G. (2009a). Reward value of intromissions and morphine in male rats evaluated by conditioned place preference *Physiol. Behav.* **98**(5), 602–607.

Camacho, F. J., Garcia-Horsman, P., and Paredes, R. G. (2009b). Hormonal and testing conditions for the induction of conditioned place preference by paced mating. *Horm. Behav.* **56**(4), 410–415.

Carboni, E., and Vacca, C. (2003). Conditioned place preference. A simple method for investigating reinforcing properties in laboratory animals. *Methods Mol. Med.* **79,** 481–498.

Chen, H. H., Yang, Y. K., Yeh, T. L., Cherng, C. F., Hsu, H. C., Hsiao, S. Y., and Yu, L. (2003). Methamphetamine-induced conditioned place preference is facilitated by estradiol pretreatment in female mice. *Chin. J. Physiol.* **46**(4), 169–174.

Chester, J. A., and Cunningham, C. L. (1998). Modulation of corticosterone does not affect the acquisition or expression of ethanol-induced conditioned place preference in DBA/2 J mice. *Pharmacol. Biochem. Behav.* **59**(1), 67–75.

Clark, A. S., and Henderson, L. P. (2003). Behavioral and physiological responses to anabolic-androgenic steroids. *Neurosci. Biobehav. Rev.* **27**(5), 413–436.

Crowder, W. F., and Hutto, C. W., Jr. (1992). Operant place conditioning measures examined using two nondrug reinforcers. *Pharmacol. Biochem. Behav.* **41**(4), 817–824.

Davis, K. W., Wellman, P. J., and Clifford, P. S. (2007). Augmented cocaine conditioned place preference in rats pretreated with systemic ghrelin. *Regul. Pept.* **140**(3), 148–152.

de Beun, R., Geerts, N. E., Jansen, E., Slangen, J. L., and van de Poll, N. E. (1991a). Luteinizing hormone releasing hormone-induced conditioned place-preference in male rats. *Pharmacol. Biochem. Behav.* **39**(1), 143–147.

de Beun, R., Jansen, E., Smeets, M. A., Niesing, J., Slangen, J. L., and van de Poll, N. E. (1991b). Estradiol-induced conditioned taste aversion and place aversion in rats: Sex- and dose-dependent effects. *Physiol. Behav.* **50**(5), 995–1000.

De Beun, R., Jansen, E., Geerts, N. E., Slangen, J. L., and Van de Poll, N. E. (1992). Temporal characteristics of appetitive stimulus effects of luteinizing hormone-releasing hormone in male rats. *Pharmacol. Biochem. Behav.* **42**(3), 445–450.

de Beun, R., Geerts, N. E., van de Poll, N. E., Slangen, J. L., and Vreeburg, J. T. M. (2004). Sex differences in luteinizing hormone releasing hormone-induced conditioned place preference in the rat. *Drug Dev. Res.* **16**(2–4), 375–383.

Der-Avakian, A., Will, M. J., Bland, S. T., Deak, T., Nguyen, K. T., Schmid, M. J., Spencer, R. L., Watkins, L. R., and Maier, S. F. (2005). Surgical and pharmacological suppression of glucocorticoids prevents the enhancement of morphine conditioned place preference by uncontrollable stress in rats. *Psychopharmacology (Berl.)* **179**(2), 409–417.

Di Chiara, G., and Bassareo, V. (2007). Reward system and addiction: What dopamine does and doesn't do. *Curr. Opin. Pharmacol.* **7**(1), 69–76.

Dietz, D., Wang, H., and Kabbaj, M. (2007). Corticosterone fails to produce conditioned place preference or conditioned place aversion in rats. *Behav. Brain Res.* **181**(2), 287–291.

Dong, Z., Han, H., Wang, M., Xu, L., Hao, W., and Cao, J. (2006). Morphine conditioned place preference depends on glucocorticoid receptors in both hippocampus and nucleus accumbens. *Hippocampus* **16**(10), 809–813.

Erskine, M. S. (1989). Solicitation behavior in the estrous female rat: A review. *Horm. Behav.* **23**(4), 473–502.

Finn, D. A., Phillips, T. J., Okorn, D. M., Chester, J. A., and Cunningham, C. L. (1997). Rewarding effect of the neuroactive steroid 3alpha-hydroxy-5alpha-pregnan-20-one in mice. *Pharmacol. Biochem. Behav.* **56**(2), 261–264.

Fitchett, A. E., Barnard, C. J., and Cassaday, H. J. (2006). There's no place like home: Cage odours and place preference in subordinate CD-1 male mice. *Physiol. Behav.* **87**(5), 955–962.

Frye, C. A., and Rhodes, M. E. (2006). Administration of estrogen to ovariectomized rats promotes conditioned place preference and produces moderate levels of estrogen in the nucleus accumbens. *Brain Res.* **1067**(1), 209–215.

Frye, C. A., Bayon, L. E., Pursnani, N. K., and Purdy, R. H. (1998). The neurosteroids, progesterone and 3alpha, 5alpha-THP, enhance sexual motivation, receptivity, and proceptivity in female rats. *Brain Res.* **808**(1), 72–83.

Frye, C. A., Park, D., Tanaka, M., Rosellini, R., and Svare, B. (2001). The testosterone metabolite and neurosteroid 3alpha-androstanediol may mediate the effects of testosterone on conditioned place preference. *Psychoneuroendocrinology* **26**(7), 731–750.

Galea, L. A., Wide, J. K., Paine, T. A., Holmes, M. M., Ormerod, B. K., and Floresco, S. B. (2001). High levels of estradiol disrupt conditioned place preference learning, stimulus response learning and reference memory but have limited effects on working memory. *Behav. Brain Res.* **126**(1–2), 115–126.

Garcia Horsman, P., and Paredes, R. G. (2004). Dopamine antagonists do not block conditioned place preference induced by paced mating behavior in female rats. *Behav. Neurosci.* **118**(2), 356–364.

Garcia-Horsman, S. P., Agmo, A., and Paredes, R. G. (2008). Infusions of naloxone into the medial preoptic area, ventromedial nucleus of the hypothalamus, and amygdala block conditioned place preference induced by paced mating behavior. *Horm. Behav.* **54**(5), 709–716.

Gonzalez-Flores, O., Camacho, F. J., Dominguez-Salazar, E., Ramirez-Orduna, J. M., Beyer, C., and Paredes, R. G. (2004). Progestins and place preference conditioning after paced mating. *Horm. Behav.* **46**(2), 151–157.

Harding, S. M., and McGinnis, M. Y. (2004). Androgen receptor blockade in the MPOA or VMN: Effects on male sociosexual behaviors. *Physiol. Behav.* **81**(4), 671–680.

Higgins, G. A., Nguyen, P., and Sellers, E. M. (1992). Morphine place conditioning is differentially affected by CCKA and CCKB receptor antagonists. *Brain Res.* **572**(1–2), 208–215.

Hughes, A. M., Everitt, B. J., and Herbert, J. (1990). Comparative effects of preoptic area infusions of opioid peptides, lesions and castration on sexual behaviour in male rats: Studies of instrumental behaviour, conditioned place preference and partner preference. *Psychopharmacology (Berl.)* **102**(2), 243–256.

Hull, C. L. (1943). Principles of behavior: An introduction to behavior theory. Appleton-Century-Crofts, New York.

King, B. E., Packard, M. G., and Alexander, G. M. (1999). Affective properties of intra-medial preoptic area injections of testosterone in male rats. *Neurosci. Lett.* **269**(3), 149–152.

Kippin, T. E., and van der Kooy, D. (2003). Excitotoxic lesions of the tegmental pedunculopontine nucleus impair copulation in naive male rats and block the rewarding effects of copulation in experienced male rats. *Eur. J. Neurosci.* **18**(9), 2581–2591.

Knutson, B., Burgdorf, J., and Panksepp, J. (2002). Ultrasonic vocalizations as indices of affective states in rats. *Psychol. Bull.* **128**(6), 961–977.

Kohlert, J. G., and Olexa, N. (2005). The role of vaginal stimulation for the acquisition of conditioned place preference in female Syrian hamsters. *Physiol. Behav.* **84**(1), 135–139.

Kudwa, A. E., Dominguez-Salazar, E., Cabrera, D. M., Sibley, D. R., and Rissman, E. F. (2005). Dopamine D5 receptor modulates male and female sexual behavior in mice. *Psychopharmacology (Berl.)* **180**(2), 206–214.

Lett, B. T., Grant, V. L., Byrne, M. J., and Koh, M. T. (2000). Pairings of a distinctive chamber with the aftereffect of wheel running produce conditioned place preference. *Appetite* **34**(1), 87–94.

Lett, B. T., Grant, V. L., and Koh, M. T. (2001). Naloxone attenuates the conditioned place preference induced by wheel running in rats. *Physiol. Behav.* **72**(3), 355–358.

Lu, L., Zhang, B., Liu, Z., and Zhang, Z. (2002). Reactivation of cocaine conditioned place preference induced by stress is reversed by cholecystokinin-B receptors antagonist in rats. *Brain Res.* **954**(1), 132–140.

Martinez, I., and Paredes, R. G. (2001). Only self-paced mating is rewarding in rats of both sexes. *Horm. Behav.* **40**(4), 510–517.

Mattson, B. J., Williams, S., Rosenblatt, J. S., and Morrell, J. I. (2001). Comparison of two positive reinforcing stimuli: Pups and cocaine throughout the postpartum period. *Behav. Neurosci.* **115**(3), 683–694.

Mattson, B. J., Williams, S. E., Rosenblatt, J. S., and Morrell, J. I. (2003). Preferences for cocaine- or pup-associated chambers differentiates otherwise behaviorally identical postpartum maternal rats. *Psychopharmacology (Berl.)* **167**(1), 1–8.

McClintock, M. K., and Adler, N. T. (1978). The role of the female during copulation in wild and domestic Norway rats (*Rattus norvegicus*). *Behaviour* **67,** 67–96.

McClintock, M. K., and Anisko, J. J. (1982). Group mating among Norway rats. I. Sex differences in the pattern and neuroendocrine consequences of copulation. *Anim. Behav.* **30,** 398–409.

McClintock, M. K., Anisko, J. J., and Adler, N. T. (1982). Group mating among Norway rats. II. The social dynamics of copulation: Competition, cooperation, and mate choice. *Anim. Behav.* **30,** 410–425.

Meerts, S. H., and Clark, A. S. (2007). Female rats exhibit a conditioned place preference for nonpaced mating. *Horm. Behav.* **51**(1), 89–94.

Meerts, S. H., and Clark, A. S. (2009a). Artificial vaginocervical stimulation induces a conditioned place preference in female rats. *Horm. Behav.* **55**(1), 128–132.

Meerts, S. H., and Clark, A. S. (2009b). Conditioned place preference for mating is preserved in rats with pelvic nerve transection. *Behav. Neurosci.* **123**(3), 539–546.

Meerts, S. H., and Clark, A. S. (2009c). Lesions of the medial preoptic area interfere with the display of a conditioned place preference for vaginocervical stimulation in rats. *Behav. Neurosci.* **123**(4), 752–757.

Mehrara, B. J., and Baum, M. J. (1990). Naloxone disrupts the expression but not the acquisition by male rats of a conditioned place preference response for an oestrous female. *Psychopharmacology (Berl.)* **101**(1), 118–125.

Miller, R. L., and Baum, M. J. (1987). Naloxone inhibits mating and conditioned place preference for an estrous female in male rats soon after castration. *Pharmacol. Biochem. Behav.* **26**(4), 781–789.

Minerly, A. E., Russo, S. J., Kemen, L. M., Nazarian, A., Wu, H. B., Weierstall, K. M., Akhavan, A., Jenab, S., and Quinones-Jenab, V. (2008). Testosterone plays a limited role in cocaine-induced conditioned place preference and locomotor activity in male rats. *Ethn. Dis.* **18**(2 Suppl. 2), S2–200-4.

Packard, M. G., Cornell, A. H., and Alexander, G. M. (1997). Rewarding affective properties of intra-nucleus accumbens injections of testosterone. *Behav. Neurosci.* **111**(1), 219–224.

Packard, M. G., Schroeder, J. P., and Alexander, G. M. (1998). Expression of testosterone conditioned place preference is blocked by peripheral or intra-accumbens injection of alpha-flupenthixol. *Horm. Behav.* **34**(1), 39–47.

Pankevich, D. E., Cherry, J. A., and Baum, M. J. (2006). Accessory olfactory neural Fos responses to a conditioned environment are blocked in male mice by vomeronasal organ removal. *Physiol. Behav.* **87**(4), 781–788.

Panksepp, J. (2007). Neuroevolutionary sources of laughter and social joy: Modeling primal human laughter in laboratory rats. *Behav. Brain Res.* **182**(2), 231–244.

Panksepp, J. B., Jochman, K. A., Kim, J. U., Koy, J. J., Wilson, E. D., Chen, Q., Wilson, C. R., and Lahvis, G. P. (2007). Affiliative behavior, ultrasonic communication and social reward are influenced by genetic variation in adolescent mice. *PLoS ONE* **2**(4), e351.

Parada, M., Chamas, L., Censi, S., Coria-Avila, G., and Pfaus, J. G. (2010). Clitoral stimulation induces conditioned place preference and Fos activation in the rat. *Horm. Behav.* **57**(2), 112–118.

Paredes, R. G. (2003). Medial preoptic area/anterior hypothalamus and sexual motivation. *Scand. J. Psychol.* **44**(3), 203–212.

Paredes, R. G. (2009). Evaluating the neurobiology of sexual reward. *ILAR J.* **50**(1), 15–27.

Paredes, R. G., and Agmo, A. (2004). Has dopamine a physiological role in the control of sexual behavior? A critical review of the evidence. *Prog. Neurobiol.* **73**(3), 179–226.

Paredes, R. G., and Alonso, A. (1997). Sexual behavior regulated (paced) by the female induces conditioned place preference. *Behav. Neurosci.* **111**(1), 123–128.

Paredes, R. G., and Baum, M. J. (1997). Role of the medial preoptic area/anterior hypothalamus in the control of masculine sexual behavior. *Annu. Rev. Sex Res.* **8,** 68–101.

Paredes, R. G., and Fernández-Guasti, A. (2008). Rewarding properties of mating. *In* "Neural Mechanisms of Drugs of Abuse and Natural Reinforcers" (M. Méndez and R. Mondragón-Ceballos, Eds.), pp. 159–170.

Paredes, R. G., and Martinez, I. (2001). Naloxone blocks place preference conditioning after paced mating in female rats. *Behav. Neurosci.* **115**(6), 1363–1367.

Paredes, R. G., and Vazquez, B. (1999). What do female rats like about sex? Paced mating. *Behav. Brain Res.* **105**(1), 117–127.

Piazza, P. V., and Le Moal, M. (1997). Glucocorticoids as a biological substrate of reward: Physiological and pathophysiological implications. *Brain Res. Brain Res. Rev.* **25**(3), 359–372.

Pierman, S., Tirelli, E., Douhard, Q., Baum, M. J., and Bakker, J. (2006). Male aromatase knockout mice acquire a conditioned place preference for cocaine but not for contact with an estrous female. *Behav. Brain Res.* **174**(1), 64–69.

Popik, P., Wrobel, M., Rygula, R., Bisaga, A., and Bespalov, A. Y. (2003). Effects of memantine, an NMDA receptor antagonist, on place preference conditioned with drug and nondrug reinforcers in mice. *Behav. Pharmacol.* **14**(3), 237–244.

Qi, J., Yang, J. Y., Wang, F., Zhao, Y. N., Song, M., and Wu, C. F. (2009). Effects of oxytocin on methamphetamine-induced conditioned place preference and the possible role of glutamatergic neurotransmission in the medial prefrontal cortex of mice in reinstatement. *Neuropharmacology* **56**(5), 856–865.

Robitaille, J. A., and Bouvet, J. (1976). Field observations on the social behavior of the Norway rat, *Rattus norvegicus* (Berkenhout). *Biol. Behav.* **1,** 289–308.

Rossi, N. A., and Reid, L. D. (1976). Affective states associated with morphine injections. *Physiol. Psychol.* **4**(3), 269–274.

Russo, S. J., Festa, E. D., Fabian, S. J., Gazi, F. M., Kraish, M., Jenab, S., and Quinones-Jenab, V. (2003). Gonadal hormones differentially modulate cocaine-induced conditioned place preference in male and female rats. *Neuroscience* **120**(2), 523–533.

Russo, S. J., Sun, W. L., Minerly, A. C., Weierstall, K., Nazarian, A., Festa, E. D., Niyomchai, T., Akhavan, A., Luine, V., Jenab, S., and Quinones-Jenab, V. (2008). Progesterone attenuates cocaine-induced conditioned place preference in female rats. *Brain Res.* **1189,** 229–235.

Sato, S. M., Schulz, K. M., Sisk, C. L., and Wood, R. I. (2008). Adolescents and androgens, receptors and rewards. *Horm. Behav.* **53**(5), 647–658.

Schechter, M. D., and Calcagnetti, D. J. (1993). Trends in place preference conditioning with a cross-indexed bibliography; 1957-1991. *Neurosci. Biobehav. Rev.* **17**(1), 21–41.

Schildein, S., Agmo, A., Huston, J. P., and Schwarting, R. K. (1998). Intraaccumbens injections of substance P, morphine and amphetamine: Effects on conditioned place preference and behavioral activity. *Brain Res.* **790**(1–2), 185–194.

Schroeder, J. P., and Packard, M. G. (2000). Role of dopamine receptor subtypes in the acquisition of a testosterone conditioned place preference in rats. *Neurosci. Lett.* **282**(1–2), 17–20.

Schultz, W. (2006). Behavioral theories and the neurophysiology of reward. *Annu. Rev. Psychol.* **57,** 87–115.

Skinner, B. F. (1938). The behavior of organisms: An experimental analysis. B. F. Skinner Foundation, Cambridge, Massachusetts.

Tenk, C. M., Wilson, H., Zhang, Q., Pitchers, K. K., and Coolen, L. M. (2009). Sexual reward in male rats: Effects of sexual experience on conditioned place preferences associated with ejaculation and intromissions. *Horm. Behav.* **55**(1), 93–97.

Tyhon, A., Lakaye, B., Adamantidis, A., and Tirelli, E. (2008). Amphetamine- and cocaine-induced conditioned place preference and concomitant psychomotor sensitization in mice with genetically inactivated melanin-concentrating hormone MCH(1) receptor. *Eur. J. Pharmacol.* **599**(1–3), 72–80.

Tzschentke, T. M. (2007). Measuring reward with the conditioned place preference (CPP) paradigm: Update of the last decade. *Addict. Biol.* **12**(3–4), 227–462.

Valverde, O., Fournie-Zaluski, M. C., Roques, B. P., and Maldonado, R. (1996). The CCKB antagonist PD-134, 308 facilitates rewarding effects of endogenous enkephalins but does not induce place preference in rats. *Psychopharmacology (Berl.)* **123**(2), 119–126.

Walf, A. A., Rhodes, M. E., Meade, J. R., Harney, J. P., and Frye, C. A. (2007). Estradiol-induced conditioned place preference may require actions at estrogen receptors in the nucleus accumbens. *Neuropsychopharmacology* **32**(3), 522–530.

Walker, Q. D., Nelson, C. J., Smith, D., and Kuhn, C. M. (2002). Vaginal lavage attenuates cocaine-stimulated activity and establishes place preference in rats. *Pharmacol. Biochem. Behav.* **73**(4), 743–752.

Wang, J., Fang, Q., Liu, Z., and Lu, L. (2006). Region-specific effects of brain corticotropin-releasing factor receptor type 1 blockade on footshock-stress- or drug-priming-induced reinstatement of morphine conditioned place preference in rats. *Psychopharmacology (Berl.)* **185**(1), 19–28.

White, N. M. (1989). Reward or reinforcement: What's the difference? *Neurosci. Biobehav. Rev.* **13**(2–3), 181–186.

Wood, R. I. (2008). Anabolic-androgenic steroid dependence? Insights from animals and humans. *Front. Neuroendocrinol.* **29**(4), 490–506.

Yahyavi-Firouz-Abadi, N., Tahsili-Fahadan, P., Ghahremani, M. H., and Dehpour, A. R. (2007). Melatonin enhances the rewarding properties of morphine: Involvement of the nitric oxidergic pathway. *J. Pineal Res.* **42**(4), 323–329.

Young, P. T. (1959). The role of affective processes in learning and motivation. *Psychol. Rev.* **66**(2), 104–125.

Young, P. T., and Christensen, K. R. (1962). Algebraic summation of hedonic processes. *J. Comp. Physiol. Psychol.* **55**, 332–336.

CHAPTER FOURTEEN

# Sex Steroids and Acetylcholine Release in the Hippocampus

Dai Mitsushima

## Contents



## Abstract

The hippocampus is processing contextual and spatial information in behaving animals. Various environmental stimuli, such as exploration, stress, or learning, increase acetylcholine (ACh) release to activate hippocampal functions. In this chapter, we review the *in vivo* ACh release property and the physiological function in the hippocampus. By monitoring 24-h ACh release profile in the hippocampus, we found a sex-specific and time-dependent ACh release property, providing a neural basis of sex-specific and time-dependent hippocampal function. In both sexes of rats, gonadectomy, known to impair the hippocampal function, dissociated the appropriate timing of the ACh release and attenuated the release amount. However, testosterone in gonadectomized males or estradiol in gonadectomized females restored the timing and maintained the amount of ACh release. Although the contributing sex hormone was sex-specific, neonatal $17\beta$-estradiol treatment in female pups successfully masculinized the contributing hormone. Moreover, some rearing conditions affected the sex-specific ACh release, since the influence was sex-specific. These results

Department of Physiology, Yokohama City University Graduate School of Medicine, Kanazawa-ku, Yokohama, Japan

*Vitamins and Hormones*, Volume 82
ISSN 0083-6729, DOI: 10.1016/S0083-6729(10)82014-X

suggest that (1) estrogen receptor mediates neonatal sexual differentiation of septohippocampal cholinergic system and (2) time of day, circulating sex hormone, and rearing conditions affect the sex-specific ACh release in the hippocampus. 

# I. Role of Acetylcholine in the Hippocampus

A number of studies suggest that acetylcholine (ACh) plays a key role in orchestrating major hippocampal functions (Fig. 14.1). In the behavioral studies, the ACh release increase during learning (Hironaka *et al.*, 2001; Ragozzino *et al.*, 1996; Stancampiano *et al.*, 1999) and are positively correlated with learning performance (Gold, 2003; Parent and Baxter, 2004). Because bilateral injections of scopolamine into the dorsal hippocampus impair spatial learning ability (Herrera-Morales *et al.*, 2007), muscarinic ACh receptors appear to mediate the consolidation of spatial memory. At the network level, ACh generates a theta rhythm (Lee *et al.*, 1994) that modulates the induction of long-term potentiation (LTP) in hippocampal CA1 neurons (Hyman *et al.*, 2003). At the cellular level, both pyramidal and nonpyramidal neurons in the hippocampal CA1 area receive direct cholinergic afferents mediated by the muscarinic receptors (Cole and Nicoll, 1983; Markram and Segal, 1990; Widmer *et al.*, 2006). *In vitro* studies showed that bath application of carbachol, a cholinergic agonist, induces LTP in CA1 pyramidal neurons without electrical tetanus stimulus, suggesting that ACh in the hippocampus plays a principal role in the synaptic plasticity of the CA1 pyramidal neurons (Auerbach and Segal, 1996). A recent study further

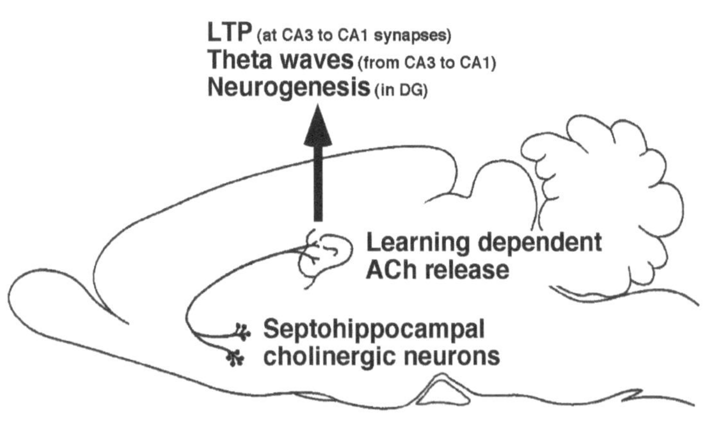


**Figure 14.1** Schematic illustration of septohippocampal cholinergic neurons. The released ACh activates major hippocampal functions. ACh, acetylcholine; LTP, long-term potentiation; DG, dentate gyrus.

revealed that focal activation of muscarinic ACh receptors in one CA1 pyramidal neuron induces $Ca^{2+}$ release from inositol 1,4,5-trisphosphate-sensitive stores to induce LTP (Fernández de Sevilla *et al.*, 2008). Moreover, genetic deficiency of muscarinic ACh receptors ($M_1$ or $M_2$) impairs LTP in CA1 region (Seeger *et al.*, 2004; Shinoe *et al.*, 2005). Not only is the synaptic plasticity, ACh release in the hippocampus is also responsible for neurogenesis in the dentate gyrus (Kotani *et al.*, 2006; Mohapel *et al.*, 2005).

## II. ACh Release in the Hippocampus Is Sex-Specific

Cholinergic neurons within the basal forebrain provide the major projection to the neocortex and hippocampus (Mesulam *et al.*, 1983). Cortical regions receive cholinergic inputs mainly from the nucleus basalis magnocellularis (NBM) or the diagonal band of Broca, whereas the hippocampus receives cholinergic inputs mostly from the medial septum (MS) and horizontal limb of diagonal band of Broca (Mesulam *et al.*, 1983). Because the widespread cholinergic projections are necessary to maintain learning and memory (Perry *et al.*, 1999, Sarter and Parikh, 2005), we hypothesized that a sex-specific ACh release in specific brain areas may underlie the sex difference in learning and memory. To estimate ACh release, we performed *in vivo* microdialysis study in freely moving rats (Fig. 14.2). Briefly, a microdialysis probe with semipermeable membrane (1.0 mm in length) was inserted into a specific brain area via surgically preimplanted guide cannula. We perfused inside of the membrane with artificial cerebrospinal fluid and assayed ACh in dialysates, using a high-performance liquid chromatography system. As a result, we are successful in finding sex-specific ACh release in the specific brain areas in behaving rats.

Using this *in vivo* measuring system, we found a sex-specific ACh release in the hippocampus. Gonadally intact male rats consistently show a greater stress response of ACh release in the hippocampus compared with diestrous or proestrous female rats, suggesting that the response of the septohippocampal cholinergic system to environmental stimuli is sexually dimorphic (Mitsushima *et al.*, 2003a). Moreover, we found that the sex-specific ACh release involves time-dependent 24-h profile: ACh release in the hippocampus was relatively similar in the light phase, but consistently lower in female rats than males in the dark phase (Masuda *et al.*, 2005). Although the ACh release clearly showed a daily rhythm in female rats, females exhibited smaller amplitude of daily change than did males. Interestingly, this daily change is quite similar to the daily rhythm in hippocampal MAPK activity and cAMP: phospholirated ERK protein, GTP-bound Ras protein, and cAMP in the hippocampus show clear daily changes in male mice

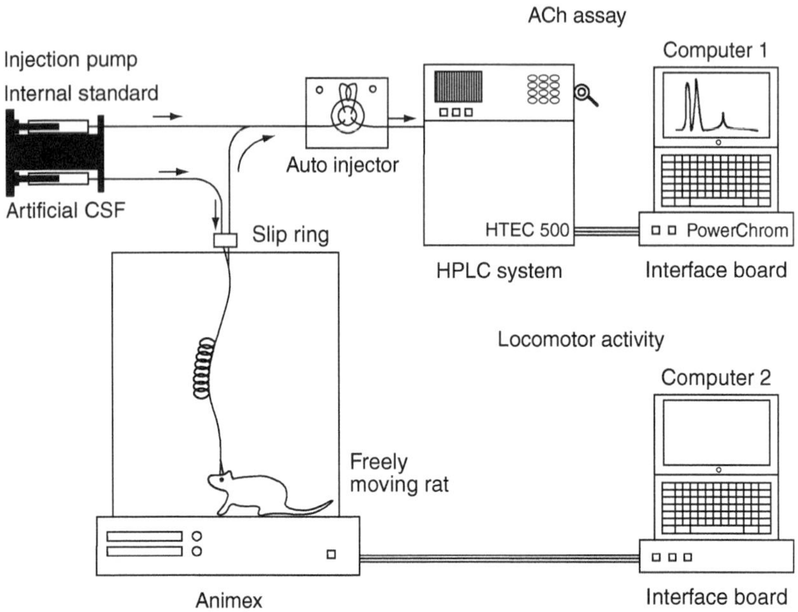


**Figure 14.2** Experimental setup of *in vivo* microdialysis system. In order to evaluate activational effect of sex hormones on the ACh release, we simultaneously measured the spontaneous locomotor activity in the same subject.

(Eckel-Mahan *et al.*, 2008). Although the changes are unknown in females, it would be of interest to elucidate the sex difference in the intracellular signaling.

In the neuroanatomical study, stereological analysis showed the number of choline acetyltransferase immunoreactive (ChAT-ir) cells in the medial septum or horizontal limb of diagonal band is not sex-specific (Takase *et al.*, 2009). Since the number of septohippocampal cholinergic neurons does not appear to be involved in the sex difference in ACh release in the hippocampus, we hypothesized that sex-specific neural circuits or substance(s) may control the endogenous release.

## III. Neural Control of Septohippocampal Cholinergic Neurons

Neurotransmitters may be involved in the expression of the sex difference in the ACh release. For instance, dopaminergic neurons in the ventral tegmental area (A10) have been shown to control septohippocampal cholinergic neurons through the A10-septal dopaminergic pathway in male rats (Nilsson *et al.*, 1992; Swanson, 1982; Yanai *et al.*, 1993). A neuroanatomical study suggested that $D_2$ receptors rather than $D_1$ receptors mediate the dopaminergic control of septohippocampal cholinergic neurons (Weiner *et al.*, 1991). It has been shown that opiatergic neurons also control septohippocampal cholinergic neurons in male rats (Mizuno and Kimura, 1996); the injection of naloxone into the MS markedly increased the ACh release in the hippocampus, while an opioid $\mu$ receptor agonist decreased the release (Mizuno and Kimura, 1996). In contrast, GABA seems to inhibit septohippocampal cholinergic neurons; the injection of muscimol into the medial septum decreased the ACh release in the hippocampus, while the injection of bicuculline increased it (Moor *et al.*, 1998). Although the neural systems are still unknown for female rats, neural control of septohippocampal cholinergic neurons should be involved in the expression of sex difference in the ACh release.

## IV. Circulating Sex Steroids Activate ACh Release

Not only neurotransmitters but also circulating sex steroids may regulate cholinergic neurons. In fact, neuroanatomical studies demonstrated in intact male and female rats that a number of dopaminergic neurons in the A10 have androgen receptor immunoreactivity (Kritzer, 1997) and 45–60% of cholinergic neurons in the MS have estrogen receptor $\alpha$ immunoreactivity (Miettinen *et al.*, 2002; Mufson *et al.*, 1999). Together with fact that female rats show a greater circulating estrogen concentration than male rats (Mitsushima *et al.*, 2003b; Shors *et al.*, 2001) and male rats show much greater circulating androgen concentration than female rats (Falvo *et al.*, 1974; Rush and Blake, 1982), it is possible that cholinergic neurons are affected by sex steroids differently in male and female rats.

The activational effects of sex steroids on cholinergic neurons have been suggested in previous neuroanatomical and neurochemical findings. For example, male gonadectomy decreases the density of cholinergic fibers in the dorsal hippocampus, while testosterone replacement in gonadectomized male rats maintains fiber density (Nakamura *et al.*, 2002). Also, estradiol increases the induction of choline acetyltransferase in the basal forebrain in gonadectomized female rats (Luine *et al.*, 1986; McEwen and Alves, 1999). A previous *in vitro* study demonstrated that estradiol treatment increases both high-affinity choline uptake and ACh synthesis in basal forebrain neurons (Pongrac *et al.*, 2004). Furthermore, we recently reported an activational effect of sex steroids on the maintenance of stress-induced ACh release in the dorsal hippocampus in immobilized rats (Mitsushima *et al.*, 2008). Despite all of this evidence suggesting the activational effect of sex steroids on ACh release in the dorsal hippocampus, conclusive evidence such as dynamic ACh changes under physiological conditions has not been presented in behaving animals. Since spontaneous movement increases ACh release (Day *et al.*, 1991; Mitsushima *et al.*, 1996), we simultaneously analyzed ACh release and spontaneous locomotor activity to determine the precise effect of sex steroids. In the recent study, we found that gonadectomy impaired ACh release without affecting spontaneous locomotor activity levels. Moreover, the activational effect on ACh release was clear especially during active period, although it is not clear during resting period (Fig. 14.3; Mitsushima *et al.*, 2009). Our results provide the first evidence that the sex-specific 24-h profile of ACh release is highly dependent on the presence of sex steroids.

## V. Activational Effect Is Sex-Specific

We found that the activational effect of sex steroids is sex-specific (Fig. 14.3). Testosterone replacement in gonadectomized female rats failed to increase ACh release to those seen in gonadectomized testosterone-primed male rats. Similarly, estradiol replacement was unable to restore ACh release in

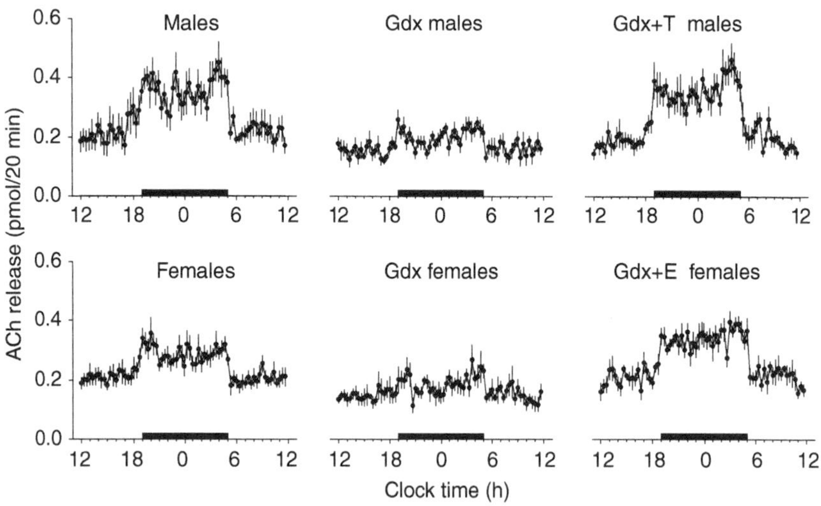


**Figure 14.3** ACh release in the hippocampus is time-dependent, sex-specific, and hormone-dependent. In freely moving condition, ACh release is high during active period (ie, dark phase) but low during resting period. Although gonadally intact males show higher levels than females, gonadal steroids are required for the difference. In gonadectomized (Gdx) groups, experiments were performed 2 weeks after the gonadectomy or steroid replacement. Horizontal black bars indicate the dark phase. +T, testosterone-priming; +E, estradiol-priming. The number of animals was 6–8 in each group (see Mitsushima *et al.*, 2009).

gonadectomized male rats (Mitsushima *et al.*, 2009). Moreover, estradiol consistently increases *N*-methyl-D-aspartate receptor binding and spine density in the CA1 area of gonadectomized female rats, although the treatment fails to increase these same parameters in gonadectomized male rats (Parducz *et al.*, 2006; Romeo *et al.*, 2005). These results suggest that sex-specific steroids are important for maintaining hippocampal functions. Based on our data, we hypothesized that the action of sex-specific steroids is due to neonatal sexual differentiation rather than the activational effects of sex steroids in adult rats. Moreover, in the latest study, we found that neonatal androgenization in females increased ACh release to resemble that of normal males without affecting spontaneous activity levels (Mitsushima *et al.*, 2009). These results indicate an organizational effect on sex-specific ACh release in behaving rats, and support currently accepted theories of sexual differentiation.

## VI. Sexual Differentiation Fates the Sex-Specific Activational Effect

Because testosterone can be aromatized to estradiol in the forebrain, neonatal sex steroids activate both estrogen and androgen receptors (McEwen, 1981). In our study, both testosterone and estradiol treatment in neonatal female pups masculinized ACh release profile in adults, suggesting an estrogen receptor-mediated masculinization of septohippocampal cholinergic systems (Mitsushima *et al.*, 2009). These results are consistent with the previous finding that testosterone or estradiol treatment in neonatal female pups improves their adult spatial performance, whereas neonatal gonadectomy in male pups impairs the performance (Williams and Meck, 1991). By contrast, dihydrotestosterone treatment failed to masculinize ACh release profile. Although dihydrotestosterone has been classically considered as a prototypical androgen receptor agonist, a metabolite of dihydrotestosterone, 3$\beta$-diol, has higher affinity for estrogen receptor $\beta$ (Lund *et al.*, 2006). Therefore, dihydrotestosterone and its metabolites may stimulate both androgen receptor and estrogen receptor $\beta$, whereas estradiol stimulates estrogen receptor $\alpha$ and $\beta$. Considering the action of sex steroids and their metabolites, estrogen receptor $\alpha$ may mediate the organizational effect on septohippocampal cholinergic system.

## VII. Interaction with Environmental Conditions

Various environmental conditions seem to interact with the effects of sex steroids. Although sex steroids did not show activational effects on the baseline levels of ACh release, sex steroids clearly activated immobility stress-induced ACh release response. In addition, contributing sex hormone to maintain the ACh release response was sex-specific: testosterone enhanced the ACh release response in male rats, while estradiol maintain the response in females (Mitsushima *et al.*, 2008). Moreover, daily light/dark cycle affects the activational effects on the ACh release in the hippocampus. Although sex steroids slightly enhanced the ACh release during the light phase, the activational effects were much stronger during the dark phase (Fig. 14.3). Considering that the time-dependent activational effect was also sex-specific and hormone-dependent, these results suggest complicated interactions with environmental conditions (Mitsushima *et al.*, 2009).

Some other environmental effects may affect the basal forebrain cholinergic system. Environmental conditions, such as complex or restricted (Brown, 1968; Smith, 1972), enriched or impoverished (Greenough *et al.*, 1972), social or isolated conditions (Hymovitch, 1952; Juraska *et al.*, 1984; Seymoure *et al.*, 1996), seem to affect spatial learning ability with sex-specific manner. For example, male rats exhibited superior performance in learning maze tests compared with female rats if they were housed socially (Einon, 1980). But if they were housed in isolation, female rats exhibited performance superior to that of male rats (Einon, 1980). Although few studies were performed on the relation between the sex-specific environmental effects and ACh release in the brain, we have reported that 4-day housing in a small cage attenuates the daily ACh release in the hippocampus in male rats (Mitsushima *et al.*, 1998), but not in female rats (Masuda *et al.*, 2005). Taken together, these results suggest that some environmental conditions may contribute to the sex difference in spatial learning ability.

Feeding conditions after weaning also affect spatial learning ability. If fed pelleted diet (i.e., standard laboratory diet), male rats show performance superior to that of female rats (Beatty, 1984; Williams and Meck, 1991). But when fed powdered diet, female rats, but not male rats, showed improved performance (Endo *et al.*, 1994; Takase *et al.*, 2005a). In our study, it was found that feeding with powdered diet after weaning increased the ACh release in the hippocampus in female rats, while it did not in male rats (Takase *et al.*, 2005b). The 24-h ACh release in female rats fed powdered diet was as high as that in male rats fed either powdered or pelleted diet, showing no sex difference. Since feeding with powdered diet improved the performance in spatial learning ability in female rats (Endo *et al.*, 1994), the increase in the ACh release in the hippocampus in female rats fed powdered diet may partly contribute to the improvement of spatial learning ability in female rats. So far, some studies have suggested that environmental conditions affect the sex difference in the spatial learning performance (Einon, 1980; Endo *et al.*, 1994). Our findings provide further evidence that environmental conditions affect the sex difference in the hippocampal function.

## VIII. Interaction with Spontaneous Behaviors

ACh release episodically changes with spontaneous movement (Day *et al.*, 1991; Mitsushima *et al.*, 1998; Mizuno *et al.*, 1991). As described above, ACh release increase during learning or exploratory behaviors (Hironaka *et al.*, 2001; Ragozzino *et al.*, 1996; Stancampiano *et al.*, 1999), which stimulate electrical activity of cholinergic neurons in the basal forebrain (Buzsáki *et al.*, 1988). Moreover, voluntary running enhances neurogenesis, spatial learning and synaptic plasticity in mice (van Praag *et al.*, 1999). By contrast, a restriction of exploratory behavior not only reduces ACh release (Mitsushima *et al.*, 1998) but also impaired spatial learning (Mitsushima *et al.*, 2001). Therefore, to analyze the precise effects of sex steroids on ACh release, we simultaneously analyzed ACh release and spontaneous locomotor activity to determine the precise effect of sex steroids. We found that gonadectomy severely impaired ACh release without changes in spontaneous locomotor activity.

Moreover, after gonadectomy, the positive correlation between ACh release and locomotor activity levels was severely impaired, suggesting that hippocampal function may not always be activated at low sex steroid levels (Mitsushima *et al.*, 2009). Learning impairment in gonadectomized rats (Daniel *et al.*, 1997; Gibbs and Pfaff, 1992; Kritzer *et al.*, 2001; Luine *et al.*, 2003; Markowska and Savonenko, 2002) may be due to insufficient activation of hippocampus at the appropriate time. Because the replacement of sex-specific steroids restored the high positive correlation between ACh release and activity levels, the correlation appears to depend on the presence of sex steroids. These results suggest that circulating sex steroids strengthen the coupling between spontaneous behaviors and ACh release (Mitsushima *et al.*, 2009).

## IX. Activational Effects and Alzheimer's Disease

In humans, circulating levels of gonadal steroids decline with age. Moreover, a reduction in ACh synthesis is known as a common feature of Alzheimer's disease (Coyle *et al.*, 1983), afflicting more than 18 million people worldwide (Ferri *et al.*, 2005; Mount and Downtown 2006). The disease is the most common form of dementia (Cummings 2004) and is frequently accompanied by insomnia, poor concentration, or day/night confusion (McCurry *et al.*, 2004; Starkstein *et al.*, 2005). Centrally active acetylcholinesterase inhibitor (donepezil) is effective in not only mild but also moderate-to-severe cases (Petersen *et al.*, 2005; Winblad *et al.*, 2006), proving the importance of endogenous ACh in humans. In addition, women are twice as likely to develop the disease (Swaab and Hofman 1995), and estradiol seems to play a protective role (Norbury *et al.*, 2007; Zandi *et al.*, 2002). A recent study using single photon emission tomography showed that estrogen replacement therapy in healthy women increases muscarinic $M_1/M_4$ receptor binding in the hippocampus (Norbury *et al.*, 2007). Conversely in men, testosterone but not estradiol seems to play a protective role (Moffat *et al.*, 2004; Rosario *et al.*, 2004) and testosterone supplementation clearly improved hippocampus-dependent learning deficits in men with Alzheimer's disease (Cherrier *et al.*, 2005). These results suggest a sex-specific activational effect of gonadal steroids on the cholinergic system in humans. Thus, there are many similarities between the rat model and the human studies, supporting the idea that the gonadal steroids replacement or an increase in their bioavailability is necessary when there is a subthreshold level of the hormone. Based on the neonatal sexual differentiation of the septohippocampal cholinergic system, we may have to search for sex-specific clinical strategies for Alzheimer's disease.

## X. Conclusions

Gonadally intact male rats consistently show a greater ACh release in the hippocampus compared with diestrous or proestrous female rats. Activational effect of sex steroids are important for the sex-specific ACh release in the hippocampus, since impaired the ACh release in gonadectomized rats does not show the sex-specific levels. Neonatal treatment of either testosterone or estradiol clearly increased ACh release in female rats, suggesting neonatal sex differentiation of septohippocampal cholinergic systems. Moreover, environmental effects on the basal forebrain cholinergic system seem to be sex-specific; housing in the small cage attenuated the ACh release only in male rats, but feeding with powdered diet after sexual maturation increases the ACh release only in female rats. These results indicate that (i) sex-specific circulating sex steroids are necessary for the sex-specific ACh release, (ii) neonatal activation of estrogen receptors is sufficient to mediate masculinization of the septohippocampal cholinergic system, and (iii) sex-specific effect of environmental condition may suggest interaction with the effect of sex hormones.

## *Acknowledgments*

This work was supported by Grant-in-Aid 18590219 from the Ministry of Education, Culture, Sports, Science, and Technology of Japan (to D. M.).

## *REFERENCES*


Auerbach, J. M., and Segal, M. (1996). Muscarinic receptors mediating depression and long-term potentiation in rat hippocampus. *J. Physiol.* **492,** 479–493.

Beatty, W. W. (1984). Hormonal organization of sex differences in play fighting and spatial behavior. *Prog. Brain Res.* **61,** 315–330.

Brown, R. T. (1968). Early experience and problem-solving ability. *J. Comp. Physiol. Psychol.* **65,** 433–440.

Buzsáki, G., Bickford, R. G., Ponomareff, G., Thal, L. J., Mandel, R., and Gage, F. H. (1988). Nucleus basalis and thalamic control of neocortical activity in the freely moving rat. *J. Neurosci.* **8,** 4007–4026.

Cherrier, M. M., Matsumoto, A. M., Amory, J. K., Asthana, S., Bremner, W., Peskind, E. R., Raskind, M. A., and Craft, S. (2005). Testosterone improves spatial memory in men with Alzheimer disease and mild cognitive impairment. *Neurology* **64,** 2063–2068.

Cole, A. E., and Nicoll, R. A. (1983). Acetylcholine mediates a slow synaptic potential in hippocampal pyramidal cells. *Science* **221,** 1299–1301.

Coyle, J. T., Price, D. L., and DeLong, M. R. (1983). Alzheimer's disease: A disorder of cortical cholinergic innervation. *Science* **219,** 1184–1190.

Cummings, J. L. (2004). Alzheimer's disease. *N. Engl. J. Med.* **351,** 56–67.

Daniel, J. M., Fader, A. J., Spencer, A. L., and Dohanich, G. P. (1997). Estrogen enhances performance of female rats during acquisition of a radial arm maze. *Horm. Behav.* **32,** 217–225.

Day, J., Damsma, G., and Fibiger, H. C. (1991). Cholinergic activity in the rat hippocampus, cortex and striatum correlates with locomotor activity: An in vivo microdialysis study. *Pharmacol. Biochem. Behav.* **38,** 723–729.

Eckel-Mahan, K. L., Phan, T., Han, S., Wang, H., Chan, G. C., Scheiner, Z. S., and Storm, D. R. (2008). Circadian oscillation of hippocampal MAPK activity and cAMP: Implications for memory persistence. *Nat. Neurosci.* **11,** 1074–1082.

Einon, D. (1980). Spatial memory and response strategies in rats: Age, sex and rearing differences in performance. *Q. J. Exp. Psychol.* **32,** 473–489.

Endo, Y., Mizuno, T., Fujita, K., Funabashi, T., and Kimura, F. (1994). Soft-diet feeding during development enhances later learning abilities in female rats. *Physiol. Behav.* **56,** 629–633.

Falvo, R. E., Buhl, A., and Nalbandov, A. V. (1974). Testosterone concentrations in the peripheral plasma of androgenized female rats and in the estrous cycle of normal female rats. *Endocrinology* **95,** 26–29.

Fernández de Sevilla, D., Núñez, A., Borde, M., Malinow, R., and Buño, W. (2008). Cholinergic-mediated $IP_3$-receptor activation induces long-lasting synaptic enhancement in CA1 pyramidal neurons. *J. Neurosci.* **28,** 1469–1478.

Ferri, C. P., Prince, M., Brayne, C., Brodaty, H., Fratiglioni, L., Ganguli, M., Hall, K., Hasegawa, K., Hendrie, H., Huang, Y., Jorm, A., Mathers, C., *et al.* (2005). Global prevalence of dementia: A Delphi consensus study. *Lancet* **366,** 2112–2117.

Gibbs, R. B., and Pfaff, D. W. (1992). Effects of estrogen and fimbria / fornix transaction on p75NGFR and ChAT expression in the medial septum and diagonal band of Broca. *Exp. Neurol.* **116,** 23–39.

Gold, P. E. (2003). Acetylcholine modulation of neural systems involved in learning and memory. *Neurobiol. Learn. Mem.* **80,** 194–210.

Greenough, W. T., Madden, T. C., and Fleischmann, T. B. (1972). Effects of isolation, daily handling, and enriched rearing on maze learning. *Psychon. Sci.* **27,** 279–280.

Herrera-Morales, W., Mar, I., Serrano, B., and Bermúdez-Rattoni, F. (2007). Activation of hippocampal postsynaptic muscarinic receptors is involved in long-term spatial memory formation. *Eur. J. Neurosci.* **25,** 1581–1588.

Hironaka, N., Tanaka, K., Izaki, Y., Hori, K., and Nomura, M. (2001). Memory-related acetylcholine efflux from the rat prefrontal cortex and hippocampus: A microdialysis study. *Brain Res.* **901,** 143–150.

Hyman, J. M., Wyble, B. P., Goyal, V., Rossi, C. A., and Hasselmo, M. E. (2003). Stimulation in hippocampal region CA1 in behaving rats yields long-term potentiation when delivered to the peak of theta and long-term depression when delivered to the trough. *J. Neurosci.* **23,** 11725–11731.

Hymovitch, B. (1952). The effects of experimental variations on problem solving in the rat. *J. Comp. Physiol. Psychol.* **45,** 313–321.

Juraska, J. M., Henderson, C., and Muller, J. (1984). Differential rearing experience, gender, and radial maze performance. *Dev. Psychobiol.* **17,** 209–215.

Kotani, S., Yamauchi, T., Teramoto, T., and Ogura, H. (2006). Pharmacological evidence of cholinergic involvement in adult hippocampal neurogenesis in rats. *Neuroscience* **142,** 505–514.

Kritzer, M. F. (1997). Selective colocalization of immunoreactivity for intracellular gonadal hormone receptors and tyrosine hydroxylase in the ventral tegmental area, substantia nigra, and retrorubral fields in the rat. *J. Comp. Neurol.* **379,** 247–260.

Kritzer, M. F., McLaughlin, P. J., Smirlis, T., and Robinson, J. K. (2001). Gonadectomy impairs T-maze acquisition in adult male rats. *Horm. Behav.* **39,** 167–174.

Lee, M. G., Chrobak, J. J., Sik, A., Wiley, R. G., and Buzsáki, G. (1994). Hippocampal theta activity following selective lesion of the septal cholinergic system. *Neuroscience* **62,** 1033–1047.

Luine, V. N., Renner, K. J., and McEwen, B. S. (1986). Sex-dependent differences in estrogen regulation of choline acetyltransferase are altered by neonatal treatments. *Endocrinology* **119,** 874–878.

Luine, V., Jacome, L. F., and MacLusky, N. J. (2003). Rapid enhancement of visual and place memory by estrogens in rats. *Endocrinology* **144,** 2836–2844.

Lund, T. D., Hinds, L. R., and Handa, R. J. (2006). The androgen 5$\alpha$-dihydrotestosterone and its metabolite 5$\alpha$-androstan-3$\beta$, 17$\beta$-diol inhibit the hypothalamo-pituitary-adrenal response to stress by acting through estrogen receptor $\beta$-expressing neurons in the hypothalamus. *J. Neurosci.* **26,** 1448–1456.

Markowska, A. J., and Savonenko, A. V. (2002). Effectiveness of estrogen replacement in restoration of cognitive function after long-term estrogen withdrawal in aging rats. *J. Neurosci.* **22,** 10985–10995.

Markram, H., and Segal, M. (1990). Long-lasting facilitation of excitatory postsynaptic potentials in the rat hippocampus by acetylcholine. *J. Physiol.* **427,** 381–393.

Masuda, J., Mitsushima, D., Funabashi, T., and Kimura, F. (2005). Sex and housing conditions affect the 24-h acetylcholine release profile in the hippocampus in rats. *Neuroscience* **132,** 537–542.

McCurry, S. M., Logsdon, R. G., Vitiello, M. V., and Teri, L. (2004). Treatment of sleep and nighttime disturbances in Alzheimer's disease: A behavior management approach. *Sleep Med.* **5,** 373–377.

McEwen, B. S. (1981). Neural gonadal steroid actions. *Science* **211,** 1303–1311.

McEwen, B. S., and Alves, S. E. (1999). Estrogen actions in the central nervous system. *Endocr. Rev.* **20,** 279–307.

Mesulam, M. M., Mufson, E. J., Wainer, B. H., and Levey, A. I. (1983). Central cholinergic pathways in the rat: An overview based on an alternative nomenclature (Ch1-Ch6). *Neuroscience* **10,** 1185–1201.

Miettinen, R. A., Kalesnykas, G., and Koivisto, E. H. (2002). Estimation of the total number of cholinergic neurons containing estrogen receptor-a in the rat basal forebrain. *J. Histochem. Cytochem.* **50,** 891–902.

Mitsushima, D., Mizuno, T., and Kimura, F. (1996). Age-related changes in diurnal acetylcholine release in the prefrontal cortex of male rats as measured by microdialysis. *Neuroscience* **72,** 429–434.

Mitsushima, D., Yamanoi, C., and Kimura, F. (1998). Restriction of environmental space attenuates locomotor activity and hippocampal acetylcholine release in male rats. *Brain Res.* **805,** 207–212.

Mitsushima, D., Funabashi, T., Shinohara, K., and Kimura, F. (2001). Impairment of maze learning in rats by restricting environmental space. *Neurosci. Lett.* **297,** 73–76.

Mitsushima, D., Masuda, J., and Kimura, F. (2003a). Sex differences in the stress-induced release of acetylcholine in the hippocampus and corticosterone from the adrenal cortex in rats. *Neuroendocrinology* **78,** 234–240.

Mitsushima, D., Tin-Tin-Win-Shwe, and Kimura, F. (2003b). Sexual dimorphism in the GABAergic control of gonadotropin release in intact rats. *Neurosci. Res.* **46,** 399–405.

Mitsushima, D., Takase, K., Funabashi, T., and Kimura, F. (2008). Gonadal steroid hormones maintain the stress-induced acetylcholine release in the hippocampus: Simultaneous measurements of the extracellular acetylcholine and serum corticosterone levels in the same subjects. *Endocrinology* **149,** 802–811.

Mitsushima, D., Takase, K., Funabashi, T., and Kimura, F. (2009). Gonadal steroids maintain 24-h acetylcholine release in the hippocampus: Organizational and activational effects in behaving rats. *J. Neurosci.* **29,** 3808–3815.

Mizuno, T., and Kimura, F. (1996). Medial septal injection of naloxone elevates acetylcholine release in the hippocampus and induces behavioral seizures in rats. *Brain Res.* **713,** 1–7.

Mizuno, T., Endo, Y., Arita, J., and Kimura, F. (1991). Acetylcholine release in the rat hippocampus as measured by the microdialysis method correlates with motor activity and exhibits a diurnal variation. *Neuroscience* **44,** 607–612.

Moffat, S. D., Zonderman, A. B., Metter, E. J., Kawas, C., Blackman, M. R., Harman, S. M., and Resnick, S. M. (2004). Free testosterone and risk for Alzheimer disease in older men. *Neurology* **62,** 188–193.

Mohapel, P., Leanza, G., Kokaia, M., and Lindvall, O. (2005). Forebrain acetylcholine regulates adult hippocampal neurogenesis and learning. *Neurobiol. Aging* **26,** 939–946.

Moor, E., DeBoer, P., and Westerink, B. H. C. (1998). GABA receptors and benzodiazepine binding sites modulate hippocampal acetylcholine release in vivo. *Eur. J. Pharmacol.* **359,** 119–126.

Mount, C., and Downtown, D. (2006). Alzheimer disease: Progress or profit? *Nat. Med.* **12,** 780–784.

Mufson, E. J., Cai, W. J., Jaffar, S., Chen, E., Stebbins, G., Sendera, T., and Kordower, J. H. (1999). Estrogen receptor immunoreactivity within subregions of the rat forebrain: Neuronal distribution and association with perikarya containing choline acetyltransferase. *Brain Res.* **849,** 253–274.

Nakamura, N., Fujita, H., and Kawata, M. (2002). Effects of gonadectomy on immunoreactivity for choline acetyltransferase in the cortex, hippocampus, and basal forebrain of adult male rats. *Neuroscience* **109,** 473–485.

Nilsson, O. G., Leanza, G., and Bjorklund, A. (1992). Acetylcholine release in the hippocampus: Regulation by monoaminergic afferents as assessed by in vivo microdialysis. *Brain Res.* **584,** 132–140.

Norbury, R., Travis, M. J., Erlandsson, K., Waddington, W., Ell, P. J., and Murphy, D. G. M. (2007). Estrogen therapy and brain muscarinic receptor density in healthy females: A SPET study. *Horm. Behav.* **51,** 249–257.

Parducz, A., Hajszan, T., Maclusky, N. J., Hoyk, Z., Csakvari, E., Kurunczi, A., Prange-Kiel, J., and Leranth, C. (2006). Synaptic remodeling induced by gonadal hormones: Neuronal plasticity as a mediator of neuroendocrine and behavioral responses to steroids. *Neuroscience* **138,** 977–985.

Parent, M. B., and Baxter, M. G. (2004). Septohippocampal acetylcholine: Involved in but not necessary for learning and memory? *Learn. Mem.* **11,** 9–20.

Perry, E., Walker, M., Grace, J., and Perry, R. (1999). Acetylcholine in mind: A neurotransmitter correlate of consciousness? *Trend Neurosci.* **22,** 273–280.

Petersen, R. C., Thomas, R. G., Grundman, M., Bennett, D., Doody, R., Ferris, S., Galasko, D., Jin, S., Kaye, J., Levey, A., Pfeiffer, E., Sano, M., *et al.* (2005). Vitamin E and donepezil for the treatment of mild cognitive impairment. *N. Engl. J. Med.* **352,** 2379–2388.

Pongrac, J. L., Gibbs, R. B., and Defranco, D. B. (2004). Estrogen-mediated regulation of cholinergic expression in basal forebrain neurons requires extracellular signal-regulated kinase activity. *Neuroscience* **124,** 809–816.

Ragozzino, M. E., Unick, K. E., and Gold, P. E. (1996). Hippocampal acetylcholine release during memory testing in rats: Augmentation by glucose. *Proc. Natl. Acad. Sci. USA* **93,** 4693–4698.

Romeo, R. D., McCarthy, J. B., Wang, A., Milner, T. A., and McEwen, B. S. (2005). Sex differences in hippocampal estradiol-induced N-methyl-D-aspartic acid binding and ultrastructural localization of estrogen receptor-α. *Neuroendocrinology* **81,** 391–399.

Rosario, E. R., Chang, L., Stanczyk, F. Z., and Pike, C. J. (2004). Age-related testosterone deplation and the development of Alzheimer disease. *JAMA* **292,** 1431–1432.

Rush, M. E., and Blake, C. A. (1982). Serum testosterone concentrations during the 4-day estrous cycle in normal and adrenalectomized rats. *Proc. Soc. Exp. Biol. Med.* **169,** 216–221.

Sarter, M., and Parikh, V. (2005). Choline transporters, cholinergic transmission and cognition. *Nat. Neurosci.* **6,** 48–56.

Seeger, T., Fedorova, I., Zheng, F., Miyakawa, T., Koustova, E., Gomeza, J., Basile, A. S., Alzheimer, C., and Wess, J. (2004). $M_2$ muscarinic acetylcholine receptor knock-out mice show deficits in behavioral flexibility, working memory, and hippocampal plasticity. *J. Neurosci.* **24,** 10117–10127.

Seymoure, P., Dou, H., and Juraska, J. M. (1996). Sex differences in radial arm maze performance: Influence of rearing environment and room cues. *Psychobiology* **24,** 33–37.

Shinoe, T., Matsui, M., Taketo, M. M., and Manabe, T. (2005). Modulation of synaptic plasticity by physiological activation of $M_1$ muscarinic acetylcholine receptors in the mouse hippocampus. *J. Neurosci.* **25,** 11194–11200.

Shors, T. J., Chua, C., and Falduto, J. (2001). Sex differences and opposite effects of stress on dendritic spine density in the male versus female hippocampus. *J. Neurosci.* **21,** 6292–6297.

Smith, H. V. (1972). Effects of environmental enrichment on open-field activity and Hebb–Williams problem solving in rats. *J. Comp. Physiol. Psychol.* **80,** 163–168.

Stancampiano, R., Cocco, S., Cugusi, C., Sarais, L., and Fadda, F. (1999). Serotonin and acetylcholine release response in the rat hippocampus during a spatial memory task. *Neuroscience* **89,** 1135–1143.

Starkstein, S. E., Jorge, R., Mizrahi, R., and Robinson, R. G. (2005). The construct of minor and major depression in Alzheimer's disease. *Am. J. Psychiatry* **162,** 2086–2093.

Swaab, D. F., and Hofman, M. A. (1995). Sexual differentiation of the human hypothalamus in relation to gender and sexual orientation. *Trend Neurosci.* **18,** 264–270.

Swanson, L. W. (1982). The projections of the ventral tegmental area and adjacent regions: A combined fluorescent retrograde tracer and immunofluorescence study in the rat. *Brain Res. Bull.* **9,** 321–353.

Takase, K., Funabashi, T., Mogi, K., Mitsushima, D., and Kimura, F. (2005a). Feeding with powdered diet after weaning increases visuospatial ability in association with increases in the expression of *N*-methyl-D-aspartate receptors in the hippocampus of female rats. *Neurosci. Res.* **53,** 169–175.

Takase, K., Mitsushima, D., Masuda, J., Mogi, K., Funabashi, T., Endo, Y., and Kimura, F. (2005b). Feeding with powdered diet after weaning affects sex difference in acetylcholine release in the hippocampus in rats. *Neuroscience* **136,** 593–599.

Takase, K., Kimura, F., Yagami, T., and Mitsushima, D. (2009). Sex-specific 24-h acetylcholine release profile in the medial prefrontal cortex: Simultaneous measurement of spontaneous locomotor activity in behaving rats. *Neuroscience* **159,** 7–15.

van Praag, H., Christie, B. R., Sejnowski, T. J., and Gage, F. H. (1999). Running enhances neurogenesis, learning and long-term potentiation in mice. *Proc. Natl. Acad. Sci. USA* **96,** 13427–13431.

Weiner, D. M., Levey, A. I., Sunahara, R. K., Niznik, H. B., O'Dowd, B. F., Seeman, P., and Brann, M. R. (1991). $D_1$ and $D_2$ dopamine receptor mRNA in rat brain. *Proc. Natl. Acad. Sci. USA* **88,** 1859–1863.

Widmer, H., Ferrigan, L., Davies, C. H., and Cobb, S. R. (2006). Evoked slow muscarinic acetylcholinergic synaptic potentials in rat hippocampal interneurons. *Hippocampus* **16,** 617–628.

Williams, C. L., and Meck, W. H. (1991). The organizational effects of gonadal steroids on sexually dimorphic spatial ability. *Psychoneuroendocrinology* **16,** 155–176.

Winblad, B., Kilander, L., Eriksson, S., Minthon, L., Båtsman, S., Wetterholm, A. L., Jansson-Blixt, C., and Haglund, A. (2006). Donepezil in patients with severe Alzheimer's disease: Double-blind, parallel-group, placebo-controlled study. *Lancet* **367,** 1057–1065.

Yanai, J., Rogel-Fuchs, Y., Pick, C. G., Slotkin, T., Seidler, F. J., Zahalka, E. A., and Newman, M. E. (1993). Septohippocampal cholinergic changes after destruction of the A10-septal dopaminergic pathways. *Neuropharmacology* **32,** 113–117.

Zandi, P. P., Carlson, M. C., Plassman, B. L., Welsh-Bohmer, K. A., Mayer, L. S., Steffens, D. C., and Breitner, J. C. S. (2002). Hormone replacement therapy and incidence of Alzheimer disease in older women. *JAMA* **288,** 2123–2129.

CHAPTER FIFTEEN

# Estradiol and GABAergic Transmission in the Hippocampus

Tomasz Wójtowicz* *and* Jerzy W. Mozrzymas*,†

## Contents



## Abstract

Estradiol is synthesized in the hippocampus and is known to increase intrinsic hippocampal excitability and capacity for synaptic plasticity. A picture emerges that at least part of these effects are due to a complex modulation of GABAergic system in developing and adult hippocampus. During development, GABAergic system undergoes profound alterations and is particularly prone to modulation. During this period, estradiol could modulate both phasic and tonic GABAergic currents and promote excitatory GABA actions. In contrast, in adult hippocampus, estradiol-induced formation of new dendritic spines in pyramidal cells is paralleled with a reduction in GABAergic drive to these neurons. Such estradiol actions could be mediated primarily through interneurons expressing estrogen receptors. In this chapter, we provide an overview of the *in vitro* and *in vivo*

* Laboratory of Neuroscience, Department of Biophysics, Wroclaw Medical University, Wroclaw, Poland
† Department of Animal Molecular Physiology, Wroclaw University, Wroclaw, Poland

*Vitamins and Hormones*, Volume 82
ISSN 0083-6729, DOI: 10.1016/S0083-6729(10)82015-1

**studies addressing the role of estradiol in regulating the GABAergic system in the hippocampal formation during development and in the adulthood. Although the mechanisms underlying such a regulation remain largely unknown, we make an attempt to present the major hypotheses and concepts related to this issue.** 

## I. Introduction

Estrogens are common cholesterol-derived steroids which easily diffuse across the cell membrane and may produce various genomic and nongenomic effects. Since the discovery of the enzyme aromatase that synthesizes 17$\beta$-estradiol in the hippocampus (Abdelgadir *et al.*, 1994; Garcia-Segura *et al.*, 1999; Sanghera *et al.*, 1991; Wehrenberg *et al.*, 2001), the role of this steroid in functioning of the hippocampal circuitry has attracted an increasing attention. The endogenous pool of the 17$\beta$-estradiol, called estradiol in the latter part of this chapter, has been recently demonstrated to be particularly rich in the hippocampus (Hojo *et al.*, 2004). In adult rodents, concentration of estradiol in this brain region highly exceeds that found in the blood due to the synthesis of this hormone in the pyramidal neurons of CA3–CA1 region as well as in the dentate gyrus (Hojo *et al.*, 2004). Thus, locally synthesized estradiol might act in a paracrine or autocrine fashion and affect neurons and local neural network functioning. Indeed, the discovery of estradiol-dependent process of new spine generation and formation of excitatory synapses on principal neurons in the adult hippocampus (Gould *et al.*, 1990; Woolley and McEwen, 1992) has shed a new light on the role of estradiol in structural plasticity. This discovery reinforced the view that dendritic spine plasticity is not limited to alterations in its morphology and function but provided direct evidence that in response to appropriate stimuli these structures can be created *de novo*. Importantly, involvement of estradiol in formation of new excitatory synapses has been implicated as an important factor affecting the excitability of hippocampal network. Several aspects related to this issue have been recently discussed in a review by Woolley (2007). In particular, estradiol increases intrinsic hippocampal excitability and capacity for synaptic plasticity (Woolley, 2007), facilitate kindled seizures in the hippocampus (Buterbaugh and Hudson, 1991) and epilepsy (Backstrom, 1976). Additional evidence for an important role of estradiol in the hippocampus came from behavioral experiments. It has been shown that intrahippocampal estradiol administration improved hippocampus-dependent spatial memory (Packard and Teather, 1997; Packard *et al.*, 1996; Sandstrom and Williams, 2001). Interestingly, a population of GABAergic interneurons in the hippocampus express estrogen receptors (ERs), and thus, could mediate, at least in part, estradiol actions on the local hippocampal circuits (Hart *et al.*, 2001; Weiland *et al.*, 1997). However, while complex estradiol actions on glutamatergic system have been extensively described (Woolley, 2007) its role in regulating GABAergic system is less understood. To date, only limited number of reports addressed the impact of estradiol on the GABAergic transmission in the adult hippocampus and even less is known about the role of estradiol during development. Importantly, during early developmental stages, the first functional synapses on principal neurons of the hippocampal formation are the GABAergic ones (reviewed in Ben-Ari *et al.*, 2007). Moreover, GABAergic system undergoes profound developmental changes, including a switch from depolarizing to hyperpolarizing mode (Ben-Ari *et al.*, 2007; Cherubini and Conti, 2001). The scope of this review is to summarize the current knowledge and opinions on the role of estradiol in modulating the GABAergic transmission in hippocampal neurons during development and adulthood.

## II. Estradiol and $GABA_AR$ Function

Estradiol is known to act genomically through nuclear ERs but also a broad spectrum of acute nongenomic actions have been recently reviewed (McCarthy, 2008; Woolley, 2007). For example, in the hippocampal CA1 region, exposure to estradiol, lasting for several minutes, depolarized neurons, induced spontaneous firing (Wong and Moss, 1991), and suppressed the afterhyperpolarization that followed the action potential (Kumar and Foster, 2002). Moreover, estradiol was found to increase the amplitude of dendritic excitatory postsynaptic potential (EPSP) in various regions of the hippocampus (Foy *et al.*, 1999; Kim *et al.*, 2006) and was also found to have a crucial impact on long-term potentiation (Foy *et al.*, 1999; Good *et al.*, 1999; Warren *et al.*, 1995) and long-term depression (Desmond *et al.*, 2000; Good *et al.*, 1999). Such enhancement of neuronal excitability and facilitation of synaptic plasticity could be in part due to a suppression of GABAergic drive in these cells through a direct modulation of $GABA_ARs$. These receptors posses the binding site for neurosteroids (Hosie *et al.*, 2007; Majewska *et al.*, 1986) and are readily modulated by many of them either positively or negatively (Majewska, 2007). For this reason, a direct effect of estradiol on $GABA_ARs$ has been tested in several studies. In a study by Wong and Moss (1992), acute application of estradiol (1 $\mu$M) on CA1 pyramidal neurons did not influence the extent of membrane potential hyperpolarization induced by iontophoretically applied GABA pulses *in vivo* (Wong and Moss, 1992). More recently, the possibility that estradiol exerts an acute effect on $GABA_A$ receptors has been tested in our laboratory (Asimiadou *et al.*, 2005). For this purpose, we have investigated the effect of estradiol on the current responses to ultrafast GABA applications. This experimental model offers a very high temporal resolution (adequate to describe the receptor gating properties) and is particularly suitable for pharmacological studies in which acute or direct effects of a modulator are considered. Using this approach, we have tested the effect of estradiol on currents elicited by application of GABA (10 mM) to somatic membrane patches excised from cultured hippocampal neurons. Within the concentration range 1–10 $\mu$M, estradiol did not affect either the amplitude or kinetics of GABA-evoked currents (Asimiadou *et al.*, 2005). We also tested the effect of estradiol on miniature inhibitory postsynaptic currents (mIPSCs) which originate from activation of synaptic $GABA_AR$s due to spontaneous exocytosis of a single vesicle of neurotransmitter. Recordings of mIPSCs performed from presumably pyramidal hippocampal neurons (visually recognized) revealed that acute application of estradiol (1 $\mu$M) did not affect either the amplitude or kinetics of mIPSCs in neurons cultured for 6–16 days *in vitro* (DIV) (Pytel *et al.*, 2007). Thus, our experiments based on the analysis of current responses and mIPSCs argue against any marked acute effect of estradiol on $GABA_AR$s at least in the considered model. This conclusion is consistent with results of Murphy *et al.* (1998a,b) who reported that acute exposure to estradiol (0.1 $\mu$g/ml) did not cause any systematic change in mIPSCs amplitude or frequency evoked by application of hyperosmotic medium (Murphy *et al.*, 1998a). Thus, on the basis of these studies, it is reasonable to assume that estradiol does not directly interfere with $GABA_A$-receptor gating and that estradiol acute actions on neuronal excitability cannot be attributed to rapidly occurring modulation of GABAergic transmission.

## III. Estradiol and GABAergic Transmission During Development

In adult animals, GABA is the key inhibitory neurotransmitter in the brain. However, during the neonatal period, the equilibrium potential for chloride is positive relative to the resting membrane potential (Rivera *et al.*, 1999, 2005) implying that chloride passive membrane transport would be depolarizing at this developmental stage. Indeed, in neurons from young animals, activation of $GABA_AR$s leads to chloride efflux from neurons that may induce sufficient membrane depolarization to open voltage-sensitive calcium channels (VSCCs; Leinekugel *et al.*, 1995) as well as to relieve the magnesium block of the *N*-methyl-D-aspartate (NMDA) receptors (Ben-Ari *et al.*, 1997). Thus, especially in early developmental stages, GABAergic signaling fulfills the criteria of the excitatory transmission. Moreover, GABA-induced increase in intracellular calcium mediates trophic effects on the developing brain (Ben-Ari *et al.*, 1994; Fiszman and Schousboe, 2004). The depolarizing effect of GABA in the developing hippocampus is particularly important in the initiation of so-called giant depolarizing potentials (GDPs), during which GABAergic transmission was proposed to act as coincidence detector (Kasyanov *et al.*, 2004). GDPs that spread through the hippocampal neural network have been shown to be accompanied by synchronous enhancements in the intracellular calcium concentrations (Leinekugel *et al.*, 1997). Thus, unlike in adults, GABAergic transmission during development plays an important role in providing excitatory drive and trophic actions in hippocampal circuit (Ben-Ari *et al.*, 2007; Cherubini *et al.*, 1991). During early development, hippocampus is also exposed to estradiol that is present at high concentrations during gestation and birth and then gradually declines to adult levels (McCarthy, 2008). High concentrations of estradiol was found in the immature rat hippocampus immediately postpartum *in vivo* (Amateau *et al.*, 2004) and it has been demonstrated that young neurons developing in slice cultures and cultures of dissociated neurons are capable of synthesizing estradiol *de novo in vitro* (Ikeda *et al.*, 2006; Kretz *et al.*, 2004). Importantly, the presence of estradiol is paralleled by a marked expression of ERs during development of the hippocampal circuit (Hosli and Hosli, 1999; Solum and Handa, 2001; Weiland *et al.*, 1997). Thus, while depolarizing GABAergic drive is a marker of early development, hippocampal neurons are capable to synthesize estradiol that is known to affect neuronal circuitry. It seemed thus particularly appealing to address the issue of a potential impact of estradiol on the GABAergic system during development. Although, as mentioned above, the early studies indicated the lack of any direct acute estradiol effect on the $GABA_A$ receptors, a possibility remained that estradiol could affect GABAergic system in a para-/autocrine manner. The possible functional significance of the GABAergic transmission modulation by estradiol in the developing hippocampus has stimulated us and other groups to investigate this issue in *in vitro* and *in vivo* models, which we briefly outline below.

## A. *In vitro* studies

In general, GABAergic transmission consists of phasic and tonic mode. Fast (phasic) inhibitory synaptic currents are fundamental for rapid and precisely targeted information transfer in the neural networks. Rapid GABAergic synaptic transmission is involved in generation of rhythmic activities of neuronal networks and is believed to be particularly important in a strict control of its timing and propagation. Following synaptic activity, GABA spills over from the synapse and builds up an ambient concentration that reflects the balance between synaptically released neurotransmitter and the effectiveness of the uptake system. The ambient GABA, that is believed to reach submicromolar or, at most, micromolar concentrations, can activate peri- and extrasynaptic $GABA_A$Rs leading to an increase in the membrane conductance and so-called tonic current that affects the neuronal excitability (Farrant and Nusser, 2005). Such GABAergic tonic current is mediated by $GABA_A$Rs characterized by the kinetic properties and pharmacology different from those described for synaptic receptors. Most interestingly, $GABA_A$ receptors containing the $\delta$ subunit occur exclusively extrasynaptically and are endowed with a very high agonist affinity that renders them very efficient in mediating the tonic currents (Farrant and Nusser, 2005). To study the impact of estradiol on both forms of GABAergic transmission during development, we have employed a model of hippocampal neurons obtained from neonatal (postnatal days 2–3) rats and cultured for 6–16 DIV (Pytel *et al.*, 2007). As already mentioned, during brain development *in vivo*, GABAergic system undergoes profound alterations (e.g., Cherubini and Conti, 2001) and it was important to check whether or not key developmental processes are maintained in the simple *in vitro* model. Over time, neonatal hippocampal neurons *in vitro* are forming a dense net of synaptic connections enabling to study GABAergic synaptic transmission. We and others have shown that hippocampal as well as cerebellar neurons developing *in vitro* retain key developmental features of GABAergic system maturation (Barberis *et al.*, 2005; Ortinski *et al.*, 2004; Pytel *et al.*, 2007). By dividing the developing hippocampal or cerebellar neurons into 3 age groups (6–8, 9–11, and 12–16 DIV) it is possible to follow up the GABAergic maturation *in vitro* (Barberis *et al.*, 2005; Ortinski *et al.*, 2004; Pytel *et al.*, 2007). Notably, the kinetics of GABAergic mIPSCs is clearly accelerated between 6 and 12 DIV and such a change was followed by monotonic increase in mIPSCs frequency (Barberis *et al.*, 2005; Ortinski *et al.*, 2004; Pytel *et al.*, 2007) similarly as observed *in vivo* (Cohen *et al.*, 2000; Hollrigel and Soltesz, 1997). It is generally accepted that the developmental acceleration of mIPSC kinetics is associated with a switch in $GABA_A$R subunit composition. As expected, age-dependent increase in expression of $\alpha 1$ subunit observed *in vivo* (Overstreet Wadiche *et al.*, 2005) is also seen in hippocampal and cerebellar cultures (Ortinski *et al.*, 2004; Pytel *et al.*, 2007), suggesting that the developmental pattern of expression of this key $GABA_A$R subunit is reproduced *in vitro*.

In the above-mentioned model both tonic and phasic GABAergic transmission was directly measured in developing hippocampal neurons chronically exposed to pharmacological doses of estradiol (Pytel *et al.*, 2007). We have found that prolonged (i.e., over 48 h) exposure of hippocampal neurons to exogenous estradiol (1 $\mu$M) resulted in an increase in mIPSCs amplitude and prolongation of their average decay time constant but this effect was restricted only to the specific time window 9–11 DIV (i.e., to the "intermediate" age group) (Pytel *et al.*, 2007). In addition, 48 h but not 24 h treatment increased mIPSCs amplitude without altering the kinetics in the 6–11 DIV group and has not been observed for older neurons (Pytel *et al.*, 2007). Such a modulation of GABAergic transmission was not accompanied by a change in mIPSCs frequency (Pytel *et al.*, 2007). Thus, long-term exposure to estradiol was shown to modulate the GABAegic transmission but this effect was restricted to specific time windows occurring only in "young" or "intermediate" age groups but not in "old" neuronal cultures. This suggests that immature phenotype of GABAergic transmission is more susceptible to modulation by estradiol.

As mentioned above, GABA spillover from synapses can activate peri- and extrasynaptic $GABA_A$Rs inducing the tonic conductance (Farrant and Nusser, 2005). Since tonic inhibition is attracting increasing attention as a major mechanism regulating neuronal excitability, we have checked whether or not estradiol treatment affected the tonic current in the considered *in vitro* developmental model. To this end, we have recorded endogenous tonic current as well as responses to exogenous application of low (0.1–1 $\mu$M) concentration of GABA ("tonic"–like GABAergic currents) in cultures treated with estradiol (Pytel *et al.*, 2007). We have found that long-term treatment with estradiol decreased the amplitude of both endogenous and evoked tonic current amplitudes but this effect was restricted to the 9–11 DIV age group. In addition, a similar trend was observed in neurons aged 6–8 DIV; however, in the latter case the change was on the border of significance (Pytel *et al.*, 2007). This suggests that estradiol profoundly influenced intrinsic membrane properties of hippocampal neurons and that, similar to synaptic currents, these changes were restricted to developing neurons. However, at the current stage one cannot discriminate between pre- and postsynaptic mechanisms of estradiol modulation of mIPSCs *in vitro* (see also Section V).

Since $GABA_A$R activation leads to activation of L-type VSCCs (Cherubini *et al.*, 1991; Leinekugel *et al.*, 1995, 1997, 1999) an efficient method to visualize the impact of estradiol on GABAergic drive *in vitro* is to monitor intracellular calcium level. This approach has been applied by Nunez *et al.* (2005) on primary hippocampal cultures of both sexes from gestational day 18 or 22 rats. By using cell permeant fluorescent probe Fura-2-AM and by applying $GABA_A$Rs-agonist muscimol it has been shown that chronic exposure to physiological estradiol concentration resulted in an increase in the percentage of neurons showing calcium transients in response to muscimol application (neurons cultured 5–10 DIV) (Nunez *et al.*, 2005). This finding is consistent with proposal that estradiol treatment might promote a delay of the developmental switch from excitatory to inhibitory mode of the GABAergic system. In addition, in estradiol-treated neurons, the recovery of intracellular calcium to the basal level was significantly prolonged suggesting that the depolarization induced by muscimol application could be more efficient in evoking calcium transients than in untreated neurons.

Sexual differentiation of the rodent brain depends on hormonal exposure during a "critical period" beginning in late gestation and ending in early neonatal life (Davis *et al.*, 1996). Various parameters of the GABAergic system including glutamic acid decarboxylase (GAD) mRNA synthesis (Davis *et al.*, 1996) and GABA production (Davis *et al.*, 1999) have been reported to be augmented in the brain of neonate male rats relative to females only during the critical period. Of note, changes in GAD mRNA levels have been positively correlated with changes in stationary GABA levels (McCarthy, 1995) and GABA turnover (Grattan *et al.*, 1996). During a perinatal critical period, steroid hormones, in particular estradiol, modulate neuronal network to differentiate male and female brains (McCarthy, 2008). In agreement with this proposal and based on calcium transients recordings, Nunez and McCarthy (2007) have shown that the time when GABA is excitatory is prolonged in male hippocampal neurons when compared to female hippocampal neurons *in vitro*. Further studies have revealed that sex differences in depolarizing GABA responses are due to *de novo* estradiol synthesis by female neurons, whereas the sex difference in resting calcium is independent of steroids (Nunez and McCarthy, 2009).

These observations corroborate with studies performed in baboon primary hippocampal cultures obtained from hippocampi of fetuses exposed to estradiol *in utero* (Nunez *et al.*, 2008). Nunez *et al.* (2008) reported that exogenous estradiol treatment (administered *in utero*) augmented the magnitude of the intracellular calcium response and increased the percentage of neurons responding to the GABA-agonist muscimol with intracellular calcium transients, indicative of depolarizing GABA, similarly as in rodent models. Thus it has been concluded that estradiol may act as a naturally occurring modulator of cellular excitation in the developing baboon and rodent brains (Nunez and McCarthy, 2009).

## B. *In vivo* studies

As described above, *in vitro* studies indicated that the GABAergic system is particularly susceptible to modulation by estrogen during development. Since developmental processes *in vitro* and *in vivo* could show profound differences, it was important to verify these observations in an *in vivo* model especially because the impact of estradiol on GABAergic transmission in the developing hippocampus *in vivo* has not been studied in detail. We have investigated the GABAergic transmission in CA1 of the hippocampi of developing rats (in the age of 7–40 days postnatally, P7–P40) injected daily with pharmacological doses of estradiol starting from P2 and throughout development (Wojtowicz *et al.*, 2008). In such a model, it is expected that exogenous estradiol circulating in the blood enriches the endogenous brain pool of this steroid and thus would rescale estradiol-induced effects. The age of the animals used for experiments (P7–P40) was chosen to cover the maturation period including its most important stages (Wojtowicz *et al.*, 2008). Thus, between P7 and P13 it is expected that a switch from depolarizing to hyperpolarizing GABA occurs (Ben-Ari *et al.*, 2007; Cherubini *et al.*, 1991; Khazipov *et al.*, 2004). By P21, GABAergic synaptic currents reach adult phenotype characterized by a rapid deactivation kinetics (Cohen *et al.*, 2000; Hollrigel and Soltesz, 1997). In these studies we limited experiments to P40, the age at which the impact of other hormones present during onset of sexual maturation remains still negligible (Germain *et al.*, 1978; Laws *et al.*, 2003). We have found that long-term treatment with exogenous estradiol affected both synaptic and tonic GABAergic transmission recorded in CA1 pyramidal cells in acute brain slices (Wojtowicz *et al.*, 2008). In particular, estradiol induced an acceleration of the mIPSCs rise-time and decay kinetics, while the mIPSCs amplitude and frequency increased (Wojtowicz *et al.*, 2008). Interestingly, the most prominent effects were restricted to the youngest considered age group (P7–P13) while no effect were observed in animals aged over P22 (Wojtowicz *et al.*, 2008). In addition, the amplitude of GABAergic tonic current, mediated by extrasynaptic $GABA_A$Rs, was strongly upregulated and, again, this effect was restricted to P7–P13 animals (Wojtowicz *et al.*, 2008). This result demonstrates that tonic inhibition might be affected by estradiol during development and therefore the impact of this hormone on the GABAergic drive is not restricted to synaptic currents. The upregulation of tonic current induced by estradiol differs from results obtained *in vitro*, where tonic current was downregulated after estradiol treatment (Pytel *et al.*, 2007; Wojtowicz *et al.*, 2008). The reason for such a discrepancy is not clear but it may be attributed to different experimental conditions in these two models (e.g., in cultures GABA is applied exogenousely at low concentrations while in slices, tonic GABA is endogenous). Nevertheless, similar to our observations in the model of cultured neurons, there is clearly a time window localized in early developmental stages when the sensitivity of GABAegic transmission to modulation by estradiol is high. Moreover, it is worth emphasizing that marked susceptibility of hippocampal neurons to estradiol overlaps, at least partially, with increased level of this hormone during development.

## IV. Estradiol and GABAergic Transmission in Mature Animals

It is widely accepted that the phenomenon of estradiol-dependent formation of new spines (Gould *et al.*, 1990; Woolley and McEwen, 1992) requires modification of GABAergic transmission to CA1 pyramidal neurons in the hippocampus. First, it has been demonstrated that exposure of 2.5–3 weeks old hippocampal cultures *in vitro* (i.e., mature in term of GABAergic phenotype) to estradiol led to a twofold increase in spine number observed 2–3 days later (Murphy and Segal, 1996). In parallel, an increase in presynaptic marker synaptophysin was observed, suggesting a *de novo* formation of synapses *in vitro* (Murphy and Segal, 1996). In this model, 24–48 h exposure of cultured neurons to estradiol (0.1 $\mu$g/ml) did not influence either the inhibitory effect of GABA on action potential discharges and the magnitude or reversal potential of the response to pulse application of GABA (10 $\mu$M) in pyramidal neurons (Murphy *et al.*, 1998a). Although the overall reactivity to GABA was not changed, GABAergic transmission to these neurons mediated by interneurons was clearly affected. Notably, 24 h estradiol treatment resulted in a reduction in mIPSCs amplitude and frequency (Murphy *et al.*, 1998a). Interestingly, pharmacological blockade of $GABA_A$Rs mediated currents with picrotoxin and bicuculline increased dendritic spine density *in vitro* (Papa and Segal, 1996). Thus estradiol-induced decrease in GABAergic transmission to pyramidal cells could effectively increase excitatory drive and thus favor formation of new dendritic spines (Murphy *et al.*, 1998a).

In agreement with *in vitro* studies, enrichment of endogenous estradiol brain pool with exogenous estradiol injected to ovariectomized female rats *in vivo* induced a reduction in amplitude and frequency of synaptically evoked IPSC 24 h after injection (Rudick and Woolley, 2001; Rudick *et al.*, 2003). In addition, the change in evoked IPSCs amplitude was accompanied by a prolongation in deactivation kinetics of these currents (Rudick and Woolley, 2001; Rudick *et al.*, 2003). Recording of mIPSCs in the above-mentioned system have shown that estradiol did not change the mIPSC amplitude but their frequency was diminished (Rudick and Woolley, 2001; Rudick *et al.*, 2003). Surprisingly, these estradiol effects were transient as another injection of this hormone 24 h after the first one resulted in restoration of evoked IPSC amplitude and mIPSC frequency to control values observed before estradiol administrations. It seems thus that the second estradiol injection resulted in a recovery of the balance between $GABA_A$R-mediated inhibition and excitation in CA1 pyramidal neurons (Rudick and Woolley, 2001). Interestingly, maintenance and formation of new dendritic spines was found to be activity-dependent (Harris, 1999; Smart and Halpain, 2000) and new spines were observed 24 h after estradiol injection to ovariectomized animals (Woolley and McEwen, 1993). Thus, as suggested by Rudick and Woolley (2001), transient estradiol-induced disinhibition of CA1 pyramidal cells would increase neural excitability and facilitate dendritic spine formation although it is clear that disinhibition alone cannot fully explain the increased spine formation *in vivo*.

It is tempting to ask about the locus of estradiol action on GABAergic transmission in the hippocampus. Hart *et al.* (2001) have found that ER$\alpha$-positive neurons show a discrete distribution in the hippocampus. Namely, the largest number of GABAergic neurons that are ER$\alpha$ positive is found in the dorsal hippocampus while in the ventral part there are fewer GABAergic neurons positive for ERα but instead many non-GABAergic neurons express this receptor (Hart *et al.*, 2001). Thus, estradiol is likely to affect directly a population of GABAergic cells in the dorsal hippocampus while having impact on some GABAergic and many non-GABAergic neurons in the ventral hippocampus (Hart *et al.*, 2001). Interestingly, estradiol can regulate synaptic inhibition of CA1 pyramidal cells by acting directly on ER-expressing cells within the hippocampus and/or by acting on ER-expressing cells in brain regions that project to the hippocampus, such as the medial septum in the basal forebrain (Rudick *et al.*, 2003; Shughrue *et al.*, 2000). Blockade of these neurons by 195IgG-Sap antibody interferes with estradiol effects on IPSCs and mIPSCs (Rudick *et al.*, 2003) demonstrating that the basal forebrain cholinergic system may play a role in estrogen-induced disinhibition of hippocampal CA1 pyramidal cells (Rudick *et al.*, 2003).

Since in hippocampus a variety of interneurons are present (Klausberger and Somogyi, 2008), there is a question regarding the cell specifity of estradiol action. Extracellular ERα has been found in presynaptic vesicles of only cholecystokinin (CCK) and neuropeptide Y (NPY) but not parvalbumin (PV)-expressing neurons (Hart *et al.*, 2007). Since CCK basket cells integrate local and subcortical inputs to fine-tune pyramidal cell activity (Freund, 2003) it seems likely that estradiol actions mediated via these interneurons may have an important impact on functioning of hippocampal circuits (Hart *et al.*, 2007). However, since only a small subset of interneurons express ERα (Hart *et al.*, 2001; Weiland *et al.*, 1997) it has been discussed how such a small subpopulation of interneurons could mediate widespread effects of estradiol at the network level? First, because each interneuron innervates over 1000 CA1 pyramidal cells (Buhl *et al.*, 1994; Freund and Buzsaki, 1996; Sik *et al.*, 1995), interneurons have heavy impact on CA1 excitability. In addition, Hart *et al.* (2001) reported a population of ERα-expressing interneurons concentration at the border between the *stratum radiatum* and *stratum lacunosum-moleculare* (Hart *et al.*, 2001). As noted by Rudick and Woolley (2001) the above-mentioned cluster of interneurons project to other interneurons that could eventually innervate pyramidal cells (Banks *et al.*, 2000; Kunkel *et al.*, 1988; Lacaille and Schwartzkroin, 1988). Thus estradiol could mediate its multiplicative actions on extensively innervated pyramidal cells through "local" and/or "peripheral" net of estradiol-sensitive interneurons.

## V. Mechanisms of Estradiol Action

In a recent review, Spencer *et al.* (2008) have emphasized that there is a number of signaling pathways which could mediate estradiol actions in the brain. These involve AKT, MAPK/ERK, BDNF, opioids, and other elements (Spencer *et al.*, 2008). However, only limited factors and mechanisms have been described so far in the context of estradiol-mediated modulation of GABAergic transmission. Below, we have highlighted some of them.

## A. ERs

Nuclear ERs appear key mediators in the modulation of GABAergic transmission by estradiol. The suppression of GABAergic transmission observed *in vivo* at the level of IPSCs and mIPSCs (Rudick *et al.*, 2003) was dependent on nuclear ER, since the application of tamoxifen, nuclear ER-antagonist, abolished the effect of estradiol (Rudick *et al.*, 2003). Consistently with this observation, estradiol required at least 24–48 h and nuclear ERs to exert its actions on mIPSCs during development *in vitro* since in the presence of tamoxifen the estradiol effects were abolished (Murphy and Segal, 1996; Pytel *et al.*, 2007). ERs are also an important target for estradiol itself since it has been reported that ER$\alpha$ is upregulated by estradiol *in vitro* (Prange-Kiel *et al.*, 2003). In addition, besides ER$\alpha$ and ER$\beta$, a novel plasma membrane ER-X has been described (Toran-Allerand, 2005; Toran-Allerand *et al.*, 2002) suggesting a new target to explore in the context of mechanism of estradiol actions in the brain.

## B. NKCC1

As mentioned above, the net effect of GABAergic transmission on membrane potential critically depends on the transmembrane chloride gradient, regulated by chloride cotransporters (Delpire, 2000). It has been shown that NKCC1 ($Na^+K^+2C1^-$ cotransporter) promotes chloride transport into the cell along with sodium and potassium, and its expression is high in neonatal brain but declines with age. On the contrary, KCC2 ($K^+C1^-$ cotransporter) promotes chloride efflux, and its expression is low at birth but increases during development becoming high by the second postnatal week (Plotkin *et al.*, 1997a,b). Due to such a developmental switch in expression of NKCC1 and KCC2 transporters, by the end of the second postnatal week, GABA function changes from depolarizing to hyperpolarizing as the reversal potential for chloride becomes negative with respect to the resting membrane potential (Ben-Ari *et al.*, 2007).

It has been shown that exposure of hippocampal neurons to physiological doses of estradiol increased the phoshorylation and hence the activity of NKCC1 along with L-type voltage-dependent calcium channels containing the $\alpha$1C subunit, while KCC1 remained unchanged (Nunez *et al.*, 2005). Similar observation was reported for primate hippocampal neurons exposed to estradiol during development (Nunez *et al.*, 2008). In addition, Nakamura *et al.* (2004) have reported that NKCC1 mRNA levels increased within 24 h after treatment with exogenous estradiol, while KCC2 mRNA was not changed. Thus, estradiol-induced upregulation of NKCC1 activity would favor prolonged excitatory GABA actions during development.

## C. $GABA_AR$s subunit expression

Another mechanism of estradiol-induced modulation of GABAergic transmission could be due to modulation of expression of $GABA_AR$s subunits. It has been shown *in vitro* that prolonged treatment with estradiol did not affect the expression of $\alpha1$, $\alpha2$, $\alpha3$, $\gamma2$, $GABA_AR$ subunit, while $\alpha4$ and $\delta$ were at the detection limit (Pytel *et al.*, 2007). In agreement with this finding, amplitude and kinetics of currents evoked by exogenous GABA applications were not different from control values (Pytel *et al.*, 2007). In addition, $\alpha1$, $\alpha2$, $\beta1$, $\beta2$, and $\gamma2$ subunits mRNA expression was not changed after 48 h estradiol treatment *in vivo* (Weiland and Orchinik, 1995). It has been also shown that there was no effect of exogenous estradiol treatment on the $GABA_A$ receptor subunit composition or the amount of $GABA_A$ receptor binding in the perinatal rat hippocampus (Davis *et al.*, 2000). Altogether, changes in $GABA_AR$ subunit expression cannot be solely attributed to estradiol-induced modulation of GABAergic transmission. However, it is important to note that combined estradiol and progesterone treatment was reported to upregulate $\delta$ subunit expression and coexpression of $\alpha4$ subunit *in vivo* (Shen *et al.*, 2005).

## D. GAD

GAD is critical in GABA synthesis. Estradiol effects on expression of this enzyme have been described both *in vitro* and *in vivo* studies. For instance, *in vitro* exposure of 2.5–3 weeks hippocampal cultures to estradiol for 24 h caused a marked decrease (by up to 80%) in the GAD content of the interneurons (Murphy *et al.*, 1998b). Prolonged exposure to estradiol resulted in a reduction in GAD65 (but not GAD67 isoform) puncta in *stratum oriens* and *stratum radiatum* of the hippocampus *in vivo* (Rudick and Woolley, 2001). Interestingly, the inhibition of either aromatase or ERs downregulated GAD65 expression level in cultures devoided of estradiol (serum and estradiol free medium) suggesting that neurons produce estradiol to maintain the level of this GABA-synthesizing enzyme (Ikeda *et al.*, 2006). Importantly, as mentioned above, changes in GAD mRNA levels have been positively correlated with changes in stationary levels of GABA (McCarthy, 1995) and GABA turnover (Grattan *et al.*, 1996). Thus, estradiol could affect GABAergic transmission through a regulation of GABA synthesis and thus its availability. Importantly, such estradiol effect would be expected to underlie a presynaptic modulation of GABAergic synaptic currents.

## E. Synapse reorganization

As already mentioned above, growing evidence indicates that estradiol may affect number and density of dendritic spines and excitatory synapses (Gould *et al.*, 1990; Leranth *et al.*, 2000; MacLusky *et al.*, 2005; McLaughlin *et al.*, 2008; Woolley and McEwen, 1992; Yankova *et al.*, 2001). Moreover, estradiol was found to increase the divergence of input from individual presynaptic boutons to multiple postsynaptic CA1 pyramidal cells (Yankova *et al.*, 2001). It seemed thus plausible that changes in GABAergic transmission might also involve structural modification of GABAergic synapses. In this section, we present examples of studies addressing this issue.

First, basing on reduction of evoked IPSCs and mIPSCs amplitude and frequency it has been suggested that estradiol could decrease the number of functional GABAergic synapses on CA1 pyramidal cells (Rudick and Woolley, 2001). However, studies done so far indicate that unlike excitatory synapses, the number of GABAergic axosomatic synapses on CA1 pyramidal cell in the hippocampus were not modified after estradiol supplementation (Ledoux and Woolley, 2005). Mechanism of estradiol action in hippocampus appears thus different than in the hypothalamus, where estradiol decreased the density of inhibitory inputs to arcuate neurons (Perez *et al.*, 1993).

Further studies addressed the impact of estradiol on the paired-pulse depression in GABAergic synapses (a phenomenon of the reduction of the second response to the same stimulus due to depletion of presynaptic GABA). Notably, paired-pulse depression was reduced after estradiol treatment indicating a downregulation of presynaptic GABA release probability (Rudick *et al.*, 2003). This mechanism is further supported by finding that estradiol reduced the number of GABA-containing vesicles adjacent to presynaptic membrane (Ledoux and Woolley, 2005) while in a subset of ER-containing synapses, GABA-containing vesicles were mobilized toward synapse (Hart *et al.*, 2007). Since the time course of synaptic currents may strongly depend on the geometry of the synaptic cleft (that largely determines the neurotransmitter clearance kinetics), it was important to check whether or not estradiol treatment affected the synapse geometry. However, basing on the data published so far, application of estradiol to ovariectomized female rats did not affect GABAergic axosomatic boutons (tested by assessment of bouton volume) on CA1 pyramidal cells (Hart *et al.*, 2007; Ledoux and Woolley, 2005), vesicle density, and presynaptic density area (Ledoux and Woolley, 2005). In addition, clusters of vesicles in inhibitory presynaptic boutons in the hippocampal CA1 cell body layer were recently found to be ER$\alpha$ positive (Hart *et al.*, 2007). Interestingly, these vesicles were mobilized toward the synapse after estradiol treatment *in vivo* showing a crucial role of extranuclear ERs as mediators of estradiol action in the hippocampus (Hart *et al.*, 2007).

### F. Receptor clusterization

Besides modulation of mIPSCs amplitude, we have reported that chronic exposure to estradiol could lead to prolongation of mIPSCs kinetics (Pytel *et al.*, 2007; Wojtowicz *et al.*, 2008). In addition, while estradiol treatment had no effect on mIPSC amplitude enhancement by flurazepam, treatment with this steroid gave rise to a larger BDZ-induced prolongation of mIPSC decaying phase in the youngest animal group (P7–P13) (Wojtowicz *et al.*, 2008). Although the mechanisms underlying such changes in mIPSC time course are not clear, a hypothesis that still awaits verification was proposed. In our recent work on the mechanism of IPSC modulation by BDZ, we have suggested that BDZ-induced prolongation of mIPSCs could involve activation of perisynaptic $GABA_ARs$ by GABA spilling over from the synapse (Mozrzymas *et al.*, 2007). Taking this into account, our observation that estradiol treatment enhances BDZ-induced prolongation of mIPSCs could suggest that estradiol might enhance the clusterization of $GABA_ARs$ within the synapse and in the perisynaptic region (Wojtowicz *et al.*, 2008). Such a possibility appears supported by the observed estradiol-induced increase in GABAergic tonic current, although the major component of this current is mediated by $GABA_ARs$ insensitive to BDZs (Farrant and Nusser, 2005). On the other hand, scaffold protein gephyrin, which is important for stabilization of $GABA_ARs$ clusters (Kneussel *et al.*, 1999; Yu *et al.*, 2007) is not regulated by estradiol in the hippocampus (Sassoe-Pognetto *et al.*, 2007). This may indicate that postsynaptic $GABA_ARs$ clusters are not a target for estradiol actions. More detailed studies are necessary to clarify that issue.

## VI. Conclusions and Future Directions

In conclusion, estradiol emerged as a potent modulator of GABAergic transmission in the hippocampus. The observed estradiol actions crucially depend on time duration of interaction between this steroid and neuronal tissue being the most effective after 24 h. At current stage of research it appears that during development, when dendritic growth, spinogenesis as well as synaptogenesis are particularly intense in the rodent brains (Ben-Ari *et al.*, 2007), estradiol would increase the impact of excitatory GABAergic transmission to hippocampal neurons. In adult brain instead, estradiol was shown to suppress GABAergic inhibitory transmission thus favoring increased excitability in CA1 and facilitating the formation of new dendritic spines and local circuit connectivity. The commonly favored mechanism of estradiol action in adult hippocampus would involve decrease in a number of vesicles available for release at inhibitory synapses and reduced probability of release, resulting in fewer GABA release sites. The estradiol effects on different elements of GABAergic system are complex and the molecular basis underlying these effects remains to be elucidated. Several important potential mechanisms emerge in this context, including recruitment of ERs, modulation of expression of NKCC1 (during development), and GAD and reorganization of the ultrastructure of the GABAergic synapses. Most interestingly, estradiol can downregulate GABAergic drive to CA1 pyramidal cells and increase their excitability, but such a decrease in inhibitory synaptic function is only transient, since at 72 h after estradiol delivery, inhibition is restored to control levels (Rudick and Woolley, 2001). This recovery may indicate that homeostatic mechanisms are involved to prevent excitotoxicity or seizures; however, the molecular basis for such process remains unknown. Altogether, basing on the above-mentioned studies, it appears that estradiol should no longer be recognized solely as a steroid hormone but also as an important signaling molecule in the hippocampus.

## ACKNOWLEDGMENTS

Supported by funds from the Polish Ministry of Science and Higher Education Grant No. PBZ-MIN-001/P05/28 and N401 028 32/0664. Partially supported by the Polish Ministry for Science and Higher Education grant (PN/030/2006).

## REFERENCES

Abdelgadir, S. E., Resko, J. A., Ojeda, S. R., Lephart, E. D., McPhaul, M. J., and Roselli, C. E. (1994). Androgens regulate aromatase cytochrome P450 messenger ribonucleic acid in rat brain. *Endocrinology* **135,** 395–401.

Amateau, S. K., Alt, J. J., Stamps, C. L., and McCarthy, M. M. (2004). Brain estradiol content in newborn rats: Sex differences, regional heterogeneity, and possible de novo synthesis by the female telencephalon. *Endocrinology* **145,** 2906–2917.

Asimiadou, S., Bittigau, P., Felderhoff-Mueser, U., Manthey, D., Sifringer, M., Pesditschek, S., Dzietko, M., Kaindl, A. M., Pytel, M., Studniarczyk, D., Mozrzymas, J. W., and Ikonomidou, C. (2005). Protection with estradiol in developmental models of apoptotic neurodegeneration. *Ann. Neurol.* **58,** 266–276.

Backstrom, T. (1976). Epileptic seizures in women related to plasma estrogen and progesterone during the menstrual cycle. *Acta Neurol. Scand.* **54,** 321–347.

Banks, M. I., White, J. A., and Pearce, R. A. (2000). Interactions between distinct GABA(A) circuits in hippocampus. *Neuron* **25,** 449–457.

Barberis, A., Lu, C., Vicini, S., and Mozrzymas, J. W. (2005). Developmental changes of GABA synaptic transient in cerebellar granule cells. *Mol. Pharmacol.* **67,** 1221–1228.

Ben-Ari, Y., Tseeb, V., Raggozzino, D., Khazipov, R., and Gaiarsa, J. L. (1994). gamma-Aminobutyric acid (GABA): A fast excitatory transmitter which may regulate the development of hippocampal neurones in early postnatal life. *Prog. Brain Res.* **102,** 261–273.

Ben-Ari, Y., Khazipov, R., Leinekugel, X., Caillard, O., and Gaiarsa, J. L. (1997). GABAA, NMDA and AMPA receptors: A developmentally regulated 'menage a trois'. *Trends Neurosci.* **20,** 523–529.

Ben-Ari, Y., Gaiarsa, J. L., Tyzio, R., and Khazipov, R. (2007). GABA: A pioneer transmitter that excites immature neurons and generates primitive oscillations. *Physiol. Rev.* **87,** 1215–1284.

Buhl, E. H., Halasy, K., and Somogyi, P. (1994). Diverse sources of hippocampal unitary inhibitory postsynaptic potentials and the number of synaptic release sites. *Nature* **368,** 823–828.

Buterbaugh, G. G., and Hudson, G. M. (1991). Estradiol replacement to female rats facilitates dorsal hippocampal but not ventral hippocampal kindled seizure acquisition. *Exp. Neurol.* **111,** 55–64.

Cherubini, E., and Conti, F. (2001). Generating diversity at GABAergic synapses. *Trends Neurosci.* **24,** 155–162.

Cherubini, E., Gaiarsa, J. L., and Ben-Ari, Y. (1991). GABA: An excitatory transmitter in early postnatal life. *Trends Neurosci.* **14,** 515–519.

Cohen, A. S., Lin, D. D., and Coulter, D. A. (2000). Protracted postnatal development of inhibitory synaptic transmission in rat hippocampal area CA1 neurons. *J. Neurophysiol.* **84,** 2465–2476.

Davis, A. M., Grattan, D. R., Selmanoff, M., and McCarthy, M. M. (1996). Sex differences in glutamic acid decarboxylase mRNA in neonatal rat brain: Implications for sexual differentiation. *Horm. Behav.* **30,** 538–552.

Davis, A. M., Ward, S. C., Selmanoff, M., Herbison, A. E., and McCarthy, M. M. (1999). Developmental sex differences in amino acid neurotransmitter levels in hypothalamic and limbic areas of rat brain. *Neuroscience* **90,** 1471–1482.

Davis, A. M., Penschuck, S., Fritschy, J. M., and McCarthy, M. M. (2000). Developmental switch in the expression of GABA(A) receptor subunits alpha(1) and alpha(2) in the hypothalamus and limbic system of the rat. *Brain Res. Dev. Brain Res.* **119,** 127–138.

Delpire, E. (2000). Cation-chloride cotransporters in neuronal communication. *News Physiol. Sci.* **15,** 309–312.

Desmond, N. L., Zhang, D. X., and Levy, W. B. (2000). Estradiol enhances the induction of homosynaptic long-term depression in the CA1 region of the adult, ovariectomized rat. *Neurobiol. Learn. Mem.* **73,** 180–187.

Farrant, M., and Nusser, Z. (2005). Variations on an inhibitory theme: Phasic and tonic activation of GABA(A) receptors. *Nat. Rev. Neurosci.* **6,** 215–229.

Fiszman, M. L., and Schousboe, A. (2004). Role of calcium and kinases on the neurotrophic effect induced by gamma-aminobutyric acid. *J. Neurosci. Res.* **76,** 435–441.

Foy, M. R., Xu, J., Xie, X., Brinton, R. D., Thompson, R. F., and Berger, T. W. (1999). 17beta-estradiol enhances NMDA receptor-mediated EPSPs and long-term potentiation. *J. Neurophysiol.* **81,** 925–929.

Freund, T. F. (2003). Interneuron diversity series: Rhythm and mood in perisomatic inhibition. *Trends Neurosci.* **26,** 489–495.

Freund, T. F., and Buzsaki, G. (1996). Interneurons of the hippocampus. *Hippocampus* **6,** 347–470.

Garcia-Segura, L. M., Wozniak, A., Azcoitia, I., Rodriguez, J. R., Hutchison, R. E., and Hutchison, J. B. (1999). Aromatase expression by astrocytes after brain injury: Implications for local estrogen formation in brain repair. *Neuroscience* **89,** 567–578.

Germain, B. J., Campbell, P. S., and Anderson, J. N. (1978). Role of the serum estrogen-binding protein in the control of tissue estradiol levels during postnatal development of the female rat. *Endocrinology* **103,** 1401–1410.

Good, M., Day, M., and Muir, J. L. (1999). Cyclical changes in endogenous levels of oestrogen modulate the induction of LTD and LTP in the hippocampal CA1 region. *Eur. J. Neurosci.* **11,** 4476–4480.

Gould, E., Woolley, C. S., Frankfurt, M., and McEwen, B. S. (1990). Gonadal steroids regulate dendritic spine density in hippocampal pyramidal cells in adulthood. *J. Neurosci.* **10,** 1286–1291.

Grattan, D. R., Rocca, M. S., Strauss, K. I., Sagrillo, C. A., Selmanoff, M., and McCarthy, M. M. (1996). GABAergic neuronal activity and mRNA levels for both forms of glutamic acid decarboxylase (GAD65 and GAD67) are reduced in the diagonal band of Broca during the afternoon of proestrus. *Brain Res.* **733,** 46–55.

Harris, K. M. (1999). Structure, development, and plasticity of dendritic spines. *Curr. Opin. Neurobiol.* **9,** 343–348.

Hart, S. A., Patton, J. D., and Woolley, C. S. (2001). Quantitative analysis of ER alpha and GAD colocalization in the hippocampus of the adult female rat. *J. Comp. Neurol.* **440,** 144–155.

Hart, S. A., Snyder, M. A., Smejkalova, T., and Woolley, C. S. (2007). Estrogen mobilizes a subset of estrogen receptor-alpha-immunoreactive vesicles in inhibitory presynaptic boutons in hippocampal CA1. *J. Neurosci.* **27,** 2102–2111.

Hojo, Y., Hattori, T. A., Enami, T., Furukawa, A., Suzuki, K., Ishii, H. T., Mukai, H., Morrison, J. H., Janssen, W. G., Kominami, S., Harada, N., Kimoto, T., *et al.* (2004). Adult male rat hippocampus synthesizes estradiol from pregnenolone by cytochromes P45017alpha and P450 aromatase localized in neurons. *Proc. Natl. Acad. Sci. USA* **101,** 865–870.

Hollrigel, G. S., and Soltesz, I. (1997). Slow kinetics of miniature IPSCs during early postnatal development in granule cells of the dentate gyrus. *J. Neurosci.* **17,** 5119–5128.

Hosie, A. M., Wilkins, M. E., and Smart, T. G. (2007). Neurosteroid binding sites on GABA (A) receptors. *Pharmacol. Ther.* **116,** 7–19.

Hosli, E., and Hosli, L. (1999). Cellular localization of estrogen receptors on neurones in various regions of cultured rat CNS: Coexistence with cholinergic and galanin receptors. *Int. J. Dev. Neurosci.* **17,** 317–330.

Ikeda, T., Matsuki, N., and Yamada, M. K. (2006). Estrogen produced in cultured hippocampal neurons is a functional regulator of a GABAergic machinery. *J. Neurosci. Res.* **84,** 1771–1777.

Kasyanov, A. M., Safiulina, V. F., Voronin, L. L., and Cherubini, E. (2004). GABA-mediated giant depolarizing potentials as coincidence detectors for enhancing synaptic efficacy in the developing hippocampus. *Proc. Natl. Acad. Sci. USA* **101,** 3967–3972.

Khazipov, R., Khalilov, I., Tyzio, R., Morozova, E., Ben-Ari, Y., and Holmes, G. L. (2004). Developmental changes in GABAergic actions and seizure susceptibility in the rat hippocampus. *Eur. J. Neurosci.* **19,** 590–600.

Kim, M. T., Soussou, W., Gholmieh, G., Ahuja, A., Tanguay, A., Berger, T. W., and Brinton, R. D. (2006). 17beta-Estradiol potentiates field excitatory postsynaptic potentials within each subfield of the hippocampus with greatest potentiation of the associational/commissural afferents of CA3. *Neuroscience* **141,** 391–406.

Klausberger, T., and Somogyi, P. (2008). Neuronal diversity and temporal dynamics: The unity of hippocampal circuit operations. *Science* **321,** 53–57.

Kneussel, M., Brandstatter, J. H., Laube, B., Stahl, S., Muller, U., and Betz, H. (1999). Loss of postsynaptic GABA(A) receptor clustering in gephyrin-deficient mice. *J. Neurosci.* **19,** 9289–9297.

Kretz, O., Fester, L., Wehrenberg, U., Zhou, L., Brauckmann, S., Zhao, S., Prange-Kiel, J., Naumann, T., Jarry, H., Frotscher, M., and Rune, G. M. (2004). Hippocampal synapses depend on hippocampal estrogen synthesis. *J. Neurosci.* **24,** 5913–5921.

Kumar, A., and Foster, T. C. (2002). 17beta-estradiol benzoate decreases the AHP amplitude in CA1 pyramidal neurons. *J. Neurophysiol.* **88,** 621–626.

Kunkel, D. D., Lacaille, J. C., and Schwartzkroin, P. A. (1988). Ultrastructure of stratum lacunosum-moleculare interneurons of hippocampal CA1 region. *Synapse* **2,** 382–394.

Lacaille, J. C., and Schwartzkroin, P. A. (1988). Stratum lacunosum-moleculare interneurons of hippocampal CA1 region. II. Intrasomatic and intradendritic recordings of local circuit synaptic interactions. *J. Neurosci.* **8,** 1411–1424.

Laws, S. C., Ferrell, J. M., Stoker, T. E., and Cooper, R. L. (2003). Pubertal development in female Wistar rats following exposure to propazine and atrazine biotransformation by-products, diamino-S-chlorotriazine and hydroxyatrazine. *Toxicol. Sci.* **76,** 190–200.

Ledoux, V. A., and Woolley, C. S. (2005). Evidence that disinhibition is associated with a decrease in number of vesicles available for release at inhibitory synapses. *J. Neurosci.* **25,** 971–976.

Leinekugel, X., Tseeb, V., Ben-Ari, Y., and Bregestovski, P. (1995). Synaptic GABAA activation induces $Ca^{2+}$ rise in pyramidal cells and interneurons from rat neonatal hippocampal slices. *J. Physiol.* **487**(Pt 2), 319–329.

Leinekugel, X., Medina, I., Khalilov, I., Ben-Ari, Y., and Khazipov, R. (1997). Ca2+ oscillations mediated by the synergistic excitatory actions of GABA(A) and NMDA receptors in the neonatal hippocampus. *Neuron* **18,** 243–255.

Leinekugel, X., Khalilov, I., McLean, H., Caillard, O., Gaiarsa, J. L., Ben-Ari, Y., and Khazipov, R. (1999). GABA is the principal fast-acting excitatory transmitter in the neonatal brain. *Adv. Neurol.* **79,** 189–201.

Leranth, C., Shanabrough, M., and Horvath, T. L. (2000). Hormonal regulation of hippocampal spine synapse density involves subcortical mediation. *Neuroscience* **101,** 349–356.

MacLusky, N. J., Luine, V. N., Hajszan, T., and Leranth, C. (2005). The 17alpha and 17beta isomers of estradiol both induce rapid spine synapse formation in the CA1 hippocampal subfield of ovariectomized female rats. *Endocrinology* **146,** 287–293.

Majewska, M. D. (2007). Steroids and ion channels in evolution: From bacteria to synapses and mind. Evolutionary role of steroid regulation of GABA(A) receptors. *Acta Neurobiol. Exp. (Wars)* **67,** 219–233.

Majewska, M. D., Harrison, N. L., Schwartz, R. D., Barker, J. L., and Paul, S. M. (1986). Steroid hormone metabolites are barbiturate-like modulators of the GABA receptor. *Science* **232,** 1004–1007.

McCarthy, M. M. (1995). Frank A. Beach Award. Functional significance of steroid modulation of GABAergic neurotransmission: Analysis at the behavioral, cellular, and molecular levels. *Horm. Behav.* **29,** 131–140.

McCarthy, M. M. (2008). Estradiol and the developing brain. *Physiol. Rev.* **88,** 91–124.

McLaughlin, K. J., Bimonte-Nelson, H., Neisewander, J. L., and Conrad, C. D. (2008). Assessment of estradiol influence on spatial tasks and hippocampal CA1 spines: Evidence that the duration of hormone deprivation after ovariectomy compromises 17beta-estradiol effectiveness in altering CA1 spines. *Horm. Behav.* **54,** 386–395.

Mozrzymas, J. W., Wojtowicz, T., Piast, M., Lebida, K., Wyrembek, P., and Mercik, K. (2007). GABA transient sets the susceptibility of mIPSCs to modulation by benzodiazepine receptor agonists in rat hippocampal neurons. *J. Physiol.* **585,** 29–46.

Murphy, D. D., and Segal, M. (1996). Regulation of dendritic spine density in cultured rat hippocampal neurons by steroid hormones. *J. Neurosci.* **16,** 4059–4068.

Murphy, D. D., Cole, N. B., Greenberger, V., and Segal, M. (1998a). Estradiol increases dendritic spine density by reducing GABA neurotransmission in hippocampal neurons. *J. Neurosci.* **18,** 2550–2559.

Murphy, D. D., Cole, N. B., and Segal, M. (1998b). Brain-derived neurotrophic factor mediates estradiol-induced dendritic spine formation in hippocampal neurons. *Proc. Natl. Acad. Sci. USA* **95,** 11412–11417.

Nakamura, N. H., Rosell, D. R., Akama, K. T., and McEwen, B. S. (2004). Estrogen and ovariectomy regulate mRNA and protein of glutamic acid decarboxylases and cation-chloride cotransporters in the adult rat hippocampus. *Neuroendocrinology* **80,** 308–323.

Nunez, J. L., and McCarthy, M. M. (2007). Evidence for an extended duration of GABA-mediated excitation in the developing male versus female hippocampus. *Dev. Neurobiol.* **67,** 1879–1890.

Nunez, J. L., and McCarthy, M. M. (2009). Resting intracellular calcium concentration, depolarizing gamma-aminobutyric acid and possible role of local estradiol synthesis in the developing male and female hippocampus. *Neuroscience* **158,** 623–634.

Nunez, J. L., Bambrick, L. L., Krueger, B. K., and McCarthy, M. M. (2005). Prolongation and enhancement of gamma-aminobutyric acid receptor mediated excitation by chronic treatment with estradiol in developing rat hippocampal neurons. *Eur. J. Neurosci.* **21,** 3251–3261.

Nunez, J. L., Aberdeen, G. W., Albrecht, E. D., and McCarthy, M. M. (2008). Impact of estradiol on gamma-aminobutyric acid- and glutamate-mediated calcium responses of fetal baboon (Papio anubis) hippocampal and cortical neurons. *Endocrinology* **149,** 6433–6443.

Ortinski, P. I., Lu, C., Takagaki, K., Fu, Z., and Vicini, S. (2004). Expression of distinct alpha subunits of GABAA receptor regulates inhibitory synaptic strength. *J. Neurophysiol.* **92,** 1718–1727.

Overstreet Wadiche, L., Bromberg, D. A., Bensen, A. L., and Westbrook, G. L. (2005). GABAergic signaling to newborn neurons in dentate gyrus. *J. Neurophysiol.* **94,** 4528–4532.

Packard, M. G., and Teather, L. A. (1997). Intra-hippocampal estradiol infusion enhances memory in ovariectomized rats. *Neuroreport* **8,** 3009–3013.

Packard, M. G., Kohlmaier, J. R., and Alexander, G. M. (1996). Posttraining intrahippocampal estradiol injections enhance spatial memory in male rats: Interaction with cholinergic systems. *Behav. Neurosci.* **110,** 626–632.

Papa, M., and Segal, M. (1996). Morphological plasticity in dendritic spines of cultured hippocampal neurons. *Neuroscience* **71,** 1005–1011.

Perez, J., Luquin, S., Naftolin, F., and Garcia-Segura, L. M. (1993). The role of estradiol and progesterone in phased synaptic remodelling of the rat arcuate nucleus. *Brain Res.* **608,** 38–44.

Plotkin, M. D., Kaplan, M. R., Peterson, L. N., Gullans, S. R., Hebert, S. C., and Delpire, E. (1997a). Expression of the Na(+)-K(+)-2C1- cotransporter BSC2 in the nervous system. *Am. J. Physiol.* **272,** C173–C183.

Plotkin, M. D., Snyder, E. Y., Hebert, S. C., and Delpire, E. (1997b). Expression of the Na-K-2C1 cotransporter is developmentally regulated in postnatal rat brains: A possible mechanism underlying GABA's excitatory role in immature brain. *J. Neurobiol.* **33,** 781–795.

Prange-Kiel, J., Wehrenberg, U., Jarry, H., and Rune, G. M. (2003). Para/autocrine regulation of estrogen receptors in hippocampal neurons. *Hippocampus* **13,** 226–234.

Pytel, M., Wojtowicz, T., Mercik, K., Sarto-Jackson, I., Sieghart, W., Ikonomidou, C., and Mozrzymas, J. W. (2007). 17 beta-estradiol modulates GABAergic synaptic transmission and tonic currents during development in vitro. *Neuropharmacology* **52,** 1342–1353.

Rivera, C., Voipio, J., Payne, J. A., Ruusuvuori, E., Lahtinen, H., Lamsa, K., Pirvola, U., Saarma, M., and Kaila, K. (1999). The K+/C1 co-transporter KCC2 renders GABA hyperpolarizing during neuronal maturation. *Nature* **397,** 251–255.

Rivera, C., Voipio, J., and Kaila, K. (2005). Two developmental switches in GABAergic signalling: The K+- C1- cotransporter KCC2 and carbonic anhydrase CAVII. *J. Physiol.* **562,** 27–36.

Rudick, C. N., and Woolley, C. S. (2001). Estrogen regulates functional inhibition of hippocampal CA1 pyramidal cells in the adult female rat. *J. Neurosci.* **21,** 6532–6543.

Rudick, C. N., Gibbs, R. B., and Woolley, C. S. (2003). A role for the basal forebrain cholinergic system in estrogen-induced disinhibition of hippocampal pyramidal cells. *J. Neurosci.* **23,** 4479–4490.

Sandstrom, N. J., and Williams, C. L. (2001). Memory retention is modulated by acute estradiol and progesterone replacement. *Behav. Neurosci.* **115,** 384–393.

Sanghera, M. K., Simpson, E. R., McPhaul, M. J., Kozlowski, G., Conley, A. J., and Lephart, E. D. (1991). Immunocytochemical distribution of aromatase cytochrome P450 in the rat brain using peptide-generated polyclonal antibodies. *Endocrinology* **129,** 2834–2844.

Sassoe-Pognetto, M., Follesa, P., Panzanelli, P., Perazzini, A. Z., Porcu, P., Sogliano, C., Cherchi, C., and Concas, A. (2007). Fluctuations in brain concentrations of neurosteroids are not associated to changes in gephyrin levels. *Brain Res.* **1169,** 1–8.

Shen, H., Gong, Q. H., Yuan, M., and Smith, S. S. (2005). Short-term steroid treatment increases delta GABAA receptor subunit expression in rat CA1 hippocampus: Pharmacological and behavioral effects. *Neuropharmacology* **49,** 573–586.

Shughrue, P. J., Scrimo, P. J., and Merchenthaler, I. (2000). Estrogen binding and estrogen receptor characterization (ERalpha and ERbeta) in the cholinergic neurons of the rat basal forebrain. *Neuroscience* **96,** 41–49.

Sik, A., Penttonen, M., Ylinen, A., and Buzsaki, G. (1995). Hippocampal CA1 interneurons: An in vivo intracellular labeling study. *J. Neurosci.* **15,** 6651–6665.

Smart, F. M., and Halpain, S. (2000). Regulation of dendritic spine stability. *Hippocampus* **10,** 542–554.

Solum, D. T., and Handa, R. J. (2001). Localization of estrogen receptor alpha (ER alpha) in pyramidal neurons of the developing rat hippocampus. *Brain Res. Dev. Brain Res.* **128,** 165–175.

Spencer, J. L., Waters, E. M., Romeo, R. D., Wood, G. E., Milner, T. A., and McEwen, B. S. (2008). Uncovering the mechanisms of estrogen effects on hippocampal function. *Front. Neuroendocrinol.* **29,** 219–237.

Toran-Allerand, C. D. (2005). Estrogen and the brain: Beyond ER-alpha, ER-beta, and 17beta-estradiol. *Ann. NY Acad. Sci.* **1052,** 136–144.

Toran-Allerand, C. D., Guan, X., MacLusky, N. J., Horvath, T. L., Diano, S., Singh, M., Connolly, E. S. Jr., Nethrapalli, I. S., and Tinnikov, A. A. (2002). ER-X: A novel, plasma membrane-associated, putative estrogen receptor that is regulated during development and after ischemic brain injury. *J. Neurosci.* **22,** 8391–8401.

Warren, S. G., Humphreys, A. G., Juraska, J. M., and Greenough, W. T. (1995). LTP varies across the estrous cycle: Enhanced synaptic plasticity in proestrus rats. *Brain Res.* **703,** 26–30.

Wehrenberg, U., Prange-Kiel, J., and Rune, G. M. (2001). Steroidogenic factor-1 expression in marmoset and rat hippocampus: Co-localization with StAR and aromatase. *J. Neurochem.* **76,** 1879–1886.

Weiland, N. G., and Orchinik, M. (1995). Specific subunit mRNAs of the GABAA receptor are regulated by progesterone in subfields of the hippocampus. *Brain Res. Mol. Brain Res.* **32,** 271–278.

Weiland, N. G., Orikasa, C., Hayashi, S., and McEwen, B. S. (1997). Distribution and hormone regulation of estrogen receptor immunoreactive cells in the hippocampus of male and female rats. *J. Comp. Neurol.* **388,** 603–612.

Wojtowicz, T., Lebida, K., and Mozrzymas, J. W. (2008). 17beta-estradiol affects GABAergic transmission in developing hippocampus. *Brain Res.* **1241,** 7–17.

Wong, M., and Moss, R. L. (1991). Electrophysiological evidence for a rapid membrane action of the gonadal steroid, 17 beta-estradiol, on CA1 pyramidal neurons of the rat hippocampus. *Brain Res.* **543,** 148–152.

Wong, M., and Moss, R. L. (1992). Long-term and short-term electrophysiological effects of estrogen on the synaptic properties of hippocampal CA1 neurons. *J. Neurosci.* **12,** 3217–3225.

Woolley, C. S. (2007). Acute effects of estrogen on neuronal physiology. *Annu. Rev. Pharmacol. Toxicol.* **47,** 657–680.

Woolley, C. S., and McEwen, B. S. (1992). Estradiol mediates fluctuation in hippocampal synapse density during the estrous cycle in the adult rat. *J. Neurosci.* **12,** 2549–2554.

Woolley, C. S., and McEwen, B. S. (1993). Roles of estradiol and progesterone in regulation of hippocampal dendritic spine density during the estrous cycle in the rat. *J. Comp. Neurol.* **336,** 293–306.

Yankova, M., Hart, S. A., and Woolley, C. S. (2001). Estrogen increases synaptic connectivity between single presynaptic inputs and multiple postsynaptic CA1 pyramidal cells: A serial electron-microscopic study. *Proc. Natl. Acad. Sci. USA* **98,** 3525–3530.

Yu, W., Jiang, M., Miralles, C. P., Li, R. W., Chen, G., and de Blas, A. L. (2007). Gephyrin clustering is required for the stability of GABAergic synapses. *Mol. Cell. Neurosci.* **36,** 484–500.

CHAPTER SIXTEEN

# Transcriptional Regulation of Hypothalamic Corticotropin-Releasing Factor Gene

Kazunori Kageyama *and* Toshihiro Suda

## Contents



## Abstract

Corticotropin-releasing factor (CRF) plays a central role in regulating stress responses. Forskolin or pituitary adenylate cyclase-activating polypeptide stimulates adenylate cyclase and then increases intracellular cAMP levels in hypothalamic cells. Activation of the protein kinase A pathway leads to binding of cAMP response element (CRE)-binding protein (CREB) on the CRF promoter. Forskolin-stimulated CRF gene transcription is mediated by CRE on the CRF 5′-promoter region. Inducible cAMP-early repressor suppresses a stress response via inhibition of the cAMP-dependent CRF gene. Glucocorticoid-dependent repression of cAMP-stimulated CRF promoter activity is mediated by both nGRE and SRE in hypothalamic cells. Interleukin (IL)-6 produced in the hypothalamus stimulates the CRF gene. Suppressor of cytokine signaling-3, which is induced by a cAMP stimulant and IL-6, is involved in the negative regulation of CRF gene expression in hypothalamic cells. Such complex mechanisms would contribute to stress responses and homeostasis in the hypothalamus. 

Department of Endocrinology and Metabolism, Hirosaki University Graduate School of Medicine, Hirosaki, Aomori, Japan

*Vitamins and Hormones*, Volume 82
ISSN 0083-6729, DOI: 10.1016/S0083-6729(10)82016-3

# I. Introduction

Corticotropin-releasing factor (CRF) plays a central role in regulating stress responses (Suda *et al.*, 1985; Vale *et al.*, 1981). Moreover, CRF coordinates neuroendocrine, behavioral, autonomic, and immune responses, and controls the hypothalamic–pituitary–adrenal (HPA) axis during stressful periods. In the hypothalamic paraventricular nucleus (PVN) of the brain, CRF is produced in response to stress. CRF and arginine vasopressin (AVP) neurons in the parvocellular region of the PVN project to the external zone of the median eminence (Gonzalez-Hernandez *et al.*, 2006; Seasholtz *et al.*, 1988). Also, CRF and AVP in parvocellular PVN neurons exert synergistic effects on adrenocorticotropic hormone (ACTH) secretion from the anterior pituitary (AP; Gillies *et al.*, 1982; Mouri *et al.*, 1993). On the other hand, ACTH stimulates the release of glucocorticoids from the adrenal glands (Whitnall, 1993). Circulating glucocorticoids are critical to recovery from stress conditions because they inhibit the hypothalamic PVN production of CRF and the pituitary production of ACTH, thereby ensuring that serum levels of glucocorticoids are appropriate to the stress experienced (Whitnall, 1993).

Pituitary adenylate cyclase-activating polypeptide (PACAP) has been shown to modulate hypothalamic CRF gene expression *in vivo* (Grinevich *et al.*, 1997). PACAP induces intracellular cAMP production in hypothalamic 4B cells and stimulates CRF gene transcription via the cAMP-dependent pathway. Limbic structures, such as the extended amygdala and the bed nuclei of the stria terminalis, are identified as innervation sites of PACAP neurons, suggesting an important role in stress responses (Kozicz and Arimura, 2002; Piggins *et al.*, 1996). Indeed, nerve fibers containing PACAP connect to CRF neurons (Hannibal *et al.*, 1995; Legradi *et al.*, 1998). Other studies also suggest that PACAP stimulates the CRF gene via the cAMP/protein kinase A (PKA) signaling pathway (Agarwal *et al.*, 2005). Activation of the PKA pathway, causing phosphorylation of cAMP response element (CRE)-binding protein (CREB), acts on the CRF promoter (Itoi *et al.*, 1996; Seasholtz *et al.*, 1988; Spengler *et al.*, 1992). A functional CRE on the 5′-promoter region takes part in regulating CRF gene expression (Seasholtz *et al.*, 1988; Spengler *et al.*, 1992).

# II. Regulatory Elements on Hypothalamic CRF Gene

Computer analysis of the proximal CRF promoter reveals several possible binding sites for transcriptional factors, such as CRE, activator protein 1 (AP-1) protein (Fos/Jun) binding sites, half glucocorticoid regulatory element (GRE), and half estrogen-responsive element (ERE; Yao and Denver, 2007).

## A. Cyclic AMP (cAMP) response element-binding protein (CREB)

Forskolin or PACAP stimulates adenylate cyclase and then increases intracellular cAMP levels in hypothalamic 4B cells (Fig. 16.1). Forskolin increases CRF transcriptional activity in hypothalamic cells (Fig. 16.2) in agreement with previous studies using other cells (Seasholtz *et al.*, 1988; Spengler *et al.*, 1992; Yamamori *et al.*, 2004). Activation of the PKA pathway leads to binding of CREB to the CRE on the CRF promoter in hypothalamic cells as well as in human placental cells (Cheng *et al.*, 2000). The forskolin-stimulated activity of CRF gene transcription is reduced in 4B cells transfected with a mutant construct, CRF-233Mtluc, in which the CRE element (TGACGTCA) is mutated (TGGATCCA), or with a deletion mutant construct of the CRF gene promoter, CRF-220luc (Fig. 16.3). Therefore, the forskolin-induced CRF gene transcription is mainly mediated by CRE, which includes −220 to −233 base pairs (bp), on the CRF 5′-promoter region in hypothalamic cells. A functional CRE on the 5′ promoter region is important for increasing CRF gene expression

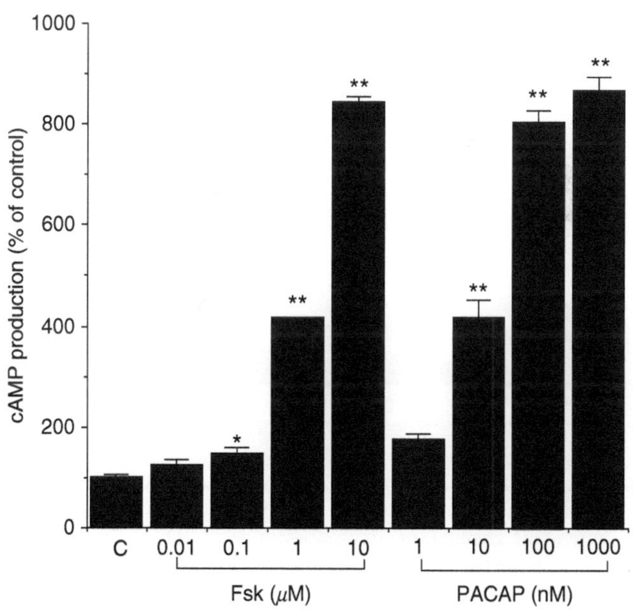


**Figure 16.1** Effects of forskolin and PACAP on cAMP production in 4B cells. * $P < 0.05$, ** $P < 0.005$ (compared with control [C]). Cells were preincubated for 20 min with medium containing 0.1 mM 3-isobutyl-1-methylxanthine, followed by the addition of forskolin (Fsk) or PACAP. The level of intracellular cAMP was measured by cAMP EIA. (Reproduction from Kageyama *et al.* (2007) with permission of the publisher.) Copyright 2007, Society for Endocrinology.

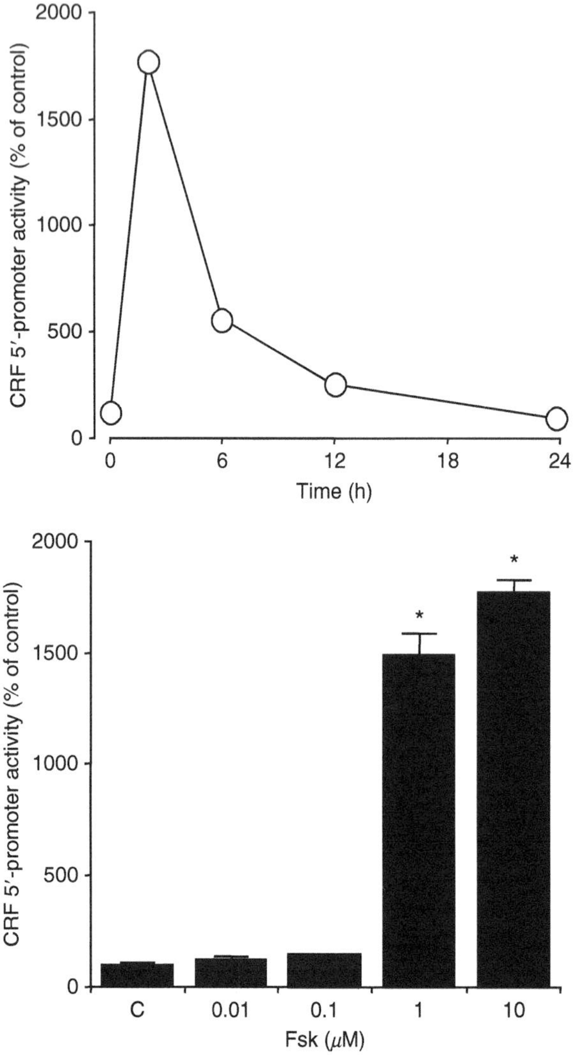


**Figure 16.2** Effects of forskolin on CRF 5′-promoter activity in 4B cells. Control cells treated with medium alone are indicated as (C). $*P < 0.05$ (compared with control [C]). Time-dependent changes in forskolin-induced CRF 5′-promoter activity (left panel). Cells were incubated with medium containing 10 $\mu$M forskolin. Dose-dependent changes in forskolin-induced CRF 5′-promoter activity (right panel). Cells were incubated for 2 h with medium containing 0.01–10 $\mu$M forskolin (Fsk). (Reproduction from Kageyama *et al.* (2008) with permission of the publisher.) Copyright 2008, Editrice Kurtis srl.

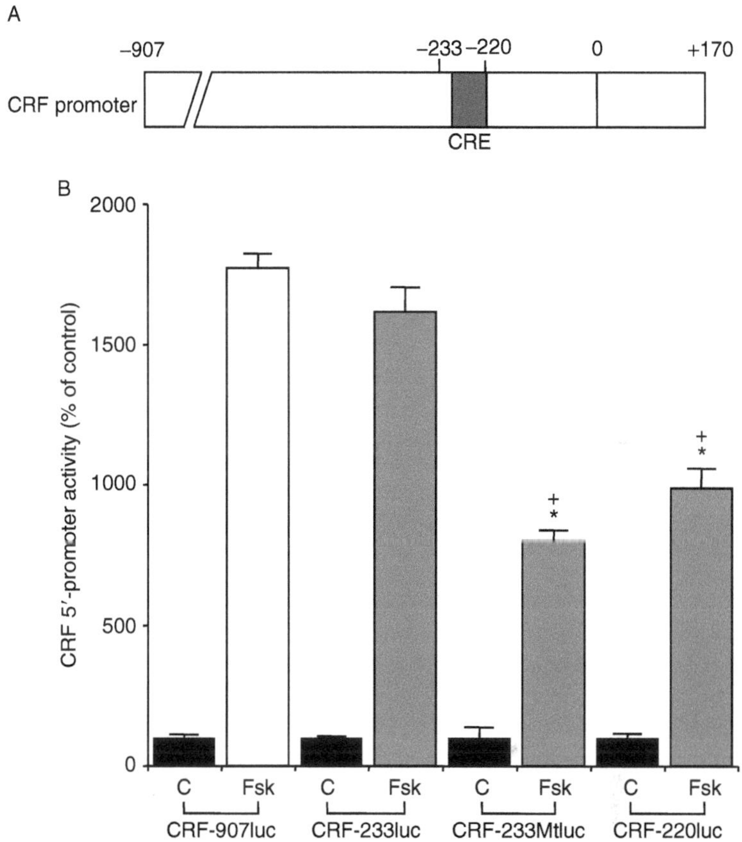


**Figure 16.3** Effects of CRE deletion on forskolin-induced CRF 5′-promoter activity in 4B cells. (A) Schematic representation of CRE in the CRF promoter. (B) Cells were transfected with full-length (CRF-907luc), deleted (CRF-220luc or CRF-233luc), or mutant (CRF-233Mtluc) promoter constructs, and then incubated for 2 h with 10 $\mu$M forskolin alone (Fsk) or vehicle (C). * $P < 0.05$ (compared with forskolin alone [Fsk] in CRF-907luc transfected cells). $^{+}$ $P < 0.05$ (compared with each control). (Reproduction from Kageyama *et al.* (2008) with permission of the publisher.) Copyright 2008, Editrice Kurtis srl.

(Seasholtz *et al.*, 1988; Spengler *et al.*, 1992). In addition, King *et al.* demonstrated a second response element by cAMP between −125 and −118 bp, a caudal-type homeobox response element on the CRF promoter (King *et al.*, 2002). Therefore, both the CRE and the caudal-type homeobox response element on the CRF promoter contribute to a cAMP-associated increase in the expression of the CRF gene (Cheng *et al.*, 2000; King *et al.*, 2002).

## B. Inducible cAMP-early repressor (ICER)

Inducible cAMP-early repressor (ICER) is a cAMP-inducible member of the CRE modulator (CREM) family and a repressor isoform of CREM (Foulkes *et al.*, 1991; Molina *et al.*, 1993). CREM, CREB, and activating transcription factor 1 (ATF-1) bind to CRE promoter elements (Lalli and Sassone-Corsi, 1994), and ICER acts as a competitive inhibitor of CRE-dependent transcription (Foulkes *et al.*, 1991). ICER may then suppress a stress response via inhibition of the cAMP-dependent CRF gene. Forskolin induces an increase in ICER protein levels in hypothalamic cells, and transfection of the ICER decreases forskolin-induced CRF 5′-promoter activity (Liu *et al.*, 2006). Therefore, considering that CRF plays a central role in controlling the HPA axis in stress, the induction of ICER may influence the suppression of a stress response via regulation of the CRF gene.

## C. Estrogen

Estrogens acting centrally, including the pituitary corticotrophs and hypothalamus, can modulate stress responses (Nakano *et al.*, 1991), and direct estrogenic regulation of CRF gene expression has been demonstrated in various tissues (Dibbs *et al.*, 1997; Vamvakopoulos and Chrousos, 1993). Estrogen regulates the HPA axis by stimulation of CRF gene expression in the hypothalamus *in vivo*, since high levels of estrogen replacement increases basal levels of CRF mRNA in the PVN of ovariectomized rats (Ochedalski *et al.*, 2007).

A physiologically relevant dose, 10 nM of estradiol (E2), stimulates both CRF gene transcription (Fig. 16.4A) and mRNA (Ogura *et al.*, 2008) expression in hypothalamic 4B cells. E2 and diarylpropionitrile (DPN), an estrogen receptor (ER) $\beta$ agonist, increase CRF gene transcriptional activity (Fig. 16.4B). Therefore, ER$\beta$ activation by estrogens induces the transcription of the CRF gene in hypothalamic cells. Treatment with both E2 and forskolin shows an additive effect on the CRF promoter activity (Fig. 16.4C). Therefore, in addition to cAMP, the presence of other signal pathways may be indicated in the activation of CRF by estrogens. ER$\beta$ often antagonizes the effect of ER$\alpha$ on gene regulation (Liu *et al.*, 2002) and stimulates CRF transcriptional activity in HeLa cells (Miller *et al.*, 2004). ERE half-sites are contained on the CRF promoter (Chen *et al.*, 2008) and are suggested to be involved in the stimulation of the CRF gene via ER$\beta$ (Vamvakopoulos and Chrousos, 1993). Therefore, direct estrogenic transcriptional regulation of the CRF gene in hypothalamic 4B cells suggests that EREs may be of functional significance via ER$\beta$ in the PVN.

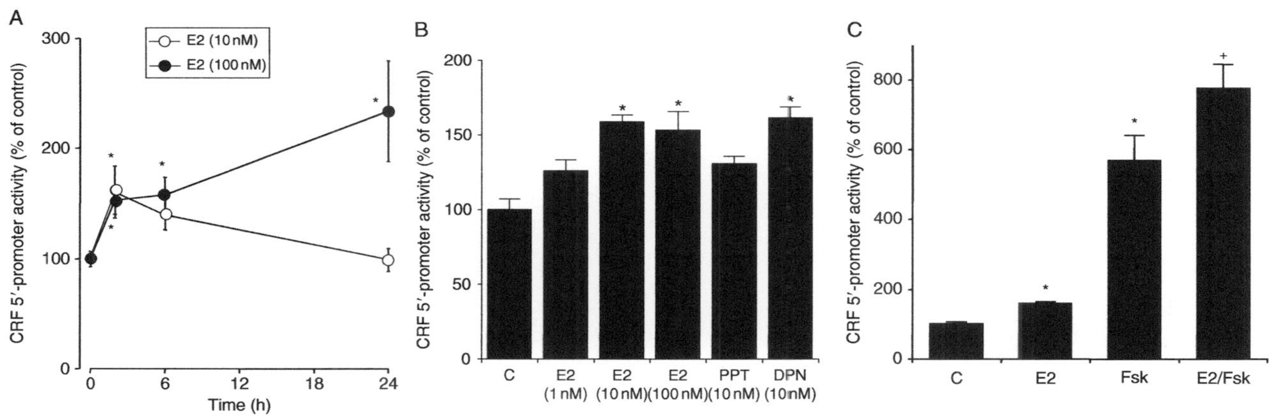


**Figure 16.4** Effects of E2 on CRF 5′-promoter activity in 4B cells. Control cells treated with medium alone are indicated as (C). * $P < 0.05$ (compared with [C]). $^{+}$ $P < 0.05$, (compared with E2 or Fsk alone). (A) Time-dependent changes in E2-induced CRF 5′-promoter activity. Cells were incubated with medium containing 10 or 100 nM E2. (B) Dose-dependent changes in E2-induced CRF 5′-promoter activity. Cells were incubated for 2 h with medium containing 1–100 nM E2, 10 nM PPT, or 10 nM DPN. (C) Effects of E2 on forskolin-induced CRF 5′-promoter activity. Cells were incubated for 2 h with medium alone (C) or medium containing 500 nM forskolin (Fsk) and/or 100 nM E2. (Reproduction from Ogura *et al.* (2008) with permission of the publisher.) Copyright 2008, Elsevier.

## D. Glucocorticoid receptor

The HPA axis is regulated by a negative feedback mechanism. Hypothalamic parvocellular neurons are known to express glucocorticoid receptors, and glucocorticoids regulate CRF gene expression directly in the hypothalamus. Repression by glucocorticoids occurs through inhibition of CRF gene transcription in a subclone of AtT-20 cells (Malkoski and Dorin, 1999). The CRF promoter does not contain a classical consensus GRE; however, there are a number of regions in the sequence where glucocorticoid receptors are able to bind (Guardiola-Diaz *et al.*, 1996). Malkoski and Dorin (1999) demonstrated glucocorticoid regulatory regions on the CRF promoter. By using a series of 5′-nested deletions, they demonstrated that dexamethasone-dependent repression of cAMP-stimulated CRF promoter activity is localized to promoter sequences between −278 and −249 bp (Fig. 16.5). High-affinity binding of the glucocorticoid receptor DNA-binding domain to this promoter region was observed with an electrophoretic mobility shift assay. Therefore, this region would contribute to the inhibition of CRF promoter activity by glucocorticoids as a negative GRE (nGRE).

We demonstrated that other promoter regions are involved in the inhibitory regulation of CRF gene expression in hypothalamic 4B cells (Fig. 16.5). The glucocorticoid suppression of cAMP-stimulated CRF promoter activity is also involved in the CRF promoter sequences between −248 and −233 bp in hypothalamic cells (Fig. 16.5). Serum response element (SRE) is included in this region and would thus contribute to the negative response to glucocorticoids, because the glucocorticoid receptor can bind to SRE and inhibit promoter activation by antagonizing the function of positive transcription (Karagianni and Tsawdaroglou, 1994). Therefore, in addition to nGRE, the glucocorticoid suppression of cAMP-stimulated CRF promoter activity may also be caused by the SRE in hypothalamic 4B cells.

## E. Activator protein 1 (fos and Jun)

In addition to the PKA pathway, protein kinase C (PKC) is involved in the regulation of forskolin-induced CRF gene expression because a PKC inhibitor inhibits forskolin-induced CRF promoter activity. Activation of the PKC pathway leads to binding of AP-1 (Fos/Jun proteins) to AP-1-binding sites on the CRF promoter.

Binding sites for both glucocorticoid receptor and AP-1 nucleoproteins have been shown at adjacent elements within the nGRE (Malkoski and Dorin, 1999) because mutations that disrupted either glucocorticoid receptor or AP-1-binding activity cause a similar loss of glucocorticoid-dependent repression. These results suggest that the nGRE functions as a composite

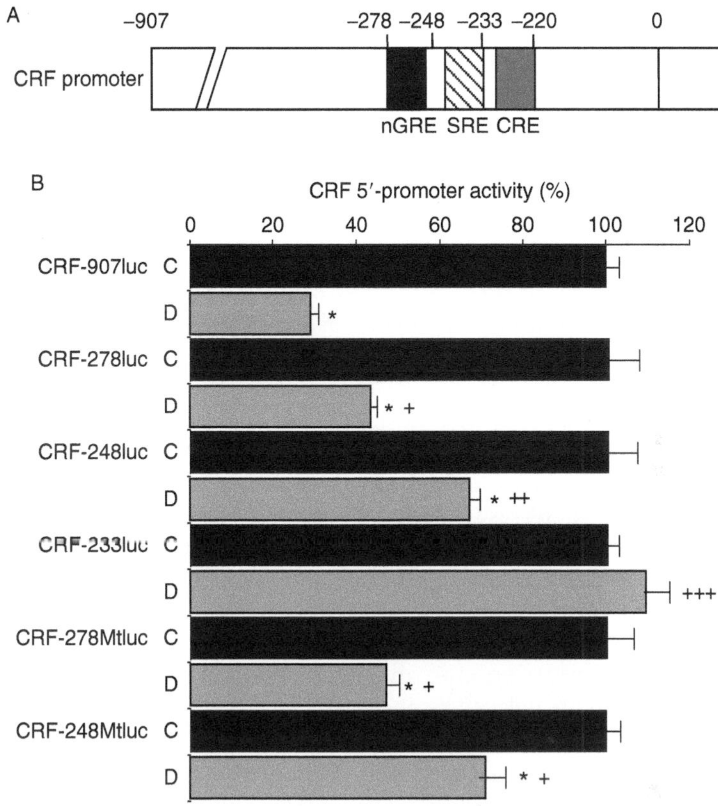


**Figure 16.5** Effects of nGRE or SRE deletion on the dexamethasone suppression of CRF 5′-promoter activity in 4B cells. (A) Schematic representation of nGRE or SRE in the CRF promoter. (B) Cells transfected with a full-length (CRF-907luc), deleted (CRF-233luc, CRF-248luc, or CRF-278luc), or mutant (CRF-278Mtluc or CRF-248Mtluc) promoter construct, were preincubated for 30 min with medium containing dexamethasone (Dex, 100 nM) or vehicle (C), followed by the addition of 10 $\mu$M forskolin for 2 h. Data are presented as relative activity, and luciferase activity in response to forskolin alone (C) was set at 100% in all transfected cells. Experiments were conducted in triplicate, and the means of three independent experiments are shown. * $P < 0.05$ (compared with forskolin alone [C] in all transfected cells). $^{+}$ $P < 0.05$ (compared with Dex in CRF-907luc). $^{++}$ $P < 0.05$ (compared with Dex in CRF-907luc and CRF-278luc). $^{+++}$ $P < 0.05$ (compared with Dex in CRF-907luc, CRF-278luc, and CRF-248luc). (Reproduction from Kageyama *et al.* (2008) with permission of the publisher.) Copyright 2008, Editrice Kurtis srl.

regulatory element, involving direct DNA binding of the glucocorticoid receptor and AP-1 nucleoproteins. King *et al.* proposed that transcription factor differences among cells cause negative or positive regulation because CREB and Fos were detected in AtT-20 while CREB and Jun were found in placental cells (King *et al.*, 2002). Furthermore, glucocorticoids can inhibit CREB and Fos in the PVN (Jacobson *et al.*, 1990; Legradi *et al.*, 1997). The modified expression of transcription factors in cells changes the binding to nGRE and/or CRE, and then the promoter activity, resulting in the suppression of CRF gene transcription.

## F. Suppressor of cytokine signaling (SOCS)-3

Following inflammatory stresses, interleukin (IL)-1 and IL-6 stimulate the HPA axis. IL-6 is an important mediator of the interaction between the neuroendocrine and immune systems. IL-6 is coexpressed with CRF and AVP in the supraoptic nucleus and PVN neurons (Ghorbel *et al.*, 2003), and plays an important role in regulating both CRF and AVP in the hypothalamus. For example, IL-6 increases CRF gene expression and secretion in the PVN (Navarra *et al.*, 1991; Vallieres and Rivest, 1999). Forskolin or PACAP increases IL-6 mRNA expression and protein levels in the hypothalamic cells. In addition, the stimulatory effects of PACAP on CRF promoter activity are significantly inhibited by treatment with anti-IL-6 monoclonal antibody in hypothalamic cells (Fig. 16.6). Therefore, endogenous IL-6 production would be involved in the PACAP-induced CRF gene transcription in an autocrine manner in hypothalamic 4B cells. Considering the delayed response to IL-6 and the partial inhibition of PACAP effects induced by anti-IL-6 antibody, IL-6 may be important for sustaining the activity of CRF and AVP genes. In fact, the CRF promoter contains multiple nuclear factor (NF)-$\kappa$B and Nurr1-binding sites in response to cytokines. Thus, it is possible that IL-6, produced in the hypothalamus, stimulates CRF and AVP gene expression.

Suppressor of cytokine signaling (SOCS)-3 acts as a potent negative regulator of cytokine signaling and suppresses cytokine-induced proopiomelanocortin gene transcription and ACTH secretion in corticotrophs (Auernhammer *et al.*, 1999; Krebs and Hilton, 2000). IL-6 stimulates the Janus kinase/signal transducers and activators of transcription (JAK/STAT) signaling pathway, while IL-6-induced SOCS-3 acts as a negative regulator and inhibits STAT phosphorylation by JAK at the receptor complex (Ram and Waxman, 1999; Schmitz *et al.*, 2000). SOCS-3 would also be involved in the negative regulation of CRF gene expression in the hypothalamus. In fact, SOCS-3 was found to be regulated by IL-6 and via the cyclic AMP/protein kinase A pathway in hypothalamic cells (Fig. 16.7). SOCS-3 knockdown increased IL-6- or forskolin-induced CRF gene transcription and mRNA levels (Fig. 16.8). Therefore, SOCS-3, which is induced by a cAMP stimulant and IL-6, is involved in the negative regulation of CRF gene expression in hypothalamic cells.

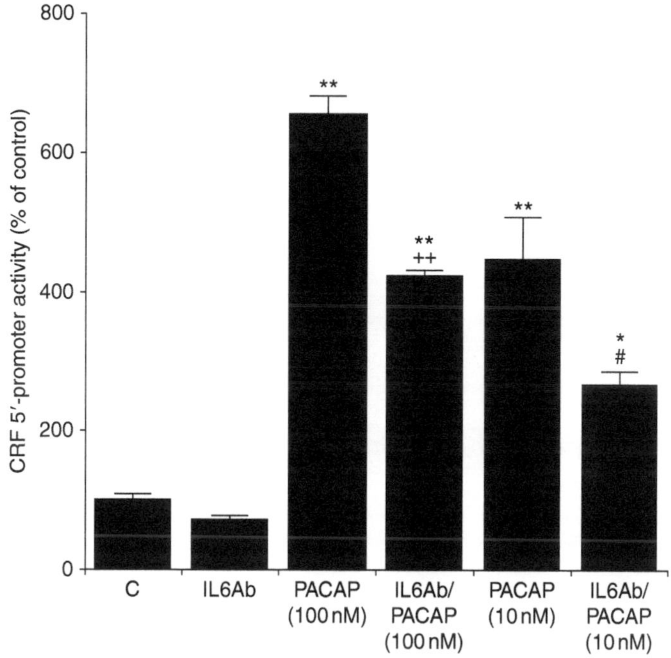


**Figure 16.6** Effects of anti-IL-6 Ab on PACAP-induced CRF 5′-promoter activity in 4B cells. $^{*}P < 0.05$, $^{**}P < 0.005$ (compared with control [C]). $^{++}P < 0.005$ (compared with 100 nM PACAP). $^{\#}P < 0.05$ (compared with 10 nM PACAP). Cells were preincubated with medium containing anti-IL-6 Ab or control IgG for 30 min, followed by the addition of 10 or 100 nM PACAP or vehicle for 2 h. Cells treated with control IgG are indicated as C. (Reproduction from Kageyama *et al.* (2007) with permission of the publisher.) Copyright 2007, Society for Endocrinology.

## III. Conclusion

Forskolin or PACAP stimulates adenylate cyclase and then increases intracellular cAMP levels in hypothalamic 4B cells. Activation of the PKA pathway leads to binding of CREB on the CRF promoter in hypothalamic cells (Fig. 16.9). Forskolin-stimulated CRF gene transcription is mediated by CRE, which included from −220 to −233, on the CRF 5′-promoter region. ICER acts as a competitive inhibitor of CRE-dependent transcription and then suppresses a stress response via inhibition of the cAMP-dependent CRF gene. E2 may enhance the activation of the CRF gene in stress. Glucocorticoid-dependent repression of cAMP-stimulated CRF promoter activity is mediated by both nGRE and SRE in hypothalamic cells. IL-6 produced in the hypothalamus stimulates CRF gene expression (Fig. 16.9). SOCS-3, induced by a cyclic AMP stimulant and IL-6, would

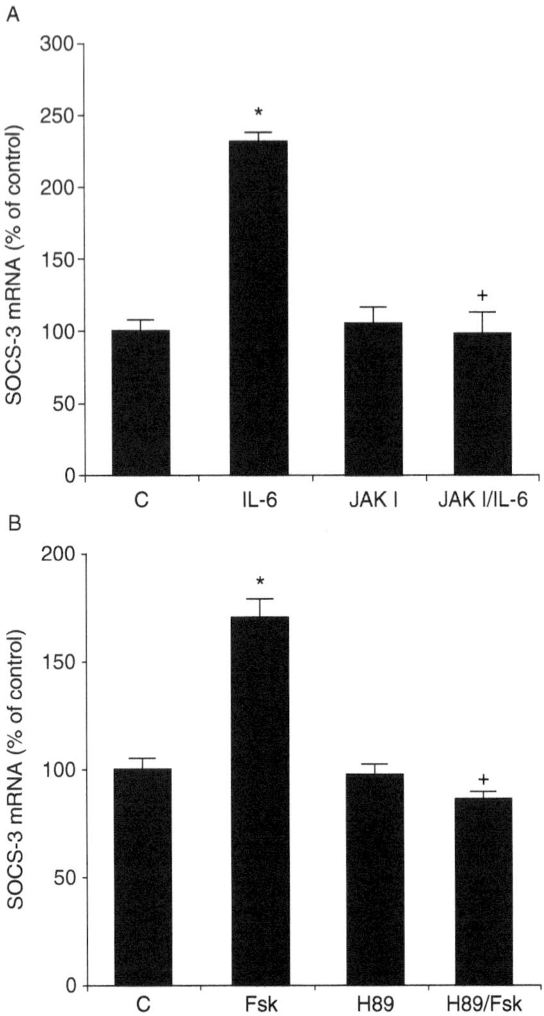


**Figure 16.7** Effects of JAK and PKA inhibitors on SOCS-3 mRNA levels. Control cells treated with medium alone are indicated as (C). Experiments were conducted in triplicate, and the means of three independent experiments are shown. Statistical analyses were performed using one-way ANOVA, followed by *post hoc* tests. $^{\star}P < 0.05$ (compared with control [C]). $^{+}P < 0.05$ (compared with forskolin [Fsk]). (A) Effects of JAK inhibitor I on IL-6-stimulated SOCS-3 mRNA levels. Cells were preincubated for 30 min with medium containing 1 $\mu$M JAK inhibitor I (JAK I) or vehicle, followed by the addition of 100 ng/ml IL-6 or vehicle for 6 h. (B) Effects of H89 on forskolin-stimulated SOCS-3 mRNA levels. Cells were preincubated for 30 min with medium containing 1 $\mu$M H89 or vehicle, followed by the addition of 10 $\mu$M forskolin (Fsk) or vehicle for 2 h. (Reproduction from Kageyama *et al.* (2009) with permission of the publisher.) Copyright 2009, Society for Endocrinology.

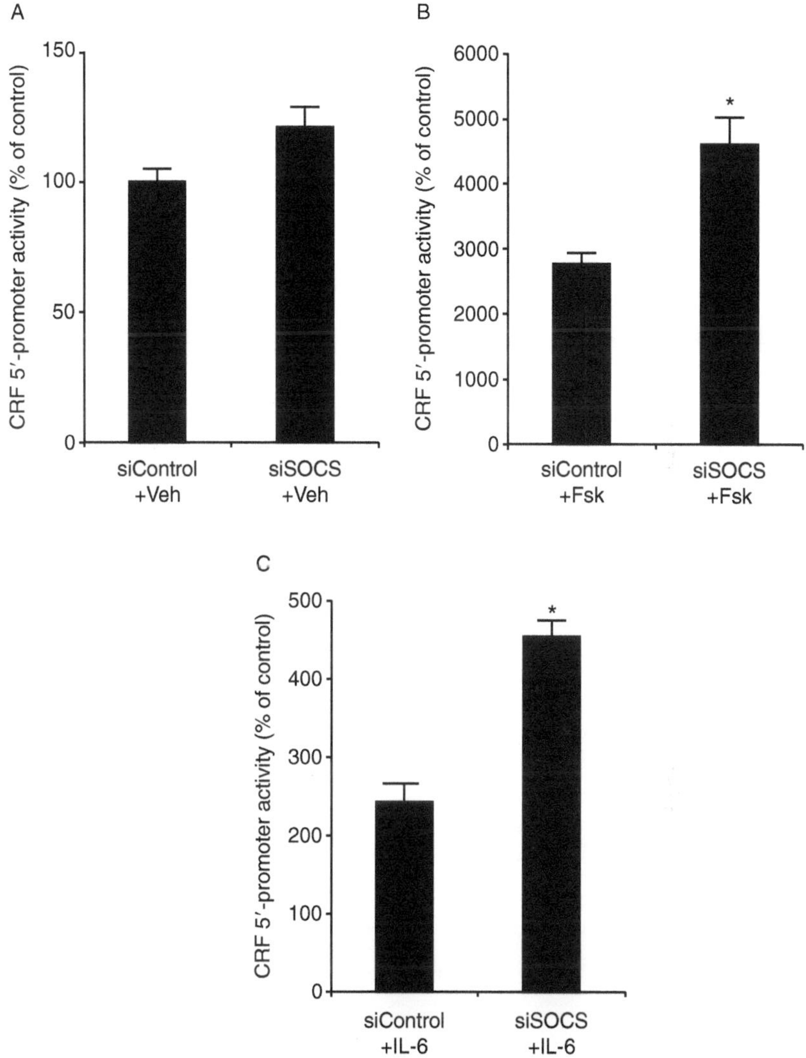


**Figure 16.8** Effects of SOCS-3 on the regulation of CRF gene in 4B cells. Experiments were conducted in triplicate, and the means of three independent experiments are shown. Statistical analyses were performed using unpaired *t*-test. * $P < 0.05$ (compared with control [siControl]). The 4B cells, seeded into 12-well plates at a density of $2 \times 10^4$ cells/well, were incubated for 24 h in 1 $\mu$l of culture medium containing siRNA for either control (siControl) or SOCS (siSOCS). (A–C) Effects of SOCS-3 on the regulation of CRF 5′-promoter activity. After transfection, the cells were retransfected with a CRF promoter construct, and then incubated with vehicle (A), 10 $\mu$M forskolin (Fsk) for 2 h (B) or 100 ng/ml IL-6 for 24 h (C). (Reproduction from Kageyama *et al.* (2009) with permission of the publisher.) Copyright 2009, Society for Endocrinology.

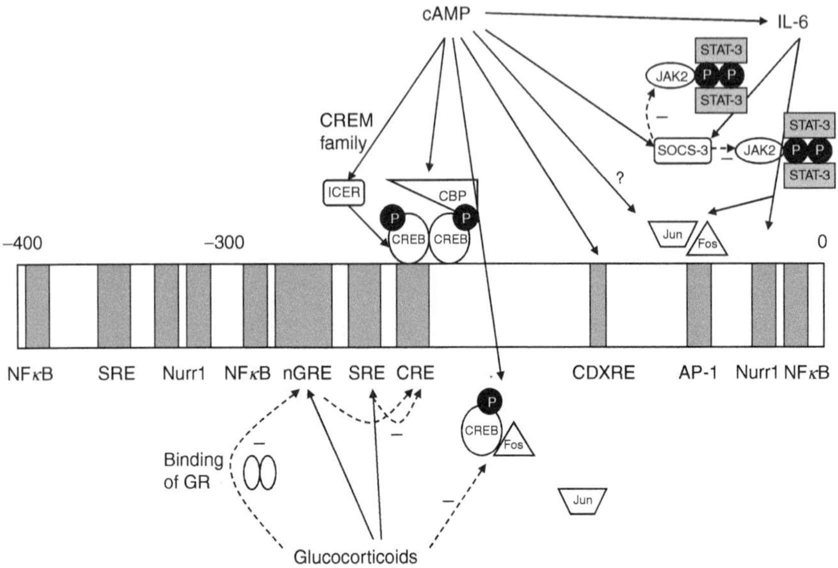


**Figure 16.9** A schematic model of transcriptional regulation of hypothalamic CRF gene. Forskolin or PACAP stimulates adenylate cyclase, and then intracellular cAMP levels in hypothalamic 4B cells. Activation of the PKA pathway leads to binding of CREB to CRE on the CRF promoter. ICER acts as a competitive inhibitor of CRE-dependent transcription, and then suppresses the stimulation of cAMP-dependent CRF gene. Glucocorticoids-dependent repression of cAMP-stimulated CRF promoter activity is mediated by both nGRE and SRE. IL-6, produced in the hypothalamus, stimulates CRF gene expression. SOCS-3, induced by a cyclic AMP stimulant and IL-6, would be involved in the negative regulation of CRF gene expression in the hypothalamus.

be involved in the negative regulation of CRF gene expression in hypothalamic cells. Such complex mechanisms would contribute to stress responses and homeostasis.

## ACKNOWLEDGMENTS

This work was supported in part by Health and Labour Science Research Grants (Research on Measures for Intractable Diseases) from the Ministry of Health, Labor, and Welfare of Japan.

## REFERENCES

Agarwal, A., Halvorson, L. M., and Legradi, G. (2005). Pituitary adenylate cyclase-activating polypeptide (PACAP) mimics neuroendocrine and behavioral manifestations of stress: Evidence for PKA-mediated expression of the corticotropin-releasing hormone (CRH) gene. *Brain Res. Mol. Brain Res.* **138,** 45–57.

Auernhammer, C. J., Bousquet, C., and Melmed, S. (1999). Autoregulation of pituitary corticotroph SOCS-3 expression: Characterization of the murine SOCS-3 promoter. *Proc. Natl. Acad. Sci. USA* **96,** 6964–6969.

Chen, X. N., Zhu, H., Meng, Q. Y., and Zhou, J. N. (2008). Estrogen receptor-alpha and -beta regulate the human corticotropin-releasing hormone gene through similar pathways. *Brain Res.* **1223,** 1–10.

Cheng, Y. H., Nicholson, R. C., King, B., Chan, E. C., Fitter, J. T., and Smith, R. (2000). Glucocorticoid stimulation of corticotropin-releasing hormone gene expression requires a cyclic adenosine 3', 5'-monophosphate regulatory element in human primary placental cytotrophoblast cells. *J. Clin. Endocrinol. Metab.* **85,** 1937–1945.

Dibbs, K. I., Anteby, E., Mallon, M. A., Sadovsky, Y., and Adler, S. (1997). Transcriptional regulation of human placental corticotropin-releasing factor by prostaglandins and estradiol. *Biol. Reprod.* **57,** 1285–1292.

Foulkes, N. S., Borrelli, E., and Sassone-Corsi, P. (1991). CREM gene: Use of alternative DNA-binding domains generates multiple antagonists of cAMP-induced transcription. *Cell* **64,** 739–749.

Ghorbel, M. T., Sharman, G., Leroux, M., Barrett, T., Donovan, D. M., Becker, K. G., and Murphy, D. (2003). Microarray analysis reveals interleukin-6 as a novel secretory product of the hypothalamo-neurohypophyseal system. *J. Biol. Chem.* **278,** 19280–19285.

Gillies, G. E., Linton, E. A., and Lowry, P. J. (1982). Corticotropin releasing activity of the new CRF is potentiated several times by vasopressin. *Nature* **299,** 355–357.

Gonzalez-Hernandez, T., Afonso-Oramas, D., Cruz-Muros, I., Barroso-Chinea, P., Abreu, P., del Mar Perez-Delgado, M., Rancel-Torres, N., and del Carmen Gonzalez, M. (2006). Interleukin-6 and nitric oxide synthase expression in the vasopressin and corticotrophin-releasing factor systems of the rat hypothalamus. *J. Histochem. Cytochem.* **54,** 427–441.

Grinevich, V., Fournier, A., and Pelletier, G. (1997). Effects of pituitary adenylate cyclase-activating polypeptide (PACAP) on corticotropin-releasing hormone (CRH) gene expression in the rat hypothalamic paraventricular nucleus. *Brain Res.* **773,** 190–196.

Guardiola-Diaz, H. M., Kolinske, J. S., Gates, L. H., and Seasholtz, A. F. (1996). Negative glucorticoid regulation of cyclic adenosine 3', 5'-monophosphate-stimulated corticotropin-releasing hormone-reporter expression in AtT-20 cells. *Mol. Endocrinol.* **10,** 317–329.

Hannibal, J., Mikkelsen, J. D., Fahrenkrug, J., and Larsen, P. J. (1995). Pituitary adenylate cyclase-activating peptide gene expression in corticotropin-releasing factor-containing parvicellular neurons of the rat hypothalamic paraventricular nucleus is induced by colchicine, but not by adrenalectomy, acute osmotic, ether, or restraint stress. *Endocrinology* **136,** 4116–4124.

Itoi, K., Horiba, N., Tozawa, F., Sakai, Y., Sakai, K., Abe, K., Demura, H., and Suda, T. (1996). Major role of 3', 5'-cyclic adenosine monophosphate-dependent protein kinase A pathway in corticotropin-releasing factor gene expression in the rat hypothalamus in vivo. *Endocrinology* **137,** 2389–2396.

Jacobson, L., Sharp, F. R., and Dallman, M. F. (1990). Induction of fos-like immunoreactivity in hypothalamic corticotropin-releasing factor neurons after adrenalectomy in the rat. *Endocrinology* **126,** 1709–1719.

Kageyama, K., Hanada, K., Iwasaki, Y., Sakihara, S., Nigawara, T., Kasckow, J., and Suda, T. (2007). Pituitary adenylate cyclase-activating polypeptide stimulates corticotropin-releasing factor, vasopressin, and interleukin-6 gene transcription in hypothalamic 4B cells. *J. Endocrinol.* **195,** 199–211.

Kageyama, K., Hanada, K., Takayasu, S., Iwasaki, Y., Sakihara, S., Nigawara, T., Kasckow, J., and Suda, T. (2008). Involvement of regulatory elements on corticotropin-releasing factor gene promoter in hypothalamic 4B cells. *J. Endocrinol. Invest.* **31,** 1079–1085.

Kageyama, K., Hanada, K., Iwasaki, Y., and Suda, T. (2009). Regulation and role of suppressor of cytokine signaling-3 in hypothalamic 4B cells. *J. Endocrinol.* **201,** 369–376.
Karagianni, N., and Tsawdaroglou, N. (1994). The c-fos serum response element (SRE) confers negative response to glucocorticoids. *Oncogene* **9,** 2327–2334.
King, B. R., Smith, R., and Nicholson, R. C. (2002). Novel glucocorticoid and cAMP interactions on the CRH gene promoter. *Mol. Cell. Endocrinol.* **194,** 19–28.
Kozicz, T., and Arimura, A. (2002). Dopamine- and cyclic AMP-regulated phosphoprotein-immunoreactive neurons activated by acute stress are innervated by fiber terminals immunopositive for pituitary adenylate cyclase-activating polypeptide in the extended amygdala in the rat. *Regul. Pept.* **109,** 63–70.
Krebs, D. L., and Hilton, D. J. (2000). SOCS: Physiological suppressors of cytokine signaling. *J. Cell. Sci.* **113,** 2813–2819.
Lalli, E., and Sassone-Corsi, P. (1994). Signal transduction and gene regulation: The nuclear response to cAMP. *J. Biol. Chem.* **269,** 17359–17362.
Legradi, G., Holzer, D., Kapcala, L. P., and Lechan, R. M. (1997). Glucocorticoids inhibit stress-induced phosphorylation of CREB in corticotropin-releasing hormone neurons of the hypothalamic paraventricular nucleus. *Neuroendocrinology* **66,** 86–97.
Legradi, G., Hannibal, J., and Lechan, R. M. (1998). Pituitary adenylate cyclase-activating polypeptide-nerve terminals densely innervate corticotropin-releasing hormone-neurons in the hypothalamic paraventricular nucleus of the rat. *Neurosci. Lett.* **246,** 145–148.
Liu, M. M., Albanese, C., Anderson, C. M., Hilty, K., Webb, P., Uht, R. M., Price, R. H. Jr., Pestell, R. G., and Kushner, P. J. (2002). Opposing action of estrogen receptors alpha and beta on cyclin D1 gene expression. *J. Biol. Chem.* **277,** 24353–24360.
Liu, Y., Kalintchenko, N., Sassone-Corsi, P., and Aguilera, G. (2006). Inhibition of corticotrophin-releasing hormone transcription by inducible cAMP-early repressor in the hypothalamic cell line, 4B. *J. Neuroendocrinol.* **18,** 42–49.
Malkoski, S. P., and Dorin, R. I. (1999). Composite glucocorticoid regulation at a functionally defined negative glucocorticoid response element of the human corticotropin-releasing hormone gene. *Mol. Endocrinol.* **13,** 1629–1644.
Miller, W. J., Suzuki, S., Miller, L. K., Handa, R., and Uht, R. M. (2004). Estrogen receptor (ER) beta isoforms rather than ERalpha regulate corticotropin-releasing hormone promoter activity through an alternate pathway. *J. Neurosci.* **24,** 10628–10635.
Molina, C. A., Foulkes, N. S., Lalli, E., and Sassone-Corsi, P. (1993). Inducibility and negative autoregulation of CREM: An alternative promoter directs the expression of ICER, an early response repressor. *Cell* **75,** 875–886.
Mouri, T., Itoi, K., Takahashi, K., Suda, T., Murakami, O., Yoshinaga, K., Andoh, N., Ohtani, H., Masuda, T., and Sasano, N. (1993). Colocalization of corticotropin-releasing factor and vasopressin in the paraventricular nucleus of the human hypothalamus. *Neuroendocrinology* **57,** 34–39.
Nakano, Y., Suda, T., Sumitomo, T., Tozawa, F., and Demura, H. (1991). Effects of sex steroids on beta-endorphin release from rat hypothalamus in vitro. *Brain Res.* **553,** 1–3.
Navarra, P., Tsagarakis, S., Faria, M. S., Rees, L. H., Besser, G. M., and Grossman, A. B. (1991). Interleukins-1 and -6 stimulate the release of corticotropin-releasing hormone-41 from rat hypothalamus in vitro via the eicosanoid cyclooxygenase pathway. *Endocrinology* **128,** 37–44.
Ochedalski, T., Subburaju, S., Wynn, P. C., and Aguilera, G. (2007). Interaction between oestrogen and oxytocin on hypothalamic-pituitary-adrenal axis activity. *J. Neuroendocrinol.* **19,** 189–197.
Ogura, E., Kageyama, K., Hanada, K., Kasckow, J., and Suda, T. (2008). Effects of estradiol on regulation of corticotropin-releasing factor gene and interleukin-6 production via estrogen receptor type beta in hypothalamic 4B cells. *Peptides* **29,** 456–464.

Piggins, H. D., Stamp, J. A., Burns, J., Rusak, B., and Semba, K. (1996). Distribution of pituitary adenylate cyclase activating polypeptide (PACAP) immunoreactivity in the hypothalamus and extended amygdala of the rat. *J. Comp. Neurol.* **376,** 278–294.

Ram, P. A., and Waxman, D. J. (1999). SOCS/CIS protein inhibition of growth hormone-stimulated STAT5 signaling by multiple mechanisms. *J. Biol. Chem.* **274,** 35553–35561.

Schmitz, J., Weissenbach, M., Haan, S., Heinrich, P. C., and Schaper, F. (2000). SOCS3 exerts its inhibitory function on interleukin-6 signal transduction through the SHP2 recruitment site of gp130. *J. Biol. Chem.* **275,** 12848–12856.

Seasholtz, A. F., Thompson, R. C., and Douglass, J. O. (1988). Identification of a cyclic adenosine monophosphate-responsive element in the rat corticotropin-releasing hormone gene. *Mol. Endocrinol.* **2,** 1311–1319.

Spengler, D., Rupprecht, R., Van, L. P., and Holsboer, F. (1992). Identification and characterization of a 3′, 5′-cyclic adenosine monophosphate-responsive element in the human corticotropin-releasing hormone gene promoter. *Mol. Endocrinol.* **6,** 1931–1941.

Suda, T., Yajima, F., Tomori, N., Demura, H., and Shizume, K. (1985). In vitro study of immunoreactive corticotropin-releasing factor release from the rat hypothalamus. *Life Sci.* **37,** 1499–1505.

Vale, W., Spiess, J., Rivier, C., and Rivier, J. (1981). Characterization of a 41-residue ovine hypothalamic peptide that stimulates secretion of corticotropin and beta-endorphin. *Science* **213,** 1394–1397.

Vallieres, L., and Rivest, S. (1999). Interleukin-6 is a needed proinflammatory cytokine in the prolonged neural activity and transcriptional activation of corticotropin-releasing factor during endotoxemia. *Endocrinology* **140,** 3890–3903.

Vamvakopoulos, N. C., and Chrousos, G. P. (1993). Evidence of direct estrogenic regulation of human corticotropin-releasing hormone gene expression. Potential implications for the sexual dimorphism of the stress response and immune/inflammatory reaction. *J. Clin. Invest.* **92,** 1896–1902.

Whitnall, M. H. (1993). Regulation of the hypothalamic corticotropin-releasing hormone neurosecretory system. *Prog. Neurobiol.* **40,** 573–629.

Yamamori, E., Asai, M., Yoshida, M., Takano, K., Itoi, K., Oiso, Y., and Iwasaki, Y. (2004). Calcium/calmodulin kinase IV pathway is involved in the transcriptional regulation of the corticotropin-releasing hormone gene promoter in neuronal cells. *J. Mol. Endocrinol.* **33,** 639–649.

Yao, M., and Denver, R. J. (2007). Regulation of vertebrate corticotropin-releasing factor genes. *Gen. Comp. Endocrinol.* **153,** 200–216.

CHAPTER SEVENTEEN

# Estrogen in the Limbic System

Gert J. ter Horst

## Contents



## Abstract

Estrogens are a group of steroid hormones that function as the primary female sex hormone. Estrogens not only have an important role in the regulation of the estrous or menstrual cycle but also control, for example, bone formation, the cardiovascular system, and cognitive functions. Estradiol (E2), the main representative of the group, is highly lipophylic and can easily pass the blood–brain barrier to modulate neuronal activity. Particularly the limbic system, a group of tightly interconnected forebrain areas controlling mood and emotion, is rich in estrogen receptors. To date two cytoplasmatic and/or nuclear estrogen receptors named ER-alpha (ERα) and ER-beta (ERβ) have been identified. In the brain, ERα plays a critical role in regulating reproductive neuroendocrine behavior and function. ERβ appears to play an important role in nonreproductive behaviors, such as learning and memory, anxiety, and mood. Five splice variants of ERβ, named Erb1, Erb2, Erb1d3, Erb2d3, and Erb1d4, have been identified with possibly different biological activities. There is evidence of a thus far not definitely characterized membrane-linked ER receptor named mER-X. In this review, the anatomy of the limbic system and the distribution of estrogen

Department of Neuroscience, Neuroimaging Center, University Medical Center Groningen, Groningen, The Netherlands

*Vitamins and Hormones*, Volume 82
ISSN 0083-6729, DOI: 10.1016/S0083-6729(10)82017-5

receptors (ERs) are described in relation to coping with stress and the higher prevalence of stress-related psychiatric disorders in women. Effects of cyclic estrogen administration and chronic stress on recovery and neuronal plasticity are illustrated with own results. 

## I. Estrogen Synthesis and Actions

The theca and granulosa cells of the ovaries are the primary sources of estrogens during reproductive life. The principal and most potent estrogen secreted is 17β-estradiol (E2) but three different naturally occurring estrogens have been recognized including estrone (E1), 17β-estradiol (E2), and estriol (E3). In nonpregnant women, E2 is the main estrogen. E1 is primarily secreted during menopause and E3 during pregnancy. To stimulate production and secretion of estrogens by the ovaries, the anterior pituitary secretes Follicle-Stimulating Hormone (FSH) and Luteinizing Hormone (LH). The production of FSH and LH in the anterior pituitary is regulated by specific cells in the hypothalamus that secrete Gonadotropin-Releasing Hormone (GnRH) in the pituitary portal veins. The hypothalamus is a critical target for the negative feedback regulation and timing of ovarian hormone release. GnRH-positive cells contain estrogen receptor beta (Erβ) and thus can respond to changes in circulating levels of estrogen (Herbison and Pape, 2001). In the ovaries, cholesterol is converted to progesterone, which is transformed to the androgens androstenedione and testosterone, which are subsequently aromatized to estradiol.

Estrogens are best known for their effect on the female reproductive organs, that is, they cause cellular proliferation and growth of the external female sex organs and the uterus. E2 also has profound effects on osteoblast and osteoclast activity. In postmenopausal women, in whom the production of E2 is reduced, the prevalence of osteoporosis significantly increases. Both positive and negative influences of postmenopausal E2 therapy on the cardiovascular system have been shown (Rosano *et al.*, 2009). Hormone replacement therapy has favorable effects on serum cholesterol levels and may reduce the incidence of cardiovascular disease. E2 administration appears to have a protective effect on atherosclerosis and lowers plasma LDL levels and triglycerides, increases plasma HDL levels, stimulates vasodilatation, and has anti-inflammatory effects (Rossouw *et al.*, 2008). However, hormone replacement with conjugated estrogens, particularly when combined with progesterone, showed an increased risk of cardiovascular disease and increased risk of blood clotting (Cushman *et al.*, 2004).

Besides actions on the reproductive tract, blood vessels, and bone formation, E2 has a major impact on the brain. Animal studies have shown that E2 can modulate the activity of different neurotransmitter systems including the serotonergic, dopaminergic, adrenergic, and cholinergic systems (Luine, 1985; McEwen and Alves, 1999; Sellix *et al.*, 2004). Changes in plasma E2 levels have been associated with differential limbic (George *et al.*, 1996) and Hypothalamic-Pituitary-Adrenal axis (HPA) activity (Carey *et al.*, 1995; Shupnik, 2002). Additionally, there is accumulating evidence that E2 has a significant impact on neuronal plasticity-related processes. Studies showed that female rats in proestrous, when E2 levels are high, have a greater density of apical dendritic spines on pyramidal neurons in the CA1 of the hippocampus than in the other stages of the estrous cycle (Shors *et al.*, 2001; Woolley *et al.*, 1990). Likewise, E2 administration to ovariectomized (OVX) females increased spine densities in the hippocampus (Murphy and Segal, 1996). Brain plasticity is critically related to the presence of growth factors and it has been demonstrated that in the hippocampus the levels of nerve growth factor (NGF) and Brain-Derived Neurotrophic Factor (BDNF) and related mRNA's change during the estrous cycle (Cavus and Duman, 2003; Gibbs, 1998, 1999; Jezierski and Sohrabji, 2000; Singh *et al.*, 1995).

## II. Estrogen Receptors

Estrogen can induce its effects by binding specific ERs that belong to the steroid hormone superfamily of nuclear receptors. Two ERs have been identified and are named ER-alpha (ER$\alpha$) and ER-beta (ER$\beta$). Both can bind E2 with high affinity. E2 is highly lipophylic and therefore can pass the blood–brain barrier and cell membranes. Classical neurotransmitter receptors are located on the cell membrane but the ERs are located in the cytoplasm and nucleus, and they translocate to the nuclear DNA domain after ligand binding (Gruber *et al.*, 2002; Koehler *et al.*, 2005). After binding estrogen the ERs attach to specific DNA sequences, the Estrogen-Response-Elements (ERE's) as homo- or heterodimers to act as transcription factors in cooperation with several cofactors (Pettersson *et al.*, 1997). Transcriptional activity of the ERs is not only confined to ERE's, because interaction with Activation-Protein 1 (AP-1) and modulation of other transcription regulators like Specificity-Protein 1 (SP-1) and Nuclear Factor-$\kappa\beta$ (NF-$\kappa\beta$) has also been demonstrated (Galien and Garcia, 1997; Paech *et al.*, 1997; Salvatori *et al.*, 2003). Binding of ligand-bound ER$\alpha$ and/or ER$\beta$ homo- or heterodimers to ERE's will lead to enhancement or repression of gene transcription and thus to alterations in protein synthesis (Gruber *et al.*, 2002). The interaction between the two ERs and the consequences for the transcriptional modulation is complicated and delicately balanced. An opposite transcriptional response can be generated by ligand-bound ER$\alpha$ and ER$\beta$, which depends on the cellular context and associated cofactors (Koehler *et al.*, 2005). Additionally, *in vitro* experiments have shown that intracellular colocalization of ERα and ERβ may specifically stimulate the formation of heterodimers after ligand binding, a mechanism by which ERβ represses the transcriptional activity of ERα (Pettersson *et al.*, 2000).

Besides rather slow effects of E2 mediated by the classical ERs, rapid effects, frequently associated with the activation of various protein-kinase cascades, have been reported (Losel and Wehling, 2003). For example, E2 administration induced both rapid and transient phosphorylation of ERK *in vivo* and *in vitro* (Kuroki *et al.*, 2000; Watters *et al.*, 1997). These rapid effects of E2 administration are most likely mediated by a putative membrane-bound ER, also described as mER-X (Toran-Allerand, 2004), and probably can explain E2-induced mobilization of intracellular calcium, stimulation of adenylate cyclase activity, and production of cyclic adenosine monophosphate (cAMP) (Aronica *et al.*, 1994; Improta-Brears *et al.*, 1999).

Both ERs are abundantly expressed in the limbic system but locally the relative distribution of ERα and ERβ may be different (Rissman, 2008; Shughrue and Merchenthaler, 2001; Shughrue *et al.*, 1997). The differential local distribution of ERs most likely underlies the different effects of local E2 administration in the brain. In addition, recent studies have identified different splice variants of ERβ with differing distribution patterns in the CNS. To date, three major splice variants of the classical ERβ receptor (Erb1), have been described in the rodent, which include a deletion of exon 3 (Erb2d3), a deletion of exon 4 (Erb2d4), and an insert between exons 5 and 6 (Erb2). It is believed that the splice variants of Erb1 may serve as low-affinity ERs (Chung *et al.*, 2007).

## III. Anatomy of the Limbic System

The limbic system is composed of a group of tightly interconnected brain areas that includes the cingulate gyrus, the anterior thalamus, the hypothalamus and mammillary bodies, the hippocampus, and the amygdala. It was first described in the first half of the previous century by Paul Broca, who coined the name "limbic system" for a part of the circuitry that is currently known as the limbic system. James Papez and shortly thereafter Paul McLean have extended the basic layout of Broca to what is currently known as the limbic system. After World War II, the introduction of neuronal tract tracing methods have significantly contributed to the understanding of the neuroanatomy of the limbic system and its internal connectivity and interactions with brainstem and spinal areas. First electrolytic lesion-based "Fink-Heimer" anterograde tracing and later various retrograde tract tracing techniques like Horse Radish Peroxidase (HRP), fluorescent Fast Blue tracing, and combinations thereof were used to study the anatomy of the limbic system. Later the use of anterogradely transported tracers like radioactive labeled amino acids and in the 1980s the introduction of antibody-based tracing with selective proteins like Phaseolus vulgaris leuco-agglutinin further increased our knowledge of the limbic neuronal network structure. It was soon recognized that the limbic system is critically involved in metabolic homeostasis, neuroendocrine activity, autonomic functions, and mood and emotion.

## IV. Distribution of Estrogen Receptors in the Limbic System

Both ER$\alpha$ and ER$\beta$ are expressed in the limbic system but expression patterns of these ERs do not overlap completely (Perez *et al.*, 2003; Shughrue and Merchenthaler, 2001; Shughrue *et al.*, 1997; Weiser *et al.* 2008). Some regions only express ER$\alpha$ like the ventromedial hypothalamic nucleus (VMH) and the subfornical organ, whereas neurons in the olfactory bulb, supraoptic nucleus, paraventricular and tuberal hypothalamic nuclei, and zona incerta exclusively express ER$\beta$. The bed nucleus of the stria terminalis (BNST), medial and cortical amygdaloid nuclei, preoptic area, suprachiasmatic nucleus (SCN) (Vida *et al.*, 2008), and lateral habenula express both ER subtypes, and also laminae IV–VI of the cerebral cortex were found to be ER$\alpha$ and ER$\beta$ positive, however with a reduced expression in the cingulate cortex. ER$\alpha$ is more abundantly expressed than ER$\beta$ in the arcuate nucleus and vice versa, ER$\beta$ is more prevalent in the hippocampus. In addition, glia cells seem to express ER$\alpha$ and ER$\beta$ (Mhyre and Dorsa, 2006) although the function of glial ERs is not yet known.

## V. Estrogen and ER$\beta$ Expression

The expression of ER$\beta$ in the limbic system seems to depend on and correlate with the estrogen levels in the blood. Pregnant and proestrous females show differential ER$\beta$ expression in the preoptic hypothalamus (POA), supraoptic hypothalamic nucleus (SON), and the medial amygdala (MA). The expression is highest in the dioestrous phase (Arteaga-Lopez *et al.* 2003). Osterlund *et al.* (1998), Patisaul *et al.* (1999), and Shima *et al.* (2003) have shown differential expression of ER$\beta$ following estrogen treatment of OVX females. Estrogen administration after ovariectomy significantly reduced ER$\beta$ expression in the olfactory bulb; entorhinal cortex; lateral septal nucleus; diagonal band; lateral, medial, and basolateral amygdala; BNST; paraventricular hypothalamic nucleus (PVN); medial amygdala; preoptic nucleus; and SCN.

Hippocampal primary cultures showed increased ERα expression but decreased ERβ expression after treatment with 17b-estradiol (Prange-Kiel *et al.*, 2003). Likewise, administration of estradiol-benzoate and also dexamethasone, a corticosterone analog, changed the level of ERβ expression and protein levels in the PVN and SON of OVX females. Dexamethasone treatment increased ERβ expression (Suzuki and Handa, 2004), whereas estradiol-benzoate administration decreased the number of ERβ-positive cells in the PVN (Suzuki and Handa, 2004; Weiser *et al.*, 2008).

Studies in ER knockout mice have shown that ERα is vital for reproductive functions (Hewitt and Korach, 2003; Ogawa *et al.*, 1998) and although ERβ is found on hypothalamic GnRH neurons, it seems not to be involved in the regulation of the preovulatory LH surge in response to rising levels of estrogen (Dorling *et al.*, 2003; Wintermantel *et al.*, 2006). Dorling *et al.* (2003) have demonstrated that these beta-receptors partially mediate the negative feedback control of anterior pituitary LH secretion.

## VI. Affective Disorders and Gender

Women in the reproductive age are more prone to developing major depression and anxiety disorders than men, the prevalence being approximately two to three times higher in women (Kessler *et al.*, 1993, 1995; Weissman and Olfson, 1995). This gender difference emerges during puberty (Bouma, 2010; Bouma *et al.*, 2008) and thereafter the prevalence rates in women remain to be higher until the menopause (Kessler, 2003). The different phases in the reproductive life of women and/or female hormonal instability seem to play a role in the gender difference. For that reason, researchers attributed the gender difference in lifetime prevalence rates of affective disorders to the cyclic release of female sex hormones, particularly estrogen (Desai and Jann, 2000; Steiner *et al.*, 2003). Indeed, puberty, pregnancy, and the perimenopause show significant changes in hormonal status and are typically associated with increased risk of occurrence of depression and anxiety (Halbreich and Kahn, 2001). Not high or low plasma levels of estrogen are implicated as causative factors for depression, but sudden changes in estrogen levels in combination with an underlying genetic vulnerability and environmental factors like chronic stress apparently make women more vulnerable. It has been proposed that postpartum depression (PPD) is precipitated by the dramatic decline in reproductive hormones that occurs just after childbirth (Stoffel and Craft, 2004). Correspondingly, the higher prevalence of affective disorders in women compared to men normalizes once menopause is established (Weissman and Olfson, 1995).

Interestingly, there has been a limited interest of preclinical researchers in the gender difference for affective disorders in the past decades. Today gender differences in depression and coping with (chronic) stress have become a booming field of research. Here we give an overview of research related to gender differences in behavior and related neurobiological and physiological changes during stress coping, focusing on the roles of estrogen.

There is extensive evidence from epidemiological studies that chronic stress or life events are an underlying cause of affective disorders and major depression (Belmaker and Agam, 2008). Preclinical research therefore has focused on behavioral and neuropathological effects of stress in rodents to model mechanisms that can lead to depression, while the pharmaceutical industry uses these animal stress models for the development and identification of modes of action of novel antidepressants. Interestingly, for a long time researchers and clinicians assumed that outcomes of studies in females would be similar to effects observed in males. Therefore, most of the commonly used stress models were developed in male rodents and the fact that women are more susceptible to stress-related psychiatric disorders like depression and anxiety was ignored. Another reason for using males is that the cyclic release of the female sex hormones, estrogen and progesterone, creates an additional and unstable endogenous factor in the models that by itself can have a significant effect on the physiology and behavior of the animals. Accordingly, the number of female rodents needed in such studies would significantly increase when one would like to include various stages of the estrous cycle. It is therefore more complicated to obtain conclusive results of stress and/or medication effects in depression models when female rodents are used. Here we review how estrogen can modulate the female stress response in view of recent evidence from different groups that the female brain is anatomically and functionally different and employs other strategies to cope with stress (Ter Horst *et al.*, 2009).

## VII. Estrogen in Hippocampus and Amygdala

The hippocampus and amygdala are putative focal points in the limbic system for E2-mediated effects on depression and anxiety. E2 administration has been shown to change the activity of the hippocampus, increasing the number of Fos-ir cells and altering hippocampal plasticity. Administration of E2 in OVX rats was associated with increased numbers of dendritic spines in the hippocampus (MacLusky *et al.*, 2005). Levels of Brain-derived Neurotrophic Factor (BDNF) fluctuate across the estrous cycle and increase following E2 administration in OVX females (Gibbs, 1998, 1999; Walf and Frye, 2006). Behavioral studies have demonstrated that performance in hippocampus-dependent cognitive tasks of OVX rats can be enhanced by E2 treatment and that direct administration of E2 in the hippocampus increased antianxiety- and antidepressant-like behavior (Walf and Frye, 2006).

Similar observations were published for effects of E2 in the amygdala. Administration of E2 increased the number of Fos-ir cells in the medial amygdala (Greco *et al.*, 2003a,b) as well as the number of synapses found on the dendrites (Nishizuka and Arai, 1982). Dendritic spine density in the medial amygdala fluctuates across the estrous cycle (Rasia-Filho *et al.*, 2004).

## VIII. Cyclic Estrogen Administration and Stress

To be able to specifically address the role of estrogen on physiological and behavioral stress coping parameters, we have introduced an artificial estrogen cycle paradigm (Gerrits *et al.*, 2005). First, we have studied how estradiol-treated female rats respond to acute and chronic stress compared to OVX rats and second, how high and low plasma estrogen levels can influence the stress response.

Cyclic E2 administration, involving subcutaneous injections every 4th day for 3 weeks, had a significant negative effect on weight gain: control OVX females gained significantly more weight than control E2-treated females. Chronic stress reduced the growth in both the OVX and E2-treated females but it was significant only in OVX rats. Basal plasma corticosterone levels were not different and both groups showed significantly increased plasma corticosterone levels after acute stress. This stress-induced increase of corticosterone gradually declined during chronic stress exposure and resulted in marginally higher plasma levels than at baseline in both OVX and E2-treated females on day 21, possibly illustrating a habituation to the stress paradigm. In addition, OVX rats showed a significant adrenal hypertrophy after 3 weeks of daily stress, whereas in E2-treated females the adrenal weight was only slightly increased.

Cyclic E2 administration significantly affected the acute and chronic stress-induced c-Fos activation in the PVN and a number of other limbic areas, including the PFC and amygdala (Gerrits *et al.*, 2006a,b). C-Fos is used as a neuroanatomical marker of neuronal activity. Effects on neuronal activity in the PVN correlated with the plasma 17$\beta$-estradiol concentration (Fig. 17.1), high plasma levels (24 h after the last E2 injection) being associated with a low number of Fos-ir positive cells in the PVN. This number increased to approximately the level found in chronic stress exposed OVX females, when the plasma estrogen levels were low at 48 h after the last E2 injection (Gerrits *et al.*, 2005). These levels of PVN activity corroborate a reduced plasma corticosterone release, that is, reduced HPA activity after 3 weeks of stress (Walf and Frye, 2006). Evidence was provided that E2 binding to ER$\beta$ drives promoter activity *in vitro* for Corticotropin Releasing Hormone (CRH) and Vasopressin (AVP) (Miller *et al.*, 2004; Pak *et al.*, 2007; Shipiro *et al.*, 2000), and that subtype selective ligands alter

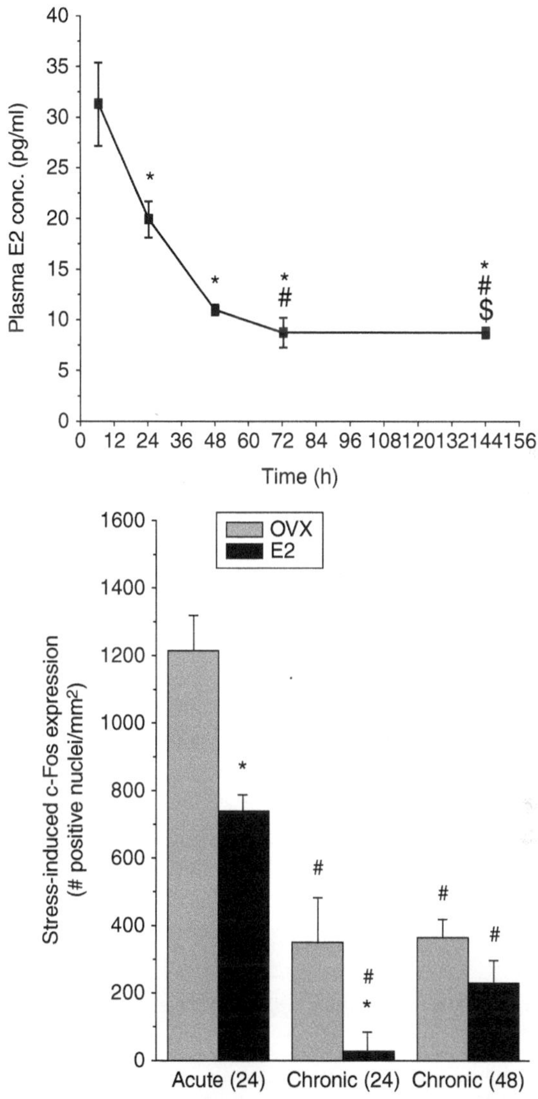


**Figure 17.1** Plasma 17-beta-estradiol concentration (pg/ml) induced by a single s.c. injection of 17-beta-estradiol-benzoate (10 $\mu$g/250 g in peanut oil) in ovariectomized female rats. (*$p < 0.05$ compared to 6 h, #$p < 0.05$ compared to 24 h, and \$$p < 0.05$ compared to 48 h). Conditioned stimulus-induced activation of the PVN measured by Fos expression after acute and chronic stress in OVX and E2-treated female rats. The rats were sacrificed 24 or 48 h after the last E2 injection (for details see Gerrits *et al.*, 2005).

CRH, adrenocorticotropin hormone (ACTH), and corticosterone responses to a stressor (Weiser *et al.*, 2008). Moreover, it was found that the number of ER$\beta$ protein expressing PVN cells was increased after chronic stress in both OVX and cyclic E2-treated females: an effect that was not found in cyclic E2-treated control females (Gerrits *et al.*, 2005).

The PFC of OVX and cyclic E2-treated females did not show significant differences in neuronal activity after chronic stress. However, both these groups showed a significantly increased Fos-ir compared to the appropriate control groups (Gerrits *et al.*, 2006a,b). In the medial amygdala, a region with an abundant ER expression (Ostlund *et al.*, 2003), the number of Fos-ir cells was higher in artificial cyclic than in OVX females after acute stress. Chronic daily stress exposure led, in both groups, to a small increase in the number of Fos-ir positive cells. The neuronal activity in the central amygdala correlated with the treatment (OVX or E2) and the duration of the protocol (Gerrits *et al.*, 2006a,b). Control OVX females that were treated with three E2-cycles showed significantly more Fos-ir cells in the central amygdala than untreated OVX females. However, after a treatment of eight E2-cycles, this basal activity increased and was significantly higher than in animals subjected to a treatment duration of three E2-cycles. Stress, either acute or chronic, had no significant additional impact on the neuronal activity in the central amygdala. We hypothesize that the observed differences in basal activity of the central amygdala after three and eight E2-cycles are related to physiological recovery after the OVX surgery. Surgery, for example, affects various cardiovascular parameters in rats, which show recovery to baseline values after 8–10 days (Greene *et al.*, 2007). The central amygdala participates in neuronal circuitry for cardiovascular regulation (Ter Horst *et al.*, 1996) as well as in many other systems that can request participation of the autonomic nervous system. The recovery of cardiovascular functions is most likely reflected in changes in basal neuronal activity. Future studies using artificial E2-cycles should consider this effect of E2 treatment duration after OVX on basal limbic activity.

## IX. Estrogen, Stress, and Neuroplasticity

Estrogens have an effect on neuronal plasticity and this is deemed critical to a healthy brain and coping with chronic stress (Pittenger and Duman, 2008). Various hypotheses have been defined for the pathobiology of depression, mostly involving changes in monoamine levels in the brain, but recently theories were defined that assume compromised neuroplasticity to be the underlying cause of depression (Duman, 2002; Duman *et al.*, 1997). Since estrogens can improve neuronal plasticity, it has become important to understand the mechanisms that mediate effects of estrogens on neuronal plasticity to be able to develop novel nonestrogen-dependent treatments for depression (Ter Horst *et al.*, 2009).

Phosphorylation of ERK1/2, an intracellular signal transduction protein, has been implicated in regulation of neuronal plasticity and survival (Bonni *et al.*, 1999; Sweatt, 2001) and this protein is most likely a critical intracellular intermediate for neuroplasticity effects mediated by estrogens. In a series of experiments, we have demonstrated that in the limbic system activation/phosphorylation of ERK1/2 indeed is part of intracellular signaling cascades that are stimulated by the presence of estrogen. A transient increase of pERK1/2 is associated with enhanced activation of cAMP-Response-Element-Binding protein (CREB) and increased transcription of different prosurvival genes (Wu *et al.*, 2001). However, persistent ERK1/2 activation can lead to inhibition of CREB-mediated gene expression (Wang *et al.*, 2003) and has been associated with neuropathological mechanisms (Colucci-D'Amato *et al.*, 2003). Acute stress caused a decrease in the number of pERK1/2 immunoreactive cells in the PFC of OVX rats (Gerrits *et al.*, 2006a,b; Kuipers *et al.*, 2003) (Fig. 17.2). Conversely, after chronic stress the number of pERK1/2 expressing neurons increased, in the PFC and hippocampus (Gerrits *et al.*, 2006a,b; Kuipers *et al.*, 2003; Pardon *et al.*, 2005; Trentani *et al.*, 2002). It should be noted that the phosphorylation of ERK1/2 is a biphasic event: a rapid inactivation after the initial burst of neuronal activation is followed by a second prolonged activation phase of 2–4 h (Pouyssegur *et al.*, 2002). Transient increases in the activation of ERK1/2 are associated with increased transcription of prosurvival genes by enhanced CREB activity (Wu *et al.*, 2001). Persistent activation of pERK1/2 can inactivate CREB by formation of complexes of CREB-binding protein and Rsk, which in turn inhibits the CREB-mediated gene expression (Wu *et al.*, 2001). Indeed, in our hands chronic stress induced a dissociation of pERK1/2 hyperphosphorylation and CREB activation in the PFC *in vivo* (Kuipers *et al.*, 2003; Trentani *et al.*, 2002). Neuronal pERK1/2 expression in the PFC of female OVX rats was increased after chronic stress. However, cyclic E2 treatment prevented this stress-induced pERK1/2 accumulation (Gerrits *et al.*, 2006a,b), which may possibly increase the neuronal viability and enhance PFC functioning under stressful conditions.

## X. Recovery After Chronic Stress; Effect of Estrogen and Antidepressants

Chronic stress-induced aberrations in OVX females were attenuated by cyclic estradiol treatment during 3 weeks (Gerrits, 2006; Gerrits *et al.*, 2005, 2006a,b), and estrogen replacement therapy has been demonstrated to be an effective treatment for depressive symptoms in menopausal women (Carranza-Lira and Valentino-Figueroa, 1999; Soares *et al.*, 2001). Chances

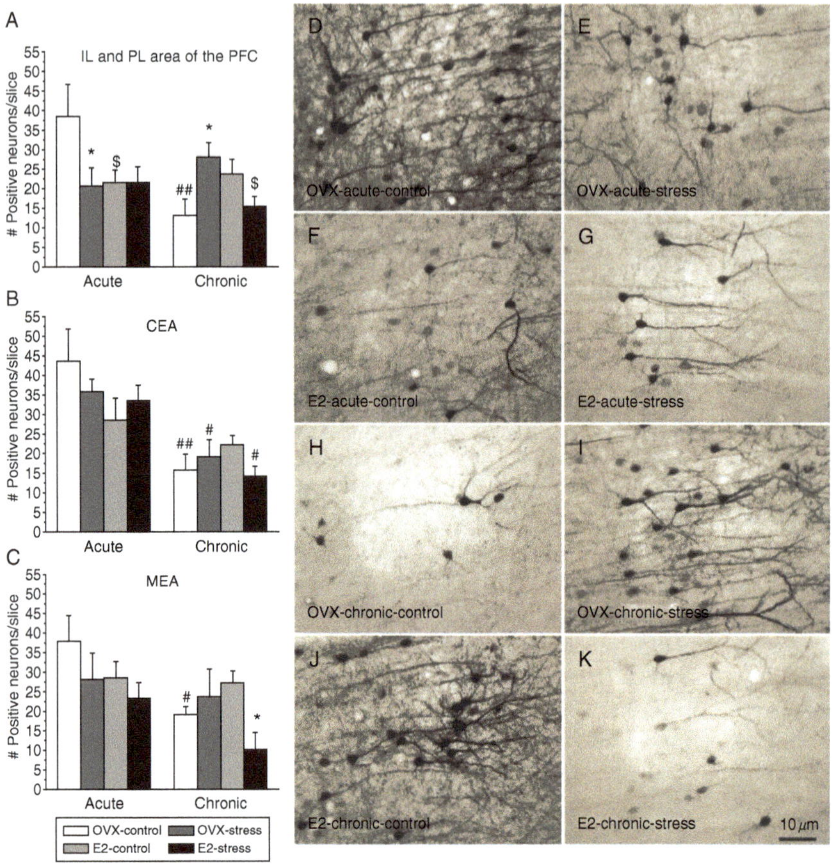


**Figure 17.2** Number of pERK1/2 immunoreactivity neurons per slice (±SEM) in the (A) PFC, (B) CeA, and (C) MeA after 3 (acute) or 22 (chronic) days of stress exposure. Photomicrographs of pERK1/2 expression in the mPFC after (D + F) acute control, (E + G) acute stress, (H + J) chronic control, and (I + K) chronic stress in OVX and E2-treated rats. (*$p < 0.05$ compared to control; #$p < 0.05$, ##$p < 0.001$ compared to acute: \$$p < 0.05$ compared to OVX; for details see Gerrits *et al.*, 2006a,b).

of recurrence of depression are significant (Kendler *et al.*, 2005) and the use of antidepressants can prevent or delay the onset of a new depressive episode. To study estrogen-induced neurobiological mechanisms of recovery after chronic stress, we have subjected OVX females to cyclic estrogen treatment and mirtazapine. Mirtazapine is a noradrenergic and specific serotonergic antidepressant, which mediates its effects by blockade of adrenergic $\alpha 2$ autoreceptors and $\alpha 2$ heteroreceptors, and 5HT2 and 5HT3 receptors. These receptors are abundantly expressed in the limbic system. Mirtazapine has also strong anxiolytic effects (Asnis *et al.*, 2004; Goodnick *et al.*, 1999). In the forced swimming test, a commonly used model for testing antidepressant activity of novel drugs, mirtazapine administration caused increased swimming behavior only after chronic treatments (Reneric *et al.*, 2002). In contrast to most pharmacological studies, which use healthy rats, in our design the normal stress response was disturbed. Prior to the start of the treatment, all OVX females were subjected to 3 weeks of chronic mild stress. Thereafter, a chronic pharmacological intervention with mirtazapine and/or cyclic estradiol was started. We have used groups of female OVX rats in which the stress exposure was either discontinued or continued during the 3 weeks of estradiol and mirtazapine treatment. Recovery females were reexposed to the stressful environment on day 42 and terminated after this final sham stress session. Using the above paradigm, sensitization of the neuronal activity could be demonstrated in several limbic areas that express ERs (Fig. 17.3), including the PFC, DG, medial and central amygdala, and the PVN (Gerrits, 2006; Gerrits *et al.*, 2006a,b). Cyclic E2 administration prevented a reexposure-induced sensitization in the PVN and central amygdala, and reduced the number of Fos-ir cells in these regions. In contrast, mirtazapine treatment increased the number of c-Fos positive cells in the central amygdala and DG. Combined administration of E2 and mirtazapine has demonstrated that cyclic E2 administration cannot augment effects of mirtazapine or vice versa in this paradigm.

The PVN and central amygdala are both output regions of the limbic system that mediate the autonomic and endocrine responses (Luiten *et al.*, 1985; Roozendaal *et al.*, 1991) and play a critical role in anxiety behavior. Suppression of an exaggerated reexposure-induced Fos response in the PVN and CeA after cyclic E2 administration may illustrate that ERs in the PVN and CeA participate in processes that inhibit excessive HPA axis and cardiovascular activity. This may provide a neurobiological basis for the anxiolytic effects of E2 (Bowman *et al.*, 2002; Marcondes *et al.*, 2001; Rachman *et al.*, 1998). However, when the mild chronic stress paradigm was continued during the E2 treatment in the recovery phase between days 22 and 43, E2 no longer had an effect on the c-Fos response in the PVN and CeA. Then, probably other not yet identified stress-related factors overrule the positive effects of the cyclic E2 administration.

## XI. Conclusion

There has been evidence that ER$\alpha$ and ER$\beta$ are differentially expressed in the subregions of the limbic system. Hormonal, neuronal plasticity-related, and behavioral responses associated with (chronic) stress are modulated by the presence of estrogen and particularly by ER$\beta$-mediated mechanisms. The hippocampus and amygdala, and also the PFC

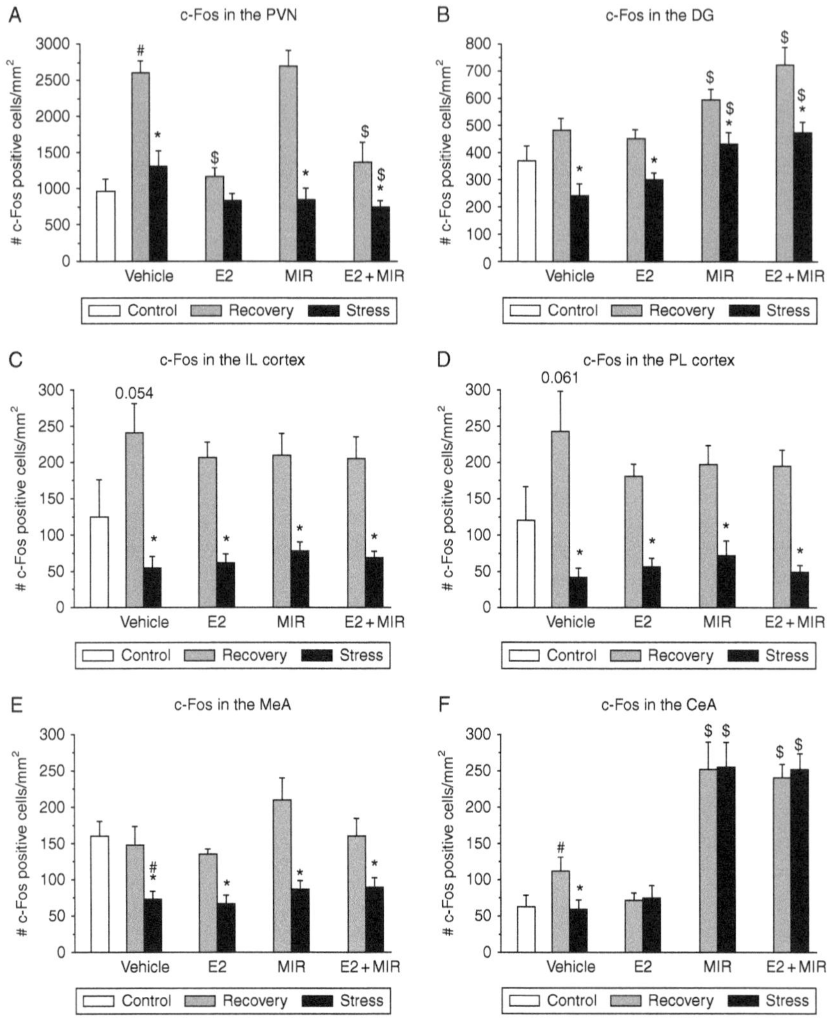


**Figure 17.3** Number of c-Fos-positive cells/mm$^2$ (+SEM) in the (A) Paraventricular hypothalamic nucleus (PVN), (B) Dentate Gyrus (DG), (C) Infralimbic (IL), and (D) prelimbic (PL) cortex, (E) medial (MeA) and (F) central (CeA) amygdala after 3 weeks of foot shock stress followed by 3 weeks of recovery or 3 weeks of continued foot shock stress. OVX females were treated with Vehicle, E2 (once every 4 days), 10 mg/kg mirtazapine MIR daily i.p., or E2 + mirtazapine E2 + (MIR), all starting in week 4 after the initial chronic stress exposure. On the last day all animals were exposed to the conditioned stimulus in the foot shock box and terminated after 2 h. (see for details Gerrits *et al.*, 2006a,b; #$p < 0.05$ compared to control; *$p < 0.05$ compared to recovery; \$$p < 0.05$ compared to vehicle).

and PVN are most likely primary target areas for the effects of estrogen on antianxiety- and antidepression-related behavior. E2-induced changes of neuroplasticity may be mediated by various intracellular signal transduction proteins, including ERK1/2.

## *REFERENCES*


Aronica, S. M., Kraus, W. L., and Katzenellenbogen, B. S. (1994). Estrogen action via the cAMP signaling pathway, stimulation of adenylate cyclase and cAMP-regulated gene transcription. *Proc. Natl. Acad. Sci. USA* **91,** 8517–8521.

Arteaga-Lopez, P. R., Dominguez, R., Cerbon, M. A., Mendoza-Rodriguez, C. A., and Cruz, M. E. (2003). Differential mRNA expression of alpha and beta estrogen receptor isoforms and GnRH in the left and right side of the preoptic and anterior hypothalamic area during the estrous cycle of the rat. *Endocrine* **21,** 251–260.

Asnis, G. M., Kohn, S. R., Henderson, M., and Brown, N. L. (2004). SSRIs versus non-SSRIs in post-traumatic stress disorder; an update with recommendations. *CNS Drugs* **64,** 383–404.

Belmaker, R. H., and Agam, G. (2008). Major depressive disorder. *N. Engl. J. Med.* **358,** 55–68.

Bonni, A., Brunet, A., West, A. E., Datta, S. R., Takasu, M. A., and Greenberg, M. E. (1999). Cell survival promoted by the Ras-MAPK signaling pathway by transcription-dependent and independent mechanisms. *Science* **286,** 1358–1362.

Bouma, E. M. (2010). The Sensitive Sex: Depressive Symptoms in Adolescents; the Role of Gender, Polymorphic Genes and Psycho-Physiological Stress Responses. Thesis, University Groningen, pp. 3–129.

Bouma, E. M., Ormel, J., Verhulst, F. C., and Oldehinkel, A. J. (2008). Stressful life events and depressive problems in early adolescent boys and girls, the influence of parental depression, temperament and family environment. *J. Affect. Disord.* **105,** 185–193.

Bowman, R. E., Ferguson, D., and Luine, V. N. (2002). Effects of chronic restraint stress and estradiol on open field activity, spatial memory, and monoaminergic neurotransmitters in ovariectomized rats. *Neuroscience* **113,** 401–410.

Carey, M. P., Deterd, C. H., de Koning, J., Helmerhorst, F., and de Kloet, E. R. (1995). The influence of ovarian steroids on hypothalamic-pituitary-adrenal regulation in the female rat. *J. Endocrinol.* **144,** 311–321.

Carranza-Lira, S., and Valentino-Figueroa, M. L. (1999). Estrogen therapy for depression in postmenopausal women. *Int. J. Gynaecol. Obstet.* **65,** 35–38.

Cavus, I., and Duman, R. S. (2003). Influence of estradiol, stress, and 5-HT2A agonist treatment on brain-derived neurotrophic factor expression in female rats. *Biol. Psychiatry* **54,** 59–69.

Chung, W. C. J., Pak, T. R., Suzuki, S., Pouliot, W. A., Andersen, M. E., and Handa, R. J. (2007). Detection and localization of an estrogen receptor beta splice variant protein (ERbeta2) in the adult female rat forebrain and midbrain regions. *J. Comp. Neurol.* **505,** 249–267.

Colucci-D´Amato, L., Perrone-Capano, C., and di Porzio, U. (2003). Chronic activation of ERK and neurodegenerative diseases. *Bioessays* **25,** 1085–1095.

Cushman, M., Kuller, L. H., Prentice, R., Rodabough, R. J., Psaty, B. M., Stafford, R. S., Sidney, S., and Rosendaal, F. R. (2004). Women's Health Initiative Investigators. Estrogen plus progestin and risk of venous thrombosis. *JAMA* **292,** 1573–1580.

Desai, H. D., and Jann, M. W. (2000). Major depression in women, a review of the literature. *J. Am. Pharm. Assoc. (Wash.)* **40,** 525–537.

Dorling, A. A., Todman, M. G., Korach, K. S., and Herbison, A. E. (2003). Critical role for estrogen receptor alpha in negative feedback regulation of gonadotropin-releasing hormone mRNA expression in the female mouse. *Neuroendocrinology* **78,** 204–209.

Duman, R. S. (2002). Pathophysiology of depression, the concept of synaptic plasticity. *Eur. Psychiatry* **17**(Suppl. 3), 306–310.

Duman, R. S., Heninger, G. R., and Nestler, E. J. (1997). A molecular and cellular theory of depression. *Arch. Gen. Psychiatry* **54,** 597–606.

Galien, R., and Garcia, T. (1997). Estrogen receptor impairs interleukin-6 expression by preventing protein binding on the NF-kappa beta site. *Nucleic Acids Res.* **25,** 2424–2429.

George, M. S., Ketter, T. A., Parekh, P. I., Herscovitch, P., and Post, R. M. (1996). Gender differences in regional cerebral blood flow during transient self-induced sadness and happiness. *Biol. Psychiatry* **40,** 859–871.

Gerrits, M. (2006). Stress and the female brain; the effects of estradiol on the neurobiological reactions to chronic stress. Thesis, University Groningen, pp. 7–140.

Gerrits, M., Grootkarzijn, H. A., Bekkering, B. F., Bruinsma, M., Den Boer, J. A., and Ter Horst, G. J. (2005). Cyclic estradiol replacement attenuates stress-induced c-fos expression in the PVN of ovariectomized rats. *Brain Res. Bull.* **67,** 147–155.

Gerrits, M., Bakker, P. L., Koch, T., and Ter Horst, G. J. (2006a). Stress-induced sensitization of the limbic system in ovariectomized rats is partially restored by cyclic 17beta-estradiol administration. *Eur. J. Neurosci.* **23,** 1747–1756.

Gerrits, M., Westenbroek, C., Koch, T., Grootkarzijn, A., and Ter Horst, G. J. (2006b). Increased limbic phosphorylated extracellular-regulated kinase 1 and 2 expression after chronic stress is reduced by cyclic 17beta-estradiol administration. *Neuroscience* **142,** 1293–1302.

Gibbs, R. B. (1998). Levels of trkA and BDNF mRNA, but not NGF mRNA, fluctuate across the estrous cycle and increase in response to acute hormone replacement. *Brain Res.* **787,** 259–268.

Gibbs, R. B. (1999). Treatment with estrogen and progesterone affects relative levels of brain-derived neurotrophic factor mRNA and protein in different regions of the adult rat brain. *Brain Res.* **844,** 20–27.

Goodnick, P. J., Puig, A., DeVane, C. L., and Freund, B. V. (1999). Mirtazapine in major depression with comorbid generalized anxiety disorder. *J. Clin. Psychiatry* **60,** 446–448.

Greco, B., Blasberg, M. E., Kosinski, E. C., and Blaustein, J. D. (2003a). Response of ERalpha-IR and ERbeta-IR cells in the forebrain of female rats to mating stimuli. *Horm. Behav.* **43,** 444–453.

Greco, B., Lubbers, L. S., and Blaustein, J. D. (2003b). Estrogen receptor beta messenger ribonucleic acid expression in the forebrain of proestrous, pregnant, and lactating female rats. *Endocrinology* **144,** 1869–1875.

Greene, A. N., Clapp, S. L., and Alper, R. H. (2007). Timecourse of recovery after surgical intraperitoneal implantation of radiotelemetry transmitters in rats. *J. Pharmacol. Toxicol. Methods* **56,** 218–222.

Gruber, C. J., Tschugguel, W., Schneeberger, C., and Huber, J. C. (2002). Production and actions of estrogens. *N. Engl. J. Med.* **346,** 340–352.

Halbreich, U., and Kahn, L. S. (2001). Role of estrogen in the aetiology and treatment of mood disorders. *CNS Drugs* **15,** 797–817.

Herbison, A. E., and Pape, J. R. (2001). New evidence for estrogen receptors in gonadotropin-releasing hormone neurons. *Front. Neuroendocrinol.* **22,** 292–308.

Hewitt, S. C., and Korach, K. S. (2003). Oestrogen receptor knockout mice, roles for oestrogen receptors alpha and beta in reproductive tissues. *Reproduction* **125,** 143–149.

Improta-Brears, T., Whortom, A. R., Codazzi, F., York, J. D., Meyer, T., and McDonnell, D. P. (1999). Estrogen-induced activation of mitogen-activated protein kinase requires mobilization of intracellular calcium. *Proc. Natl. Acad. Sci. USA* **96,** 4686–4691.

Jezierski, M. K., and Sohrabji, F. (2000). Region- and peptide-specific regulation of the neurotrophins by estrogen. *Brain Res. Mol. Brain Res.* **85,** 77–84.

Kendler, K. S., Myers, J., and Prescott, C. A. (2005). Sex differences in the relationship between social support and risk for major depression: A longitudinal study of opposite-sex twin pairs. *Am. J. Psychiatry* **162,** 250–256.

Kessler, R. C. (2003). Epidemiology of women and depression. *J. Affect. Disord.* **74,** 5–13.

Kessler, R. C., Mc Gonagle, K. A., Swartz, M., Blazer, D. G., and Nelson, C. B. (1993). Sex and depression in the National Comorbidity Survey, Lifetime prevalence, chronicity and recurrence. *J. Affect. Disord.* **29,** 85–96.

Kessler, R. C., Sonnega, A., Bromet, E., Hughes, M., and Nelson, C. B. (1995). Posttraumatic stress disorder in the National Comorbidity Survey. *Arch. Gen. Psychiatry* **52,** 1048–1060.

Koehler, K. F., Helguero, L. A., Haldosen, L. A., Warner, M., and Gustafsson, J. A. (2005). Reflections on discovery and significance of estrogen receptor beta. *Endocr. Rev.* **26,** 465–478.

Kuipers, S. D., Trentani, A., Den Boer, J. A., and Ter Horst, G. J. (2003). Molecular correlates of impaired prefrontal plasticity in response to chronic stress. *J. Neurochem.* **85,** 1312–1323.

Kuroki, Y., Fukushima, K., Kanda, Y., Mizuno, K., and Wantanabe, Y. (2000). Putative membrane-bound estrogen receptors possibly stimulate mitogen-activated protein kinase in the rat hippocampus. *Eur. J. Pharmacol.* **400,** 205–209.

Losel, R., and Wehling, M. (2003). Nongenomic actions of steroid hormones. *Nat. Rev. Mol. Cell Biol.* **4,** 46–56.

Luine, V. N. (1985). Estradiol increases choline acetyltransferase activity in specific basal forebrain nuclei and projection areas in female rats. *Exp. Neurol.* **89,** 484–490.

Luiten, P. G. M., Ter Horst, G. J., Karst, H., and Steffens, A. B. (1985). The course of paraventricular hypothalamic efferents to autonomic structures in medulla and spinal cord. *Brain Res.* **329,** 374–378.

MacLusky, N. J., Luine, V. N., Hajszan, T., and Leranth, C. (2005). The 17alpha and 17beta isomers of estradiol both induce rapid spine synapse formation in the CA1 hippocampal subfield of ovariectomized female rats. *Endocrinology* **146,** 435–440.

Marcondes, F. K., Miguel, K. J., Melo, L. L., and Spadari-Bratfisch, R. C. (2001). Estrous cycle influences the response of female rats in the elevated plus-maze test. *Physiol. Behav.* **74,** 435–440.

McEwen, B. S., and Alves, S. E. (1999). Estrogen actions in the central nervous system. *Endocr. Rev.* **20,** 279–307.

Mhyre, A. J., and Dorsa, D. M. (2006). Estrogen activates rapid signaling in the brain, role of estrogen receptor alpha and estrogen receptor beta in neurons and glia. *Neuroscience* **138,** 851–858.

Miller, W. J., Suzuki, S., Miller, M. K., Handa, R., and Uht, R. M. (2004). Estrogen receptor (ER)beta isoforms rather than ERalpha regulate corticotrophin-releasing hormone promoter activity through an alternative pathway. *J. Neurosci.* **24,** 10628–10635.

Murphy, D. D., and Segal, M. (1996). Regulation of dendritic spine density in cultured rat hippocampal neurons by steroid hormones. *J. Neurosci.* **16,** 4059–4068.

Nishizuka, M., and Arai, Y. (1982). Synapse formation in response to estrogen in the medial amygdala developing in the eye. *Proc. Natl. Acad. Sci. USA* **79,** 7024–7026.

Ogawa, S., Eng, V., Taylor, J., Lubahn, D. B., Korach, K. S., and Pfaff, D. W. (1998). Roles of estrogen receptor-alpha gene expression in reproduction-related behaviors in female mice. *Endocrinology* **139,** 5070–5081.

Osterlund, M., Kuiper, C. G., Gustafsson, J. A., and Hurd, Y. L. (1998). Differential distribution and regulation of estrogen receptor-alpha and –beta mRNA within the female rat brain. *Brain Res. Mol. Brain Res.* **54,** 175–180.

Ostlund, H., Keller, E., and Hurd, Y. L. (2003). Estrogen receptor gene expression in relation to neuropsychiatric disorders. *Ann. N.Y. Acad. Sci.* **1007,** 54–63.

Paech, K., Webb, P., Kuiper, C. G., Nilsson, S., Gustafsson, J., Kushner, P. J., and Scanlan, T. S. (1997). Differential ligand activation of estrogen receptors ERalpha and ERbeta at AP1 sites. *Science* **277,** 1508–1510.

Pak, T. R., Chung, W. C., Hinds, L. R., and Handa, R. J. (2007). Estrogen receptor-beta mediates dihydrotestosterone-induced stimulation of the arginine vasopressin promoter in neuronal cells. *Endocrinology* **148,** 3371–3382.

Pardon, M. C., Roberts, R. E., Marden, C. A., Bianchi, M., Latif, M. L., and Duxon, M. S. (2005). Social threat and novel cage stress-induced sustained extracellular-regulated kinase 1/2 phosphorylation but differential modulation of brain-derived neurotrophic factor (BDNF) expression in the hippocampus of NMR1 mice. *Neuroscience* **132,** 561–574.

Patisaul, H. B., Whitten, P. L., and Young, L. J. (1999). Regulation of estrogen receptor beta mRNA in the brain, opposite effects of 17beta-estradiol and the phytoestrogen, coumestrol. *Brain Res. Mol. Brain Res.* **67,** 165–171.

Perez, S. E., Chen, E.-Y., and Mufson, E. J. (2003). Distribution of estrogen receptor alpha and beta immunoreactive profiles in the postnatal rat brain. *Develop. Brain Res.* **145,** 117–139.

Pettersson, K., Grandien, K., Kuiper, G. C., and Gustafsson, J. A. (1997). Mouse estrogen receptor beta forms estrogen response element-binding heterodimers with estrogen receptor alpha. *Mol. Endocrinol.* **11,** 1186–1196.

Pettersson, K., Delaunay, F., and Gustafsson, J. A. (2000). Estrogen receptor beta acts as a dominant regulator of estrogen signaling. *Oncogene* **19,** 4970–4978.

Pittenger, C., and Duman, R. S. (2008). Stress, depression, and neuroplasticity, a convergence of mechanisms. *Neuropsychopharmacology* **33,** 88–109.

Pouyssegur, J., Volmat, V., and Lenormand, P. (2002). Fidelity and spatio-temporal control in MAP kinase (ERKs) signaling. *Biochem. Pharmacol.* **64,** 755–763.

Prange-Kiel, J., Wehrenberg, U., Jarry, H., and Rune, G. M. (2003). Para/autocrine regulation of estrogen receptors in hippocampal neurons. *Hippocampus* **13,** 226–234.

Rachman, I. M., Unnerstall, J. R., Pfaff, D. W., and Cohen, R. S. (1998). Estrogen alters behavior and forebrain c-fos expression in ovariectomized rats subjected to the forced swim test. *Proc. Natl. Acad. Sci. USA* **95,** 13941–13946.

Rasia-Filho, A. A., Fabian, C., Rigoti, K. M., and Achaval, M. (2004). Influence of sex, estrous cycle and motherhood on dendritic spine density in the rat medial amygdala revealed by Golgi method. *Neuroscience* **126,** 839–847.

Reneric, J. P., Bouvard, M., and Stinus, L. (2002). In the rat forced swimming test, chronic but not sub-acute administration of dual 5-HT/NA antidepressant treatments may produce greater effects than selective drugs. *Behav. Brain Res.* **136,** 521–532.

Rissman, E. F. (2008). Roles of oestrogen receptors alpha and beta in behavioural neuroendocrinology, beyond yin/yang. *J. Neuroendocrinol.* **20,** 873–879.

Roozendaal, B., Koolhaas, J. M., and Bohus, B. (1991). Attenuated cardiovascular, neuroendocrine, and behavioral responses after a single footshock in central amygdaloid lesioned male rats. *Physiol. Behav.* **50,** 771–775.

Rosano, G. M., Vitale, C., and Fini, M. (2009). Cardiovascular aspects of menopausal hormone replacement therapy. *Climacteric* **12**(Suppl. 1), 41–60.

Rossouw, J. E., Cushman, M., Greenland, P., Lloyd-Jones, D. M., Bray, P., Kooperberg, C., Pettinger, M., Robinson, J., Hendrix, S., and Hsia, J. (2008). Inflammatory, lipid, thrombotic, and genetic markers of coronary heart disease risk in the women's health initiative trials of hormone therapy. *Arch. Intern. Med.* **168,** 2245–2253.

Salvatori, L., Pallante, P., Ravenna, L., Chinzari, P., Frati, L., Russo, M. A., and Petrangeli, E. (2003). Oestrogens and selective oestrogen receptor (ER) modulators

regulate EGF receptor gene expression through human ER alpha and beta subtypes via an Sp1 site. *Oncogene* **22,** 4875–4881.
Sellix, M. T., Egli, M., Henderson, R. P., and Freeman, M. E. (2004). Ovarian steroid hormones modulate circadian rhythms of neuroendocrine dopaminergic neuronal activity. *Brain Res.* **1005,** 164–181.
Shima, N., Yamaguchi, Y., and Yuri, K. (2003). Distribution of estrogen receptor beta mRNA-containing cells in ovariectomized and estrogen-treated female rat brain. *Anat. Sci. Int.* **78,** 85–97.
Shipiro, R. A., Xu, C., and Dorsa, D. M. (2000). Differential transcriptional regulation of rat vasopressin gene expression by estrogen receptor alpha and beta. *Endocrinology* **141,** 4056–4064.
Shors, T. J., Chua, C., and Falduto, J. (2001). Sex differences and opposite effects of stress on dendritic spine density in the male versus female hippocampus. *J. Neurosci.* **21,** 6292–6297.
Shughrue, P. J., and Merchenthaler, I. (2001). Distribution of estrogen receptor beta immunoreactivity in the rat central nervous system. *J. Comp. Neurol.* **436,** 64–81.
Shughrue, P. J., Lane, M. V., and Merchenthaler, I. (1997). Comparative distribution of estrogen receptor-alpha and beta mRNA in the rat central nervous system. *J. Comp. Neurol.* **388,** 507–525.
Shupnik, M. A. (2002). Oestrogen receptors, receptor variants and oestrogen actions in the hypothalamic-pituitary axis. *J. Neuroendocrinol.* **14,** 85–94.
Singh, M., Meyer, E. M., and Simpkins, J. W. (1995). The effect of ovariectomy and estradiol replacement on brain-derived neurotrophic factor messenger ribonucleic acid expression in cortical and hippocampal brain regions in female Spargue-Dawley rats. *Endocrinology* **136,** 2320–2324.
Soares, C. N., Almeida, O. P., Joffe, H., and Cohen, L. S. (2001). Efficacy of estradiol for the treatment of depressive disorders in perimenopausal women: A double-blind, randomized, placebo-controlled trial. *Arch. Gen. Psychiatry* **58,** 529–534.
Steiner, M., Dunn, E., and Born, L. (2003). Hormones and mood, from menarche to menopause and beyond. *J. Affect. Disord.* **74,** 67–83.
Stoffel, E. C., and Craft, R. M. (2004). Ovarian hormone withdrawal-induced 'depression' in female rats. *Physiol. Behav.* **83,** 505–513.
Suzuki, S., and Handa, R. J. (2004). Regulation of estrogen receptor-beta expression in the female rat hypothalamus, differential effects of dexamethasone and estradiol. *Endocrinology* **145,** 3658–3670.
Sweatt, J. D. (2001). The neuronal MAP kinase cascade, a biochemical signal integration system subserving synaptic plasticity and memory. *J. Neurochem.* **76,** 1–10.
Ter Horst, G. J., Hautvast, R. W., de Jongste, M. J., and Korf, J. (1996). Neuroanatomy of cardiac-activity regulating circuitry, a transneuronal retrograde viral labeling study in the rat. *Eur. J. Neurosci.* **8,** 2029–2041.
Ter Horst, G. J., Wichmann, R., Gerrits, M., Westenbroek, C., and Lin, Y. (2009). Sex differences in stress responses, focus on ovarian hormones. *Physiol. Behav.* **97,** 239–249.
Toran-Allerand, C. D. (2004). Estrogen and the brain, beyond ER-alpha and ER-beta. *Exp. Gerontol.* **39,** 1579–1587.
Trentani, A., Kuipers, S. D., Ter Horst, G. J., and Den Boer, J. A. (2002). Selective chronic stress-induced in vivo ERK1/2 hyperphosphorylation in medial prefrontocortical dendrites, implications for stress-related cortical pathology. *Eur. J. Neurosci.* **15,** 1681–1691.
Vida, B., Hrabovszky, E., Kalamatianos, T., Coen, C. W., Liposits, Z., and Kallo, I. (2008). Oestrogen receptor alpha and beta immunoreactive cells in the suprachiasmatic nucleus of mice, distribution, sex differences and regulation by gonadal hormones. *J. Neuroendocrinol.* **20,** 1270–1277.

Walf, A. A., and Frye, C. A. (2006). A review and update of mechanisms of estrogen in the hippocampus and amygdala for anxiety and depression behavior. *Neuropsychopharmacology* **31,** 1097–1111.

Wang, Z., Zhang, B., Wang, M., and Carr, B. L. (2003). Persistent ERK phosphorylation negatively regulates cAMP response element-binding protein activity via recruitment of CREB-binding protein to pp 90RSK. *J. Biol. Chem.* **278,** 11138–11344.

Watters, J. J., Campbell, J. S., Cunningham, M. J., Krebs, E. G., and Dorsa, D. M. (1997). Rapid membrane effects of steroids in neuroblastoma cells, effects of estrogen on mitogen activated protein kinase signaling cascade and c-fos immediate early gene transcription. *Endocrinology* **138,** 4030–4033.

Weiser, M. J., Foradori, C. D., and Handa, R. J. (2008). Estrogen receptor beta in the brain, from form to function. *Brain Res. Rev.* **57,** 309–320.

Weissman, M. M., and Olfson, M. (1995). Depression in women, implications for health care research. *Science* **269,** 799–801.

Wintermantel, T. M., Campbell, R. E., Porteous, R., Bock, D., Grone, H. J., Todman, M. G., Korach, K. S., Greiner, E., Perez, C. A., Schutz, G., and Herbison, A. E. (2006). Definition of estrogen receptor pathway critical for estrogen positive feedback to gonadotropin-releasing hormone neurons and fertility. *Neuron* **52,** 271–280.

Woolley, C. S., Gould, E., Frankfurt, M., and McEwen, B. S. (1990). Naturally occurring fluctuation in dendritic spine density on adult hippocampal pyramidal neurons. *J. Neurosci.* **10,** 4035–4039.

Wu, G. Y., Deisseroth, K., and Tsien, R. W. (2001). Activity-dependent CREB phosphorylation, convergence of fast, sensitive calmodulin kinase pathway and a slow, less sensitive mitogen-activated protein kinase pathway. *Proc. Natl. Acad. Sci. USA* **98,** 2808–2813.

CHAPTER EIGHTEEN

# Corticotropin-Releasing Hormone and Arginine Vasopressin in Depression: Focus on the Human Postmortem Hypothalamus

Ai-Min Bao* *and* Dick F. Swaab†

## Contents



## Abstract

The neuropeptides corticotropin-releasing hormone (CRH) and arginine vasopressin (AVP) are crucially involved in the pathogenesis of depression. The close correlation between the etiology of depression and dysregulation of the stress responses is based upon a hyperactivity of the hypothalamo–pituitary–adrenal (HPA) axis. CRH neurons in the paraventricular nucleus are the motor of the HPA-axis. Centrally released CRH, AVP, and increased levels of cortisol all contribute

* Department of Neurobiology, Institute of Neuroscience, Zhejiang University School of Medicine, Hangzhou, Zhejiang, People's Republic of China
† Netherlands Institute for Neuroscience, An Institute of the Royal Netherlands Academy of Arts and Sciences, Amsterdam, The Netherlands

*Vitamins and Hormones*, Volume 82
ISSN 0083-6729, DOI: 10.1016/S0083-6729(10)82018-7

to the signs and symptoms of depression. Single-nucleotide polymorphisms in the CRH and AVP receptor genes are associated with the risk for depression. Activation of the HPA-axis is generally regarded to be the final common pathway of the pathogenesis of depression. Sex hormones are crucially involved in the regulation of CRH gene expression. The decreased activity of the biological clock, the suprachiasmatic nucleus, as indicated by its lower AVP expression, is the basis for the disturbed rhythms in depression. Both similarities and differences are found in the activity changes in the CRH and AVP systems in depressive disorders and depression in Alzheimer's disease. 

# I. Introduction

## A. Neuroendocrine cells in the human hypothalamus

There are three types of neuroendocrine cells in the human hypothalamus that play a role in the integration of endocrine, autonomic, and higher brain functions: (i) the small parvocellular neuroendocrine neurons, for example, in the paraventricular nucleus (PVN), that release their peptides, such as corticotropin-releasing hormone (CRH) and thyroxine-releasing hormone (TRH), into the portal capillaries and further convey to the anterior lobe of the pituitary; (ii) the large magnocellular neurons of the supraoptic nucleus (SON) and the PVN, which produce arginine vasopressin (AVP) and oxytocin (OXT), both transported by their axons to the neurohypophysis; and (iii) a part of the peptidergic cells, that project to other neurons, where their peptides, such as CRH, AVP, and OXT, are synaptically released and act as neurotransmitters/neuromodulators, influencing a great variety of functions such as sexual behavior, social behavior, the regulation of stress, and metabolism (Swaab, 2004a,b). In this way, the centrally projecting peptidergic neurons also regulate the autonomic nervous system. An example of the centrally projecting peptidergic hypothalamic neurons is the biological clock, the suprachiasmatic nucleus (SCN). AVP and vasoactive intestinal polypeptide containing SCN fibers terminate throughout the hypothalamus (Dai *et al.*, 1997, 1998a,b). The SCN also regulates the pineal gland in this way, via a polysynaptic autonomic pathway (Buijs and Kalsbeek, 2001). Although in the human brain, CRH is produced in the PVN and not in the SON and SCN, AVP is produced in all these three nuclei. Neurons of the PVN and SCN are projecting centrally (Swaab, 2003). It should be noted, however, that so far we still lack the methods to establish whether the "third" type of neurons, that is, the centrally projecting neurons, are a unique subtype or that they are part of the population of parvocellular neurons that also release peptides into the bloodstream of the median eminence or neurohypophysis.

Because of the use of selective serotonin reuptake inhibitors (SSRIs) in clinics, the focus on the pathogenesis of depression has been on the monoaminergic, especially serotonergic, system during the last four decades. While the monoamine theory, that postulates dysfunctional noradrenergic and serotonergic systems as the underlying cause of depression, has been valuable in the development of antidepressants that are thought to act by reversing these dysfunctional states, recent clinical and experimental studies have questioned this reductionist view of depression (Leonard, 2007). It has now become clear that dysregulation of the monoaminergic systems is neither a sufficient nor a necessary cause of depression. On the other hand, a wealth of data indicates that neuropeptides, such as CRH, AVP, OXT, TRH, and neuropeptide Y (NPY), are crucially involved in the clinical signs and symptoms of depression (Bao *et al.*, 2008; Mathe *et al.*, 2007; Swaab, 2004a,b). This review focuses on the two classical neurohormones, CRH and AVP, which also have central projections and are now considered to play a leading role in the pathophysiology of depression.

## B. From stress response to depression

The rationale to make CRH the focus of depression studies was due to the correlation that was found between the dysregulation of the stress response and mood disorders, especially depression by Hans Selye in the 1930s (Selye, 1998). It is well known now that although the stress response is necessary to maintain homeostasis of the body, long-term activation of the stress system brings hazardous or even lethal effects, and increases the risk of depression (Selye, 1998). Abnormalities in the "stress system" have been documented in at least a subset of depressed individuals (Holsboer, 2001; Young *et al.*, 1991, 2003). We have learned that stress is not simply a trigger that can be neatly separated from the disease process. The neural network that encodes and evaluates the stressful event is the very same brain circuit that, when it is hyperactive due to a combination of a genetic background, developmental sequelae and life events, underlies negative emotions and moods (Akil, 2005; Swaab, 2004a,b). Major depression (MD) thus arises from the interaction of vulnerability genes and developmental and environmental factors (Swaab *et al.*, 2005). Single-nucleotide polymorphisms (SNPs) in a series of genes, such as in the CRH receptor1 (CRHR1) gene, or in the AVP receptor (AVPR) V1b gene, have been found to be either associated with increased susceptibility to MD (Liu *et al.*, 2006b) or to protect against recurrent MD (van West *et al.*, 2004). Prenatal environmental stressors, such as placental insufficiency, food shortage, or nicotine exposure due to smoking by the pregnant mother, may sensitize the child for developing depression in later life (Clark, 1998; Swaab, 2004a, b). Psychosocial stress, such as early maternal separation, child abuse or neglect, death or loss of a spouse, personal injury or illness, forms risk factors, or triggers the early episodes of depression (Bao *et al.*, 2008; Swaab *et al.*, 2005).

## II. The HPA-Axis as the Final Common Pathway in Depression

Brain areas that are closely involved in a stress response include the brainstem, which is the first relay station for many physiological stressors; the amygdala, which processes fear and anxiety responses; the hippocampus, which mediates learning and memory and encodes the importance of a stimulus to the organism; and the prefrontal cortex (PFC), which holds cognitive and executive functions (Drevets *et al.*, 2008; Swaab *et al.*, 2000). The stress-induced neural activation converges on the hypothalamus, and the hypothalamo–pituitary–adrenal (HPA) axis is regarded to be the final common pathway in the mediation of the stress response (Bao *et al.*, 2008; Swaab, 2004a,b). When activated, the HPA-axis stimulates the synthesis and release of corticosteroids—cortisol in humans and corticosterone in rodents, which normally have broad biological effects throughout the body that are adaptive but can become damaging when chronically elevated. Cortisol exerts negative feedback at the pituitary and the hypothalamic level via two types of receptors, that is, the mineralocorticoid receptor (MR) and the glucocorticoid receptor (GR), to turn off the stress responses after the threat has passed. The termination of the stress response is as important as its initiation, since superfluous cortisol can cause broad endocrine dysregulation and may lead to neurodegeneration (Swaab *et al.*, 2005). Moreover, quite a few observations have pointed to a causal role for glucocorticoids in at least some subset of mood disorders, that is, atypical depression (Bao *et al.*, 2008).

Environmental and genetic risk factors for depression correlate with increased HPA-axis activity in adulthood. Small size at birth is associated with an alteration in the set-point of the HPA-axis, with an increased cortisol responsiveness, and with risk of depression in adulthood (Phillips 2001; Thompson *et al.*, 2001). Childhood abuse may also predispose individuals to adult onset depression accompanied by a permanent hyperactivity of the HPA system (Tarullo and Gunnar 2006). HPA-axis hyperactivity in depression is clear at all levels. Elevated cortisol levels, disturbed dexamethasone suppression, altered corticosteroid receptor function, enhanced adrenal response to adrenocorticotropin (ACTH), blunted pituitary ACTH response to CRH as well as adrenal and pituitary enlargement have been found in patients suffering from depression. Studies have also shown that abnormalities in HPA-axis function already exist prior to the onset of the clinical symptoms in high-risk probands of MD patients, suggesting that such abnormalities not only correlate but can also precipitate depressive episodes (Holsboer, 2000). Increased ACTH secretion occurred in depressed in-patients regardless of their cortisolemic status, confirming the presence of a central HPA-axis-overdrive in severe depression (Carroll *et al.*, 2007). Studies in patients with psychotic depression also support this concept, since treatment aimed directly at interfering with the consequences of HPA-axis abnormalities, for instance, by high doses of GR antagonist, can reverse the clinical symptoms (Belanoff *et al.*, 2002). When patients or animal models for depression are treated with antidepressants or electroconvulsive therapy, or when patients show spontaneous remission, the HPA-axis function returns to normal (Nemeroff, 1996).

## A. CRH in depression

CRH-expressing neurons in the hypothalamic PVN are the central driving force in regulating the activity of the HPA-axis. CRH is widely believed to mediate stress-induced behaviors and to play a broader, integrative role in the psychological stress response and in depression. The number of CRH neurons, the number of CRH neurons coexpressing AVP, and the amount of CRH mRNA in the PVN, are significantly increased in depressed subjects (Raadsheer *et al.*, 1994a, 1995; Wang *et al.*, 2008). That hyperactive CRH neurons get involved in the etiology of depression is demonstrated not only by the activation of the HPA-axis, which results in the hypersecretion of glucocorticoids, but also via the central CRH effects, including cardiovascular regulation, respiration, appetite control, stress-related behavior and mood, cerebral blood flow regulation, and stress-induced analgesia (for review, see Swaab, 2003). Similar symptoms, such as decreased food intake, decreased sexual activity, disturbed sleep and motor behavior, and increased anxiety, can all be induced in experimental animals by the intracerebroventricular injection of CRH (Holsboer 2001). During postnatal development of the stress system, CRH controls the activity of the HPA-axis and mediates the effects of early disturbances such as maternal deprivation through the CRH-R1 (Schmidt *et al.*, 2006). Both the basic and the clinical studies suggest that disrupting the CRH signaling through CRH-R1 can ameliorate stress-related clinical conditions. It was found in transgenic mouse models that CRH overexpression in the entire central nervous system resulted in stress-induced hypersecretion of stress hormones and increased active stress-coping behavior. These changes were related to the acute effects of overexpressed CRH as they were normalized by CRH-R1 antagonist treatment (Lu *et al.*, 2008a,b). In addition, SNPs in the CRHR1 gene were found to be associated with increased susceptibility to MD (Liu *et al.*, 2006b). It is also of interest to notice that mice lacking the CRH gene exhibit normal stress-induced behavior that is specifically blocked by a CRH-R1 antagonist. Since the other known mammalian ligand for CRH receptors is urocortin, while in these animals, urocortin mRNA is absent from regions known to mediate stress-related behaviors, an unidentified alternative CRH-like molecule other than CRH or urocortin, was proposed to act through the CRH receptors in brain regions to mediate stress-induced behaviors, either alone or in concert with CRH (Weninger *et al.*, 1999). This possibility has to be investigated further.

Antidepressant drugs were found to attenuate the synthesis of CRH by stimulation and/or upregulation of GR expression, which renders the HPA system more susceptible to feedback inhibition by cortisol (Barden 1996). The clinical observation that patients with treatment-resistant depression noticed a significant improvement in mood after receiving dexamethasone while remaining on their antidepressant (sertraline or fluoxetine) further supports the concept that hyperactive CRH neurons play a causal role in the symptomatology of depression (Dinan *et al.*, 1997). Moreover, the CRH concentrations in cerebrospinal fluid (CSF) in healthy volunteers and depressed patients decrease due to antidepressant drugs (Heuser *et al.*, 1998), although it should be noted that CSF-CRH is also derived from other brain areas, such as the thalamus (Bao *et al.*, 2005; Hsu *et al.*, 2001). Lastly, significantly increased CRH mRNA levels in the PVN of the depressed patients were found to be accompanied by a significantly increased expression of genes involved in the activation of CRH neurons, such as CRHR1, MR, estrogen receptor-alpha (ER-$\alpha$), and AVPR1a, and with significantly decreased expression of genes involved in the inhibition of CRH neurons, such as the androgen receptor (AR) mRNA (Wang *et al.*, 2008). These findings further raise the possibility that a disturbed receptor balance in the PVN may contribute to the activation of the HPA-axis in depression. Together, the aforementioned arguments have led to the *CRH-hypothesis of depression*, that is, the hyperactivity of CRH neurons, and thus, of the HPA-axis, is of crucial importance for the induction of the symptoms of depression.

## B. Sex difference in depression: Relationship with HPA-axis activity

It should be noted that there is also additional evidence supporting the *CRH-hypothesis of depression*. A good example is the sex difference found in depression. Indeed, studies in rodents have found that the female brain's innate strategy to handle stress differs from that of the male brain (Ter Horst *et al.*, 2009). Since MD is twice as prevalent in women of reproductive age as in men, organizing and/or activating effects of sex hormones on the HPA-axis are proposed as risk factors for depression. The possible importance of fluctuating levels of sex hormones as a risk factor for depression is underlined by the higher prevalence of premenstrual depression, antepartum, or postpartum depression, and depression during the transition to menopause (Bao *et al.*, 2004, 2008). In addition, the organizing effects of estrogens during fetal life may also be responsible for a higher prevalence of mood disorders, as appeared in children that were exposed *in utero* to diethylstilbestrol (Meyer-Bahlburg and Ehrhardt, 1987). Indeed, age-related sex differences have been found in CRH-expressing neurons in the human hypothalamic PVN, which illustrate a relationship between sex hormones and CRH: (1) From the age of 24 years onward, men had significantly more CRH neurons than women while (2) there was a significant age-related increase of CRH neurons in men, but not in women (Fig. 18.1). Moreover, abnormal hormone status, induced by castration, ovariectomy, or a sex hormone-producing tumor, was accompanied by changes in the number of CRH-expressing neurons (Fig. 18.2) (Bao and Swaab, 2007). It should be noted that sex hormones can also be synthesized locally in the brain as "neurosteroids," which may interact with circulating sex hormones by thus far unknown mechanisms. Both sources of sex hormones act by their binding to sex hormone receptors that are expressed in CRH neurons of the hypothalamus and in corticotropes of the adenohypophysis (Bao *et al.*, 2005, 2006; Pereira-Lima *et al.*, 2004; Scheithauer *et al.*,

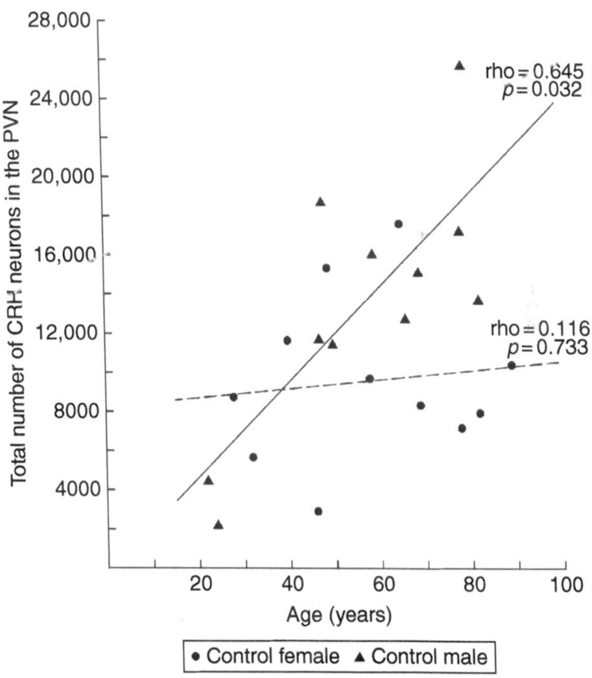


**Figure 18.1** The total number of corticotropin-releasing hormone (CRH)-immunoreactive neurons in the hypothalamic paraventricular nucleus (PVN) of control males (▲) and females (●). The control males had significantly ($p = 0.004$) more CRH neurons than control females from age 24 onward and showed a significantly positive correlation (the solid line) between age and total number of CRH neurons. The control female group did not show significant correlation (the dashed line) between age and total number of CRH neurons. (From Bao and Swaab, 2007; Fig. 2, with permission.)

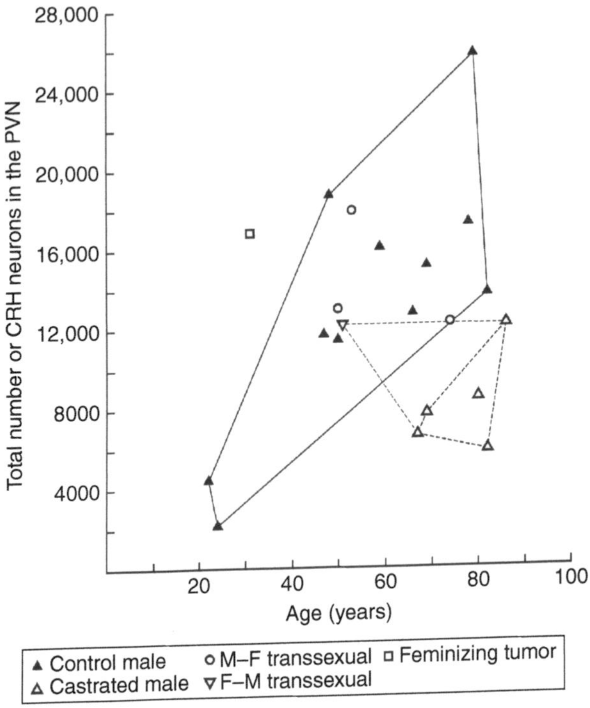


**Figure 18.2** The total number of CRH-immunoreactive neurons in the hypothalamic PVN of control males and subjects with abnormal sex hormone status. Values of control-male-group (▲, solid line), castrated-male-group (Δ, dashed line), and "extended group," that is, the castrated-male-group (Δ) plus an ovariectomized female-to-male (F–M) transsexual (∇, dashed line), are delineated by a minimum convex polygon. Note that the total number of CRH neurons in the PVN of the five castrated males ($n = 5$, age $\geq 67$) is significantly ($p = 0.008$) lower than that of the matched old control males ($n = 5$, age $\geq 66$). Such a significant difference remained ($p = 0.009$) when the ovariectomized F–M transsexual was included in the "extended group" ($n = 6$, age $\geq 51$) compared with the matched control males ($n = 6$, age $\geq 50$). The total numbers of CRH neurons in the three M–F castrated-with-estrogen-replacement M–F transsexuals (○) were significantly larger than those in the castration group ($p = 0.036$) and the "extended group" ($p = 0.024$), while there was no significant difference when compared with age-matched control males (age 50–70, $n = 6$, $p = 0.905$; or age 50–78, $n = 5$, $p = 1.000$). The 31-year-old male with an estrogen-producing adrenal tumor (□) had a very high total number (16,832) of CRH neurons in the PVN. M–F transsexual: male-to-female transsexual; F–M transsexual: female-to-male transsexual; Feminizing tumor: estrogen-producing adrenal tumor. (From Bao and Swaab, 2007; Fig. 3, with permission.)

2008; Stefaneanu *et al.*, 1994). We have found colocalization of CRH and sex hormone receptors, indicating a direct effect of sex hormones on CRH neurons. Both nuclear ERα (Bao *et al.*, 2005) (Fig. 18.3) and nuclear AR (Bao *et al.*, 2006) (Fig. 18.4) were present in CRH-expressing neurons in

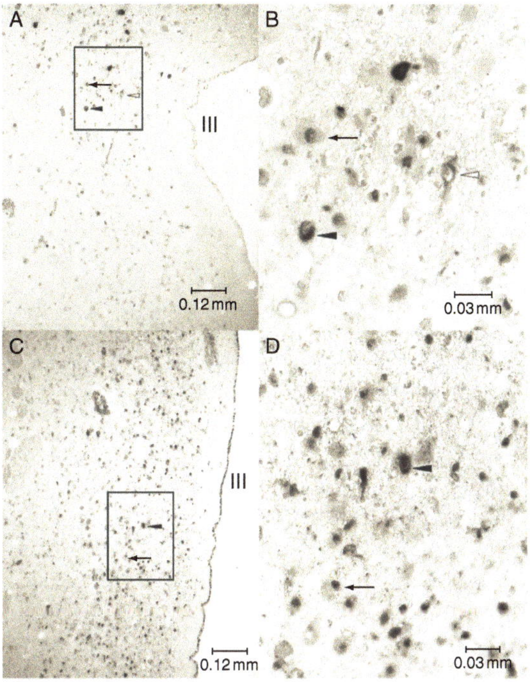


**Figure 18.3** Frontal section of the PVN in a control (C12) (A, B) and a patient with mood disorder (D10) (C, D) stained for CRH (blue) and ERa (red). (B) and (D) represent a 4× higher magnification of (A) and (C). The arrows, solid and hollow arrowheads in (A, B) and (C, D) indicate the same place in the preparation to facilitate comparison. Both sections show the central part (mid-level) of the PVN and contain the largest number of stained neurons. It is clear by comparing (A) with (C) and (B) with (D), that the number of stained neurons is markedly increased in this mood disorder patient. III: the third ventricle. The arrow points to an ERa nuclear single-staining cell; the solid arrowhead points to a cytoplasmic CRH-ERa nuclear double-staining cell and the hollow arrowhead points to a CRH single-staining cell. (From Bao *et al.*, 2005; Fig. 2, with permission.) (See Color Insert.)

the human hypothalamic PVN. In addition, there was a significantly positive correlation between the increased number of CRH neurons containing nuclear ER$\alpha$ and the increased number of CRH neurons in mood disorders, both in males and females (Bao *et al.*, 2005). It is known that the human CRH gene promoter contains five perfect, half-palindromic estrogen-responsive elements (EREs) (Vamvakopoulos and Chrousos, 1993), while animal studies have shown that estrogens stimulate CRH expression (Lund *et al.*, 2004). Thus, human brain material, animal, and cell-line studies confirm a key stimulating role of estrogens on CRH production. We have also identified an androgen-responsive element (ARE) in the CRH gene promoter region that initiates a repressing effect of AR on CRH expression

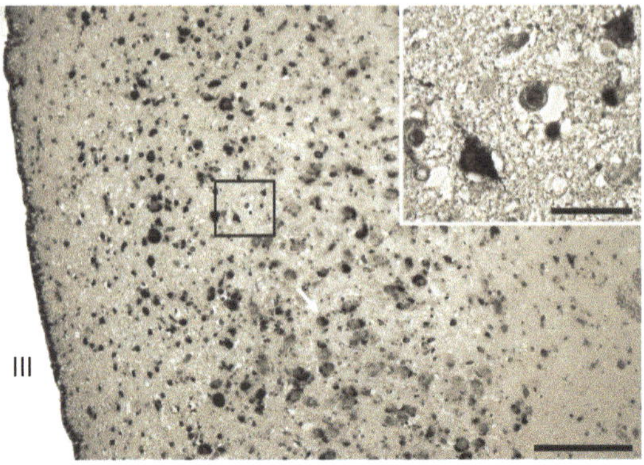


**Figure 18.4** Frontal section of the PVN in subject (#00182) stained for CRH (red) and androgen receptor (AR) (blue). III: the third ventricle. The upper-right corner represents a higher magnification of the framed field and shows cytoplasmic CRH (red)–AR (blue) nuclear double-staining neurons. The arrow points to some AR single-staining cells. Bar in the upper-right corner = 16 μm; in the lower right corner = 100 μm. (From Bao *et al.* 2006; Fig. 4, with permission.) (See Color Insert.)

(Bao *et al.*, 2006), which is in agreement with an animal study showing that androgens inhibit CRH production (Lund *et al.*, 2004). This opposite effect of estrogens and androgens on CRH neurons may be the basis for the sex difference in the prevalence of depression.

Finally, concerning the *CRH-hypothesis of depression*, it should be noted that although some negative neuroendocrine findings (Brunner *et al.*, 2001; Roy, 1996) seem to oppose the idea that CRH plays a key role in the pathogenesis of depression, one should realize that the cortisol measurements leading to this conclusion relate only to the CRH neurons that project via the median eminence to the periphery. Actually, the *CRH-hypothesis of depression* may still hold for these cases, since a subgroup of CRH neurons that project to brain areas and not to the median eminence may be activated in depression and induce the symptoms. Such a subgroup of CRH neurons may not be monitored endocrinologically in the periphery.

## III. The AVP Hypothesis of Depression

Even after decades of intense research, new data are still emerging in relation to the classic antidiuretic nonapeptide AVP. Besides their role in the regulation of osmolality, blood pressure, temperature, and corticosteroid secretion, the central vasopressinergic fibers are also involved in the stress response, cognition, paternal behavior and social attachment, and in emotionality (de Wied and van Ree, 1982; Insel, 1997; Swaab, 2004a,b). Intranasal AVP differentially affects social communication in men and women (Thompson *et al.*, 2006), polymorphisms in the AVPR1a were found to be related to the autism spectrum (Yirmiya *et al.*, 2006) and to pair-bonding behavior in men and thus, in the evolutionary development of the social brain (Walum *et al.*, 2008).

## A. Different vasopressin systems

There are at least four different vasopressinergic systems intimately involved in the signs and symptoms of depression (for reviews, see Swaab, 2003, 2004a,b): (i) AVP is produced as a neurohormone by the large magnocellular neurons of the hypothalamic SON and PVN, axons of which run to the neurohypophysis where they release AVP and OXT into the general circulation. Circulating AVP has an influence on the anterior pituitary and high circulating levels also affect mood. (ii) Small parvocellular neurons of the PVN secrete CRH and AVP also as neurohormones from their axons in the median eminence into the portal capillaries that transport them to the anterior lobe of the pituitary. AVP strongly potentiates the ACTH-releasing activity and eventually, the production of corticosteroids from the adrenal gland (Engelmann *et al.*, 2004). In addition, (iii) vasopressinergic fibers are found to project from the hypothalamus to the subregions of the hippocampus, septum, amygdala, and brainstem areas, where AVP serves as a neurotransmitter/neuromodulator via AVPR1a and AVPR1b receptors that are widely distributed (Surget and Belzung, 2008). Moreover, particularly the magnocellular hypothalamic neurons release AVP from their dendrites and somata, subsequently diffusing through the brain's extracellular fluid to act as neuromodulators on receptors at some distance from their site of release (Ludwig *et al.*, 2005). (iv) AVP is also released into the brain with a circadian rhythm by neurons of the SCN, the biological clock of the hypothalamus, which shows significant changes in depression. Once overexpressed and overreleased, AVP may contribute to hyperanxiety and depression-like behaviors, while AVP deficit, in turn, may cause signs of diabetes insipidus, hypoanxiety, (Inder *et al.*, 1997; Landgraf *et al.*, 2007; Mlynarik *et al.*, 2007) and disturbed rhythmicity (Zhou *et al.*, 2001).

## B. Chronic stress and depression

As mentioned before, a multiple misbalance of receptor genes that are involved in the regulation of the HPA-axis activity has been proposed in depression (Wang *et al.*, 2008), in which the increased AVPR1a suggests a role of enhanced somatodendritic AVP release in the PVN (Surget and Belzung, 2008). Increased co-storage of CRH and AVP in the CRH terminals was found after immobilization or social defeat stressors

(Bartanusz *et al.*, 1993; de Goeij *et al.*, 1991; Keeney *et al.*, 2006). When released together into the portal capillaries, AVP strongly potentiates the ACTH-releasing activity (Gillies *et al.*, 1982; Rivier and Vale, 1983). In addition, circulating AVP from the SON may induce ACTH release from the pituitary (Gispen-de Wied *et al.*, 1992). Transient activation of the HPA-axis following a single exposure to a stressor may induce delayed and long-lasting hyperproduction, hyperstorage, and hypersecretion of AVP, for example, from hypothalamic CRH neurons, which results in hyperresponsiveness of the HPA-axis to subsequent stimuli (Schmidt *et al.*, 1995, 1996). Such sensitization of neuronal processes is proposed to be an important feature in promoting depressive-like states. It was suggested that with each stressor experience and with each successive episode of depression, neuronal sensitization becomes more pronounced, and hence, the stressor severity necessary to elicit the neurochemical changes (and thus, to induce a depressive episode) becomes progressively smaller (Post, 1992). It is also important to note that, whereas in acute stress CRH is the main cause of increased ACTH release, animal experiments show that in chronic stress there is a switch from CRH to AVP stimulation of ACTH release (Scott and Dinan, 1998); since depression is a chronic disorder, the AVP-driven HPA-axis-hyperactivity in depression is receiving more and more attention. In case of chronic depression, AVP is persistently increased within CRH neurons, so that even minor day-to-day annoyances might trigger excessive CRH/AVP release. This, in turn, would favor the presentation of dysphoric symptoms, and might even be a factor responsible for triggering MD and dysthymia (Griffiths *et al.*, 2000). The number of CRH and AVP-colocalizing neurons is indeed increased in the human PVN in depression and during the course of aging, at least in males (Bao and Swaab, 2007; Raadsheer *et al.*, 1993, 1994b).

CRH and AVP mediate ACTH release via different second messenger systems. CRH activates G-protein-linked adenylate cyclase, leading to cAMP formation and protein-kinase-A activation. AVP works through a specific AVP receptor subtype termed AVPR1b, which is strongly expressed by pituitary corticotropes and activates only phospholipase C. The AVPR1b receptor is required for a normal pituitary and adrenal response to some acute stressful stimuli and is necessary for a normal ACTH response during chronic stress (Lolait *et al.*, 2007). AVPR1b receptor-mRNA levels and coupling of the receptor to phospholipase C are stimulated by glucocorticoids, while AVP facilitates corticotrope responsiveness in spite of the elevated levels of plasma glucocorticoids during chronic stress or depression (Rabadan-Diehl and Aguilera, 1998). In this respect, it is interesting that following a challenge with desmopressin (an analog of vasopressin that stimulates the AVPR1b receptor), the ACTH and cortisol responses were appreciably greater in MD patients than in controls, suggesting that the AVPR1b receptor was more reactive in depression (Dinan *et al.*,

2004). A major SNP haplotype of the AVPR1b receptor has been found to protect against recurrent MD (van West *et al.*, 2004), while evidence was also found for the involvement of SNPs of AVPR1b in childhood-onset mood disorders, particularly in females (Dempster *et al.*, 2007), supporting the possibility of a direct involvement of AVP in the pathogenesis of depression. A recent study on AVP-deficient Brattleboro rats has shown that AVP is involved in the development of depression-like behavior, in particular of the coping style and anhedonia. In addition, behavioral and endocrine responses were found to be dissociated, which suggested that brain vasopressinergic circuits—distinct from those regulating the HPA-axis—are involved in generating depression-like behavior (Mlynarik *et al.*, 2006). The identification of genetic polymorphisms underlying a special subtype of depression not only explains individual variation in social memory and emotionality, but may also help to characterize potential targets for therapeutic interventions (Frank and Landgraf, 2008).

Feedback of corticosteroids takes place on the PVN, SON, and SCN. Following different types of corticosteroid treatment in different disorders or during the presence of high levels of endogenous corticosteroids produced by a tumor, we found—in postmortem tissue—not only that CRH-expressing neurons are strongly downregulated, but also that AVP expression in the SON and PVN is strongly decreased. On the other hand, OXT neurons were not affected, which further illustrates that, in the human brain, selective negative feedback of corticosteroids is present in CRH cells and in cells that coexpress AVP. The glucocorticoid-induced suppression of AVP-synthesis has been proposed to occur at the posttranscriptional level (Erkut *et al.*, 1998, 2002). In the human SCN, we found a diminished AVP mRNA following the administration of corticosteroids (Liu *et al.*, 2006a), which may be one explanation for the disturbed circadian rhythms in depression.

## C. AVP in the SON and PVN in depression

In the PVN of MD patients, the number of AVP- and OXT-expressing neurons is increased (Purba *et al.*, 1996). Using radioactive *in situ* hybridization, our group determined the amount of AVP mRNA in the PVN and SON in formalin-fixed, paraffin-embedded archival postmortem brain tissues of depressed subjects and their controls (Fig. 18.5). In the SON, a 60% increase of AVP mRNA expression was found in depressed subjects as compared with control ones (Fig. 18.6A), but AVP mRNA expression was only significantly increased in both the SON and the PVN in the melancholic subgroup, compared with control subjects (Meynen *et al.*, 2006) (Fig. 18.6B). Recently, we performed laser microdissection and real-time PCR in the PVN and the SON in snap-frozen postmortem human hypothalamic tissue of depressed patients and their matched controls

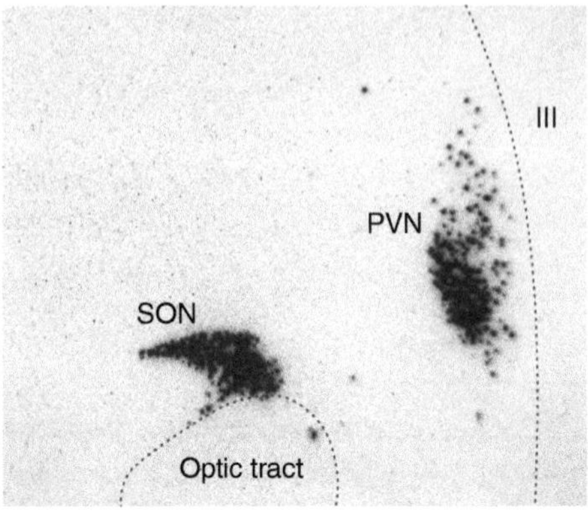


**Figure 18.5** Radioactive *in situ* hybridization vasopressin mRNA signal on film of the supraoptic nucleus (SON) and PVN of a representative patient. Note the intense signal in both nuclei (III: third ventricle). (From Meynen *et al.*, 2006; Fig. 2, with permission.)

(Fig. 18.7), while no significant difference of AVP mRNA expression was found between the SON or the PVN of controls and depressed patients (Wang *et al.*, 2008). However, it was noticed that there was only one patient who had been diagnosed with the melancholic type of depression in the latter study, which fully explained the lack of significant change of AVP mRNA in the depressed subjects (Wang *et al.*, 2008). Enhanced AVP mRNA production in the SON of depressed patients (Meynen *et al.*, 2006) leads to increased plasma levels of AVP (van Londen *et al.*, 1997, 1998b, 2001) that are also reported to be related to an enhanced suicide risk in depression (Inder *et al.*, 1997), as well as to anxious-retarded depression (motivational inhibition) (de Winter *et al.*, 2003; Goekoop *et al.*, 2006). The relationship between enhanced AVP and suicide risk, however, was not reflected in plasma neurophysin levels (Pitchot *et al.*, 2008). It has also been proposed that AVP may not only be important in mediating the psychomotor retardation but also in affecting memory processes in depressed patients, possibly by altering arousal and attention (van Londen *et al.*, 1998a,b). The possibility that circulating AVP may also induce symptoms of depression is supported by the case-story of a man displaying chronically elevated plasma AVP levels due to AVP secretion by an olfactory neuroblastoma, which induced his first episode of MD. Depressive symptoms improved markedly after surgical resection of the tumor and subsequent normalization of plasma AVP levels. In this patient, the HPA-axis was suppressed (Muller *et al.*, 2000).

In addition, in depression and suicide, centrally released AVP from the PVN may play a role, since immunoreactive AVP was elevated in the dorsomedial PFC and reduced in the dorsal vagal complex (Merali *et al.*, 2006).

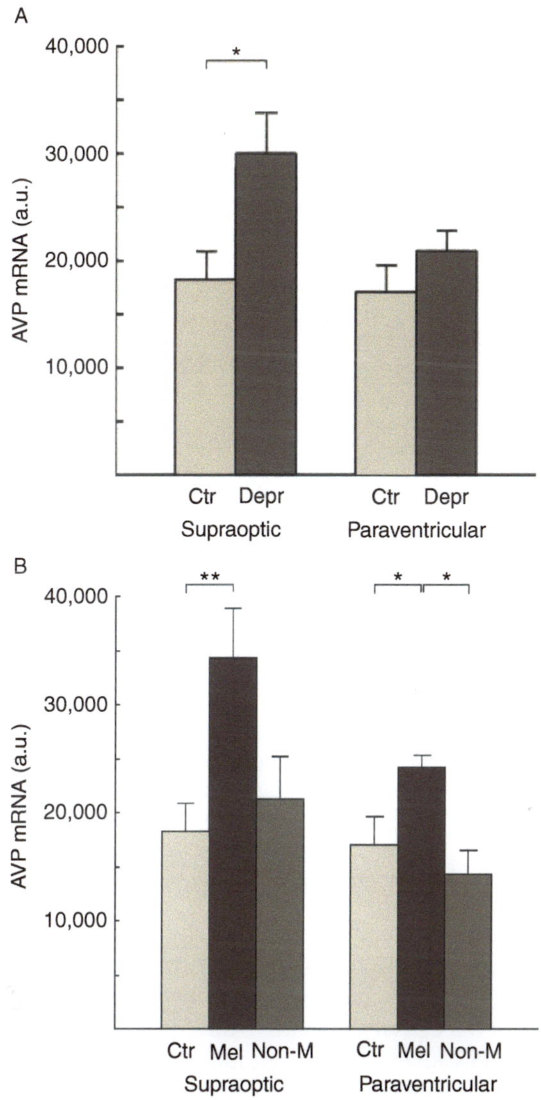


**Figure 18.6** (A) Amount of vasopressin (AVP) mRNA signal in the SON and PVN in depressed subjects (Depr, $n = 9$) and controls (Ctr). (B) Amount of AVP mRNA signal in the SON and PVN in melancholic subgroup (Mel, $n = 6$), nonmelancholic subgroup (Non-m) and controls (Ctr). *Statistically significant difference at $p < 0.05$, **$p < 0.01$. Bars show means, error bars indicate the SEM; a.u., arbitrary units. (A) From Meynen *et al.* (2006); Fig. 1, with permission. (B) is kindly offered by Dr. G. Meynen, results from the study of Meynen *et al.* (2006).

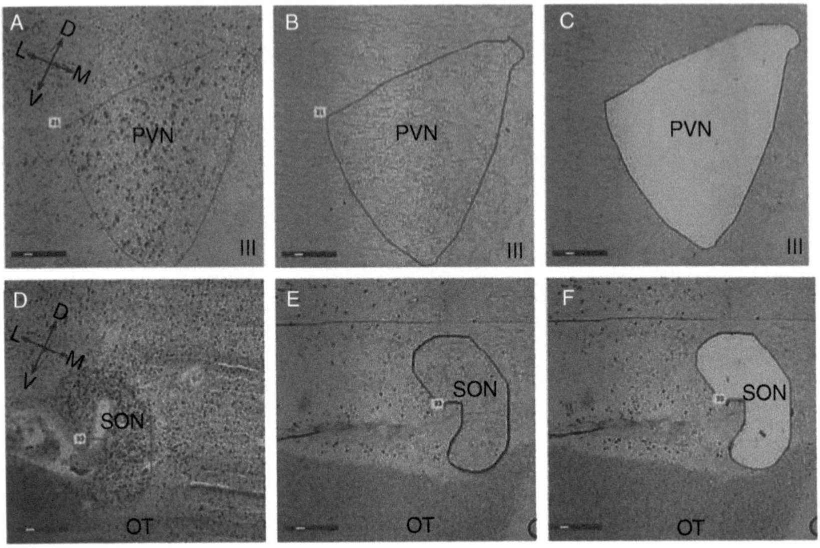


**Figure 18.7** The sections for laser microdissection (LMD). Sections of the PVN (A–C) and SON (D–F) at the right side of hypothalamus as seen under the PALM laser-dissection microscope. A thionin-stained section of the PVN for the orientation is shown in panel (A). An unstained section adjacent to panel (A) is represented in panel (B) before LMD, in which the PVN area is outlined under the microscope. Section (B) is represented in panel (C) after laser dissection. The sections of the right SON under the PALM laser-dissection microscope with the same LMD procedure as for the PVN are represented in panels (D)–(F). Bar = 300 $\mu$m. The arrows show the orientation: V, ventral; D, dorsal; M, medial; L, lateral; OT, optic tract; and III, third ventricle. (From Wang *et al.*, 2008; Fig. 1, with permission.)

## D. AVP in the circadian system in depression

The stress response is strongly influenced by the time of the day. The hypothalamic SCN, the biological clock, is responsible for the rhythmic changes of the stress system. It shows not only circadian but also circannual variations in neuronal activity (Hofman and Swaab, 1992, 1993), supposed to be related to circadian and circannual fluctuations in mood and to sleeping disturbances in depression (van Londen *et al.*, 2001). Polymorphisms in the clock genes appeared to be associated with mood disorders, susceptibility, and dysfunctional circadian rhythm (Johansson *et al.*, 2003; Kishi *et al.*, 2009; Shi *et al.*, 2008), indicating that the SCN may also play a causal role in depression. A disorder of SCN function, characterized by an increased number of AVP-expressing neurons, a decreased amount of AVP mRNA in this nucleus, and diminished circadian fluctuation of AVP mRNA, has been found in depressed patients (Zhou *et al.*, 2001) (Fig. 18.8). Decreased activity of the SCN in depression is presumed to be at least partly due to the increased circulating plasma cortisol levels, since

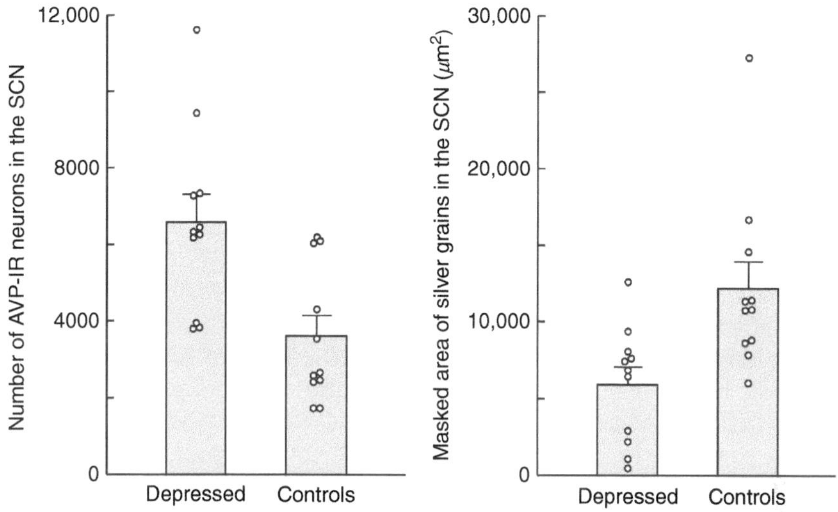


**Figure 18.8** The number of arginine vasopressin-immunoreactive (AVP-IR) neurons (A) and the mask area of silver grains of the AVP mRNA (B) in the suprachiasmatic nucleus (SCN) in control subjects ($n = 11$) and depressed subjects ($n = 11$). The error bars indicate the SD. Note the change in the balance between the presence of more AVP and less AVP mRNA in depression. There is probably a disorder of the transport of AVP that leads to the accumulation of the peptide, in spite of the decreased production rate. (From Zhou *et al.*, 2001; Fig. 2, with permission.)

corticosteroids were found to inhibit SCN function (Liu *et al.*, 2006a). Light therapy was found to be effective not only for the seasonal subtype but also for MD, while as an adjuvant to conventional antidepressants in unipolar patients, or lithium in bipolar patients, morning light hastens and potentiates the antidepressant response (Wirz-Justice *et al.*, 2005). Animal data have shown that AVP neurons of the SCN exert an inhibitory influence on CRH neurons in the PVN (Kalsbeek *et al.*, 1992). It seems thus logical to propose that in depressed patients with a deficient SCN, stress will result in a disproportionally high activity of the HPA-axis because of a deficient cortisol feedback due to the presence of glucocorticoid resistance. The increased cortisol levels cause a decrease in activity of the AVP neurons in the SCN that subsequently fail to inhibit sufficiently the CRH neurons in the PVN of depressed patients. Such an impaired negative feedback mechanism may lead to a further increase in the activity of the HPA-axis in depression. Both high CRH and cortisol levels contribute to the symptoms of depression. Light therapy may activate the SCN, directly inducing an increased synthesis and release of AVP that will inhibit the CRH neurons. However, it should be noted that human beings are diurnal creatures and might hold a different mode of interaction between SCN-AVP neurons and the PVN-CRH neurons to that of the nocturnal rat, especially because of the fact that opposite actions of hypothalamic AVP have recently been observed on the circadian corticosterone rhythm in nocturnal versus diurnal species (Kalsbeek *et al.*, 2008). The exact mechanism of action of light in depressed patients thus deserves further study.

## IV. Depression in Alzheimer's Disease

There is a very high prevalence of depression in Alzheimer's Disease (AD), affecting up to 50% of the patients. Indeed, AD is accompanied by an activated HPA-axis. CRH mRNA levels in the PVN of AD patients were markedly higher than those of comparison subjects, although they are even higher in depressed patients (Raadsheer *et al.*, 1995) (Fig. 18.9). In the group of AD patients, the mean postmortem CSF total cortisol level was 83% higher than that in the controls, especially in presenile AD patients (<65 years) (Swaab *et al.*, 1994). CRH neurons in the PVN of AD patients showed similar age-dependent increases in AVP colocalization to that of the control subjects (Raadsheer *et al.*, 1994b).

Recently, we found a positive correlation between the Cornell scale for Depression in Dementia and the number of CRH neurons in the PVN, suggesting that MD and depression in AD share, at least partly, their pathogenesis (Meynen *et al.*, 2007) (Fig. 18.10). The SCN is also clearly affected in AD (Swaab *et al.*, 1985). A diminished AVP mRNA content was already present in the SCN from the very first preclinical AD stages onward, explaining the disruption of circadian rhythms and nightly restlessness in AD (Wu *et al.*, 2006). However, no data are available to compare the SCN of depressed versus nondepressed AD patients. Our recent long-term, double-blind, placebo-controlled trials showed that a whole day of bright (±1000 lux) versus dim (±300 lux) light not only improved circadian rhythmicity and attenuated cognitive deterioration on the Mini-Mental State Examination Scale, but also ameliorated depressive symptoms on the Cornell Scale for Depression in Dementia (Riemersma-van der Lek *et al.*, 2008).

There are, however, also differences between MD and depression in AD. Although cortisol levels in AD patients were found to be more than double those of controls, no significant differences were found between depressed and nondepressed AD patients (Hoogendijk *et al.*, 2006). Moreover, recently it has been found that AD patients did not differ from controls with respect to the amount of AVP- or OXT-mRNA in the PVN or SON. Also, no differences were found between depressed and nondepressed AD patients and no relationship was found between the depression severity and AVP- or OXT-mRNA expression, all of which indicate that AVP and OXT gene expression in the PVN and SON is unchanged in depressed AD patients compared to nondepressed AD patients. This is in contrast with the

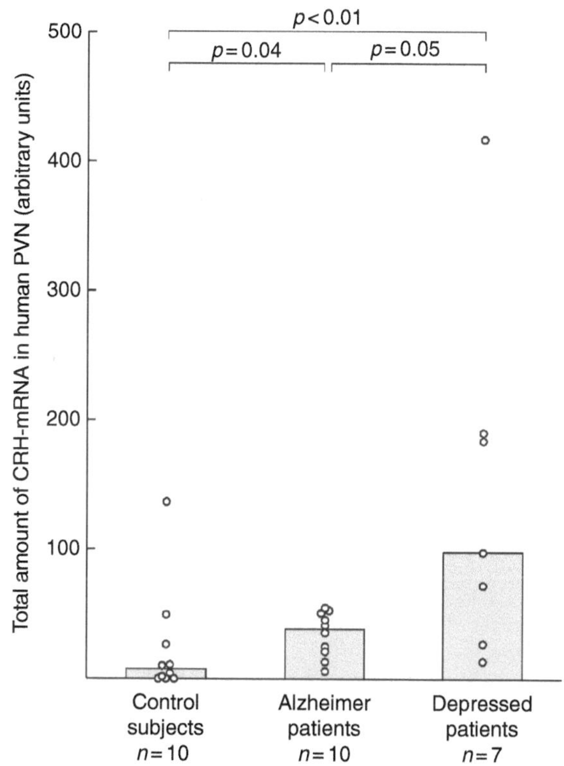


**Figure 18.9** Total hybridization signal for human CRH mRNA (arbitrary units) in the PVN. Bars indicate median values per patient group. The PVN of the Alzheimer patients ($n = 10$) contained significantly more (MW; $U = 23.0$, $W = 0.78$, $Z = -2.0$, $p = 0.04$) CRH mRNA than that of comparison subjects ($n = 10$). The amount of radioactivity in depressed patients ($n = 7$) was significantly higher than in comparison cases (MW; $U = 7.0$, $W = 91.0$, $Z = -2.7$, $p = 0.006$) and Alzheimer's disease patients (MW; $U = 23.0$, $W = 0.78$, $Z = -2.0$, $p = 0.05$). (From Raadsheer *et al.*, 1995; Fig. 2, with permission.)

enhanced AVP gene expression in MDD, suggesting a difference in pathophysiology between MDD and depression in AD (Meynen *et al.*, 2009).

## V. Conclusions

Both the HPA-axis and the hypothalamo-neurohypophysial systems are crucially involved in the stress response. The neuropeptides CRH and AVP, both secreted to the pituitary and released in other brain areas, play key roles in the etiology and pathophysiology of depression. HPA-axis

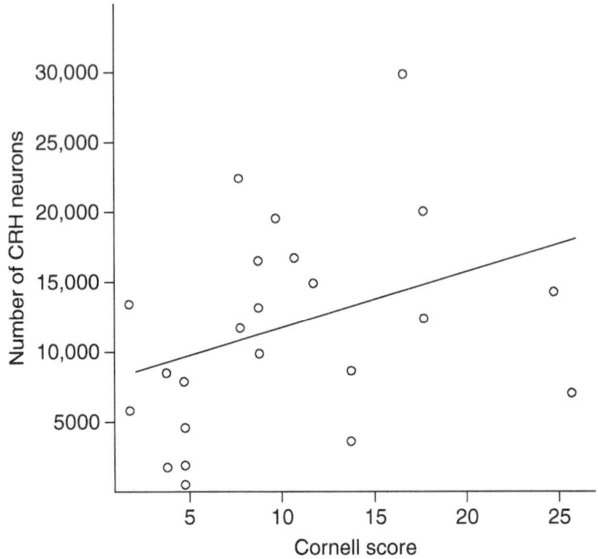


**Figure 18.10** Correlation between Cornell Scores for depression and the number of CRH-IR neurons in the PVN (rho = 0.433, $p = 0.039$). (From Meynen *et al.*, 2007; Fig. 2 with permission.)

hyperactivity is present in a majority of depressed subjects. CRH not only causes the neuroendocrine-hyperdrive, resulting in increased plasma cortisol levels, possibly causal in the development for depressive symptoms, but also induces depressive signs and symptoms by the direct release of CRH in brain areas. More recently, attention has been drawn to the important role of AVP on HPA-axis activation in depression. A clear sex difference has been indicated in the stress response and in the susceptibility, progress, and actual outcome of depression, which is closely correlated to the direct effects of sex hormones on the activity of CRH neurons. The decreased activity of the SCN, indicated by changes of AVP-expressing neurons, may be the basis for the biological rhythm disorders in depression. Similar as well as different changes in the CRH and AVP neuropeptidergic systems have been found in AD patients with depression and in idiopathic depressed patients.

## ACKNOWLEDGMENTS

The authors want to thank Mrs. Wilma Verweij for her secretarial assistance. Dr. A.-M. Bao is supported by Nature Science Foundation of China (30970928), Science Foundation of Chinese University, China Exchange Programme of the Royal Netherlands Academy of Arts and Sciences (KNAW) (project 09CDP011) and the Hersenstichting Nederland (14F06(2).07).

## REFERENCES


Akil, H. (2005). Stressed and depressed. *Nat. Med.* **11,** 116–118.

Bao, A. M., and Swaab, D. F. (2007). Gender difference in age-related number of corticotropin-releasing hormone-expressing neurons in the human hypothalamic paraventricular nucleus and the role of sex hormones. *Neuroendocrinology* **85,** 27–36.

Bao, A. M., Ji, Y. F., Van Someren, E. J., Hofman, M. A., Liu, R. Y., and Zhou, J. N. (2004). Diurnal rhythms of free estradiol and cortisol during the normal menstrual cycle in women with major depression. *Horm. Behav.* **45,** 93–102.

Bao, A. M., Hestiantoro, A., Van Someren, E. J., Swaab, D. F., and Zhou, J. N. (2005). Colocalization of corticotropin-releasing hormone and oestrogen receptor-alpha in the paraventricular nucleus of the hypothalamus in mood disorders. *Brain* **128,** 1301–1313.

Bao, A. M., Fischer, D. F., Wu, Y. H., Hol, E. M., Balesar, R., Unmehopa, U. A., *et al.* (2006). A direct androgenic involvement in the expression of human corticotropin-releasing hormone. *Mol. Psychiatry* **11,** 567–576.

Bao, A. M., Meynen, G., and Swaab, D. F. (2008). The stress system in depression and neurodegeneration: Focus on the human hypothalamus. *Brain Res. Rev.* **57,** 531–553.

Barden, N. (1996). Modulation of glucocorticoid receptor gene expression by antidepressant drugs. *Pharmacopsychiatry* **29,** 12–22.

Bartanusz, V., Jezova, D., Bertini, L. T., Tilders, F. J., Aubry, J. M., and Kiss, J. Z. (1993). Stress-induced increase in vasopressin and corticotropin-releasing factor expression in hypophysiotrophic paraventricular neurons. *Endocrinology* **132,** 895–902.

Belanoff, J. K., Rothschild, A. J., Cassidy, F., DeBattista, C., Baulieu, E. E., Schold, C., *et al.* (2002). An open label trial of C-1073 (mifepristone) for psychotic major depression. *Biol. Psychiatry* **52,** 386–392.

Brunner, J., Stalla, G. K., Stalla, J., Uhr, M., Grabner, A., Wetter, T. C., *et al.* (2001). Decreased corticotropin-releasing hormone (CRH) concentrations in the cerebrospinal fluid of eucortisolemic suicide attempters. *J. Psychiatr. Res.* **35,** 1–9.

Buijs, R. M., and Kalsbeek, A. (2001). Hypothalamic integration of central and peripheral clocks. *Nat. Rev. Neurosci.* **2,** 521–526.

Carroll, B. J., Cassidy, F., Naftolowitz, D., Tatham, N. E., Wilson, W. H., Iranmanesh, A., *et al.* (2007). Pathophysiology of hypercortisolism in depression. *Acta Psychiatr. Scand. Suppl.* 90–103.

Clark, P. M. (1998). Programming of the hypothalamo-pituitary-adrenal axis and the fetal origins of adult disease hypothesis. *Eur. J. Pediatr.* **157**(Suppl. 1), S7–S10.

Dai, J., Swaab, D. F., and Buijs, R. M. (1997). Distribution of vasopressin and vasoactive intestinal polypeptide (VIP) fibers in the human hypothalamus with special emphasis on suprachiasmatic nucleus efferent projections. *J. Comp. Neurol.* **383,** 397–414.

Dai, J., Swaab, D. F., Van der Vliet, J., and Buijs, R. M. (1998a). Postmortem tracing reveals the organization of hypothalamic projections of the suprachiasmatic nucleus in the human brain. *J. Comp. Neurol.* **400,** 87–102.

Dai, J., Van der Vliet, J., Swaab, D. F., and Buijs, R. M. (1998b). Human retinohypothalamic tract as revealed by in vitro postmortem tracing. *J. Comp. Neurol.* **397,** 357–370.

de Goeij, D. C., Kvetnansky, R., Whitnall, M. H., Jezova, D., Berkenbosch, F., and Tilders, F. J. (1991). Repeated stress-induced activation of corticotropin-releasing factor neurons enhances vasopressin stores and colocalization with corticotropin-releasing factor in the median eminence of rats. *Neuroendocrinology* **53,** 150–159.

de Wied, D., and van Ree, J. M. (1982). Neuropeptides, mental performance and aging. *Life Sci.* **31,** 709–719.

de Winter, R. F., van Hemert, A. M., DeRijk, R. H., Zwinderman, K. H., Frankhuijzen-Sierevogel, A. C., Wiegant, V. M., *et al.* (2003). Anxious-retarded depression: Relation with plasma vasopressin and cortisol. *Neuropsychopharmacology* **28,** 140–147.

Dempster, E. L., Burcescu, I., Wigg, K., Kiss, E., Baji, I., Gadoros, J., *et al.* (2007). Evidence of an association between the vasopressin V1b receptor gene (AVPR1B) and childhood-onset mood disorders. *Arch. Gen. Psychiatry* **64,** 1189–1195.

Dinan, T. G., Lavelle, E., Cooney, J., Burnett, F., Scott, L., Dash, A., *et al.* (1997). Dexamethasone augmentation in treatment-resistant depression. *Acta Psychiatr. Scand.* **95,** 58–61.

Dinan, T. G., O'Brien, S., Lavelle, E., and Scott, L. V. (2004). Further neuroendocrine evidence of enhanced vasopressin V3 receptor responses in melancholic depression. *Psychol. Med.* **34,** 169–172.

Drevets, W. C., Price, J. L., and Furey, M. L. (2008). Brain structural and functional abnormalities in mood disorders: Implications for neurocircuitry models of depression. *Brain Struct. Funct.* **213,** 93–118.

Engelmann, M., Landgraf, R., and Wotjak, C. T. (2004). The hypothalamic-neurohypophysial system regulates the hypothalamic-pituitary-adrenal axis under stress: An old concept revisited. *Front. Neuroendocrinol.* **25,** 132–149.

Erkut, Z. A., Pool, C., and Swaab, D. F. (1998). Glucocorticoids suppress corticotropin-releasing hormone and vasopressin expression in human hypothalamic neurons. *J. Clin. Endocrinol. Metab.* **83,** 2066–2073.

Erkut, Z. A., Gabreels, B. A., Eikelenboom, J., van Leeuwen, F. W., and Swaab, D. F. (2002). Glucocorticoid treatment is associated with decreased expression of processed AVP but not of proAVP, neurophysin or oxytocin in the human hypothalamus: Are PC1 and PC2 involved? *Neuro Endocrinol. Lett.* **23,** 33–44.

Frank, E., and Landgraf, R. (2008). The vasopressin system—From antidiuresis to psychopathology. *Eur. J. Pharmacol.* **583,** 226–242.

Gillies, G. E., Linton, E. A., and Lowry, P. J. (1982). Corticotropin releasing activity of the new CRF is potentiated several times by vasopressin. *Nature* **299,** 355–357.

Gispen-de Wied, C. C., Westenberg, H. G., Koppeschaar, H. P., Thijssen, J. H., and van Ree, J. M. (1992). Stimulation of the pituitary-adrenal axis with a low dose [Arg8]-vasopressin in depressed patients and healthy subjects. *Eur. Neuropsychopharmacol.* **2,** 411–419.

Goekoop, J. G., de Winter, R. P., de Rijk, R., Zwinderman, K. H., Frankhuijzen-Sierevogel, A., and Wiegant, V. M. (2006). Depression with above-normal plasma vasopressin: Validation by relations with family history of depression and mixed anxiety and retardation. *Psychiatry Res.* **141,** 201–211.

Griffiths, J., Ravindran, A. V., Merali, Z., and Anisman, H. (2000). Dysthymia: A review of pharmacological and behavioral factors. *Mol. Psychiatry* **5,** 242–261.

Heuser, I., Bissette, G., Dettling, M., Schweiger, U., Gotthardt, U., Schmider, J., *et al.* (1998). Cerebrospinal fluid concentrations of corticotropin-releasing hormone, vasopressin, and somatostatin in depressed patients and healthy controls: Response to amitriptyline treatment. *Depress. Anxiety* **8,** 71–79.

Hofman, M. A., and Swaab, D. F. (1992). Seasonal changes in the suprachiasmatic nucleus of man. *Neurosci. Lett.* **139,** 257–260.

Hofman, M. A., and Swaab, D. F. (1993). Diurnal and seasonal rhythms of neuronal activity in the suprachiasmatic nucleus of humans. *J. Biol. Rhythms* **8,** 283–295.

Holsboer, F. (2000). The corticosteroid receptor hypothesis of depression. *Neuropsychopharmacology* **23,** 477–501.

Holsboer, F. (2001). Stress, hypercortisolism and corticosteroid receptors in depression: Implications for therapy. *J. Affect. Disord.* **62,** 77–91.

Hoogendijk, W. J., Meynen, G., Endert, E., Hofman, M. A., and Swaab, D. F. (2006). Increased cerebrospinal fluid cortisol level in Alzheimer's disease is not related to depression. *Neurobiol. Aging* **27,** 780 e781 780–e782.

Hsu, D. T., Lombardo, K. A., Herringa, R. J., Bakshi, V. P., Roseboom, P. H., and Kalin, N. H. (2001). Corticotropin-releasing hormone messenger RNA distribution and stress-induced activation in the thalamus. *Neuroscience* **105,** 911–921.

Inder, W. J., Donald, R. A., Prickett, T. C., Frampton, C. M., Sullivan, P. F., Mulder, R. T., *et al.* (1997). Arginine vasopressin is associated with hypercortisolemia and suicide attempts in depression. *Biol. Psychiatry* **42,** 744–747.

Insel, T. R. (1997). A neurobiological basis of social attachment. *Am. J. Psychiatry* **154,** 726–735.

Johansson, C., Willeit, M., Smedh, C., Ekholm, J., Paunio, T., Kieseppa, T., *et al.* (2003). Circadian clock-related polymorphisms in seasonal affective disorder and their relevance to diurnal preference. *Neuropsychopharmacology* **28,** 734–739.

Kalsbeek, A., Buijs, R. M., van Heerikhuize, J. J., Arts, M., and van der Woude, T. P. (1992). Vasopressin-containing neurons of the suprachiasmatic nuclei inhibit corticosterone release. *Brain Res.* **580,** 62–67.

Kalsbeek, A., Verhagen, L. A., Schalij, I., Foppen, E., Saboureau, M., Bothorel, B., *et al.* (2008). Opposite actions of hypothalamic vasopressin on circadian corticosterone rhythm in nocturnal versus diurnal species. *Eur. J. NeuroSci.* **27,** 818–827.

Keeney, A., Jessop, D. S., Harbuz, M. S., Marsden, C. A., Hogg, S., and Blackburn-Munro, R. E. (2006). Differential effects of acute and chronic social defeat stress on hypothalamic-pituitary-adrenal axis function and hippocampal serotonin release in mice. *J. Neuroendocrinol.* **18,** 330–338.

Kishi, T., Kitajima, T., Ikeda, M., Yamanouchi, Y., Kinoshita, Y., Kawashima, K., *et al.* (2009). Association study of clock gene (CLOCK) and schizophrenia and mood disorders in the Japanese population. *Eur. Arch. Psychiatry Clin. Neurosci.* **259,** 293–297.

Landgraf, R., Kessler, M. S., Bunck, M., Murgatroyd, C., Spengler, D., Zimbelmann, M., *et al.* (2007). Candidate genes of anxiety-related behavior in HAB/LAB rats and mice: Focus on vasopressin and glyoxalase-I. *Neurosci. Biobehav. Rev.* **31,** 89–102.

Leonard, B. E. (2007). Psychopathology of depression. *Drugs Today (Barc.)* **43,** 705–716.

Liu, R. Y., Unmehopa, U. A., Zhou, J. N., and Swaab, D. F. (2006a). Glucocorticoids suppress vasopressin gene expression in human suprachiasmatic nucleus. *J. Steroid Biochem. Mol. Biol.* **98,** 248–253.

Liu, Z., Zhu, F., Wang, G., Xiao, Z., Wang, H., Tang, J., *et al.* (2006b). Association of corticotropin-releasing hormone receptor1 gene SNP and haplotype with major depression. *Neurosci. Lett.* **404,** 358–362.

Lolait, S. J., Stewart, L. Q., Jessop, D. S., Young, W. S. 3rd., and O'Carroll, A. M. (2007). The hypothalamic-pituitary-adrenal axis response to stress in mice lacking functional vasopressin V1b receptors. *Endocrinology* **148,** 849–856.

Lu, A., Steiner, M. A., Whittle, N., Vogl, A. M., Walser, S. M., Ableitner, M., *et al.* (2008a). Conditional CRH overexpressing mice: An animal model for stress-elicited pathologies and treatments that target the central CRH system. *Mol. Psychiatry* **13,** 989.

Lu, A., Steiner, M. A., Whittle, N., Vogl, A. M., Walser, S. M., Ableitner, M., *et al.* (2008b). Conditional mouse mutants highlight mechanisms of corticotropin-releasing hormone effects on stress-coping behavior. *Mol. Psychiatry* **13,** 1028–1042.

Ludwig, M., Bull, P. M., Tobin, V. A., Sabatier, N., Landgraf, R., Dayanithi, G., *et al.* (2005). Regulation of activity-dependent dendritic vasopressin release from rat supraoptic neurones. *J. Physiol.* **564,** 515–522.

Lund, T. D., Munson, D. J., Haldy, M. E., and Handa, R. J. (2004). Androgen inhibits, while oestrogen enhances, restraint-induced activation of neuropeptide neurones in the paraventricular nucleus of the hypothalamus. *J. Neuroendocrinol.* **16,** 272–278.

Mathe, A. A., Husum, H., El Khoury, A., Jimenez-Vasquez, P., Gruber, S. H., Wortwein, G., *et al.* (2007). Search for biological correlates of depression and mechanisms of action of antidepressant treatment modalities. Do neuropeptides play a role? *Physiol. Behav.* **92,** 226–231.

Merali, Z., Kent, P., Du, L., Hrdina, P., Palkovits, M., Faludi, G., *et al.* (2006). Corticotropin-releasing hormone, arginine vasopressin, gastrin-releasing peptide, and neuromedin

B alterations in stress-relevant brain regions of suicides and control subjects. *Biol. Psychiatry* **59,** 594–602.
Meyer-Bahlburg, H. F. L., and Ehrhardt, A. A. (1987). A prenatal-hormone hypothesis for depression in adults with a history of fetal DES exposure. *In* "Hormones and Depression" (U. Halbreich, Ed.), pp. 325–338. Raven Press, New York, USA.
Meynen, G., Unmehopa, U. A., van Heerikhuize, J. J., Hofman, M. A., Swaab, D. F., and Hoogendijk, W. J. (2006). Increased arginine vasopressin mRNA expression in the human hypothalamus in depression: A preliminary report. *Biol. Psychiatry* **60,** 892–895.
Meynen, G., Unmehopa, U. A., Hofman, M. A., Swaab, D. F., and Hoogendijk, W. J. (2007). Relation between corticotropin-releasing hormone neuron number in the hypothalamic paraventricular nucleus and depressive state in Alzheimer's disease. *Neuroendocrinology* **85,** 37–44.
Meynen, G., Unmehopa, U. A., Hofman, M. A., Swaab, D. F., and Hoogendijk, W. J. (2009). Hypothalamic vasopressin and oxytocin mRNA expression in relation to depressive state in Alzheimer's disease: A difference with major depressive disorder. *J. Neuroendocrinol.* **21,** 722–729.
Mlynarik, M., Zelena, D., Bagdy, G., Makara, G. B., and Jezova, D. (2006). Signs of attenuated depression-like behavior in vasopressin deficient Brattleboro rats. *Horm. Behav.* .
Mlynarik, M., Zelena, D., Bagdy, G., Makara, G. B., and Jezova, D. (2007). Signs of attenuated depression-like behavior in vasopressin deficient Brattleboro rats. *Horm. Behav.* **51,** 395–405.
Muller, M. B., Landgraf, R., and Keck, M. E. (2000). Vasopressin, major depression, and hypothalamic-pituitary-adrenocortical desensitization. *Biol. Psychiatry* **48,** 330–333.
Nemeroff, C. B. (1996). The corticotropin-releasing factor (CRF) hypothesis of depression: New findings and new directions. *Mol. Psychiatry* **1,** 336–342.
Pereira-Lima, J. F., Marroni, C. P., Pizarro, C. B., Barbosa-Coutinho, L. M., Ferreira, N. P., and Oliveira, M. C. (2004). Immunohistochemical detection of estrogen receptor alpha in pituitary adenomas and its correlation with cellular replication. *Neuroendocrinology* **79,** 119–124.
Phillips, D. I. (2001). Fetal growth and programming of the hypothalamic-pituitary-adrenal axis. *Clin. Exp. Pharmacol. Physiol.* **28,** 967–970.
Pitchot, W., Scantamburlo, G., Pinto, E., Hansenne, M., Reggers, J., Ansseau, M., *et al.* (2008). Vasopressin-neurophysin and DST in major depression: Relationship with suicidal behavior. *J. Psychiatr. Res.* **42,** 684–688.
Post, R. M. (1992). Transduction of psychosocial stress into the neurobiology of recurrent affective disorder. *Am. J. Psychiatry* **149,** 999–1010.
Purba, J. S., Hoogendijk, W. J., Hofman, M. A., and Swaab, D. F. (1996). Increased number of vasopressin- and oxytocin-expressing neurons in the paraventricular nucleus of the hypothalamus in depression. *Arch. Gen. Psychiatry* **53,** 137–143.
Raadsheer, F. C., Sluiter, A. A., Ravid, R., Tilders, F. J., and Swaab, D. F. (1993). Localization of corticotropin-releasing hormone (CRH) neurons in the paraventricular nucleus of the human hypothalamus; age-dependent colocalization with vasopressin. *Brain Res.* **615,** 50–62.
Raadsheer, F. C., Hoogendijk, W. J., Stam, F. C., Tilders, F. J., and Swaab, D. F. (1994a). Increased numbers of corticotropin-releasing hormone expressing neurons in the hypothalamic paraventricular nucleus of depressed patients. *Neuroendocrinology* **60,** 436–444.
Raadsheer, F. C., Tilders, F. J., and Swaab, D. F. (1994b). Similar age related increase of vasopressin colocalization in paraventricular corticotropin-releasing hormone neurons in controls and Alzheimer patients. *J. Neuroendocrinol.* **6,** 131–133.
Raadsheer, F. C., van Heerikhuize, J. J., Lucassen, P. J., Hoogendijk, W. J., Tilders, F. J., and Swaab, D. F. (1995). Corticotropin-releasing hormone mRNA levels in the

paraventricular nucleus of patients with Alzheimer's disease and depression. *Am. J. Psychiatry* **152,** 1372–1376.

Rabadan-Diehl, C., and Aguilera, G. (1998). Glucocorticoids increase vasopressin V1b receptor coupling to phospholipase C. *Endocrinology* **139,** 3220–3226.

Riemersma-van der Lek, R. F., Swaab, D. F., Twisk, J., Hol, E. M., Hoogendijk, W. J., and Van Someren, E. J. (2008). Effect of bright light and melatonin on cognitive and noncognitive function in elderly residents of group care facilities: A randomized controlled trial. *JAMA* **299,** 2642–2655.

Rivier, C., and Vale, W. (1983). Interaction of corticotropin-releasing factor and arginine vasopressin on adrenocorticotropin secretion in vivo. *Endocrinology* **113,** 939–942.

Roy, A. (1996). HPA axis function and temperament in depression: A negative report. *Biol. Psychiatry* **39,** 364–366.

Scheithauer, B. W., Kovacs, K., Zorludemir, S., Lloyd, R. V., Erdogan, S., and Slezak, J. (2008). Immunoexpression of androgen receptor in the nontumorous pituitary and in adenomas. *Endocr. Pathol.* **19,** 27–33.

Schmidt, E. D., Janszen, A. W., Wouterlood, F. G., and Tilders, F. J. (1995). Interleukin-1-induced long-lasting changes in hypothalamic corticotropin-releasing hormone (CRH)—Neurons and hyperresponsiveness of the hypothalamus-pituitary-adrenal axis. *J. Neurosci.* **15,** 7417–7426.

Schmidt, E. D., Binnekade, R., Janszen, A. W., and Tilders, F. J. (1996). Short stressor induced long-lasting increases of vasopressin stores in hypothalamic corticotropin-releasing hormone (CRH) neurons in adult rats. *J. Neuroendocrinol.* **8,** 703–712.

Schmidt, M. V., Deussing, J. M., Oitzl, M. S., Ohl, F., Levine, S., Wurst, W., *et al.* (2006). Differential disinhibition of the neonatal hypothalamic-pituitary-adrenal axis in brain-specific CRH receptor 1-knockout mice. *Eur. J. Neurosci.* **24,** 2291–2298.

Scott, L. V., and Dinan, T. G. (1998). Vasopressin and the regulation of hypothalamic-pituitary-adrenal axis function: Implications for the pathophysiology of depression. *Life Sci.* **62,** 1985–1998.

Selye, H. (1998). A syndrome produced by diverse nocuous agents. 1936. *J. Neuropsychiatry Clin. Neurosci.* **10,** 230–231.

Shi, J., Wittke-Thompson, J. K., Badner, J. A., Hattori, E., Potash, J. B., Willour, V. L., *et al.* (2008). Clock genes may influence bipolar disorder susceptibility and dysfunctional circadian rhythm. *Am. J. Med. Genet. B Neuropsychiatr. Genet.* **147B,** 1047–1055.

Stefaneanu, L., Kovacs, K., Horvath, E., Lloyd, R. V., Buchfelder, M., Fahlbusch, R., *et al.* (1994). In situ hybridization study of estrogen receptor messenger ribonucleic acid in human adenohypophysial cells and pituitary adenomas. *J. Clin. Endocrinol. Metab.* **78,** 83–88.

Surget, A., and Belzung, C. (2008). Involvement of vasopressin in affective disorders. *Eur. J. Pharmacol.* **583,** 340–349.

Swaab, D. F., Fliers, E., and Partiman, T. S. (1985). The suprachiasmatic nucleus of the human brain in relation to sex, age and senile dementia. *Brain Res.* **342,** 37–44.

Swaab, D. F., Raadsheer, F. C., Endert, E., Hofman, M. A., Kamphorst, W., and Ravid, R. (1994). Increased cortisol levels in aging and Alzheimer's disease in postmortem cerebrospinal fluid. *J. Neuroendocrinol.* **6,** 681–687.

Swaab, D. F., Fliers, E., Hoogendijk, W. J., Veltman, D. J., and Zhou, J. N. (2000). Interaction of prefrontal cortical and hypothalamic systems in the pathogenesis of depression. *Prog. Brain Res.* **126,** 369–396.

Swaab, D. F. (2003). The human hypothalamus. Basic and clinical aspects. Part I: Nuclei of the hypothalamus. *In* "Handbook of Clinical Neurology" (M. J. Aminoff, F. Boller, and D. F. Swaab, Eds.), Vol. 79. Elsevier, Amsterdam.

Swaab, D. F. (2004a). Neuropeptides in hypothalamic neuronal disorders. *Int. Rev. Cytol.* **240,** 305–375.

Swaab, D. F. (2004b). The human hypothalamus. Basic and clinical aspects. Part II: Neuropathology of the hypothalamus and adjacent brain structures. *In* "Handbook of Clinical Neurology" (M. J. Aminoff, F. Boller, and D. F. Swaab, Eds.), Vol. 80. Elsevier, Amsterdam.

Swaab, D. F., Bao, A. M., and Lucassen, P. J. (2005). The stress system in the human brain in depression and neurodegeneration. *Ageing Res. Rev.* **4,** 141–194.

Tarullo, A. R., and Gunnar, M. R. (2006). Child maltreatment and the developing HPA axis. *Horm. Behav.* **50,** 632–639.

Ter Horst, G. J., Wichmann, R., Gerrits, M., Westenbroek, C., and Lin, Y. (2009). Sex differences in stress responses: Focus on ovarian hormones. *Physiol. Behav.* **97,** 239–249.

Thompson, C., Syddall, H., Rodin, I., Osmond, C., and Barker, D. J. (2001). Birth weight and the risk of depressive disorder in late life. *Br. J. Psychiatry* **179,** 450–455.

Thompson, R. R., George, K., Walton, J. C., Orr, S. P., and Benson, J. (2006). Sex-specific influences of vasopressin on human social communication. *Proc. Natl. Acad. Sci. USA* **103,** 7889–7894.

Vamvakopoulos, N. C., and Chrousos, G. P. (1993). Evidence of direct estrogenic regulation of human corticotropin-releasing hormone gene expression. Potential implications for the sexual dimophism of the stress response and immune/inflammatory reaction. *J. Clin. Invest.* **92,** 1896–1902.

van Londen, L., Goekoop, J. G., van Kempen, G. M., Frankhuijzen-Sierevogel, A. C., Wiegant, V. M., van der Velde, E. A., *et al.* (1997). Plasma levels of arginine vasopressin elevated in patients with major depression. *Neuropsychopharmacology* **17,** 284–292.

van Londen, L., Goekoop, J. G., Zwinderman, A. H., Lanser, J. B., Wiegant, V. M., and De Wied, D. (1998a). Neuropsychological performance and plasma cortisol, arginine vasopressin and oxytocin in patients with major depression. *Psychol. Med.* **28,** 275–284.

van Londen, L., Kerkhof, G. A., van den Berg, F., Goekoop, J. G., Zwinderman, K. H., Frankhuijzen-Sierevogel, A. C., *et al.* (1998b). Plasma arginine vasopressin and motor activity in major depression. *Biol. Psychiatry* **43,** 196–204.

van Londen, L., Goekoop, J. G., Kerkhof, G. A., Zwinderman, K. H., Wiegant, V. M., and De Wied, D. (2001). Weak 24-h periodicity of body temperature and increased plasma vasopressin in melancholic depression. *Eur. Neuropsychopharmacol.* **11,** 7–14.

van West, D., Del-Favero, J., Aulchenko, Y., Oswald, P., Souery, D., Forsgren, T., *et al.* (2004). A major SNP haplotype of the arginine vasopressin 1B receptor protects against recurrent major depression. *Mol. Psychiatry* **9,** 287–292.

Walum, H., Westberg, L., Henningsson, S., Neiderhiser, J. M., Reiss, D., Igl, W., *et al.* (2008). Genetic variation in the vasopressin receptor 1a gene (AVPR1A) associates with pair-bonding behavior in humans. *Proc. Natl. Acad. Sci. USA* **105,** 14153–14156.

Wang, S. S., Kamphuis, W., Huitinga, I., Zhou, J. N., and Swaab, D. F. (2008). Gene expression analysis in the human hypothalamus in depression by laser microdissection and real-time PCR: The presence of multiple receptor imbalances. *Mol. Psychiatry* **13**(786–799), 741.

Weninger, S. C., Dunn, A. J., Muglia, L. J., Dikkes, P., Miczek, K. A., Swiergiel, A. H., *et al.* (1999). Stress-induced behaviors require the corticotropin-releasing hormone (CRH) receptor, but not CRH. *Proc. Natl. Acad. Sci. USA* **96,** 8283–8288.

Wirz-Justice, A., Benedetti, F., Berger, M., Lam, R. W., Martiny, K., Terman, M., *et al.* (2005). Chronotherapeutics (light and wake therapy) in affective disorders. *Psychol. Med.* **35,** 939–944.

Wu, Y. H., Fischer, D. F., Kalsbeek, A., Garidou-Boof, M. L., van der Vliet, J., van Heijningen, C., *et al.* (2006). Pineal clock gene oscillation is disturbed in Alzheimer's disease, due to functional disconnection from the "master clock". *FASEB J.* **20,** 1874–1876.

Yirmiya, N., Rosenberg, C., Levi, S., Salomon, S., Shulman, C., Nemanov, L., *et al.* (2006). Association between the arginine vasopressin 1a receptor (AVPR1a) gene and autism in a family-based study: Mediation by socialization skills. *Mol. Psychiatry* **11,** 488–494.
Young, E. A., Haskett, R. F., Murphy-Weinberg, V., Watson, S. J., and Akil, H. (1991). Loss of glucocorticoid fast feedback in depression. *Arch. Gen. Psychiatry* **48,** 693–699.
Young, E. A., Lopez, J. F., Murphy-Weinberg, V., Watson, S. J., and Akil, H. (2003). Mineralocorticoid receptor function in major depression. *Arch. Gen. Psychiatry* **60,** 24–28.
Zhou, J. N., Riemersma, R. F., Unmehopa, U. A., Hoogendijk, W. J., van Heerikhuize, J. J., Hofman, M. A., *et al.* (2001). Alterations in arginine vasopressin neurons in the suprachiasmatic nucleus in depression. *Arch. Gen. Psychiatry* **58,** 655–662.

CHAPTER NINETEEN

# Postnatal Ontogeny of the Glucocorticoid Receptor in the Hippocampus

Anastasia Galeeva,*,† Markku Pelto-Huikko,† Svetlana Pivina,* *and* Natalia Ordyan*

## Contents



## Abstract

Corticosteroid hormones are important intrinsic factors that not only mediate the response to stress but also largely contribute to the main physiological processes. The biological actions of these steroids involve, first of all, the activation of specific receptors, namely mineralocorticoid (MR) and glucocorticoid (GR) receptors. These two receptor types govern a flexible and well-balanced mechanism that leads to the often opposing changes in the cell. The hippocampus is the central part of the extrahypothalamic feedback loop in the control of the hypothalamic–pituitary–adrenal (HPA) axis activity. The coexpression of both MR and GR in the hippocampus serves a coordinated response to corticosteroids in the hippocampal neurons, thereby mediating the neuronal excitability, stress response, and behavioral adaptation. Each receptor type reveals distinct ontogenetic pattern over the postnatal period. This review addresses the issues relating to postnatal development of the HPA axis and especially the hippocampal expression of the GR proteins in intact and prenatally stressed rats. © 2010 Elsevier Inc.

* Pavlov Institute of Physiology, Russian Academy of Science, Makarova 6, Saint-Petersburg, Russia
† Medical School of Tampere University, University of Tampere, Tampere, Finland

*Vitamins and Hormones*, Volume 82
ISSN 0083-6729, DOI: 10.1016/S0083-6729(10)82019-9

# I. Introduction

Hypothalamic–pituitary–adrenal (HPA) axis is one of the important hormonal systems of our organism, especially in the sense of an individual's ability to cope with stress; and corticosteroid hormones thus have widely extended regulatory functions. These hormones govern multiple aspects of a mammal's physiology, such as metabolic reactions, circadian activities, and response to stress. Corticosteroids appear also to be very important players at a developmental stage. They are required for the normal maturation of a vast majority of peripheral tissues as well as for the regulation of cellular differentiation of neural tissue (for review see Meyer, 1985). Direct effect of corticosteroids has been described for both central and peripheral nervous systems, including adrenal medulla (for review see Doupe and Patterson, 1982). While recent evidences suggest that corticosteroids may produce rapid nongenomic cellular effects (Atkinson *et al.*, 2008) via putative membrane-bound receptors, corticosteroids are classically considered to exert their biological action through two ligand-activated transcription factors, mineralocorticoid receptors (MRs) and glucocorticoid receptors (GRs). Corticosteroids (cortisol in human, and corticosterone in rat and mouse) acting through MRs and GRs, which are abundantly coexpressed in the neurons of limbic structures (Herman *et al.*, 2003), control the stress-related processes. Thus, MR is implicated in the onset of the stress response. GR, in turn, terminates the stress reactions, mobilizes the energy resources required for this purpose and facilitates recovery, being activated only by high concentrations of corticosteroids. In other words, the activation of MR by corticosteroids leads to the onset of a mechanism that initiates the stress response, that is, before GR-mediated mechanisms develop and start to lessen the initial stress reaction, thereby facilitating recovery and adaptation.

The involvement of the hippocampus in the regulation of the HPA function has been extensively documented and reviewed (De Kloet, 2003; De Kloet *et al.*, 1998; Feldman and Weidenfeld, 1999; Herman and Cullinan, 1997; Herman *et al.*, 1989; Jacobson and Sapolsky, 1991; Sapolsky and Meaney, 1986). Lesion studies of the hippocampus revealed that it plays an important role in the inhibition of secretion of glucocorticoids and in shutting off their secretion after stress (Herman *et al.*, 1989), whereas the magnitude of hippocampal inhibition on the HPA activity is dependent on the duration and type of stress to which an animal is exposed. Though the role of the hippocampus in extrahypothalamic negative feedback in adults is well defined, it is still unclear whether the hippocampus may take part in regulation of neonatal HPA hyporesponsiveness. There is evidence that glucocorticoids play an essential role in neurogenesis in some brain region, especially in the hippocampus (Sousa *et al.*, 2008). Since the biological action of glucocorticoids is mediated by the intracellular GR and MR, knowledge of the distribution patterns of these receptors in the developing brain is essential.

In this review, we once more consider a pattern of postnatal expression of GRs in the hippocampus of rats as well as the effect of prenatal stress on the changes in the HPA axis during stress-hyporesponsive period (SHRP). A separate subchapter is also concerned with postnatal ontogeny of MRs in the hippocampus.

## II. Postnatal Development of the HPA Axis and the Stress-Hyporesponsive Period

The proper prenatal development and neonatal maturation of the HPA axis represent events that are extremely important in adaptation to extrauterine life. During neonatal ontogeny, the HPA axis shows a regulation profoundly different from that found in the adult. As it has already been shown in many studies, basal levels of corticosterone in rats are high at the time of birth. Following a surge in perinatal glucocorticoid levels, there is SHRP, in which rodents have a reduced capacity for secretion of adrenocorticotropic hormone (ACTH) and corticosterone in response to several mild stress stimuli (De Kloet *et al.*, 1988; Sapolsky and Meaney, 1986; Schmidt *et al.*, 2003; Walker *et al.*, 1986, 1991). However, simultaneously they retain the capability to demonstrate an adult-like HPA axis response with elevated corticosterone levels to life-threatening situations such as cold, infection, exposure to noxious agents, hypoxia, and long-term maternal deprivation (Levine, 2001; Levine *et al.*, 1994; O'Grady *et al.*, 1993; van Oers *et al.*, 1998; Vazquez and Akhil, 1993; Yi and Baram, 1994; Zelena *et al.*, 2008). The reduced adrenal production of corticosterone in response to exogenous ACTH during SHRP has been demonstrated (Arai and Widmaier, 1993; Yoshimura *et al.*, 2003). Between postnatal days (PDs) 3 and 14, a period of adrenal hyporesponsiveness to the stressors has been well documented in rodents. This is not unique to rodents since the adrenal hyporesponsiveness has also been shown to occur in guinea pigs (Jones and Roebuck, 1980), fetal sheep (Répérant and Durand, 1997), and in rabbits during the first days of life (Devaskar *et al.*, 1980). In humans, HPA hyporesponsiveness appears to develop gradually during the first year of life, but its duration remains unknown (Gunnar and Donzella, 2002; Watterberg and Scott, 1995).

During SHRP, corticosterone remains at very low levels until around the end of the second postnatal week and then demonstrates a sustained rise to adult levels. From day 14, plasma ACTH and corticosterone levels rise significantly after exposure to various stressors (Campagne *et al.*, 2009;

Galeeva *et al.*, 2006; Walker *et al.*, 1991). Although infant rats are capable of responding to stressful stimuli, their circadian rhythmicity and feedback regulation are not fully developed until much later (Levin and Levine, 1975; Vazquez and Akhil, 1993). The same pattern has been reported for ontogenetic changes in plasma levels of corticosteroid-binding globulin (CBG) (Viau *et al.*, 1996).

Although several aspects of stress hyporesponsiveness involved in the HPA regulatory system have been elucidated, a complete explanation of SHRP is still unknown. It has been suggested that various components of the HPA axis contribute to maintaining low circulating corticosterone concentrations. The HPA system does not develop uniformly; different components of the system have different ontogenetic patterns. At the hypothalamic level, the portal vessel system and the neuronal network which innervates corticotropin releasing hormone (CRH) neurons in the parvocellular subdivision of the paraventricular nucleus (PVN) are not fully developed (Daikoku *et al.*, 1981, 1984; Ugrumov and Mitskevich, 1992; Ugrumov *et al.*, 1985). Nevertheless, in the anterior pituitary corticotropes starting from the embryonic day 15, the expression of pro-opiomelanocortin (POMC), its processing to ACTH and $\beta$-endorphin as well as the expression of CRH has been reported (Chatelain and Dupouy, 1981; Grino *et al.*, 1989; Hindelang *et al.*, 1990; Keegan *et al.*, 1994). The SHRP phenomenon might be due to immaturity in the regulatory system at the level of adrenocortical secretagogues. The maximal steroidogenic response to ACTH in collagenase-dispersed adrenocortical cells significantly varied in the following order: adult > PD 1 > day 10 (Arai and Widmaier, 1993). Other factors such as a reduction in the expression of cytochrome P450 enzyme (Arai and Widmaier, 1993), a decline in enzymatic activity of P450C21 (Nagaya *et al.*, 1995), or some changes in adrenal sympathetic innervation (Walker, 1995) might be responsible for the neonatal adrenal hyporesponsivity. Walker and colleagues (2004) proposed one more mechanism whereby an exposure to high circulating levels of leptin, such as those found in suckling neonates (Ahima *et al.*, 1998), could maintain low adrenal responsivity during this critical period of neonatal development.

From PD 4 until approximately 10, CBG transiently disappears from the circulation (Sakly and Koch, 1983). Simultaneously, low activity of 11$\beta$-hydroxysteroid dehydrogenase type 2, an enzyme that converts the bioactive glucocorticoids into their inactive forms, has been reported (Brown *et al.*, 1996). These findings suggest that even though the amount of circulating glucocorticoid hormone is much reduced during SHRP, glucocorticoids bioavailability to brain and peripheral organs is strongly enhanced. Such developmental specificity could lead to extensive occupancy of corticosteroid receptors and an enhanced feedback action of hormone. It has been suggested that a glucocorticoid negative-feedback system in the pituitary and resultant inhibition of responsiveness to CRH play a major role in mediating the SHRP (Halasz *et al.*, 1997; Sapolsky and Meaney, 1986; Walker *et al.*, 1986).

Also it has been proposed that certain aspects of rodent maternal behavior, such as licking and feeding, might play a significant role in regulating the neonatal's HPA axis activity (Campagne and Meaney, 2001; Rosenfeld *et al.*, 1991; van Oers *et al.*, 1998). There are experimental data showing that prolonged separation of the pups during SHRP disinhibits the stress hyporesponsiveness of the HPA, increasing basal levels of corticosterone and adrenal sensitivity to ACTH (Levine, 1994; Okimoto *et al.*, 2002; Rosenfeld *et al.*, 1991; Schmidt *et al.*, 2004). At the same time, basal levels of CRH gene expression in the PVN have been reported to be decreased following maternal deprivation (Schmidt *et al.*, 2004; van Oers *et al.*, 1998). Furthermore, expression levels of GRs and MRs are downregulated in the hippocampus of maternally deprived pups (Avishai-Eliner *et al.*, 1999; Schmidt *et al.*, 2004). In addition, following maternal deprivation, both the ACTH and corticosterone response to some stressors, such as novelty exposure and a saline injection, is markedly elevated (Enthoven *et al.*, 2008; Rosenfeld *et al.*, 1991; Suchecki *et al.*, 1993). Taken together, this data suggest that the hyporesponsiveness of the HPA axis during early neonatal period of the development is not absolute and disappears under situations that threaten health or survival.

Generally, SHRP may be considered as a mechanism that protects non-matured neurons from fatal impact of circulating corticosteroids, which, if not stopped in time, may lead to the inhibition of neuronal cell division, glial development, alteration of neurofibril density, and as result to an retardation of brain growth (Antonow-Schlorke *et al.*, 2003; Bohn and Friedrich, 1982; Howard and Benjamins, 1975; Kapellou *et al.*, 2003). Depending on the developmental stage, glucocorticoids or stress exposure may lead to long-term effects on neuroendocrine, behavioral, emotional, and cognitive functions, which in turn are associated with increased risks of developing some metabolic, cardiovascular, and brain disorders in later life (Barker, 2007; Gluckman *et al.*, 2005; Seckl and Meaney, 2006). Such developmental "programming" of adult phenotype may occur not only as a result of the postnatal manipulations. Maternal stress during pregnancy also produces HPA axis activation in neonatal offspring that results in the vast array of neuroendocrine (e.g., HPA axis stress-reactivity) and behavioral (e.g., anxiety-like and depressive behaviors) alterations observed in adults (Dodic *et al.*, 2003; Seckl and Holmes, 2007; Weinstock, 2005, 2008).

There are numerous data indicating the existence of sex-differences in the HPA axis activity in adult rodents. Thus, the basal levels of corticosterone secretion and stress-induced response of the HPA are much greater in female than in male rats (Critchlow *et al.*, 1963; Le Mevel *et al.*, 1979;

Seale *et al.*, 2004). An activational effect of estrogens and inhibitory effect of androgens on stress-induced HPA response clearly contribute to this difference (Figueiredo *et al.*, 2007; Romeo *et al.*, 2004; Viau and Meaney, 1991; Williamson and Viau, 2008). In ovariectomized female rats, estradiol not only potentiates the corticosterone response to numerous stressors (Viau and Meaney, 1991), but also enhances basal level of corticosterone (Figueiredo *et al.*, 2007). Moreover, it has been shown that gonadal steroids produced organizational effects on the HPA axis activity during critical period of perinatal life. Exposure of neonatal female rats to a single injection of testosterone propionate disrupted the development of the characteristic female pattern of corticosterone secretion as well as the normal female HPA response to stress, resulting in a pattern similar to that seen in males (Seale *et al.,* 2005). The same effects of neonatal gonadectomy on adult male HPA axis activity has been demonstrated recently (Bingham and Viau, 2008). In addition, adult male rats who had been exposed to perinatal flutamide or the aromatase inhibitors 1,4,6-androstatriene-3,17-dione (ATD) demonstrated all parameters of increased HPA activity that resembled the normal physiological state of the intact adult female (Seale *et al.*, 2005).

It has also been shown that susceptibility of the HPA axis to neonatal programming is sexually dimorphic (Halasz *et al.*, 1997; Knuth and Etgen, 2005; Papaioannou *et al.*, 2002; Renard *et al.*, 2007). Taking together all the data described above, we can speculate that some sex-differences in the HPA axis activity might be occurring in rats during the postnatal life. Our own study revealed a profound difference in the corticosterone levels between male and female pups over a period of time (Fig. 19.1). In fact, it is interesting to note that plasma corticosterone values in males and females change in opposite directions along the course of development. Indeed, during the first two weeks of life, the plasma corticosterone in females remains at significantly lower levels than in males, and then starting from PD 18 rises considerably above the values which could be obtained in male subjects. Hence, the adult-like pattern of basal corticosterone secretion in female rats is set already during the juvenile period of life, a phenomenon that appears to be caused by the functional maturation of the pituitary–gonadal axis and the biological activity of estradiol. Although rat ovaries contain estradiol already in the neonatal period, its biological activity is extremely low at this time (Sokka and Huhtaniemi, 1995). The main factor reducing the natural activity of estradiol is α-fetoprotein (AFP), which actively binds the hormone in juvenile rats (Meijs-Roelofs and Kramer, 1979; Raynaud *et al.*, 1971). The blood level of AFP declines in female rats by the end of the third week of life, and at the same time the basal activity of the HPA axis gradually increases.

In summary, the principles that govern the activity of the HPA axis in neonates are very different from those in adults. The animal experimental data also clearly show sex-dependent differences in the regulation of the HPA axis prenatal development and postnatal maturation.

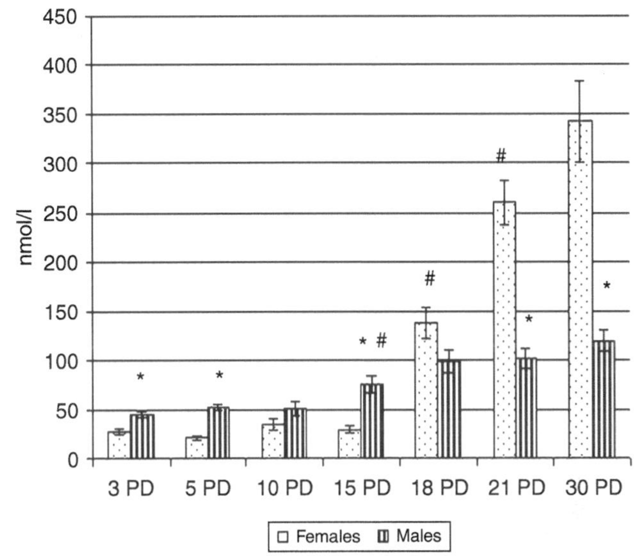


**Figure 19.1** Sex-specific postnatal pattern of the plasma corticosterone levels. Intact females (dotted bars) and males (vertically stripped bars) of selected postnatal days (3, 5, 10, 15, 18, 21, 30 PD) were used for the study. The serum corticosterone concentrations were measured by radioimmunoassay as described by Ordyan *et al.* (2001). Each data bar represents the mean ± SEM ($n = 8$).

# III. Postnatal Expression of Corticosteroid Receptors in the Rat Hippocampus

## A. Ontogeny of the glucocorticoid receptor

Corticosteroids are classically considered to exert their biological effects by binding in a ligand-dependent manner to specific transcription factors, MR and GR, previously more referred to as type I and type II receptors with different affinities to corticosterone, namely very high affinity ($K_d \sim 0.5$ nM) for type I (MR) and relatively low affinity (($K_d \sim 5.0$ nM) for type II (GR) (Reul and de Kloet, 1985). The ligand binding results in the translocation of the ligand–receptor complex to the nucleus, where it binds to glucocorticoid response element (GRE) in the promoter region of the target gene to influence gene transcription. Mapping studies in the brain of adult subjects have revealed that while GRs are rather widely distributed, MRs have a more discrete distribution (for review see De Kloet and Reul, 1987; De Kloet *et al.*, 1998). Thus, MR expression is mainly restricted to the paraventricular hypothalamic nucleus, certain hippocampal subfields, and the septum. These patterns of distribution as well as the pharmacological profiles of each receptor may, to some extent, explain the wide spectrum of homeostatic and psychophysiological functions that are subject to regulation by corticosteroids.

The lower affinity of GR allows them to respond largely in phase with the diurnal rhythm. Pulsatility thus has particular significance on occupancy of the lower affinity GR receptors. Hence, the GR becomes progressively activated during stress- and circadian-induced increases in the frequency and amplitude of corticosteroid secretory bursts (Young *et al.*, 2004).

The accumulated data in the literature suggest that during the perinatal period of development, central and peripheral parts of the HPA axis mature mostly independent of one another and lack the direct regulation of GR expression by endogenous or/and exogenous glucocorticoids (Ghosh *et al.*, 2000; Kalinyak *et al.*, 1989; Noorlander *et al.*, 2006). There are even data indicating that the administration of glucocorticoids during fetal development may instead upregulate the GR expression (Sweezey *et al.*, 1998). Additionally it has been shown that the GRs are capable of interacting with endogenous glucocorticoids during first two weeks of life, although HPA axis is rather in quiescence during this neonatal period. (Meaney *et al.*, 1985). Since the authors found the GRs to be relatively unsusceptible to manipulations with corticosterone level at PD 10, they concluded that the neonatal GR system develops independently of circulating corticosterone. At the same time, it was hypothesized that the most feasible reason for the persistence of the neonatal hyporesponsiveness of the HPA axis is an enhanced GR-mediated negative feedback at the level of the anterior pituitary (Sakly and Koch, 1983; Sapolsky and Meaney, 1986; Walker *et al.*, 1986) due to adult-like levels of GRs in the anterior pituitary over the postnatal period of life (Sakly and Koch, 1981). Moreover, this assumption is also corroborated by the data demonstrating that GRs are in fact expressed in most of the hypothalamic nuclei (especially in the PVN) just after birth over the course of postnatal development (Rosenfeld *et al.*, 1988a,b). In the light of new fact that GR mutants die about 1 week after birth, displaying a fulminant increase in plasma corticosterone (Erdmann *et al.*, 2008), this interpretation now seems to be reasonably convincing. Indeed, the capability of GR to execute their biological functions during neonatal ontogeny, at least at the pituitary level, may serve to maintain the very low postnatal activity of the HPA axis, protecting the immature system from probable corticosterone-induced impact damage.

Although there are a number of studies describing the brain expression and functioning of GR across ontogeny, the available data are still inconsistent. Analysis of a spatiotemporal expression of the receptors in the postnatal rat brain made by the concurrent binding assay revealed the following pattern of the expression: binding of GR proteins was detected already at PD 1, following which the binding demonstrated substantial increase approaching an adult level by day 35 of life (Meaney *et al.*, 1985).

Concomitantly, it was found that the level of GR in the hippocampus of 15-day pups significantly exceeded that of adult animals. Studies of the postnatal ontogeny of corticosteroid receptor expression in the rat hippocampus by using an analysis of mRNAs levels have shown a gradual increase in GR mRNA expression from birth onward. The expression of GR transcripts achieved adult-like levels by PD 16. The authors also found that the expression of GR mRNAs has a regional specificity within the hippocampal formation (Bohn *et al.*, 1994; Van Eekelen *et al.*, 1991). The levels of GR transcripts were significantly lower in the CA3 field of the hippocampus in both adults and pups.

Thus, most studies implying regulation of the GR expression in the fetal and neonatal brains were based on a ligand binding or mRNA analysis. Both these approaches, however, have major shortcomings in methodology. First, a ligand binding assay is influenced by circulating glucocorticoids, requiring interventions such as adrenalectomy (Reul *et al.*, 1989). Second, GR protein is regulated in part by posttranscriptional events, meaning that mRNA measurements may not reflect the actual amount of GR protein present in the cell (Dong *et al.*, 1988). In our studies, we used densitometric quantification of immunoblots, following equalization against actin, expressed in units of optical density. The expression of the main GR protein isoform (94 kDa) in the hippocampal region (Fig. 19.2) was found to be

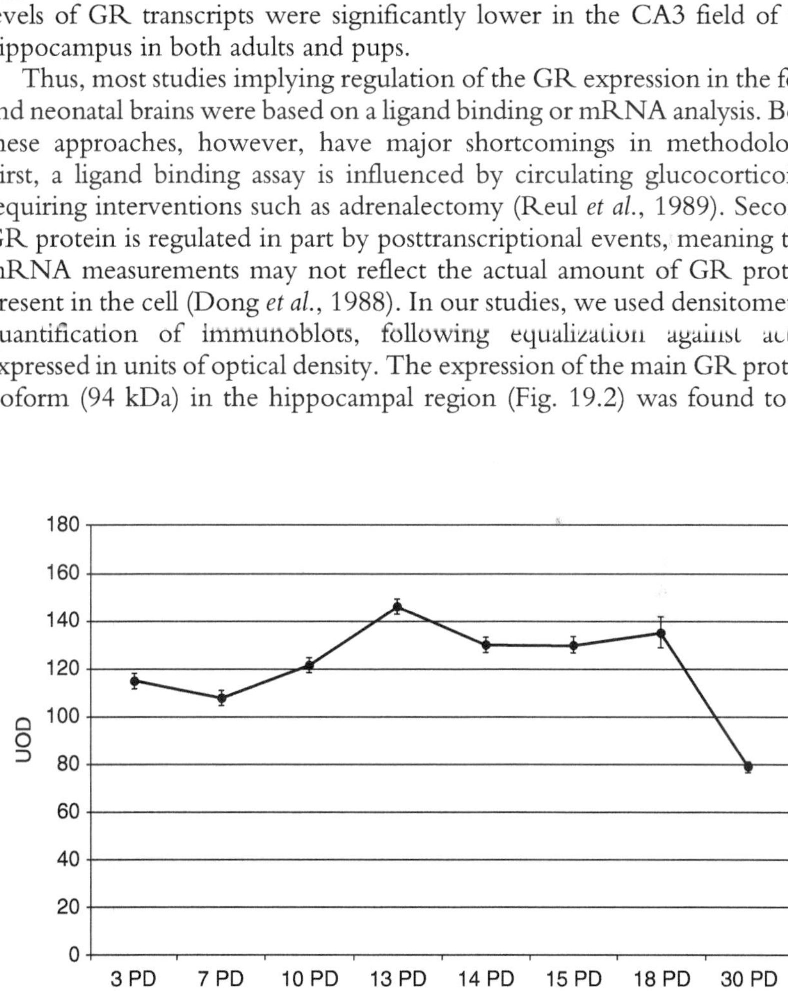


**Figure 19.2** Postnatal ontogeny of glucocorticoid receptor protein expression in the hippocampus of rats. Data represent densitometric quantification of immunoblots, following equalization against actin, expressed in units of optical density (UOD). Each data point depicts the mean ± SEM of four determinations at each postnatal day. The whole procedure was performed as described previously by Galeeva *et al.* (2006).

resembling with the pattern that previously has already been described by means of using the protein-binding assay. The protein was detected at a rather high level already at PD 3, reached the maximum by PD 13, and then slightly decreased after PD 18. At the same time, regional distribution of the GR protein within the hippocampus was rather similar to the distribution of the GR transcripts distribution (Fig. 19.3): a very low level of the protein is seen in the CA3 and the DG subfields, and fairly high levels in the CA1 and the CA2 subfields. In early studies of the hippocampal GR-immunoreactivity (GRir) distribution in the postnatal period, it was described that GRir had the U-shaped intensity curve over the course of the postnatal development (Lawson *et al.,* 1991; Rosenfeld *et al.*, 1988b), showing a remarkable decline in the GRir around PD 10–12. The studies have been performed by using a monoclonal antibody that recognizes preferably the activated form of the receptor. This may explain the discrepancy between this and the earlier reports. The drastically different results we got might simply have occurred by the biological nature of a polyclonal antibody, which also recognizes other protein form of the receptor (Galeeva *et al.*, 2006).

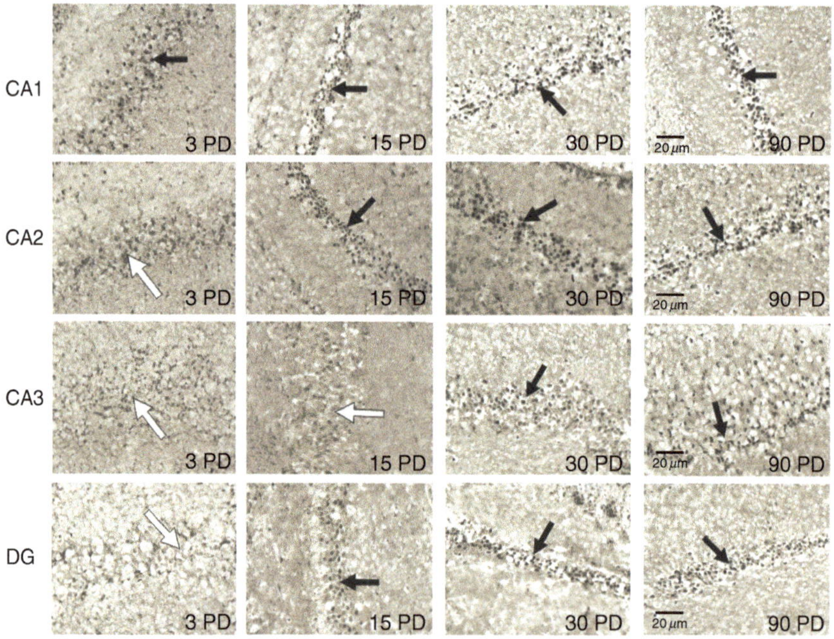


**Figure 19.3** Photomicrographs showing immunohistochemical staining detected by a polyclonal antibody against the glucocorticoid receptor in the different hippocampal regions. Each row represent the one area of the rat hippocampus at selected postnatal days (3, 15, 30, 90 PD). The brains were processed as described by Galeeva *et al.* (2006). Empty arrows depict the negative signal. CA1, CA2, and CA3, pyramidal cell fields of the Ammon's horn; DG, dentate gyrus.

Nowadays, the polymorphic nature of the GR gene has been described for many mammal species (mouse, rat, human) (for review see Duma *et al.*, 2006; Ju *et al.*, 2009). In fact, the two phenotypically distinct GR$\alpha$ and GR$\beta$ isoforms have been cloned from human tissues as far back as 1985 by Weinberger *et al.* (1985). Recently, an existence of multiple isoforms of GR$\alpha$ was described (Lu and Cidlowski, 2005). It has been shown that four main functional GR$\alpha$ isoforms are alternatively translated from different start codons placed in the N-terminal region. These codons are highly conserved within such species as human, monkey, rat, and mouse. Their corresponding proteins, weighing 94, 91, 82, and 54 kDa were assigned as follows: GR-A, GR-B, GR-C, and GR-D respectively. All reported isoforms were found in both rat and mouse species with different abundances among various tissues (Lu and Cidlowski, 2005). The experiments show that GR$\alpha$ N-terminal isoforms represent a functional receptor with its own distinct ability to confer a glucocorticoid response to a given cell. All existing isoforms exhibit a similar affinity for glucocorticoids and the most of them undergo ligand-induced nuclear localization (Lu and Cidlowski, 2005). For some of the isoforms, the specific activities have been delineated. Indeed, GR$\alpha$-C3 displayed the highest transcriptional activity, whereas others (A, B, C1 and C2) were intermediate. Expression of the GR$\alpha$-C3 correlated with increased sensitivity to glucocorticoid-induced apoptosis and the relatively inactive GR$\alpha$-D3 was associated with resistance to glucocorticoid-induced apoptosis (Lu and Cidlowski, 2005). During postnatal period of the brain development, high levels of GR are presented in brain structures that show intensive cell proliferation and restructuring. In this connection, the previous data on the pattern of GR proteins expression over the postnatal ontogeny (Galeeva *et al.*, 2006) may have particular meaning, which needs to be studied in more detail.

## B. Ontogeny of the mineralocorticoid receptor

The activation of the MR is required to maintain neuronal integrity and a stable excitatory tone, at least in the hippocampus (Joëls *et al.*, 2007). The excitatory output from the hippocampus to inhibitory interneurons in the hypothalamus can enhance inhibitory input to the CRH-producing cells in the PVN (Herman *et al.*, 2003). These findings may explain the inhibitory tone exerted via hippocampal MRs on basal and stress-induced HPA activity in rodents and and humans (Kellner and Wiedemann, 2008; Reul *et al.*, 2000). Hippocampal MRs are also involved in cognitive processes underlying the novelty assessment and the flexibility in the selection of an appropriate behavioral response to a challenging situation (Oitzl *et al.*, 1994).

In contrast to the well-defined expression of the hippocampal GRs described above, perinatal detection of MRs in the hippocampus was earlier regarded as rather restricted due to technical detection limits. Binding capacity of MR increased rapidly achieving adult values somewhere between the first and the second postnatal week (Cirini *et al.*, 1985; Rosenfeld *et al.*, 1988a; Rosenfeld *et al.*, 1990). These authors also found that prior to PD 8, MRs are not extracted from the nuclear fraction unless a high-ionic strength is used (Rosenfeld *et al.*, 1990). By using a selective GR agonist (RU 28362) to discriminate between MR and GR sites, it has also been shown that MRs are not detectable until PD 8 (Rosenfeld *et al.*, 1988a), but from this day on, MR demonstrated adult-like affinity and capacity.

The study of the MR protein expression in the rat hippocampus over the period (Fig. 19.4) of the postnatal development revealed a sustained rise of the intensity signal starting from PD 9 onwards. Simultaneously, the intensity of the protein band was under the detection limits up to this time.

Altogether, the findings suggest that the MR proteins are expressed in the hippocampus already perinatally, but at a rather low level.

In conclusion, both the experimental data and the theoretical considerations lead to the confirmation of the capability of the hippocampal receptors to bind ligands already in the infant. This also tells of the own specific role that hippocampal receptors could play throughout the postnatal ontogeny.

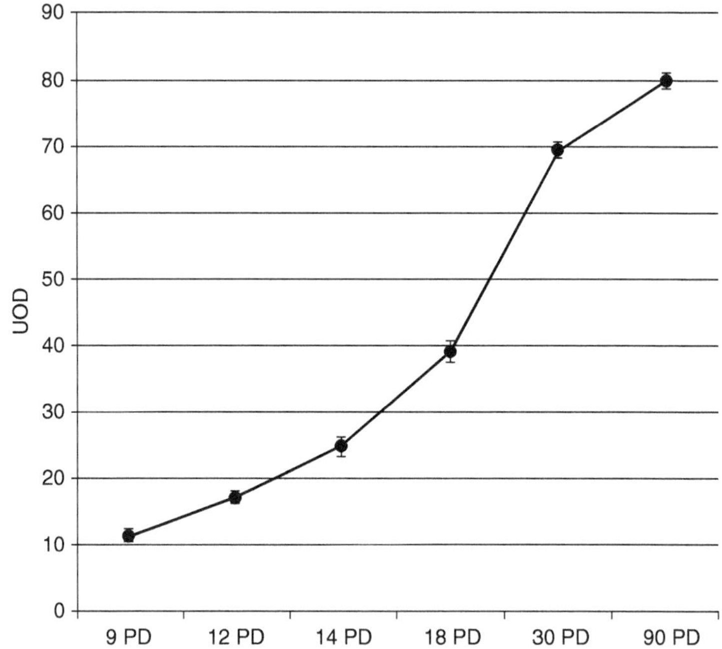


**Figure 19.4** Dynamic of postnatal expression of the mineralocorticoid receptor in the rat hippocampus. The mineralocorticoid receptor (MR) proteins in the hippocampal tissue extracts were first detected only at about PD 9, and since that time gradual increase in the expression is seen up to PD 14. From around the beginning of the second postnatal week, the expression of the MR starts to increase markedly.

## IV. Effect of Prenatal Stress on the Expression of the GR in the Hippocampus During Postnatal Development

A late gestational rise in endogenous plasma glucocorticoid levels occurs in the fetuses of many species and is necessary for tissue and organ maturation as well as for the initiation of parturition (Liggins, 1994). This trend corresponds to a widely observed increase in the HPA axis activity in the late gestation, but this also apparently requires a decrease of negative regulatory control of the HPA axis activity. While elevated levels of glucocorticoids during late pregnancy are required for normal fetal development (Liggins, 1994), it has been previously shown that exposure to high levels of endogenous or exogenous glucocorticoids leads to a number of negative consequences for fetuses, including fetal growth restriction (Newnham *et al.*, 1999), reduced brain growth (Huang *et al.*, 1999), reduced neuronal myelination (Quinlivan *et al.*, 1999), and fetal HPA axis hyperactivity near term (Sloboda *et al.*, 2000). Other studies have also shown that early developmental influences affecting HPA axis function may have profound long-term programming effects at the level of hippocampus (Levitt *et al.*, 1996; Meaney and Aitken 1985; Meaney *et al.*, 1989; Seckl and Meaney, 2006).

It is generally accepted that prenatal stress alters an individual's ability to cope with stress in adult life. Thus, it has been reported that rats that have experienced prenatal stress exhibit behavioral abnormalities under stressful conditions in adult life, such as increased anxiety (Vallée *et al.*, 1997) or emotionality (Fride *et al.*, 1986; Joffe, 1977; Thompson, 1957; Wakshlak and Weinstock, 1990). In humans, prenatal stress also causes severe abnormalities in infants after birth (Beversdorf *et al.*, 2005; for review see Nelson, 2000), as well as metabolic and neuroendocrine disorders such as hypertension, type 2 diabetes, ischemic heart disease, and different types of cognitive and behavioral disorders in adulthood (Seckl, 1998).

Although different animal models of perinatal stress have been studied since 1957 (Alonso *et al.*, 1991; Dahlof *et al.*, 1978; Fride and Weinstock, 1984; Maccari *et al.*, 1995; Peters, 1982; Seckl, 1998; Takahashi and Kalin, 1991; Thompson, 1957; Ward, 1972), there are still obscure sides to the question of the individual's maladaptation and the circumstances of its development from birth into adulthood.

Many authors have reported that the pathophysiological consequences of prenatal stress exposure are gender-dependent, but particular data supporting this outcome are still inconsistent. Thus, it has been described that prenatal restrain stress affects more female than male offspring, inducing elevated levels of blood ACTH, corticosterone, and CBG only in females

(McCormick *et al.*, 1995). Simultaneously, authors found no effect of prenatal stress on the GR density (measured by protein-binding assay) in the hippocampus. For mice, it was also found that female offspring are more sensitive to maternal stress, which expressed increased risk of depression development (Alonso *et al.*, 2000). However, there are other studies showing contrary evidences that males are more liable to be affected by prenatal stress (Ward, 1972; Weinstock, 2007), in particular they are more prone to develop cognitive retardation and, related to this, impaired neurogenesis in the hippocampus. In part, such contradiction can be referred to a genetic component that has also been implicated in mediation of the effects of maternal stress (Stöhr *et al.*, 1998). However, most researchers fail to take into account the early postnatal development of HPA axis, when they examine the psychophysiologic consequences of prenatal stress on adult offspring. From our point of view, however, it is extremely important to study in detail, first, the entire period of postnatal development, whereas all three postnatal weeks are crucial for the proper establishment of HPA axis functioning in the adulthood.

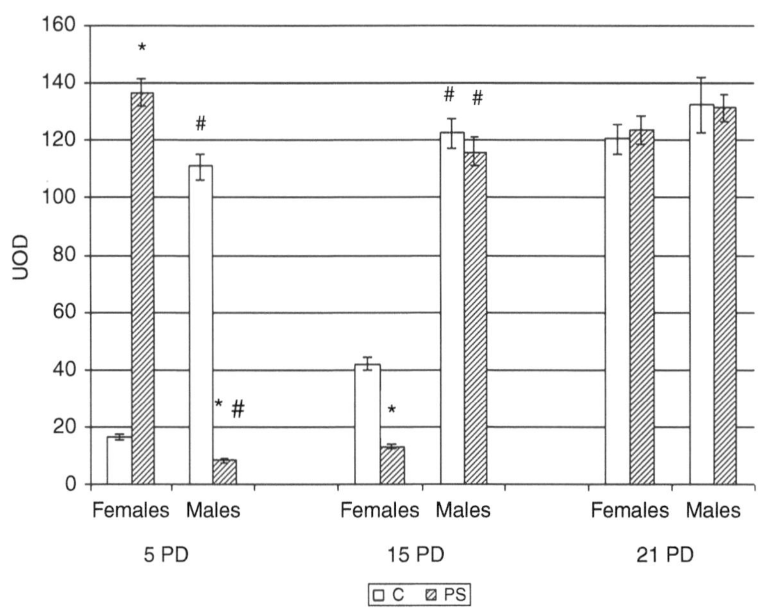


**Figure 19.5** Developmental changes in the GR protein expression in the hippocampus of control (empty bar) and prenatally stressed (hatched bars) rats (n = 8 (control) + n = 8 (prenatal stress) per day) (*) Inter-group statistically significant differences; (#) statistically significant differences between various days in each group. Data bars are mean ± SEM of six separate experiments. The data indicate that during first two weeks males and females are significantly different from each other, whereas by PD 21 the differences disappear.

Animal model of maternal stress, such as prenatal restrain stress used in our laboratory, is widely accepted as a particularly useful paradigm. Our data show that developing HPA axis of both male and female offspring was influenced by maternal stress, and the pattern of changes occurred in a sex-specific manner. Thus, the amount of GR protein in the hippocampus of prenatally stressed males was expressed 10 times less than in the hippocampus of control males on the fifth PD (Fig. 19.5), but this enormous difference disappeared completely by PD 15. However, at the adrenal level, the effect of prenatal stress on plasma corticosterone (Fig. 19.6) in the male offspring appeared to be more influential only at the end of the second postnatal week, whereas prenatal stress led to a significant increase in plasma corticosterone in females only at the fifth PD. Simultaneously, the substantial changes elicited by prenatal stress in the expression of the GR protein in the hippocampus of the female offspring were found at both times of measurement, though the changes were opposite. It is interesting to note that all the effects of the prenatal stress described above disappeared completely from PD 21 onwards, that is, from the beginning of the prepubertal period.

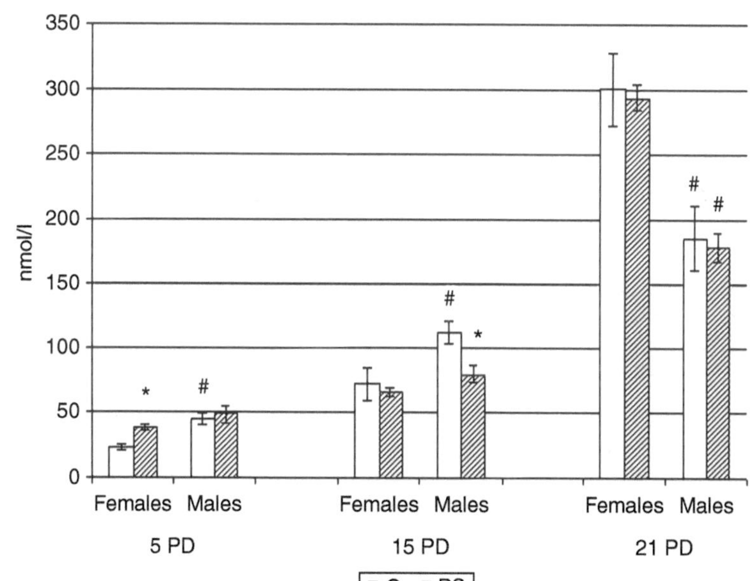


**Figure 19.6** Developmental changes in the plasma corticosterone levels of control (empty bar) and prenatally stressed (hatched bars) rats at selected postnatal days (the same groups as represented on Fig. 5). (*) Inter-group statistically significant differences; (#) statistically significant differences between various days in each group.

In summary, taking into account our data and the data available from the literature, we can speculate that in the adult the HPA system pays for the extra efforts that have been made in the early postnatal life to correct the negative consequences of the prenatal stress.

## REFERENCES


Ahima, R. S., Prabakaran, D., and Flier, J. S. (1998). Postnatal leptin surge and regulation of circadian rhythm of leptin by feeding. Implications for energy homeostasis and neuroendocrine function. *J. Clin. Invest.* **101,** 1020–1027.

Alonso, S. J., Arevalo, R., Afonso, D., and Rodriguez, M. (1991). Effects of maternal stress during pregnancy on forced swimming test behavior of the offspring. *Physiol. Behav.* **50,** 511–517.

Alonso, S. J., Damas, C., and Navarro, E. (2000). Behavioral despair in mice after prenatal stress. *J. Physiol. Biochem.* **56,** 77–82.

Antonow-Schlorke, I., Schwab, M., Li, C., and Nathanielsz, P. W. (2003). Glucocorticoid exposure at the dose used clinically alters cytoskeletal proteins and presynaptic terminals in the fetal baboon brain. *J. Physiol.* **547,** 117–123.

Arai, M., and Widmaier, E. P. (1993). Steroidogenesis in isolated adrenocortical cells during development in rats. *Mol. Cell. Endocrinol.* **92,** 91–97.

Atkinson, H. C., Wood, S. A., Castrique, E. S., Kershaw, Y. M., Wiles, C. C., and Lightman, S. L. (2008). Corticosteroids mediate fast feedback of the rat hypothalamic–pituitary–adrenal axis via the mineralocorticoid receptor. *Am. J. Physiol. Endocrinol. Metab.* **294,** E1011–E1022.

Avishai-Eliner, S., Hatalski, C. G., Tabachnik, E., Eghbal-Ahmadi, M., and Baram, T. Z. (1999). Differential regulation of glucocorticoid receptor messenger RNA (GR-mRNA) by maternal deprivation in immature rat hypothalamus and limbic regions. *Brain Res. Dev. Brain Res.* **114,** 265–268.

Barker, D. J. (2007). The origins of the developmental origins theory. *J. Int. Med.* **261,** 412–417.

Beversdorf, D. Q., Manning, S. E., Hillier, A., Anderson, S. L., Nordgren, R. E., Walters, S. E., Nagaraja, H. N., Cooley, W. C., Gaelic, S. E., and Bauman, M. L. (2005). Timing of prenatal stressors and autism. *J. Autism Dev. Disord.* **35,** 471–478.

Bingham, B., and Viau, V. (2008). Neonatal gonadectomy and adult testosterone replacement suggest an involvement of limbic arginine vasopressin and androgen receptors in the organization of the hypothalamic-pituitary-adrenal axis. *Endocrinology* **149,** 3581–3591.

Bohn, M. C., and Friedrich, V. L. Jr. (1982). Recovery of myelination in rat optic nerve after developmental retardation by cortisol. *J. Neurosci.* **2,** 1292–1298.

Bohn, M. C., Dean, D., Hussain, S., and Guiliano, R. (1994). Development of mRNAs for glucocorticoid and mineralocorticoid receptors in rat hippocampus. *Dev. Brain Res.* **77,** 157–162.

Brown, R. W., Diaz, R., Robson, A. C., Kotelevtsev, Y. V., Mullins, J. J., Kaufman, M. H., and Seckl, J. R. (1996). The ontogeny of 11 beta-hydroxysteroid dehydrogenase type 2 and mineralocorticoid receptor gene expression reveal intricate control of glucocorticoid action in development. *Endocrinology* **137,** 794–797.

Campagne, D. L., and Meaney, M. J. (2001). Like mother, like daughter: Evidence for nongenomic transmission of parental behavior and stress responsivity. *Prog. Brain Res.* **133,** 287–302.

Campagne, D. L., de Kloet, E. R., and Joels, M. (2009). Fundamental aspects of the impact of glucocorticoids on the (immature) brain. *Semin. Fetal Neonatal Med.* **14,** 136–142.

Chatelain, A., and Dupouy, J. P. (1981). Activity of the pituitary-adrenal system in rat fetuses subjected to encephalectomy in early or late stages of pregnancy. *Neuroendocrinology* **33,** 148–152.

Cirini, H., Magarinos, A. M., De Nicola, A. F., Rainbow, T. C., and McEwen, B. S. (1985). Further studies of brain aldosterone binding sites employing new mineralocorticoid and glucocorticoid receptor markers in vitro. *Brain Res.* **361,** 212–216.

Critchlow, V., Liebelt, R. A., Bar-Sela, M., Mountcastl, E. W., and Lipscomb, H. S. (1963). Sex difference in resting pituitary-adrenal function in the rat. *Am. J. Physiol.* **205,** 807–815.

Dahlof, L. G., Hard, E., and Larsson, K. (1978). Influence of maternal stress on the development of the fetal genital system. *Physiol. Behav.* **20,** 193–195.

Daikoku, S., Kawano, H., Abe, K., and Yoshinada, K. (1981). Topographical appearance of adenohypophyseal cells with special reference to the development of the portal system. *Arch. Histol. Jpn.* **44,** 103–116.

Daikoku, S., Okamura, Y., Kawano, H., Tsuruo, Y., Malgawa, M., and Shibasaki, T. (1984). Immunohistochemical study on the development of CRF-containing neurons in the hypothalamus of the rat. *Cell Tissue Res.* **238,** 539–544.

De Kloet, E. R. (2003). Hormones, brain and stress. *Endocr. Regul.* **37,** 51–68.

De Kloet, E. R., and Reul, J. M. (1987). Feedback action and tonic influence of corticosteroids on brain function: A concept arising from the heterogeneity of brain receptor systems. *Psychoneuroendocrinology* **12,** 83–105.

De Kloet, E. R., Rosenfeld, P., Van Eekelen, J. A. M., Sutanto, W., and Levine, S. (1988). Stress, glucocorticoids and development. *In* "Progress in Brain Research" (G.J. Boer, M. G.P. Feenstra, M. Mirmiran, and D.F. Swaab, Eds.), pp. 101–120. Elsevier Science, Amsterdam.

De Kloet, E. R., Vreugdenhil, E., Oitzl, M. S., and Joels, M. (1998). Brain corticosteroid balance in the health and disease. *Endocr. Rev.* **19,** 269–301.

Devaskar, U., Magyar, D., Fridshal, D., Buster, J., and Nathanielsz, P. W. (1980). Development of responsiveness of dispersed rabbit adrenocortical cells to synthetic adrenocorticotropic hormone [ACTH-(1-24)] and alpha melanocyte-stimulating hormone. *Endocrinology* **107,** 809–815.

Dodic, M., Moritz, K., and Wintour, E. M. (2003). Prenatal exposure to glucocorticoids and adult disease. *Arch. Physiol. Biochem.* **111,** 61–69.

Dong, Y., Poellinger, L., Gustafsson, J.-A., and Okret, S. (1988). Regulation of glucocorticoids receptor expression: Evidence for transcriptional and posttranslational mechanisms. *Mol. Endocrinol.* **2,** 1254–1256.

Doupe, A. J., and Patterson, P. H. (1982). Glucocorticoids and the developing nervous system. *In* "Current Topics in Neuroendocrinology" (D. Ganten and D. Pfaff, Eds.), pp. 23–43. Springer, Berlin, Heidelberg.

Duma, D., Jewell, C. M., and Cidlowski, J. A. (2006). Multiple glucocorticoid receptor isoforms and mechanisms of post-translational modification. *J. Steroid Biochem. Mol. Biol.* **102,** 11–21.

Enthoven, L., Oitzl, M. S., Koning, N., van der Mark, M., and de Kloet, E. R. (2008). Hypothalamic-pituitary-adrenal axis activity of newborn mice rapidly desensitizes to repeated maternal absence, but becomes highly responsive to novelty. *Endocrinology* **149,** 6366–6377.

Erdmann, G., Schütz, G., and Berger, S. (2008). Loss of glucocorticoid receptor function in the pituitary results in early postnatal lethality. *Endocrinology* **149,** 3446–3451.

Feldman, S., and Weidenfeld, J. (1999). Glucocorticoid receptor antagonists in hippocampus modify the negative feedback following neural stimuli. *Brain Res.* **821,** 33–37.

Figueiredo, H. F., Ulrich-Lai, Y. M., Choi, D. C., and Herman, J. P. (2007). Estrogen potentiates adrenocortical responses to stress in female rats. *Am. J. Physiol. Endocrinol. Metab.* **292,** E1173–E1182.

Fride, E., Dan, Y., Feldon, J., Halevy, G., and Weinstock, M. (1986). Effects of prenatal stress on vulnerability to stress in prepubertal and adult rats. *Physiol. Behav.* **37,** 681–687.

Fride, E., and Weinstock, M. (1984). The effects of prenatal exposure to predictable or unpredictable stress on early development in the rat. *Dev. Psychobiol.* **17,** 651–660.

Galeeva, A., Ordyan, N., Pivina, S., and Pelto-Huikko, M. (2006). Expression of glucocorticoid receptors in the hippocampal region of the rat brain during postnatal development. *J. Chem. Neuroanat.* **31,** 216–225.

Ghosh, B., Wood, C. R., Held, G. A., Abbott, B. D., and Lau, C. (2000). Glucocorticoid receptor regulation in the rat embryo: A potential site for developmental toxicity? *Toxicol. Appl. Pharmacol.* **164,** 221–229.

Gluckman, P. D., Cutfield, W., Hofman, P., and Hanson, M. A. (2005). The fetal, neonatal, and infant environments—The long-term consequences for disease risk. *Early Hum. Dev.* **81,** 51–59.

Grino, M., Young, W. S. III, and Burgunder, J.-M. (1989). Ontogeny of expression of the corticotropin-releasing factor gene in the hypothalamic paraventricular nucleus and of the proopiomelanocortin gene. *Psychoneuroendocrinology* **124,** 60–68.

Gunnar, M. R., and Donzella, B. (2002). Social regulation of the cortisol level in early human development. *Psychoneuroendocrinology* **27,** 199–220.

Halasz, I., Rittenhouse, P. A., Zorrila, E. P., and Redei, E. (1997). Sexually dimorphic effects of maternal adrenalectomy on the hypothalamic corticotrophin-releasing factor, glucocorticoid receptor and anterior pituitary POMC mRNA levels in rat neonates. *Dev. Brain Res.* **100,** 198–204.

Herman, J. P., and Cullinan, W. E. (1997). Neurocircuitry of stress: Central control of the hypothalamo-pituitary-adrenocortical axis. *Trends Neurosci.* **20,** 78–84.

Herman, J. P., Schafer, M. K., Young, E. A., Thompson, R., Douglass, J., Akil, H., and Watson, S. J. (1989). Evidence for hippocampal regulation of neuroendocrine neurons of the hypothalamo-pituitary-adrenocortical axis. *J. Neurosci.* **9,** 3072–3082.

Herman, J. P., Figueiredo, H., Mueller, N. K., Ulrich-Lai, Y., Ostrander, M. M., Choi, D. C., and Cullinan, W. E. (2003). Central mechanisms of stress integration: Hierarchical circuitry controlling hypothalamo–pituitary–adrenocortical responsiveness. *Front. Neuroendocrinol.* **24,** 151–180.

Hindelang, C., Felix, J. M., Laurent, F. M., Klein, M. J., and Stoeckel, M. E. (1990). Ontogenesis of proopiomelanocortin gene expression and regulation in the rat pituitary intermediate lobe. *Mol. Cell. Endocrinol.* **70,** 225–235.

Howard, E., and Benjamins, J. A. (1975). DNA, ganglioside and sulfatide in brains of rats given corticosterone in infancy, with an estimate of cell loss during development. *Brain Res.* **92,** 73–87.

Huang, W. L., Beazley, L. D., Quinlivan, J. A., Evans, S., Newnham, J., and Dunlop, S. A. (1999). Effect of corticosteroids on brain growth in fetal sheep. *Obstet. Gynecol.* **94,** 213–218.

Jacobson, L., and Sapolsky, R. M. (1991). The role of the hippocampus in feedback regulation of the hypothalamo-pituitary-adrenocortical axis. *Endocr. Rev.* **12,** 118–134.

Joffe, J. M. (1977). Modification of prenatal stress effects in rats by dexamethasone and adrenocorticotrophin. *Physiol. Behav.* **19,** 601–606.

Joëls, M., Karst, H., Krugers, H. J., and Lucassen, P. J. (2007). Chronic stress: Implications for neuronal morphology, function and neurogenesis. *Front. Neuroendocrinol.* **28,** 72–96.

Jones, C. T., and Roebuck, M. M. (1980). The development of the pituitary-adrenal axis in the guinea-pig. *Acta Endocrinol.* **94,** 107–116.

Ju, H., Wang, X., Liu, Z., Liu, P., Zhao, Y., Xiong, R., Zhou, Y., and Chen, X. (2009). Identification and tissue distribution of a novel rat glucocorticoid receptor splice variant. *Acta Biochim. Pol.* **56,** 109–113.

Kalinyak, J. E., Griffin, C. A., Hamilton, R. W., Bradshaw, J. G., Perlman, A. J., and Hoffman, A. R. (1989). Development and hormonal regulation of glucocorticoid receptor messenger RNA in the rat. *J. Clin. Invest.* **84,** 1843–1848.

Kapellou, O., Ajaye-Obe, M., Kennea, N., Counsell, S., Allsop, J., Saeed, N., Duggan, P., Maalouf, E., Laroche, S., Cowan, F., Rutherford, M., and Edwards, A. D. (2003). Quantitation of brain development in preterm infants treated with corticosteroids. *Early Hum. Dev.* **73,** 111–123.

Keegan, C. E., Herman, J. P., Karolyi, I. J., O'Shea, K. S., Camper, S. A., and Seasholtz, A. F. (1994). Differential expression of corticotropin-releasing hormone in developing mouse embryos and adult brain. *Endocrinology* **34,** 2547–2555.

Kellner, M., and Wiedemann, K. (2008). Mineralocorticoid receptors in brain, in health and disease: Possibilities for new pharmacotherapy. *Eur. J. Pharmacol.* **583,** 372–378.

Knuth, E. D., and Etgen, A. M. (2005). Corticosterone secretion induced by chronic isolation in neonatal rats is sexually dimorphic and accompanied by elevated ACTH. *Horm. Behav.* **47,** 65–75.

Lawson, A., Ahima, R., Krozowski, Z., and Harlan, R. (1991). Postnatal development of corticosteroid receptor immunoreactivity in the rat hippocampus. *Dev. Brain Res.* **62,** 69–79.

Le Mevel, J. C., Abitbol, S., Beraud, G., and Maniey, J. (1979). Temporal changes in plasma adrenocorticotropin concentration after repeated neurotropic stress in male and female rats. *Endocrinology* **105,** 812–817.

Levin, R., and Levine, S. (1975). Development of circadian periodicity in base and stress levels of corticosterone. *Am. J. Physiol.* **229,** 1397–1399.

Levine, S. (1994). The ontogeny of the hypothalamic-pituitary-adrenal axis. The influence of maternal factors. *Ann. NY Acad. Sci.* **764,** 275–288.

Levine, S. (2001). Primary social relationships influence the development of the hypothalamic-pituitary-adrenal axis in the rat. *Physiol. Behav.* **73,** 255–260.

Levine, S., Berkenbosch, F., Suchecki, D., and Tilders, F. J. D. (1994). Pituitary-adrenal and interleukin-6 responses to recombinant interleukin-1 in neonatal rats. *Psychoneuroendocrinology* **19,** 143–153.

Levitt, N. S., Lindsay, R. S., Holmes, M. C., and Seckl, J. R. (1996). Dexamethasone in the last week of pregnancy attenuates hippocampal glucocorticoid receptor gene expression and elevates blood pressure in the adult offspring in the rat. *Neuroendocrinology* **64,** 412–419.

Liggins, G. C. (1994). The role of cortisol in preparing the fetus for birth. *Reprod. Fertil. Dev.* **6,** 141–150.

Lu, N. Z., and Cidlowski, J. A. (2005). Translational regulatory mechanisms generate N-terminal glucocorticoid receptor isoforms with unique transcriptional target genes. *Mol. Cell* **18,** 331–342.

Maccari, S., Piazza, P. V., Kabbaj, M., Barbazanges, A., Simon, H., and Le Moal, M. (1995). Adoption reverses the long-term impairment in glucocorticoid feedback induced by prenatal stress. *J. Neurosci.* **15,** 110–116.

McCormick, C. M., Smythe, J. W., Sharma, S., and Meaney, M. J. (1995). Sex-specific effects of prenatal stress on hypothalamicpituitary-adrenal responses to stress and brain glucocorticoid receptor density in adult rats. *Dev. Brain Res.* **84,** 55–61.

Meaney, M. J., and Aitken, D. H. (1985). The effects of early postnatal handling on hippocampal glucocorticoid receptor concentrations: Temporal parameters. *Dev. Brain Res.* **22,** 301–304.

Meaney, M. J., Sapolsky, R. M., and McEwen, B. S. (1985). The development of the glucocorticoid receptor system in the rat limbic brain. I. Ontogeny and autoregulation. *Brain Res.* **350,** 159–164.

Meaney, M. J., Aitken, D. H., Viau, V., Sharma, S., and Sarrieau, A. (1989). Neonatal handling alters adrenocortical negative feedback sensitivity and hippocampal type II glucocorticoid receptor binding in the rat. *Neuroendocrinology* **50,** 597–604.

Meijs-Roelofs, H. M. A., and Kramer, P. (1979). Maturation of the inhibitory feedback action of oestrogen on follicle-stimulating hormone secretion in the immature female rat: A role for the alpha-foetoprotein. *J. Endocrinol.* **81,** 199–208.

Meyer, J. S. (1985). Biochemical effects of corticosteroids on neural tissues. *Physiol. Rev.* **65,** 946–1021.

Nagaya, M., Arai, M., and Widmaier, E. P. (1995). Ontogeny of immunoreactive and bioactive microsomal steroidogenic enzymes during adrenocortical development in rats. *Mol. Cell. Edocrinol.* **114,** 27–34.

Nelson, C. A. (2000). The effects of early adversity on neurobehavioral development. Minnesota symposia on child psychology. *Lawrence Erlbaum Associates.* Vol. **31**, Mahwah, NJ, 345.

Newnham, J. P., Evans, S. F., Godfrey, M., Huang, W., Ikegami, M., and Jobe, A. (1999). Maternal, but not fetal, administration of corticosteroids restricts fetal growth. *J. Matern. Fetal Med.* **8,** 81–87.

Noorlander, C. W., De Graan, P. N. E., Middeldorp, J., van Beers, J. J. B. C., and Visser, G. H. A. (2006). Ontogeny of hippocampal corticosteroid receptors: Effects of antenatal glucocorticoids in human and mouse. *J. Comp. Neurol.* **499,** 924–932.

O'Grady, M. P., Hall, N. R. S., and Menzies, R. A. (1993). Interleukin-1$\beta$ stimulates adrenocorticotropin and corticosterone release in 10-day-old rat pups. *Psychoneuroendocrinology* **18,** 241–247.

Oitzl, M. S., Fluttert, M., and de Kloet, E. R. (1994). The effect of corticosterone on reactivity to spatial novelty is mediated by central mineralocorticosteroid receptors. *Eur. J. Neurosci.* **6,** 1072–1079.

Okimoto, D. K., Blaus, A., Schmidt, M., Gordon, M. K., Dent, G. W., and Levine, S. (2002). Differential expression of c-fos and tyrosine hydroxylase mRNA in the adrenal gland o the infant rats: Evidence for an adrenal hyporesponsive period. *Endocrinology* **143,** 1717–1725.

Ordyan, N. E., Pivina, S. G., Rakitskaya, V. V., and Shalyapina, V. G. (2001). The neonatal glucocorticoid treatment-produced long-term changes of the pituitary-adrenal function and brain corticosteroid receptors in rats. *Steroids* **66,** 883–888.

Papaioannou, A., Gerozissis, K., Prokopiou, A., Bolaris, S., and Stylianopoulou, F. (2002). Sex differences in the effects of neonatal handling on the animal's response to stress and the vulnerability for depressive behaviour. *Behav. Bran Res.* **129,** 131–139.

Peters, D. A. (1982). Prenatal stress: Effects on brain biogenic amine and plasma corticosterone levels. *Pharmacol. Biochem. Behav.* **17,** 721–725.

Quinlivan, J. A., Dunlop, S. A., Newnham, J., Evans, S. F., and Beazley, L. D. (1999). Repeated, but not single, maternal administration of corticosteroids delays myelination in the brain of fetal sheep. *Prenat. Neonatal Med.* **4,** 47–55.

Raynaud, J.-P., Mercier-Bobard, C., and Baulieu, E. E. (1971). Rat estradiol binding plasma protein (EBP). *Steroids* **18,** 767–788.

Renard, G. M., Rivarola, M. A., and Suarez, M. M. (2007). Sexual dimorphism in rats: Effects of early maternal separation and variable chronic stress on pituitary-adrenal axis and behavior. *Int. J. Dev. Neurosci.* **25,** 373–379.

Répérant, N. E., and Durand, P. (1997). The development of the ovine fetal adrenal gland and its regulation. *Reprod. Nutr. Dev.* **37,** 81–95.

Reul, J. M., and de Kloet, E. R. (1985). Two receptor systems for corticosterone in rat brain: Microdistribution and differential occupation. *Endocrinology* **117,** 2505–2511.

Reul, J. M. H. M., Pearce, P. T., Funder, J. W., and Krozowski, Z. S. (1989). Type I and type II corticosteroid receptor gene expression in the rat: Effect of adrenalectomy and dexamethasone administration. *Mol. Endocrinol.* **3,** 1674–1680.

Reul, J. M., Gesing, A., Droste, S., Stec, I. S., Weber, A., Bachmann, C., Bilang-Bleuel, A., Holsboer, F., and Linthorst, A. C. (2000). The brain mineralocorticoid receptor: Greedy for ligand, mysterious in function. *Eur. J. Pharmacol.* **405,** 235–249.

Romeo, R. D., Lee, S. J., and McEwen, B. S. (2004). Differential stress reactivity in intact and ovariectomized prepubertal and adult female rats. *Neuroendocrinology* **80,** 387–393.

Rosenfeld, P., Sutanto, W., Levine, S., and de Kloet, E. R. (1988a). Ontogeny of type I and type II corticosteroid receptors in the rat hippocampus. *Dev. Brain Res.* **42,** 113–118.

Rosenfeld, P., van Eekelen, J. A. M., Levine, S., and de Kloet, E. R. (1988b). Ontogeny of the type 2 glucocorticoid receptor in discrete rat brain region: An immunocytochemical study. *Dev. Brain Res.* **42,** 119–127.

Rosenfeld, P., Sutanto, W., Levine, S., and de Kloet, E. R. (1990). Ontogeny of mineralocorticoid (type 1) receptors in brain and pituitary: an in vivo autoradiographical study. *Brain Res. Dev. Brain Res.* **52,** 57–62.

Rosenfeld, P., Gutierrez, Y. A., Martin, A. M., Mallett, H. A., Alleva, E., and Levine, S. (1991). Maternal regulation of the adrenocortical response in preweanling rats. *Physiol. Behav.* **50,** 661–671.

Sakly, M., and Koch, B. (1981). Ontogenesis of glucocorticoid receptors in anterior pituitary gland: Transient dissociation among cytoplasmic receptor density, nuclear uptake, and regulation of corticotropic activity. *Endocrinology* **108,** 591–596.

Sakly, M., and Koch, B. (1983). Ontogenetical variations of transcortin modulate glucocorticoids receptor function and corticotropic activity in the pituitary gland. *Horm. Metab. Res.* **15,** 92–96.

Sapolsky, R. M., and Meaney, M. J. (1986). Maturation of the adrenocortical stress response: Neuroendocrine control mechanisms and the stress hyporesponsiveness period. *Brain Res. Rev.* **11,** 65–76.

Schmidt, M., Enthoven, L., van der Mark, M., Levine, S., de Kloet, E. R., and Oitzl, M. S. (2003). The postnatal development of the hypothalamic-pituitary-adrenal axis in the mouse. *Int. J. Dev. Neurosci.* **21,** 125–132.

Schmidt, M., Enthoven, L., van Woezik, J. H. G., Levine, S., de Kloet, E. R., and Oitzl, M. S. (2004). The dynamics of the hypothalamic-pituitary-adrenal axis during maternal deprivation. *J. Neuroendocrinol.* **16,** 52–57.

Seale, J. V., Wood, S. A., Atkinson, H. C., Bate, E., Lightman, S. L., Ingram, C. D., Jessop, D. S., and Harbuz, M. S. (2004). Gonadectomy reverses the sexually diergic patterns of circadian and stress-induced hypothalamo-pituitary-adrenal axis activity in male and female rats. *J. Neuroendocrinol.* **16,** 516–524.

Seale, J. V., Wood, S. A., Atkinson, H. C., Harbuz, M. S., and Lightman, S. L. (2005). Postnatal masculinization alters the HPA axis phenotype in the adult female rat. *J. Physiol.* **563,** 265–274.

Seckl, J. R. (1998). Physiologic programming of the fetus. *Clin. Perinatol.* **25,** 939–962 vii.

Seckl, J. R., and Holmes, M. C. (2007). Mechanisms of disease: Glucocorticoids, their placental metabolism and fetal 'programming' of adult pathophysiology. *Nat. Clin. Pract. Endocrinol. Metab.* **3,** 479–488.

Seckl, J. R., and Meaney, M. J. (2006). Glucocorticoid "programming" and PTSD risk. *Ann. NY Acad. Sci.* **1071,** 351–378.

Sloboda, D. M., Newnham, J., and Challis, J. R. G. (2000). Effects of repeated maternal betamethasone administration on growth and hypothalamic–pituitary–adrenal function of the ovine fetus at term. *J. Endocrinol.* **165,** 79–91.

Sokka, T. A., and Huhtaniemi, I. T. (1995). Functional maturation of the pituitary-gonadal axis in the neonatal female rat. *Biol. Reprod.* **52,** 1404–1409.

Sousa, N., Cergueira, J. J., and Almeida, O. F. (2008). Corticosteroid receptors and neuroplasticity. *Brain Res. Rev.* **57,** 561–570.

Stöhr, T., Schulte Wermeling, D., Szuran, T., Pliska, V., Domeney, A., Welzl, H., Weiner, I., and Feldon, J. (1998). Differential effects of prenatal stress in two inbred strains of rats. *Pharmacol. Biochem. Behav.* **59,** 799–805.

Suchecki, D., Mozaffarian, D., Gross, G., Rosenfeld, P., and Levine, S. (1993). Effects of maternal deprivation on the ACTH stress response in the infant rat. *Neuroendocrinology* **57,** 204–212.

Sweezey, N. B., Ghibu, F., Gagnon, S., Schotman, E., and Hamid, Q. (1998). Glucocorticoid receptor mRNA and protein in fetal rat lung in vivo: Modulation by glucocorticoids and androgen. *Am. J. Physiol.* **275,** L103–L109.

Takahashi, L. K., and Kalin, N. H. (1991). Early developmental and temporal characteristics of stress-induced secretion of pituitary–adrenal hormones in prenatally stressed rat pups. *Brain Res.* **558,** 75–78.

Thompson, W. R. (1957). Influence of prenatal maternal anxiety on emotionality in young rats. *Science* **125,** 698–699.

Ugrumov, M. V., and Mitskevich, M. S. (1992). Development of neuroendocrine regulation during ontogenesis. *Sov. Sci. Rev. Sect. Physiol. Gen. Biol.* **5,** 41–49.

Ugrumov, M. V., Ivanova, I. P., Mitskevich, M. S., Liposits, Zs., Setalo, G., and Flerko, B. (1985). Axovascular relationships in developing median eminence of perinatal rats with special reference to luteinizing hormone-releasing hormone projections. *Neuroscience* **16,** 897–906.

Vallée, M., Mayo, W., Darnaudéry, M., Corpéchot, C., Young, J., Koehl, M., Le Moal, M., Baulieu, E. E., Robel, P., and Simon, H. (1997). Neurosteroids: deficient cognitive performance in aged rats depends on low pregnenolone sulfate levels in the hippocampus. *Proc. Natl. Acad. Sci. USA* **94,** 14865–14870.

Van Eekelen, J. A. M., Bohn, M. C., and de Kloet, E. R. (1991). Postnatal ontogeny of mineralocorticoid and glucocorticoid receptor gene expression in regions of the rat tel- and diencephalon. *Dev. Brain Res.* **61,** 33–43.

Van Oers, H. J., de Kloet, E. R., Li, C., and Levine, S. (1998). The ontogeny of glucocorticoid negative feedback: Influence of maternal deprivation. *Endocrinology* **139,** 2838–2846.

Vazquez, D. M., and Akhil, H. (1993). Pituitary-adrenal response to ether vapor in the weanling animal: Characterization of the inhibitory effect on glucocorticoids on adrenocortical secretion. *Pediatr. Res.* **34,** 646–653.

Viau, V., and Meaney, M. J. (1991). Variation in the hypothalamic-pituitary-adrenal response to stress during the estrous cycle in the rat. *Endocrinology* **129,** 2503–2511.

Viau, V., Sharma, S., and Meaney, M. J. (1996). Changes in plasma adrenocorticotropin, corticosterone, corticosteroid-binding globulin, and hippocampal glucocorticoid receptor occupancy/translocation in rat pups in response to stress. *J. Neuroendocrinol.* **8,** 1–8.

Wakshlak, A., and Weinstock, M. (1990). Neonatal handling reverses behavioral abnormalities induced in rats by prenatal stress. *Physiol. Behav.* **48,** 289–292.

Walker, C. D. (1995). Chemical sympathectomy and maternal separation affect neonatal stress responses and adrenal sensitivity to ACTH. *Am. J. Physiol.* **268,** R1281–R1288.

Walker, C. D., Sapolsky, R. M., Meaney, M. J., Vale, W. W., and Rivier, C. L. (1986). Increased pituitary sensitivity to glucocorticoid feedback during the stress nonresponsive period in the neonatal rat. *Endocrinology* **119,** 1816–1821.

Walker, C. D., Scribner, K. A., Cascio, C. S., and Dallma, N. M. F. (1991). The pituitary-adrenocortical system of neonatal rats is responsive to stress throughout development in a time-dependent and stressor-specific fashion. *Endocrinology* **128,** 1385–1395.

Walker, C. D., Salzmann, C., Long, H., Otis, M., Roberge, C., and Gallo-Payer, N. (2004). Direct inhibitory effects of leptin on the neonatal adrenal and potential consequences for brain glucocorticoid feedback. *Endocr. Res.* **30,** 837–844.

Ward, I. L. (1972). Prenatal stress feminizes and demasculinizes the behavior of males. *Science* **175,** 82–84.

Watterberg, K. L., and Scott, S. M. (1995). Evidence of early adrenal insufficiency in babies who develop bronchopulmonary dysplasia. *Pediatrics* **95,** 120–125.

Weinberger, C., Hollenberg, S. M., Ong, E. S., Harmon, J. M., Brower, S. T., Cidlowski, J., Thompson, E. B., Rosenfeld, M. G., and Evans, R. M. (1985). Identification of human glucocorticoid receptor complementary DNA clones by epitope selection. *Science* **228,** 740–742.

Weinstock, M. (2005). The potential influence of maternal stress hormones on development and mental health of the offspring. *Brain Behav. Immun.* **19,** 296–308.

Weinstock, M. (2007). Gender differences in the effects of prenatal stress on brain development and behaviour. *Neurochem. Res.* **32,** 1730–1740.

Weinstock, M. (2008). The long-term behavioural consequences of prenatal stress. *Neurosci. Biobehav. Rev.* **32,** 1073–1086.

Williamson, M., and Viau, V. (2008). Selective contributions of the medial preoptic nucleus to testosterone-depending regulation of the paraventricular nucleus of the hypothalamus and HPA axis. *Am. J. Physiol. Regul. Intgr. Comp. Physiol.* **294,** R1020–R1030.

Yi, S. J., and Baram, T. Z. (1994). Corticotropin-releasing hormone mediates the response to cold stress in the neonatal rat without compensatory enhancement of the peptide's gene expression. *Endocrinology* **135,** 2364–2368.

Yoshimura, S., Sakamoto, S., Kudo, H., Sassa, S., Kumal, A., and Okamoto, R. (2003). Sex-differences in adrenocortical responsiveness during development in rats. *Steroids* **68,** 439–445.

Young, E. A., Abelson, J., and Lightman, S. L. (2004). Cortisol pulsatility and its role in stress regulation and health. *Front. Neuroendocrinol.* **25,** 69–76.

Zelena, D., Domokos, A., Barna, I., Mergl, Z., Halle, R. J., and Makara, G. (2008). Control of the hypothalamo-pituitary-adrenal axis in the neonatal period: Adrenocorticotropin and corticosterone stress responses dissociated in vasopressin-deficient Brattleboro rats. *Endocrinology* **149,** 2576–2583.

CHAPTER TWENTY

# Mineralocorticoid and Glucocorticoid Receptors in Hippocampus: Their Impact on Neurons Survival and Behavioral Impairment After Neonatal Brain Injury

Justyna Rogalska

## Contents



Department of Animal Physiology, Institute of General and Molecular Biology, N. Copernicus University, Torun, Poland

*Vitamins and Hormones,* Volume 82
ISSN 0083-6729, DOI: 10.1016/S0083-6729(10)82020-5

## Abstract

Glucocorticoids (GC) exert multiple effects within the central nervous system via mineralocorticoid receptors (MR) and glucocorticoid receptors (GR) activation. MR expression is associated with a neuroprotective phenotype, whereas GR activation is implicated in the induction of an endangered neural phenotype and the opposite actions are most evident in hippocampus, where these receptors are predominantly present.

Hippocampus has an overall inhibitory influence on the activity of the hypothalamic–pituitary–adrenal (HPA) axis and it has been suggested that efficient learning and adequate stress response depend on the appropriate functioning of the axis brought by coordinated activation of MR and GR in this region.

There is a growing body of evidence that perinatal asphyxia causes irreversible damage to the brain leading to neurons loss in regions vulnerable to oxygen shortage especially in hippocampus. In the present review, some aspects of recently acquired insight in the role of GC receptors in promoting neuronal death and survival after hippocampal injury are discussed. Since the unbalance of MR and GR in hippocampus creates a condition of disturbed neuroendocrine regulation their potential impact on behavioral impairment will also be reviewed. 

## I. Introduction

The limbic system and the hippocampus in particular play pivotal roles in cognition (Nyakas *et al.*, 1996) as well as in the regulation of the hypothalamic–pituitary–adrenal (HPA) axis (Jacobson and Sapolsky, 1991). Accumulating evidence shows that corticosteroids modulate hippocampal function and may underlie some aspects of the physiological and behavioral effects of neuronal damage. Corticosteroid hormones entering the brain can bind to two types of intracellular receptors that regulate the transcriptions of responsive genes, that is, mineralocorticoid receptors (MR) and glucocorticoid receptors (GR) (de Kloet *et al.*, 1998; Reul and de Kloet, 1985). Acutely or chronically unbalanced (significantly reduced or enhanced) glucocorticoids (GC) concentrations may directly threaten physiological functions, and enhance neuronal vulnerability under pathological conditions (Goodman *et al.*, 1996; McCullers *et al.*, 2002a; Sapolsky, 2000; Sapolsky and Pulsinelli, 1985). On the other hand, a fine-tuned action of GCs is essential for neural development, particularly by their control of cellular differentiation, and for maintenance of neural integrity and function during neonatal period as well as in adulthood (de Kloet *et al.*, 1998). The activation of MRs and GRs appears to have opposing effects in the hippocampus. MRs enhance neurogenesis and the differentiation of processes of cultured hippocampal neurons, whereas GRs have a suppressive effect on development of the morphology of these neurons (Fujioka *et al.*, 2006).

It is increasingly clear that perinatal events have significant and sometimes dramatic effects on the physiological phenotype of the adult, a phenomenon termed fetal or neonatal "programming" (Matthews, 2000; Sanchez *et al.*, 2001). Such programming occurs in a variety of systems and reflects the action of a factor during sensitive periods or "windows" of development to exert organizational effects that persist through life. It is accepted that prenatal programming of the HPA axis occurs, to a large part, at the level of hippocampal GRs and MRs (Matthews, 2000).

The review focuses on the hippocampus as the main locus of MRs and GRs in the brain and the target of these early-life events. Indeed, early-life experience, via complex interactions with genetic factors (Sanchez *et al.*, 2001), have been suspected to be a major determinant of hippocampal dysfunction and long-term behavioral impairment. In rodents, maturation and full differentiation of the hippocampal formation take place during early postnatal life (Gaarskjaer, 1986) and the early experience affects neuronal morphology and synaptic function in the hippocampus (Poeggel *et al.*, 2003). Therefore, it is possible that processes triggered by early-life events, such as neonatal asphyxia, may interrupt or corrupt the functional and structural maturation of the hippocampal network in an irreversible manner (Nyakas *et al.*, 1996).

There is a growing body of evidence that perinatal asphyxia, as well as ischemia, cause irreversible damage to the brain leading to neurons loss in regions vulnerable to oxygen shortage, especially in the hippocampus (Mehta *et al.*, 2007; Nyakas *et al.*, 1996). The damage produced by asphyxia is almost proportional to the duration and severity of the hypoxic episode, inducing true encephalopathy with moderate or severe neonatal symptoms (Nyakas *et al.*, 1996), or mild or absent pathological signs at birth, which might become evident later in life (Briscoe *et al.*, 2001). Damages to the hippocampus are implicated to play a role in long-term neurological and behavioral disturbances, such as cerebral dysfunction (Nyakas *et al.*, 1996), changes in brain corticosteroid receptor profiles (Boksa *et al.*, 1996; de Kloet *et al.*, 1998; Nyakas *et al.*, 1996), sympathoadrenal dysfunction (Boksa, 1997), abnormal responses to stress (Boksa *et al.*, 1998; Caputa *et al.*, 2005; Lai *et al.*, 2007; Rogalska *et al.*, 2004, 2006a), and memory impairment (Lai *et al.*, 2007; Nyakas *et al.*, 1996; Rogalska *et al.*, 2009b).

These behavioral aspects parallel an important remodeling in brain structure and function, which involves rearrangement of synapses as well as a new neuroendocrine homeostasis (Andersen, 2003). Obstetric complications and fetal and perinatal asphyxia in particular have been associated with the onset of neurodevelopmental disorder known as attention-deficit/hyperactivity disorder (ADHD). ADHD is the most prominent childhood psychiatric condition and its features or core deficits frequently persist well into adulthood (Nyakas *et al.*, 1996).

The remainder of this review is aimed at analyzing the mechanisms underlying the bidirectional action of GCs on neuronal survival. Specifically, this review will focus on the role of corticosteroid receptors in promotion of neurodegeneration or rescuing neurons there from. Moreover, coordinated actions mediated by MRs and GRs in limbic areas such as hippocampus, which are the route through which corticosteroids influence emotional and cognitive behaviors, will be discussed. To begin with, however, the characteristics of distinct classes of corticosteroid receptors have to be provided.

## II. Corticosteroid Receptors Properties

MR and GR are a ligand-dependent transcription factors belonging to the steroid hormone receptors family. Both types of receptors share a common endogenous ligand: the adrenal steroid hormone corticosterone (CORT) in rodents or cortisol in humans (de Kloet *et al.*, 1998).

Under normal conditions high affinity MRs are substantially occupied (greater than 70–90%) by physiological levels of glucocorticoids (Reul and de Kloet, 1985); in contrast the lower affinity GRs only comes into play when hormone levels rise either during the circadian peak or in response to stress (de Kloet *et al.*, 1998; Reul and de Kloet, 1985). Therefore under basal conditions MR binding is near saturated and MR signaling is near-maximal. Increased cortisol or corticosterone levels will not lead to increased MR signaling, which is limited by the availability of MR itself rather than by ligand concentration. On the other hand, under basal conditions GR binding is low and therefore increased hormone level will lead to increased GR signaling. Summarizing, MR signaling is predominantly regulated by receptor abundance whereas GR signaling is more dependent on hormone concentrations.

Classically, corticosteroid receptors are described as part of a cytoplasmic multiprotein complex, which consists of one receptor molecule and several heat shock proteins (hsp) including two molecules of hsp90, one hsp70, one hsp56, and an immunophilin (Smith and Toft, 1993). After binding of corticosterone to its receptor, chaperone proteins dissociate and the receptor–ligand complex moves to the nuclear compartment (Duma *et al.*, 2006). Finally, the receptor as a homodimer binds to recognition sites in the DNA and thus activates gene transcription. Via this genomic pathway the hormone alters properties of hippocampal cells slowly and for a prolonged period. For instance, in this way, hippocampal cell properties, like calcium influx, were found to be altered (Karst *et al.*, 2000).

Emerging evidence suggests that glucocorticoids also act by nongenomic mechanisms on cell signaling processes and in such fashion have rapid effects in brain (Karst *et al.*, 2005). As an example, recently was observed an association between increased MR mRNA and mitogen-activated protein kinases (MAPKs) in primary neuronal cultures subjected to cell stress (Kang *et al.*, 2009; Macleod *et al.*, 2003). Moreover, corticosterone rapidly and reversibly enhances the glutamate transmission in the CA1 hippocampal area (Karst *et al.*, 2005). The rapid effect is also accomplished through a nongenomic pathway involving membrane-located receptors (Karst *et al.*, 2005). Interestingly, the rapid effect depends critically on the classical mineralocorticoid receptor, as evidenced by the effectiveness of agonists, antagonists, and brain-specific inactivation of the mineralocorticoid but not the glucocorticoid receptor gene (Karst *et al.*, 2005). Primarily, MR is present in the nuclear compartment (Nishi *et al.*, 2001). However, part of the brain MRs can shuttle from the nucleus/cytoplasm to the cell membrane (Karst *et al.*, 2005). It is noteworthy, that the nongenomic pathway could be activated only when hormone levels are elevated, thus the MR could serve as a "corticosensor" in the brain, rapidly changing excitatory transmission while corticosteroid levels are enhanced (Karst *et al.*, 2005). Moreover, rapid actions of corticosterone would allow the brain to change its function within minutes after elevations of corticosteroid levels, in addition to responding later through gene-mediated signaling pathways (Karst *et al.*, 2005).

In the brain, MRs are present predominantly in hippocampus, whereas GRs are expressed ubiquitously with high concentrations in hippocampus, septum, and hypothalamus (de Kloet *et al.*, 1998). Furthermore, MRs and GRs appear to be coexpressed in the neurons of the hippocampus (Van Eekelen *et al.*, 1988). It is thought that in these neurons both types of receptors coordinate stress responsiveness, behavioral adaptation (de Kloet *et al.*, 1998), and neuronal excitability (Joëls and de Kloet, 1990). Moreover, although GR and MR share many similarities (structural homology, common agonists, hormone response elements) they have opposite actions on neuronal viability where predominant MR activation appears to have a survival effect, for instance in dentate gyrus neurons (Woolley *et al.*, 1991), while GR overactivation increases vulnerability to excitotoxicity (Sapolsky, 2000).

## III. MR and GR Under Development of the Hippocampus

An undisturbed development of the hippocampus is essential for the normal function of the organism during adulthood. In rats the hippocampus undergoes rapid development during the first 10 days of life (Gaarskjaer, 1986). However, in most fetal tissues intracellular corticosteroid receptors are expressed from early embryonal life, for instance the GR-mRNA in the rat septo-hippocampal formation is present as early as the 14th fetal day (Yi *et al.*, 1994). On the other hand, the expression of mineralocorticoid receptors is restricted to later gestational ages and only specific tissues (Brown *et al.*, 1996). It is supposed that in the fetus these receptors may play a role in neuronal development and function of the brain, trough trophic or direct membrane effects (Joëls and de Kloet, 1992).

In a number of tissues including corticolimbic brain regions, GC receptors level is the highest during the first few weeks of rats' life (Meaney *et al.*, 1985), suggesting that they may play a specific role at early development. By the first postnatal day, robust GR-gene expression is visible in CA1, with little message in CA3 or dentate gyrus (Yi *et al.*, 1994). GR density increases progressively toward adult levels at the end of the postnatal week 3 (Meaney *et al.*, 1985; Sarrieau *et al.*, 1988; Schmidt *et al.*, 2003). In contrast, MRs, which are already highly expressed following birth, reach adult levels by the end of the first postnatal week (Sarrieau *et al.*, 1988). GC receptors distribution is similar, but not identical to that seen in postnatal and adult rats (Van Eekelen *et al.*, 1988, 1991). It means that they may mediate different functions in developing contrary to mature neurons.

## IV. Detrimental Effects of Hippocampal Glucocorticoid Receptor Activation

### A. GR expression after hypoxia/ischemia

Since the hippocampus expresses high levels of adrenal steroid hormone receptors, which makes this region a prime target for glucocorticoid action (Reul and de Kloet, 1985; Reul *et al.*, 1989), it is particularly vulnerable and sensitive to neurotoxic insults (Sapolsky, 2000). Hypoxia occurs when the demand for oxygen to maintain normal cellular ATP requirements outweighs the vascular supply and it has been documented as an integral part of many pathological states, including neonatal asphyxia or ischemic disease (Mehta *et al.*, 2007; Nyakas *et al.*, 1996). The interplay between hypoxia and glucocorticoid-mediated cellular responses in physiology and disease has become increasingly appreciated over recent years. Indeed, MR and GR expression was significantly increased in the hippocampal CA1 region after ischemia/reperfusion (Hwang *et al.*, 2006). Similarly, cerebral hypoxia as well as ischemia in adult rats was associated with increased hippocampal GR expression (Leonard *et al.*, 2005; Macleod *et al.*, 2003).

However, another line of evidences suggests that the increase in MR expression in the rat's hippocampus following anoxia under hyperthermic conditions as well as in the dentate gyrus of human hippocampus following brief episodes of cerebral ischemia were not associated with changes in hippocampal GR expression (Lai *et al.*, 2009; Rogalska *et al.*, 2009a). Similarly, perinatal hypoxia (from prenatal day 19 to postnatal day 14) has no effect on the hippocampal glucocorticoid receptor mRNA levels (Raff *et al.*, 2007).

There is some evidence for autoregulation of steroid receptor expression, and it might be that increased MR expression inhibits the GR expression in hippocampus. Indeed, the transcriptional regulation of GR biosynthesis by the neuronal MR has been suggested (Herman and Spencer, 1998). Both MR antagonist spironolactone and GR antagonist RU486 increased GR mRNA levels in sham-operated animals (McCullers and Herman, 2001; McCullers *et al.*, 2002b). Moreover, in rats treated with the MR antagonist spironolactone GR mRNA expression increased in CA1 and dentate gyrus (Herman and Spencer, 1998), suggesting that MR tonically inhibits GR biosynthesis in hippocampus. Additionally, GR agonist dexamethasone (DEX) (Reul *et al.*, 1989), MR agonist aldosterone, or low MR-binding doses of CORT (Chao *et al.*, 1998) each reverse adrenalectomy-induced increases in GR mRNA. These results provide compelling evidence for the existence of regulation GR mRNA expression in hippocampus by both MR and GR. Furthermore, it cannot be ruled out that changes in GR signaling in response to hypoxic/ischemic insult are dependent on circulating glucocorticoid levels rather than on changes in gene expression. Indeed, both synthetic glucocorticoid (dexamethasone) and endogenous glucocorticoid (corticosterone) decreased GR mRNA levels in hippocampal neurons (Erdeljan *et al.*, 2001).

## B. The impact of GR on anoxia-induced excitotoxicity

Deleterious effects of glucocorticoids are mediated by low-affinity GRs. It was evidenced by the pharmacological observation that synthetic GR agonists, for example, methylprednisolone (Uhler *et al.*, 1994) and DEX (Almeida *et al.*, 2000) share the ability to enhance the vulnerability of neuronal cells, whereas non-GR ligands do not exert toxic properties in the rodent brain (Goodman *et al.*, 1996).

Inactivation of GR by glucocorticoid removal or GR blockade has been shown to improve survival of rodent hippocampal neurons following several types of neurotoxic challenge. For example, blockade of GR with RU486 (Antonawich *et al.*, 1999) and removal of endogenous glucocorticoids by adrenalectomy (Sapolsky and Pulsinelli, 1985) attenuate loss of CA1 neurons following ischemia. Additionally, GR blockade with RU486 prevents CA1 neuron loss 24 h after traumatic brain injury (McCullers *et al.*, 2002b).

It is noteworthy that excessive GR activation increases hippocampal neuronal vulnerability to neuronal insults such as excitotoxicity and ischemia (Goodman *et al.*, 1996; McCullers *et al.*, 2002a; Sapolsky, 2000; Sapolsky and Pulsinelli, 1985). These effects may occur due to direct action of glucocorticoids or they may merely exacerbate the deleterious effects of other more potent neurotoxic mediators such as excitatory amino acids and reactive oxygen species (Almeida *et al.*, 2000; Behl *et al.*, 1997; Lu *et al.*, 2003; Uhler *et al.*, 1994). Indeed, DEX-induced apoptosis in the dentate gyrus seems to involve the generation of nitric oxide and possibly other reactive oxygen species (Almeida *et al.*, 2000).

As it was described earlier, MR- and GR-mediated actions occur through activation or repression of gene transcription. The actions depend on the cellular context, which is determined in part by other agents (e.g., neurotransmitters, hormones, and cytokines) (de Kloet *et al.*, 1998). Importantly, the expression of various neuroactive substances, their release from neurons, and the density and/or affinity of their receptor subtypes are profoundly changed in hypoxic/ischemic conditions (Mehta *et al.*, 2007; Nyakas *et al.*, 1996). This indicates that when neurons are shifted from their basal condition, for example, under anoxic conditions, the expression of MRs and GRs is aimed to restoring homeostasis. Although it is difficult to isolate specific factors influencing neuronal GR expression, some results suggest a role of glutamate in a regulation of hippocampal GRs (McCullers and Herman, 2001). Extracellular glutamate levels increase after anoxia (Nyakas et al, 1996) and ischemia (Mehta *et al.*, 2007), and both types of injury modify patterns of GR mRNA expression in the hippocampus (Boksa *et al.*, 1996; Macleod *et al.*, 2003). Moreover, RU486 reverses glucocorticoid-mediated vulnerability to glutamate in a cultured murine hippocampal cell line (Behl *et al.*, 1997).

The glucocorticoids may heighten neuronal sensitivity to insult through several mechanisms. It has been reported that activated GR stimulates or represses expression of hypoxic responsive genes, including glucose transporter 3 and vascular endothelial growth factor (Kodama *et al.*, 2003; Leonard *et al.*, 2005). Moreover, elevated GRs level causes an altered metabolic state that is partially conferred by inhibition of glucose uptake in neurons (Virgin *et al.*, 1991) and increase of neuronal vulnerability to insult through energy depletion (Sapolsky, 1986). These changes subsequently endanger hippocampal neurons when challenged with a second metabolic insult such as glutamate excitoxicity or hypoxia–ischemia

(Roy and Sapolsky, 2003; Sapolsky and Pulsinelli, 1985). Except excitatory amino acid accumulation, elevated glucocorticoid concentrations can also result in disturbed modulation of inhibitory (GABAergic) neurotransmission (Abraham *et al.*, 1996). Furthermore, GR-activation may hamper neuronal function via increasing intracellular $Ca^{2+}$ concentrations (Nair *et al.*, 1998), and free radical formation (McIntosh and Sapolsky, 1996), all of which may compromise cells by triggering $Ca^{2+}$-dependent proteolysis, lipid peroxidation, oxidative stress, and mitochondrial dysfunction.

## C. Apoptosis as a mechanism of GR-mediated cell death in the hippocampus

The key GR-mediated mechanism leading to neuronal cell death seems to be the apoptosis (Almeida *et al.*, 2000; Haynes *et al.*, 2001). The regulation of apoptosis is complex, but it is now established that proteins encoded by the bcl-2 gene family are major regulatory components of the apoptotic pathway (Merry and Korsmeyer, 1997; Tsujimoto, 2003). The Bcl-2 family comprises death inducer (e.g., Bax, Bcl-xS) and death repressor (e.g., Bcl-2, Bcl-xL) proteins (Merry and Korsmeyer, 1997). These proteins are activated by physiological or injurious stimuli and appear to operate upstream of events like changes in the plasma membrane, redox potentials, mitochondrial permeability transition phenomena, and free radical generation, leading to the final execution phase of the apoptotic process, which involves the activation of cysteine proteases—the caspases (Almeida *et al.*, 2000; Beere and Green, 2001, Jacobson *et al.*, 1997; Tsujimoto, 2003). All these processes have been implicated to underlie the mechanism of neonatal hypoxic/ischemic brain damage (Nyakas *et al.*, 1996). Indeed, in response to ischemic brain injury, neurons display a rapid change in gene expression patterns which may serve to augment (e.g., Bcl-2, Bcl-xL) (Akhtar *et al.*, 2004) or inhibit (e.g., caspase 3, p53) (Chen *et al.*, 1998) neuronal survival. Interestingly, the administration of the highly specific GR agonist DEX to rats results in apoptosis in the hippocampus, associated with concomitant alterations in the ratio of pro- (Bax) to antiapoptotic (Bcl-2, Bcl-xL) molecules (Almeida *et al.*, 2000). Additionally, the exposure of primary postnatal hippocampal culture to DEX led to a significant loss of mature (MAP2-positive) neurons, an event accompanied by a significant increase in the incidence of apoptosis (Crochemore *et al.*, 2005). In line with these results there is a report demonstrating that in the mouse neural cell line HT-22, DEX leads to a marked reduction in cell proliferation, with cells being arrested in the G1 phase of the cell cycle. The process seems to be GR-mediated since it was attenuated by the GR antagonist RU38486 (Crochemore *et al.*, 2002).

A very likely candidate for mediation between GR receptors and Bcl-2 family members is the tumor suppressor protein p53. There are reports showing an interactions between GR and p53 (Sloviter *et al.*, 1989;

Woolley *et al.*, 1991), a strong inducer of the death promoter bax (Miyashita *et al.*, 1994) and at the same time being a repressor of bcl-2 (Miyashita *et al.*, 1994). Moreover, nuclear translocation of GR and p53 occurred contemporaneously, suggesting that physical interactions between these two transcriptional factors can occur (Sengupta and Wasylyk, 2001). In addition to p53, interactions between GR and a number of other transcription factors such as nuclear factor kB and activator protein 1, which themselves show complex interactions, have been reported (McKay and Cidlowski, 1998). All of these transcription factors have been implicated in cellular death and survival processes, depending on their ability to influence the transcription of death inducer or repressor genes (Miyashita *et al.*, 1994).

As the primary mediators of apoptosis, caspases cleave a variety of endogenous cytosolic and nuclear substrates (Jacobson *et al.*, 1997). Recent data showed that GMEB1 (glucocorticoid modulatory element-binding protein 1—a modulator of transactivation) effectively attenuated caspase activation and apoptosis caused by hypoxia and oxidative stress. Thus this gene could be implicated as a potent inhibitor of caspase activation and apoptosis in response to these kinds of stresses (Nakagawa *et al.*, 2008).

When a tissue is exposed to hypoxia, a variety of cellular responses is generated, leading to cell and tissue adaptation via induction of the expression of a number of genes. These hypoxic responses are controlled mainly at the level of transcription by hypoxia-inducible factor-1 (HIF-1). Interestingly, the data demonstrating HIF-1α-mediated induction of the GRs at the protein level under both normoxic and hypoxic conditions (Leonard *et al.*, 2005) indicate that this transcription factor is responsible, at least in part, for the hypoxic induction of the GRs. Ultimately, the alterations in GR expression in hypoxia not only alter GR-dependent gene expression but also influence hypoxia-dependent gene expression. It was revealed that activation of the GR enhances hypoxia-dependent gene expression and a hypoxic response element (HRE) activity (Kodama *et al.*, 2003). The cross talk between hypoxia-generated signals and the GR system may occur at multiple levels via distinct mechanisms, thereby enabling the fine-tuning of adaptive responses (Kodama *et al.*, 2003). For example, during hypoxic events, HIF-1 mediates the transcription of pro-apoptotic gene Bnip3 via interaction with HRE in the promoter region (Sowter *et al.*, 2001). Moreover, DEX treatment of postnatal rat pups increased Bnip3 expression in the hippocampus under hypoxic condition as well as the treatment with RU28362, a glucocorticoid receptor selective agonist significantly increases Bnip3 mRNA in primary cortical neurons (Sandau and Handa, 2007). Using *in vivo* approaches, Bnip3 protein levels were found to be increased following a hypoxic–ischemic insult in the adult rat hippocampus, cortex, and striatum (Althaus *et al.*, 2006). Similarly, in a differentiated oligodendroglial cell line, Bnip3 protein levels are increased in response to hypoxia (Burton *et al.*, 2006). These data, together with the report suggesting that the transcription factors reside in the same nuclear compartments, strongly support the important role of intracellular GR-HIF-1α communication in adaptive regulation of gene expression. These results reveal a novel signaling aspect responsible for the incorporation of hypoxic and glucocorticoid stimuli, which can be an important cooperative pathway for the control of gene expression observed in complex tissue microenvironments under hypoxic state.

Summarizing, the provided evidences support the existence of interactions between GR-mediated events and other mechanisms underlying excitotoxicity under hypoxia/ischemia-induced neuronal death.

## V. Neuroprotective Role of MR Overexpression in Damaged Hippocampus

### A. Injury-induced MR expression

Recent studies suggest that the MR needs to be present and functional for neuronal survival in the damaged brain regions (Table 20.1). A neuroprotective effect of a low-level occupancy MR has been recognized in granule neurons of the hippocampal dentate gyrus. These neurons, which exert high endogenous MR expression, undergo apoptosis following adrenalectomy (Sloviter *et al.*, 1989) that can be prevented by a low-dose of corticosterone through selective MR activation (Woolley *et al.*, 1991). Moreover, the

**Table 20.1** Neuroprotective effect of MR activation on neurons survival

| Model | References |
|---|---|
| Adrenalectomy-induced cell death in rats | Woolley *et al.* (1991) |
| Exposure of rats to MR-activating doses of CORT | Hassan *et al.* (1996) |
| Exposure of rats to CORT in a dose specifically activating MR | Almeida *et al.* (2000) |
| Transient global ischemia in rats | Macleod *et al.* (2003) |
| Exposure of primary rat hippocampal neurons to MR-activating doses of CORT | Crochemore *et al.* (2005) |
| Staurosporine or oxygen–glucose deprivation-induced cell death in rat pheochromocytoma PC12 cell line | Lai *et al.* (2005) |
| Transient global cerebral ischemia in mice | Lai *et al.* (2007) |
| Postmortem samples from patients who had a brief cardiac arrest, and presumed cerebral hypoxia | Lai *et al.* (2009) |

preferential MR agonist, CORT, and the specific GR agonist DEX produced opposite effects on dentate cell survival. While CORT treatment diminished the number of TUNEL-positive (apoptotic) cells, treatment with DEX induced a significant degree of apoptosis (Almeida *et al.*, 2000). In primary rat hippocampal neurons, nanomolar concentrations of corticosterone are sufficient to inhibit GR's killing activities via selective MR activation. Moreover, both spironolactone and another MR antagonist (RU28318) accentuated DEX-induced apoptosis (Crochemore *et al.*, 2005). Similarly, the presence of low corticosterone levels in culture system, which activates the MRs, prevented cortical neuron death following glucocorticoid and hypoxia treatment (Sandau and Handa, 2007). These observations are of particular interest since they indicate that chronic administration of low, MR-activating doses of CORT can reverse cell death. It was also suggested that the relative occupation of MRs and GRs also contributes to maintenance of hippocampal cell numbers, since activation of MRs counteract deleterious GC actions on neuronal survival (Almeida *et al.*, 2000; Hassan *et al.*, 1996). Additional evidence of the importance of MRs and GRs to the structure of the hippocampal formation derives from studies using transgenic mice; genetic disruption of MR leads to dentate granule cell degeneration, whereas GR−/− mice do not display signs of hippocampal degeneration (Gass *et al.*, 2001).

There is an interesting report showing the rapid upregulation of MR (mRNA and protein) in the rat primary cortical neurons undergoing age-related or staurosporine-induced apoptosis and in rat hippocampus following hypothermic transient global ischemia. In both paradigms, increased MR density was associated with increased survival of neurons and MR antagonism with spironolactone attenuated this protective effect (Macleod *et al.*, 2003). MR expression was also significantly increased 4–7 days after ischemia reperfusion in CA1 where the neuronal death occurred (Hwang *et al.*, 2006). Moreover, MR blockade causes an increase in kainic acid-induced hippocampal cell death (McCullers and Herman, 2001) whereas a shift in the balance toward MR activation following GR blockade protects neurons from ischemic insult (Krugers *et al.*, 2000). Similarly, overexpression of human MR in PC12 cells prevented staurosporine- and oxygen/glucose deprivation-induced cell death, and spironolactone attenuated that effect (Lai *et al.*, 2005). Lai *et al.* (2007) also demonstrated that, compared with wild-type mice, transgenic mice overexpressing MR specifically in their forebrain show significantly reduced neuronal death in the hippocampus after transient global cerebral ischemia. In addition, hippocampal MR density is increased in postmortem samples from patients who had a brief cardiac arrest, and presumed cerebral hypoxia, some weeks prior to death (Lai *et al.*, 2009), implying that MR expression is also increased in human brain following neuronal stress.

It has been shown that staurosporine causes apoptosis in a range of mammalian cell types including neurons in primary culture (Lobner and Choi, 1996). Paradoxically, at lower concentrations staurosporine is neuroprotective (Chopp *et al.*, 1999) and this neuroprotection is mediated by MR activation since staurosporine caused a rapid increase in MR mRNA and protein in cortical as well as hippocampal neurons in culture (Macleod *et al.*, 2003). On the other hand, *in vivo* studies have shown that high doses of staurosporine cause neuronal damage (Chopp *et al.*, 1999). It is possible that brain damage induced by high doses of staurosporine is too extensive for MR effect to be sufficient for neuroprotection. Thus, the extent of injury can determine whether the upregulation of MR receptors will be sufficient for protection.

Interestingly, the upregulation of MR mRNA expression in ischemic human brain was most pronounced in the dentate gyrus of the hippocampus (Lai *et al.*, 2009). This region normally exhibits high levels of MR (Seckl *et al.*, 1991) and in this location neuronal death is least pronounced following ischemic insult (Walton *et al.*, 1999). The greatest protection in MR-Tg mice was observed in the CA2 region, where neurons contained the highest MR levels compared to the other hippocampal regions (Lai *et al.*, 2007). Similarly, the most pronounced induction of MR mRNA was observed in CA3 hippocampal neurons in a rat model of hypothermic transient global ischemia (Macleod *et al.*, 2003), again in a region of high MR expression (Van Eekelen *et al.*, 1991) and little affected by damage (Walton *et al.*, 1999). One possible explanation of these observations may be that enhanced MR levels in response to injury contribute, at least in part, to the relative resistance to ischemic damage in these regions. Moreover, it also suggests that some threshold of MR density must be reached before protection is attained.

Taken together, these findings provide powerful evidence that increased MR expression seen *in vitro* and *in vivo* following neuronal injury is directly and casually linked to the promotion of survival in these systems. Importantly, the protective role of MR following diverse types of cellular injury suggests that increased MR density may represent a generic endogenous survival response.

## B. The mechanisms of mineralocorticoid action

The mechanism by which increased MR expression protects neurons is not clear. In the cytoplasm, MR forms a complex with other proteins including hsp and immunophilins, and this complex dissociates on ligand binding. Both hsp and immunophilins exhibit neuroprotective properties (Avramut and Achim, 2003; Beere and Green, 2001), and the consequences of neuroprotection by overexpressed MR may partly be due to the release of these chaperone proteins.

However, the most rational explanation of protective role of MR is a direct effect of increased MR expression on transcriptional activity. Indeed, MR activation is known to reduce the transcription of calcium channel subunits (L-type channels, P/Q type channels) and alter the ratio NMDA receptor subunit genes (increase NR2A:NR2B) leading to overall decreased NMDA receptor activity and reduced calcium flux (Nair *et al.*, 1998).

Importantly, MR signaling has been shown to upregulate the apoptosis inhibitory proteins Bcl-2 and Bcl-xL (Almeida *et al.*, 2000), neurotrophic factors, including nerve growth factor and basic fibroblast growth factor, and to alter the expression of the pro-apoptotic p53 protein (Almeida *et al.*, 2000; Hansson *et al.*, 2000; McCullers and Herman, 1998). Moreover, MR blockade with spironolactone decreases basal bcl-2 messenger RNA (mRNA) levels (McCullers and Herman, 1998; McCullers *et al.*, 2002a), suggesting that MR activation maintains neuroprotective bcl-2 expression and may participate in regulation of neuronal viability after injury. This could be of substantial importance in reducing delayed (mainly apoptotic) cell death after neonatal asphyxia and ischemia.

It also should be noted that the pro-apoptotic protein Bax is widely distributed in the brain (Shindler *et al.*, 1997) and importantly, its distribution in the hippocampus (the highest level in dentate gyrus and CA2 areas) corresponds closely to that of MR (Hassan *et al.*, 1999). A reciprocal distribution pattern of Bax and MR is another evidence that MR can preserve neurons against apoptosis.

## C. The effect of different neuronal stresses on transcriptional regulation of the MR gene

It should be mentioned that the rat MR gene gives rise through alternate splicing to three distinct MR mRNA transcripts (MR$\alpha$, $\beta$, and $\gamma$) which differ in their 5′-untranslated region (Kwak *et al.*, 1993) but encode the same mature protein. Each corresponding 5′ flanking region acts as a functional promoter allowing independent regulation of each mRNA species controlled by two promoters P1 and P2 (Zennaro *et al.*, 1996) which correspond to rat MR$\alpha$ and MR$\beta$. These alternate promoters appear to determine tissue- and development-specific expression of the MR gene (Vazquez *et al.*, 1998) and suggest separate mechanisms for upregulation of MR in response to different stimuli.

MR$\beta$ transcript was specifically upregulated in rat primary cortical cultures undergoing hypothermic oxygen–glucose deprivation (OGD/H) through activation of its own promoter, an effect which appears to be mediated through the ERK1/2 pathway. A specific increase in MR$\beta$ transcript expression was also found *in vivo* in hypothermic anoxic neonatal rat hippocampus (Kang *et al.*, 2009). However, the OGD/H exposure of differentiated PC12 cells did not change MR$\gamma$ activity, but enhanced MR$\alpha$

promoter activity and this was not associated with elevated transcript levels either *in vitro* or *in vivo* (Kang *et al.*, 2009). Variation in expression and differential regulation of the MR transcripts has previously been reported to occur during brain development where both the MR$\beta$ and MR$\gamma$ are expressed during development but decline in adulthood (Vazquez *et al.*, 1998). Cellular responses to brain injury often appear to recapitulate events that occur during development (Chen *et al.*, 2005) so activation of the MR$\beta$ promoter with increased MR$\beta$ expression may be a plausible response to injury. These results provide evidence for specific transcriptional regulation of the MR gene by different conditions of neuronal stress.

### D. Detrimental effect of MR activation on blood flow and perfusion after cerebral ischemia

The results confirmed the neuroprotective role of MR induction contrast with studies showing that mineralocorticoid receptor activation causes cerebral vessel remodeling and exacerbates the damage caused by cerebral ischemia (Dorrance *et al.*, 2006). Indeed, chronic MR blockade reduces cerebral infarct size (Dorrance *et al.*, 2001; Oyamada *et al.*, 2008), but this appears to be a vascular phenomenon because MR blockade increases vessel lumen diameter, which presumably increases blood flow and perfusion of the tissue to reduce ischemic damage (Rigsby *et al.*, 2005). In addition, the blockade of MR in astrocytes migrating to the ischemic core after cerebral ischemia appears to protect damaged neurons via indirect effects (Oyamada *et al.*, 2008). Although, the vascular protection afforded by MR antagonism might seem to be at odds with the results seen within the neurons, where MR activation is required for survival, it is possible that the significance of MR activation for brain protection differs in neurons and astrocytes. Since mineralocorticoid receptors have been identified in the brain cells and blood vessels, they are both a potential target for therapeutic intervention (Rigsby *et al.*, 2005).

In summary, the neuroprotective consequences of MR activation might therefore be due to a combination of direct inhibition of excitotoxic processes and promotion of an apoptosis-resisting internal milieu.

## VI. The Effect of Temperature on Hypoxia/Ischemia-Induced Changes in MR and GR Expression

Hypothermia and hyperthermia during neonatal period, both under control and anoxic conditions specifically induces MR but not GR expression in neonatal hippocampus (Kang *et al.*, 2009; Rogalska *et al.*, 2009a).

It should be mentioned that newborn mammals maintain their normal body temperature 4 °C below that of adult rats (Rogalska and Caputa, 2005). Thus, the normal physiological body temperature of newborn rats is as low as 33 °C. Accordingly, in newborn rats forced decreasing body temperature to a level of 31 °C should be referred to as hypothermia whereas forced raising body temperature to a level of 37 °C (typical for adults) should be referred as hyperthermia (Rogalska and Caputa, 2005).

Induction of MR in the hippocampus of neonates in response to injury (Kang *et al.*, 2009; Rogalska *et al.*, 2009a) is likely to exert direct prosurvival actions since the deleterious effects of both hypothermia and hyperthermia under perinatal anoxia were proved. Neonatal hypothermia (31 °C) under anoxic condition leads to acidosis (Caputa *et al.*, 2001), which could have deleterious effect on brain injury since acidosis is associated with an increased risk of encephalopathy and cerebral palsy in newborns (Impey *et al.*, 2008).

Similarly, hyperthermia during hypoxia–ischemia makes the immature brain inordinately susceptible to hypoxic–ischemic insult and causes brain injury. This can occur despite the fact that hypoxic–ischemic insult is mild and causes no or little injury by itself (Tomimatsu *et al.*, 2003). Prominent harmful effects of elevated temperature in animals subjected to hypoxia/ischemia were observed when temperature elevations occurred immediately after the insult and continued for 24 h (Reglodi *et al.*, 2000). This suggests that the neonatal brain is extremely susceptible to hyperthermia. Neonatal asphyxia in rats, incubated at an elevated ambient temperature to keep their rectal temperature at 39 °C, induced plasma hyperferremia (Caputa *et al.*, 2001) followed by iron accumulation in the frontal cortex, the hippocampus, and the corpus striatum (Rogalska *et al.*, 2006b), which creates the risk of oxidative stress under these conditions. Moreover, intraischemic hyperthermia may activate caspase-3, which leads to the escalation of apoptotic cell death and results in aggravation of neuronal injury in the immature brain (Tomimatsu *et al.*, 2003). It is also plausible that the hyperthermic neonatal exposure to anoxia causes lifelong emotional disturbances, such as stress-induced hyperactivity in juvenile rats (Rogalska *et al.*, 2004) and reduced alertness to external stimuli signaling a potential danger in adult (Caputa *et al.*, 2005) and old (Rogalska *et al.*, 2006a) rats. These animal data are strongly supported by clinical observations. Namely, relatively high temperature applied in incubators during usual care after hypoxia–ischemia is associated with increased risk of adverse outcomes (Impey *et al.*, 2008).

Intriguingly, MR upregulation in hyperthermic (39 °C) neonatal rat brain was also observed (Rogalska *et al.*, 2009a). Disturbed cell proliferation and migration by brief hyperthermia may result in permanent effects on the brain (Edwards *et al.*, 1971). This is supported by observation that exposure of neonatal rats to hyperthermia impairs their behavior till adulthood (Caputa *et al.*, 2005; Rogalska *et al.*, 2006a). The dissonance between potentially protective overexpression of MR and disturbed behavior which reflects brain injury in animals exposed neonatally to hyperthermia or hyperthermia and anoxia might be explained by more severe insult sufficient to cause damage to the brain. In spite of that, MR overexpression could determine the final extent of brain damage.

Taken together, the MR response to both hypo- and hyperthermia in neonatal brain is in line with the findings that sublethal challenge can induce neuronal MR expression (Macleod *et al.*, 2003) and may serve as a compensatory mechanism designed to limit the neuronal death. Although the consequences of increased MR level under these circumstances are not known, it is possible that this represents endogenous response to protect the brain from subsequent injury.

Since decreased brain temperature has been shown to provide a considerable neuroprotection (Corbett *et al.*, 2000), this particular level of naturally regulated core temperature (ca. 33 °C) in newborn rats could protect them against detrimental effects of parturitional asphyxia. While the exact mechanism by which the reduced temperature exerts this protective effect is not clear; changes in gene expression have been reported (Kobayashi *et al.*, 2008). Body temperature reduced to 33 °C increased hippocampal total MR mRNA expression observed in adult rats subjected to transient global cerebral ischemia (Macleod *et al.*, 2003). It appears to be of biological importance, as a pharmacological antagonism of MR with spironolactone increased neuronal death suggesting that MR induction is an endogenous neuroprotective mechanism and may mediate some of the effects of hypothermia. Moreover, hypothermia reduced the density of GRs in ischemic adult animals (Macleod *et al.*, 2003). Importantly, recently observed lack of an anoxia-induced injury under normothermic condition may be due to the fact that normal body temperature in neonates is 33 °C. This may represent an adaptive response and offer protection occurring at this temperature. Accordingly, in these conditions the increase in MR expression was not observed (Rogalska *et al.*, 2009a), which proves that MR overexpression represents endogenous response to injury.

# VII. The Long-Term Consequences of Alterations in MR and GR Receptors After Neonatal Brain Injury

## A. The influence of injury-induced MR/GR unbalance on the extent of HPA activation

There is convincing evidence now that early-life experience alters brain corticosteroid receptor profiles (Boksa *et al.*, 1996; de Kloet *et al.*, 1998; Nyakas *et al.*, 1996) leading to changes in stress response system and emotionality that persist into adulthood (de Kloet *et al.*, 1998; Nyakas *et al.*, 1996). A change in the balance of MR- and GR-mediated processes progressively creates a condition of disturbed neuroendocrine regulation and impaired behavioral adaptation (de Kloet *et al.*, 1998).

The developmental changes in GC receptor levels (described in Chapter 3) are associated with a distinct pattern of basal corticosterone secretion during the early neonatal period. In the rat pups, plasma GC levels are relatively high at birth, fall to barely detectable levels by second postnatal day, and begin to rise around 14–15th day to reach adult levels by 21st day after birth (Schroeder and Henning, 1989). Although birth complications clearly affect hippocampal MR and GR levels (Boksa *et al.*, 1996; Kang *et al.*, 2009; Rogalska *et al.*, 2009a), it is unclear whether neonatal asphyxia might have more lasting effects on the developmental profile of GC secretion during the first few weeks of life. However, recent data showed that an acute period of birth anoxia affects the developmental profile of both basal and stress-induced GC secretion. The plasma corticosterone response to a 20-min restraint stress was reduced in adult animals born by cesarean section with added period of anoxia (Boksa *et al.*, 1996). In addition, neonatal anoxia under hyperthermic conditions causes the decrease of stress-induced level of corticosterone on 14th day of life (Rogalska and Caputa, 2010). Furthermore, a prolonged, low level of corticosterone secretion was observed for at least 6 h after injury (McCullers *et al.*, 2002b). Alterations in secretory patterns may be due to mineralocorticoid receptor signaling, which is believed to play a major role in inhibiting basal HPA tone during the trough of the diurnal corticosterone rhythm (Berger *et al.*, 2006; de Kloet *et al.*, 1998). Intrahippocampal application of GR antagonist suppresses adrenal corticosterone release under conditions in which MR antagonists enhance HPA-axis activity (Van Haarst *et al.*, 1997). These findings are in agreement with electrophysiological data showing opposite effects of MR and GR activation on excitability and excitatory outflow of the hippocampus (Joëls, 2001). Since blockade of MR with spironolactone (McCullers *et al.*, 2002b) results in increased plasma corticosterone levels, it is possible that hippocampal increase of MR density subsequent to neuronal stress (McCullers *et al.*, 2002b; Rogalska *et al.*, 2009a) in combination with other factors might culminate in decreased HPA secretion. However, it should be noted that HPA hyposecretion following hippocampal damage is not observed in all studies (Nyakas *et al.*, 1994). For instance, the perinatal hypoxia augmented the corticosterone response to stress in 6-month-old rats (Raff *et al.*, 2007). The inconsistencies among experiments may be related to time after insult when assessment was performed or experimental design issues. To sum up, perinatal hypoxia changes the "programming" of the hypothalamic–pituitary–adrenal axis to respond to stress situation in adulthood.

Importantly, the characteristic of ADHD-like cognitive impairments in an animal model of ADHD was accompanied by significant changes in basal levels of hormonal indices of the pituitary–adrenal axis, particularly low level of corticosterone (King *et al.*, 2000). Dysfunction of the pituitary–adrenal axis has also been reported in children with persistent ADHD, who also have low basal cortisol level (King *et al.*, 1998).

At present, there is a lack of information whether the precise timing of the adrenal quiescent phase and the subsequent rise of plasma corticosterone may differentially affect the development of various body tissues. However, if the precise timing is critical, the prolonged early decrease in plasma corticosterone may have developmental repercussions.

## B. Behavioral effects involving MR and GR levels disturbance in hippocampus

Neonatal asphyxia in mammals causes delayed behavioral disturbances such as abnormal responses to stress and impaired learning, which persist over the entire life span (Caputa *et al.*, 2005; Nyakas *et al.*, 1996; Rogalska *et al.*, 2004, 2006a, 2009b). There is growing body of evidence that MR and GR regulation at critical stages of development can have lifelong effects on behavior (Matthews, 2000). However, it is important to realize that specific MR or GR activation does not regulate emotional behavior and physiology; rather, they induce chemical changes in particular sets of neurons, making certain behavioral and physiological outcomes more likely in a certain context in time, as a result of the strengthening or weakening of particular neural pathways (Sapolsky, 2000). For instance, GR overexpression in the forebrain may increase anxiety responses by means of alterations in the noradrenergic system, because the anxiety responses are blocked by norepinephrine-related antidepressants (Wei *et al.*, 2004).

Rats with damage to the hippocampal formation due to neonatal asphyxia show deficient regulation of the stress response (Nyakas *et al.*, 1996). One enduring view is that a loss of neurons bearing GR and MR in the hippocampus is the primary cause of disinhibited hypothalamic–pituitary–adrenal (HPA) activity. Moreover, hippocampal cell loss is likely to have repercussions on cognition and the regulation of mood and anxiety (de Kloet *et al.*, 1998). The most important consequence of neonatal asphyxia that is likely to have long-lasting impact is behavioral hyperactivity. In animal models, hypoxia/ischemia has been shown to increase emotional reactivity to a novel environment as indexed by hyperactivity in the open-field test (Nyakas *et al.*, 1996; Rogalska *et al.*, 2004). At the level of neural mechanisms, changes in open-field activity are coupled with changes in the function of the hypothalamic–pituitary–adrenal (HPA) axis (van den Buuse *et al.*, 2002). Moreover, it was suggested that the increased locomotor activity in many psychiatric diseases may derive from the interference with the development of brain regions, for example, of the hippocampus, striatum, or amygdala during a specific postnatal time window (Viggiano, 2008). In addition, alterations in most neurotransmitter systems can give rise to a hyperactive phenotype (Nyakas *et al.*, 1996; Viggiano, 2008). The status of particular neurotransmitter systems can regulate the hippocampal corticosterone receptors, which in turn, could alter an organism's ability to respond to stress. The cholinergic innervation of hippocampal neurons may play a role in the regulation of glucocorticoid receptors and HPA function, through a neural mechanism. There is an interesting report showing that young rats with selective cholinergic lesions have reduced hippocampal GR mRNA and dysregulated HPA function in response to acute stress (Han *et al.*, 2002).

It has been suggested that MRs also play a role in the evaluation of an anxiety-related behavior. In general, high hippocampal level of MRs was associated with a low anxiety trait (Berger *et al.*, 2006; Herrero *et al.* 2006). These results are in agreement with other evidence indicating that increased MRs level in the hippocampus may be associated with decreased anxiety (Catalani *et al.*, 2000). There is an interesting report showing that MR stimulation enhances exploration of novel items in an open field (Oitzl *et al.*, 1994). Additionally, the increased neuronal MR level may also be causal in mediating the reduced anxiety-related behavior in forebrain-specific MR overexpressing transgenic mice (Lai *et al.*, 2007).

Contrary to MR function in decreasing anxiety state, emerging evidences strongly suggest that GR receptors play an important role in fear potentiation. Indeed, glucocorticoid receptor antagonists reduced the enhanced anxiety state, as measured in the elevated plus-maze (Korte *et al.*, 1995). In agreement with this finding, it was shown that administration of corticosterone, at a dose sufficient to activate GRs, reduced open-arm exploration in the elevated plus-maze 24 h later (Calvo *et al.*, 1998), suggesting that GR activation produces a long-term enhanced anxiety state. Additional support comes from the finding that rats exhibiting high rates of exploratory locomotion are characterized by decreased level of hippocampal GR which may contribute to the increased novelty seeking observed in these animals (Kabbaj *et al.*, 2000). Moreover, an increase in GR expression in the forebrain leads to an increase in anxiety-like behavior as measured in plus-maze test (Wei *et al.*, 2004). Similarly, adult rats, prenatally treated with DEX, exhibited reduced exploratory behavior in an open field as well as in an elevated plus-maze (Welberg *et al.*, 2001). A direct relationship between brain glucocorticoid receptor levels and anxiety-like behavior is supported by the phenotype of transgenic mice with disrupted GR expression in the brain, which show strikingly attenuated anxiety (Tronche *et al.*, 1999).

Thus, it could be presumed that, the permanent changes in stress response observed in hyperthermic animals, both control and anoxic ones

(Caputa *et al.*, 2005; Rogalska *et al.*, 2004, 2006a) may be explained by the increased MR expression as a part of anoxia underlying processes (Rogalska *et al.*, 2009a).

Efficient learning and memory performance is thought to depend on the appropriate functioning of the HPA axis brought about by coordinated activation of MR and GR (Berger *et al.*, 2006; de Kloet *et al.*, 1998). Indeed, activation of MR is thought to be implicated in memory acquisition (appraisal of information and response selection) (Oitzl and de Kloet, 1992; Oitzl *et al.*, 1994), whereas activation of GR is thought to be related to consolidation of acquired information (Oitzl and de Kloet, 1992). Studies using both specific MR- and GR-antagonists and transgenic models of reduced MR and GR signaling indicate that chronic changes in HPA-axis activity interfere with cognitive performance (Gass *et al.*, 2001). Additionally, forebrain-specific MR overexpressing transgenic mice also display behavioral differences, showing improved spatial memory retention compared to wild-type counterparts (Lai *et al.*, 2007). Similarly, loss of limbic MR in mice with inactivated the MR gene appears primarily to cause a defect in behavioral flexibility combined with lacking control of appropriate responses, reflected in various learning-related and -unrelated deficits (Berger *et al.*, 2006).

Taken together, behavioral disturbances observed in animals exposed neonatally to anoxia (Boksa *et al.*, 1998; Caputa *et al.*, 2005; Nyakas *et al.*, 1996; Rogalska *et al.*, 2004, 2006a, 2009b) could be related in part to abnormalities in various neurotransmitter systems of the brain (Nyakas *et al.*, 1996; Viggiano, 2008) as well as the unbalance of MR/GR levels (Boksa *et al.*, 1996; Rogalska *et al.*, 2009a).

## VIII. Conclusion

From a critical analysis of the available data, the picture that emerges is that corticosterone exerts a bidirectional effect on excitotoxic damage, which became apparent in the U-shaped profile of a dose–response relationship between plasma corticosterone concentration and the extent of excitotoxic injury to hippocampal neurons. Whereas the loss of serum corticosterone and highly elevated corticosterone concentrations potentiate excitotoxicity, moderate levels of plasma corticosterone in a narrow concentration window significantly protect against excitotoxic neuronal damage. It means that the tonic activation of MR is essential for the survival of hippocampal neurons. On the other hand, GR-mediated action can lead to hippocampal cell death without the participation of other aggravating factors, so long as their effects are not masked by previously activated MR. The early-life unbalance in MR/GR-mediated actions compromises homeostatic processes in hippocampal neurons, which is thought to cause hypothalamic–pituitary–adrenal dysregulation (de Kloet and DeRijk, 2004) and may underlie changes in stress response and learning abilities observed in adolescence as well as in adulthood (Boksa *et al.*, 1998; Caputa *et al.*, 2005; Nyakas *et al.*, 1996; Rogalska *et al.*, 2004, 2006a, 2009b).

If postnatal manipulations could "overwrite" the deleterious outcome of an adverse prenatal environment, then there might be the way to exploit the MR and GR levels manipulations for therapeutic benefit to protect brain from the effects of the insult. Especially, the regulation of MR level may be an important target for reduction of brain damage after acute neuronal injury such as neonatal hypoxia/ischemia.

## REFERENCES

Abraham, I., Juhasz, G., Kekesi, K. A., and Kovacs, K. J. (1996). Effect of intrahippocampal dexamethasone on the levels of amino acid transmitters and neuronal excitability. *Brain Res.* **733,** 56–63.

Akhtar, R. S., Ness, J. M., and Roth, K. A. (2004). Bcl2 family regulation of neuronal development and neurodegeneration. *Acta Biochim. Biophys.* **1644,** 189–203.

Almeida, O. F. X., Conde, G. L., Crochmore, C., Demenieux, B. A., Fischer, D., Hassan, A. H., Meyer, M., Holsboer, F., and Michaelidis, T. M. (2000). Subtle shifts in the ratio between pro- and antiapoptotic molecules after activation of corticosteroid receptors decide neuronal fate. *FASEB J.* **14,** 779–790.

Althaus, J., Bernaudin, M., Petit, E., Toutain, J., Touzani, O., and Rami, A. (2006). Expression of the gene encoding the pro-apoptotic BNIP3 protein and stimulation of hypoxia-inducible factor-1alpha (HIF-1alpha) protein following focal cerebral ischemia in rats. *Neurochem. Int.* **48,** 687–695.

Andersen, S. L. (2003). Trajectories of brain development: point of vulnerability or window of opportunity? *Neurosci. Biobehav. Rev.* **27,** 3–18.

Antonawich, F. J., Miller, G., Rigsby, D. C., and Davis, J. N. (1999). Regulation of ischemic cell death by glucocorticoids and adrenocorticotropic hormone. *Neuroscience* **88,** 319–325.

Avramut, M., and Achim, C. L. (2003). Immunophilins in nervous system degeneration and regeneration. *Curr. Top. Med. Chem.* **3,** 1376–1382.

Beere, H. M., and Green, D. R. (2001). Stress management—Heat shock protein-70 and the regulation of apoptosis. *Trends Cell Biol.* **11,** 6–10.

Behl, C., Lezoualc'h, F., Trapp, T., Widmann, M., Skutella, T., and Holsboer, F. (1997). Glucocorticoids enhance oxidative stress-induced cell death in hippocampal neurons in vitro. *Endocrinology* **38,** 101–106.

Berger, S., Wolfer, D. P., Selbach, O., Alter, H., Erdmann, G., Reichardt, H. M., Chepkova, A. N., Welz, H., Hass, H. L., Lipp, H.-P., and Schutz, G. (2006). Loss of the limbic mineralocorticoid receptor impairs behavioural plasticity. *Proc. Natl. Acad. Sci. USA* **103,** 195–200.

Boksa, P. (1997). Early developmental profiles of plasma corticosterone are altered by birth condition in the rat: A comparison of vaginal birth, Caesarean section and Caesarean section with added anoxia. *Pediatr. Res.* **41,** 34–43.

Boksa, P., Krishnamurthy, A., and Sharma, S. (1996). Hippocampal and hypothalamic type I corticosteroid receptor affinities are reduced in adult rats born by a caesarean procedure with or without an added period of anoxia. *Neuroendocrinology* **64,** 25–34.

Boksa, P., Wilson, D., and Rochford, J. (1998). Responses to stress and novelty in adult rats born vaginally, by cesarean section, or by cesarean section with acute anoxia. *Biol. Neonate* **74,** 48–50.

Briscoe, J., Gathercole, S. E., and Marlow, N. (2001). Everyday memory and cognitive ability in children born very prematurely. *J. Child Psychol. Psychiatry* **42,** 749–754.

Brown, R. W., Diaz, R., Robson, A. C., Kotelevtsev, Y., Mullins, J. J., Kaufman, M. H., and Seckl, J. R. (1996). The ontogeny of 11-hydroxysteroid dehydrogenase type 2 and mineralocorticoid receptor gene expression reveal intricate control of glucocorticoid action in development. *Endocrinology* **137,** 794–797.

Burton, T. R., Henson, E. S., Baijal, P., Eisenstat, D. D., and Gibson, S. B. (2006). The pro-cell death Bcl-2 family member, BNIP3, is localized to the nucleus of human glial cells: Implications for glioblastoma multiforme tumor cell survival under hypoxia. *Int. J. Cancer* **118,** 1660–1669.

Calvo, N., Martijena, I. D., Molina, V. A., and Volosin, M. (1998). Metyrapone pretreatment prevents the behavioral and neurochemical sequelae induced by stress. *Brain Res.* **800,** 227–235.

Caputa, M., Rogalska, J., and Nowakowska, A. (2001). Effect of temperature on postanoxic, potentially neurotoxic changes of plasma pH and free iron level in newborn rats. *Brain Res. Bull.* **55,** 281–286.

Caputa, M., Rogalska, J., Wentowska, K., and Nowakowska, A. (2005). Perinatal asphyxia, hyperthermia and hyperferremia as factors inducing behavioural disturbances in adulthood: A rat model. *Behav. Brain Res.* **163,** 246–256.

Catalani, A., Casolini, P., Scaccianoce, S., Patacchioli, F. R., Spinozzi, P., and Angelucci, L. (2000). Maternal corticosterone during lactation permanently affects brain corticosteroid receptors, stress response and behaviour in rat progeny. *Neuroscience* **100,** 319–325.

Chao, H. M., Ma, L. Y., McEwen, B. S., and Sakai, R. R. (1998). Regulation of glucocorticoid receptor and mineralocorticoid receptor messenger ribonucleic acids by selective agonists in the rat hippocampus. *Endocrinology* **139,** 1810–1814.

Chen, J., Nagayama, T., Jin, K., Stetler, R. A., Zhu, R. L., Graham, S. L., and Simon, R. P. (1998). Induction of caspase-3 like protease may mediate delayed neuronal death in the hippocampus after transient cerebral ischaemia. *J. Neurosci.* **18,** 4914–4928.

Chen, J., Leong, S. Y., and Schachner, M. (2005). Differential expression of cell fate determinants in neurons and glial cells of adult mouse spinal cord after compression injury. *Eur. J. NeuroSci.* **22,** 1895–1906.

Chopp, M., Li, Y., and Jiang, N. (1999). Increase in apoptosis and concomitant reduction of ischemic lesion volume and evidence for synaptogenesis after transient focal cerebral ischemia in rat treated with staurosporine. *Brain Res.* **828,** 197–201.

Corbett, D., Hamilton, M., and Colbourne, F. (2000). Persistent neuroprotection with prolonged postischemic hypothermia in adult rats subjected to transient middle cerebral artery occlusion. *Exp. Neurol.* **163,** 200–206.

Crochemore, C., Michaelidis, T. M., Fischer, D., Loeffler, J. P., and Almeida, O. F. X. (2002). Enhancement of p53 activity and inhibition of neural cell proliferation by glucocorticoid receptor activation. *FASEB J.* **16,** 761–770.

Crochemore, C., Lu, J., Wu, Y., Liposits, Z., Sousa, N., Holsboer, F., and Almeida, O. F. (2005). Direct targeting of hippocampal neurons for apoptosis by glucocorticoids is reversible by mineralocorticoid receptor activation. *Mol. Psychiatry* **10,** 790–798.

de Kloet, E. R., and DeRijk, R. (2004). Signaling pathways in brain involved in predisposition and pathogenesis of stress-related disease: Genetic and kinetic affecting the MR/GR balance. *Ann. NY Acad. Sci.* **1032,** 14–34.

de Kloet, E. R., Vreugdenhil, E., Oitzl, M. S., and Joels, M. (1998). Brain corticosteroid receptor balance in health and disease. *Endocr. Rev.* **19,** 269–301.

Dorrance, A. M., Osborn, H. L., Grekin, R., and Webb, R. C. (2001). Spironolactone reduces cerebral infarct size and EGF-receptor mRNA in stroke prone rats. *Am. J. Physiol. Regul. Integr. Comp. Physiol.* **281,** R944–R950.

Dorrance, A. M., Rupp, N. C., and Noguiera, E. F. (2006). Mineralocorticoid receptor activation causes cerebral vessel remodeling and exacerbates the damage caused by cerebral ischemia. *Hypertension* **47,** 590–595.

Duma, D., Jewell, C. M., and Cidlowski, J. A. (2006). Multiple glucocorticoid receptor isoforms and mechanisms of post-translational modification. *J. Steroid Biochem. Mol. Biol.* **102,** 11–21.

Edwards, M. J., Penny, R. H., and Zevnik, I. (1971). A brain cell deficit in newborn guinea pigs following prenatal hyperthermia. *Brain Res.* **28,** 341–345.

Erdeljan, P., MacDonald, J. F., and Matthews, S. G. (2001). Glucocorticoids and serotonin alter glucocorticoid receptor (GR) but not mineralocorticoid receptor (MR) mRNA levels in fetal mouse hippocampal neurons, in vitro. *Brain Res.* **896,** 130–136.

Fujioka, A., Fujioka, T., Ishida, Y., Maekawa, T., and Nakamura, S. (2006). Differential effects of prenatal stress on the morphological maturation of hippocampal neurons. *Neuroscience* **141,** 907–915.

Gaarskjaer, F. B. (1986). The organization and development of the hippocampal mossy fiber system. *Brain Res.* **396,** 335–357.

Gass, P., Reichardt, H. M., Strekalova, T., Henn, F., and Tronche, F. (2001). Mice with targeted mutations of glucocorticoid and mineralocorticoid receptors: Models of depression and anxiety? *Physiol. Behav.* **73,** 811–825.

Goodman, Y., Bruce, A. J., Cheng, B., and Mattson, M. P. (1996). Estrogens attenuate and corticosterone exacerbates excitotoxicity, oxidative injury, and amyloid beta-peptide toxicity in hippocampal neurons. *J. Neurochem.* **66,** 1836–1844.

Han, J. S., Bizon, J. L., Chun, H. J., Maus, C. E., and Gallagher, M. (2002). Decreased glucocorticoid receptor mRNA and dysfunction of HPA axis in rats after removal of the cholinergic innervation to hippocampus. *Eur. J. Neurosci.* **16,** 1399–1404.

Hansson, A. C., Cintra, A., Belluardo, N., Sommer, W., Bhatnagar, M., Bader, M., Ganten, D., and Fuxe, F. (2000). Gluco and mineralocorticoid receptor mediated regulation of neurotrophic factor gene expression in the dorsal hippocampus and neocortex of the rat. *Eur. J. Neurosci.* **12,** 2918–2934.

Hassan, A. H., von Rosenstiel, P., Patchev, V. K., Holsboer, F., and Almeida, O. F. X. (1996). Exacerbation of apoptosis in the dentate gyrus of the aged rat by dexamethasone and the protective role of corticosterone. *Exp. Neurol.* **140,** 43–52.

Hassan, A. H. S., Patchev, V. K., von Rosenstiel, P., Holsboer, F., and Almeida, O. F. X. (1999). Plasticity of hippocampal corticosteroid receptors during aging in the rat. *FASEB J.* **13,** 115–122.

Haynes, L. E., Griffiths, M. R., Hyde, R. E., Barber, D. J., and Mitchell, I. J. (2001). Dexamethasone induces limited apoptosis and extensive sublethal damage to specific subregions of the striatum and hippocampus: Implications for mood disorders. *Neuroscience* **104,** 57–69.

Herman, J. P., and Spencer, R. (1998). Regulation of hippocampal glucocorticoid receptor gene transcription and protein expression in vivo. *J. Neurosci.* **18,** 7462–7473.

Herrero, A. I., Sandi, C., and Venero, C. (2006). Individual differences in anxiety trait are related to spatial learning abilities and hippocampal expression of mineralocorticoid receptors. *Neurobiol. Learn. Mem.* **86,** 150–159.

Hwang, I. K., Yoo, K.-Y., Namb, Y. S., Choi, J. H., Lee, S., Kwon, Y.-G., Kang, T.-C.h., Kim, Y.-S., and Won, M. H. (2006). Mineralocorticoid and glucocorticoid receptor expressions in astrocytes and microglia in the gerbil hippocampal CA1 region after ischemic insult. *Neurosci. Res.* **54,** 319–327.

Impey, L. W. M., Greenwood, C. E. L., Black, R. S., Yeh, P. S.-Y., Sheil, O., and Doyle, P. (2008). The relationship between intrapartum maternal fever and neonatal acidosis as risk factors for neonatal encephalopathy. *Am. J. Obstet. Gynecol.* **198,** 49.e1–49.e6.

Jacobson, L., and Sapolsky, R. (1991). The role of the hippocampus in feedback regulation of the hypothalamic–pituitary–adrenal axis. *Endocr. Rev.* **12,** 118–134.

Jacobson, M. D., Weil, M., and Raff, M. C. (1997). Programmed cell death in animal development. *Cell* **88,** 347–354.

Joëls, M. (2001). Corticosteroid actions in the hippocampus. *J. Neuroendocrinol.* **13,** 657–669.

Joëls, M., and de Kloet, E. R. (1990). Mineralocortieoid receptor-mediated effects on membrane properties of rat CA1 pyramidal neurons in vitro. *Proc. Natl. Acad. Sci. USA* **87,** 4495–4498.

Joëls, M., and de Kloet, E. R. (1992). Control of neuronal excitability by corticosteroid hormones. *Trends Neurosci.* **15,** 25–30.

Kabbaj, M., Devine, D. P., Savage, V. R., and Akil, H. (2000). Neurobiological correlates of individual differences in novelty-seeking behavior in the rat: Differential expression of stress-related molecules. *J. Neurosci.* **20,** 6983–6988.

Kang, P., Rogalska, J., Walter, C. A., Burke, M., Seckl, J. R., Macleod, M. R., and Lai, M. (2009). Injury-induced mineralocorticoid receptor expression involves differential promoter usage: A novel role for the rat MR$\beta$ variant. *Mol. Cell. Endocrinol.* **305,** 56–62.

Karst, H., Karten, Y. J., Reichardt, H. M., de Kloet, E. R., Schütz, G., and Joëls, M. (2000). Corticosteroid actions in hippocampus require DNA binding of glucocorticoid receptor homodimers. *Nat. Neurosci.* **3,** 977–978.

Karst, H., Berger, S., Turiault, M., Tronche, F., Schütz, G., and Joëls, M. (2005). Mineralocorticoid receptors are indispensable for nongenomic modulation of hippocampal glutamate transmission by corticosterone. *Proc. Natl. Acad. Sci. USA* **102,** 19204–19207.

King, J. A., Barkley, R. A., and Barrett, S. (1998). Attention deficit hyperactivity disorder and the stress response. *Biol. Psychiatry* **44,** 72–74.

King, J. A., Barkley, R. A., Delville, Y., and Ferris, C. F. (2000). Early androgen treatment decreases cognitive function and catecholamine innervation in an animal model of ADHD. *Behav. Brain Res.* **107,** 35–43.

Kobayashi, M. S., Asai, S., Ishikawa, K., Nishida, Y., Nagata, T., and Takahashi, Y. (2008). Global profiling of influence of intra-ischemic brain temperature on gene expression in rat brain. *Brain Res. Rev.* **58,** 171–191.

Kodama, T., Shimizu, N., Yoshikawa, N., Makino, Y., Ouchida, R., Okamoto, K., Hisada, T., Nakamura, H., Morimoto, C., and Tanaka, H. (2003). Role of the glucocorticoid receptor for regulation of hypoxia-dependent gene expression. *J. Biol. Chem.* **278,** 33384–33391.

Korte, S. M., de Boer, S. F., de Kloet, E. R., and Bohus, B. (1995). Anxiolytic-like effects of selective mineralocorticoid and glucocorticoid antagonists on fear-enhanced behavior in the elevated plus-maze. *Psychoneuroendocrinology* **20,** 385–394.

Krugers, H. J., Maslam, S., Korf, J., Joëls, M., and Holsboer, F. (2000). The corticosterone synthesis inhibitor metyrapone prevents hypoxia/ischaemic-induced loss of synaptic function in the rat hippocampus. *Stroke* **31,** 1162–1172.

Kwak, S. P., Pate, P. D., Thompson, R. C., Akil, H., and Watson, S. J. (1993). 5′-Heterogeneity of the mineralocorticoid receptor messenger ribonucleic acid: Differential expression and regulation of splice variants within rat hippocampus. *Endocrinology* **133,** 2344–2350.

Lai, M., Seckl, J., and Macleod, M. (2005). Overexpression of the mineralocorticoid receptor protects against injury in PC12 cells. *Mol. Brain Res.* **135,** 276–279.

Lai, M., Horsburgh, K., Bae, S.-E., Carter, R. N., Stenvers, D. J., Fowler, J. H., Yau, J. L., Gomez-Sanchez, C. E., Holmes, M. C., Kenyon, C. J., Seckl, J. R., and Macleod, M. R. (2007). Forebrain mineralocorticoid receptor overexpression enhances memory, reduces anxiety and attenuates neuronal loss in cerebral ischaemia. *Eur. J. Neurosci.* **25,** 1832–1842.

Lai, M., Bae, S.-E., Bell, J. E., Seckl, J. R., and Macleod, M. R. (2009). Mineralocorticoid receptor mRNA expression is increased in human hippocampus following brief cerebral ischaemia. *Neuropathol. Appl. Neurobiol.* **35,** 156–164.

Leonard, M. O., Godson, C., Brady, H. R., and Taylor, C. T. (2005). Potentiation of glucocorticoid activity in hypoxia through induction of the glucocorticoid receptor. *J. Immunol.* **174,** 2250–2257.

Lobner, D., and Choi, D. W. (1996). Preincubation with protein synthesis inhibitors protects cortical neurons against oxygen-glucose deprivation-induced death. *Neuroscience* **72,** 335–341.

Lu, J., Goula, D., Sousa, N., and Almeida, O. F. X. (2003). Ionotropic and metabotropic glutamate receptor mediation of glucocorticoid-induced apoptosis in hippocampal cells and the neuroprotective role of synaptic N-methyl-D-aspartate receptors. *Neuroscience* **121,** 123–131.

Macleod, M. R., Johansson, I. M., Soderstrom, I., Lai, M., Gido, G., Wieloch, T., Seckl, J. R., and Olsson, T. (2003). Mineralocorticoid receptor expression and increased survival following neuronal injury. *Eur. J. Neurosci.* **17,** 1549–1555.

Matthews, S. G. (2000). Antenatal glucocorticoids and programming of the developing CNS. *Pediatr. Res.* **47,** 291–300.

McCullers, D. L., and Herman, J. P. (1998). Mineralocorticoid receptors regulate bcl-2 and p53 expression in hippocampus. *Neuroreport* **9,** 3085–3089.

McCullers, D. L., and Herman, J. P. (2001). Adrenocorticosteroid receptor blockade and excitotoxic challenge regulate adrenocorticosteroid receptor mRNA levels in hippocampus. *J. Neurosci. Res.* **64,** 277–283.

McCullers, D. L., Sullivan, P. G., Scheff, S. W., and Herman, J. P. (2002a). Mifepristone protects CA1 hippocampal neurons following traumatic brain injury in rat. *Neuroscience* **109,** 219–230.

McCullers, D. L., Sullivan, P. G., Scheff, S. W., and Herman, J. P. (2002b). Traumatic brain injury regulates adrenocorticosteroid receptor mRNA levels in rat hippocampus. *Brain Res.* **947,** 41–49.

McIntosh, L. J., and Sapolsky, R. M. (1996). Glucocorticoids may enhance oxygen radical-mediated neurotoxicity. *Neurotoxicology* **17,** 873–882.

McKay, L. I., and Cidlowski, J. A. (1998). Cross-talk between nuclear factor-kappa B and the steroid hormone receptors: Mechanisms of mutual antagonism. *Mol. Endocrinol.* **12,** 45–56.

Meaney, M. J., Sapolsky, R. M., and McEwen, B. S. (1985). The development of the glucocorticoid receptor system in the rat limbic brain. II. An autoradiographic study. *Brain Res.* **350,** 165–168.

Mehta, S. R., Manhas, N., and Raghubir, R. (2007). Molecular targets in cerebral ischemia for developing novel therapeutics. *Brain Res. Rev.* **54,** 34–66.

Merry, D. E., and Korsmeyer, S. J. (1997). Bcl-2 gene family in the nervous system. *Annu. Rev. Neurosci.* **20,** 245–267.

Miyashita, T., Krajewski, S., Krajewska, M., Wang, H. G., Lin, H. K., Lieberman, D. A., Hoffman, B., and Reed, J. C. (1994). Tumor suppressor p53 is a regulator of bcl-2 and bax gene expression in vitro and in vivo. *Oncogene* **9,** 1799–1805.

Nair, S. M., Werkman, T. R., Craig, J., Finnell, R., Joëls, M., and Eberwine, J. H. (1998). Corticosteroid regulation of ion channel conductance and mRNA levels in individual hippocampal CA1 neurones. *J. Neurosci.* **18,** 2685–2696.

Nakagawa, T., Tsuruma, K., Uehara, T., and Nomura, Y. (2008). GMEB1, a novel endogenous caspase inhibitor, prevents hypoxia- and oxidative stress-induced neuronal apoptosis. *Neurosci. Lett.* **438,** 34–37.

Nishi, M., Ogawa, H., Ito, T., Matsuda, K. I., and Kawata, M. (2001). Dynamic changes in subcellular localization of mineralocorticoid receptor in living cells: In comparison with glucocorticoid receptor using dual color labeling with green fluorescent protein spectral variants. *Mol. Endocrinol.* **15,** 1077–1092.

Nyakas, C., Buwalda, B., Markel, E., Korte, S. M., and Luiten, P. G. M. (1994). Life-spanning behavioural and adrenal dysfunction induced by prenatal hypoxia in the rat is prevented by the calcium antagonist nimodipine. *Eur. J. Neurosci.* **6,** 746–753.

Nyakas, C., Buwalda, B., and Luiten, P. G. M. (1996). Hypoxia and brain development. *Prog. Neurobiol.* **49,** 1–51.

Oitzl, M. S., and de Kloet, E. R. (1992). Selective corticosteroid antagonists modify specific aspects of spatial orientation learning. *Behav. Neurosci.* **106,** 62–71.

Oitzl, M. S., Fluttert, M., and de Kloet, E. R. (1994). The effect of corticosterone on reactivity to spatial novelty is mediated by central mineralocorticosteroid receptors. *Eur. J. Neurosci.* **6,** 1072–1079.

Oyamada, N., Sone, M., Miyashita, K., Park, K., Taura, D., Inuzuka, M., Sonoyama, T., Tsujimoto, H., Fukunaga, Y., Tamura, N., Itoh, H., and Nakao, K. (2008). The role of mineralocorticoid receptor expression in brain remodeling after cerebral ischemia. *Endocrinology* **149,** 3764–3777.

Poeggel, G., Helmeke, C., Abraham, A., Schwabe, T., Friedrich, P., and Braun, K. (2003). Juvenile emotional experience alters synaptic composition in the rodent cortex, hippocampus, and lateral amygdala. *Proc. Natl. Acad. Sci. USA* **100,** 16137–16142.

Raff, H., Jacobson, L., and Cullinan, W. E. (2007). Augmented hypothalamic corticotrophin-releasing hormone mRNA and corticosterone responses to stress in adult rats exposed to perinatal hypoxia. *J. Neuroendocrinol.* **19,** 907–912.

Reglodi, D., Somogyvari-Vigh, A., Maderdrut, J. L., Vigh, S., and Arimura, A. (2000). Postischemic spontaneous hyperthermia and its effects in middle cerebral artery occlusion in the rat. *Exp. Neurol.* **163,** 399–407.

Reul, J. M., and de Kloet, E. R. (1985). Two receptor systems for corticosterone in rat brain: Microdistribution and differential occupation. *Endocrinology* **117,** 2505–2511.

Reul, J. M., Pearce, P. T., Funder, J. W., and Krozowski, Z. S. (1989). Type I and type II corticosteroid receptor gene expression in the rat: Effect of adrenalectomy and dexamethasone administration. *Mol. Endocrinol.* **3,** 1674–1680.

Rigsby, Ch.S., Cannady, W. E., and Dorrance, A. M. (2005). Aldosterone: Good guy or bad guy in cerebrovascular disease? *Trends Endocrinol. Metab.* **16,** 401–406.

Rogalska, J., and Caputa, M. (2005). Spontaneously reduced body temperature and gasping ability as a mechanism of extreme tolerance to asphyxia in neonatal rats. *J. Therm. Biol.* **30,** 360–369.

Rogalska, J., and Caputa, M. (2010). Neonatal asphyxia under hyperthermic conditions alters HPA axis function in juvenile rats. *Neurosci Lett.* **472,** 68–72.

Rogalska, J., Caputa, M., Wentowska, K., and Nowakowska, A. (2004). Stress-induced behaviour in juvenile rats: Effects of neonatal asphyxia, body temperature and chelation of iron. *Behav. Brain Res.* **154,** 321–329.

Rogalska, J., Caputa, M., Wentowska, K., and Nowakowska, A. (2006a). Stress-induced behaviour in adult and old rats: Effects of neonatal asphyxia, body temperature and chelation of iron. *J. Physiol. Pharm.* **57,** 17–34.

Rogalska, J., Danielisova, V., and Caputa, M. (2006b). Effect of neonatal body temperature on postanoxic, potentially neurotoxic iron accumulation in the rat brain. *Neurosci. Lett.* **393,** 249–254.

Rogalska, J., Kang, P., Wotherspoon, W., Macleod, M. R., and Lai, M. (2009a). Effect of hyperthermia and anoxia on glucocorticoid andmineralocorticoid receptor expression in neonatal rat hippocampus. *Neurosci. Lett.* **450,** 196–200.

Rogalska, J., Caputa, M., Piątkowska, K., and Nowakowska, A. (2009b). Neonatal asphyxia and hyperthermia and cognitive deficits in adult rats: Role of iron. *J. Therm. Biol.* **34,** 391–400.

Roy, M., and Sapolsky, R. M. (2003). The exacerbation of hippocampal excitotoxicity by glucocorticoids is not mediated by apoptosis. *Neuroendocrinology* **77,** 24–31.

Sanchez, M. M., Ladd, C. O., and Plotsky, P. M. (2001). Early adverse experience as a developmental risk factor for later psychopathology: Evidence from rodent and primate models. *Dev. Psychopathol.* **13,** 419–449.

Sandau, U. S., and Handa, R. J. (2007). Glucocorticoids exacerbate hypoxia-induced expression of the pro-apoptotic gene bnip3 in the developing cortex. *Neuroscience* **144,** 482–494.

Sapolsky, R. M. (1986). Glucocorticoid toxicity in the hippocampus: Reversal by supplementation with brain fuels. *J. Neurosci.* **6,** 2240–2244.

Sapolsky, R. M. (2000). Glucocorticoids and hippocampal atrophy in neuropsychiatric disorders. *Arch. Gen. Psychiatry* **57,** 925–930.

Sapolsky, R. M., and Pulsinelli, W. A. (1985). Glucocorticoids potentiate ischemic injury to neurons: Therapeutic implications. *Science* **229,** 1397–1400.

Sarrieau, A., Sharma, S., and Meaney, M. J. (1988). Postnatal development and environmental regulation of hippocampal glucocorticoid and mineralocorticoid receptors. *Brain Res.* **471,** 158–162.

Schmidt, M. V., Enthoven, L., van der Mark, M., Levine, S., de Kloet, E. R., and Oitzl, M. S. (2003). The postnatal development of the hypothalamic–pituitary–adrenal axis in the mouse. *Int. J. Dev. Neurosci.* **21,** 125–132.

Schroeder, R. J., and Henning, S. J. (1989). Roles of plasma clearance and corticosteroid-binding globulin in the developmental increase of circulating corticosterone in infant rats. *Endocrinology* **124,** 2612–2618.

Seckl, J., Dickson, K., Yates, C., and Fink, G. (1991). Distribution of glucocorticoid and mineralocorticoid receptor messenger RNA expression in human post-mortem hippocampus. *Brain Res.* **561,** 332–337.

Sengupta, S., and Wasylyk, B. (2001). Ligand-dependent interaction of the glucocorticoid receptor with p53 enhances their degradation by Hdm2. *Genes Dev.* **15,** 2367–2380.

Shindler, K. S., Latham, C. B., and Roth, K. A. (1997). bax deficiency prevents the increased cell death of immature neurons in bcl-x-deficient mice. *J. Neurosci.* **17,** 3112–3119.

Sloviter, R. S., Valiquette, G., Abrams, G. M., Ronk, E. C., Sollas, A. L., Paul, A., and Neubort, S. (1989). Selective loss of hippocampal granule cells in the mature rat brain after adrenalectomy. *Science* **243,** 535–538.

Smith, D. F., and Toft, D. O. (1993). Steroid receptors and their associated proteins. *Mol. Endocrinol.* **7,** 4–11.

Sowter, H. M., Ratcliffe, P. J., Watson, P., Greenberg, A. H., and Harris, A. L. (2001). HIF-1-dependent regulation of hypoxic induction of the cell death factors BNIP3 and NIX in human tumors. *Cancer Res.* **61,** 6669–6673.

Tomimatsu, T., Fukuda, H., Kanagawa, T., Mu, J., Kanzaki, T., and Murata, Y. (2003). Effects of hyperthermia on hypoxic-ischemic brain damage in the immature rat: Its influence on caspase-3-like protease. *Am. J. Obstet. Gynecol.* **188,** 768–773.

Tronche, F., Kellendonk, C., Kretz, O., Gass, P., Anlag, K., Orban, P. C., Bock, R., Klein, R., and Schutz, G. (1999). Disruption of the glucocorticoid receptor gene in the nervous system results in reduced anxiety. *Nat. Genet.* **23,** 99–103.

Tsujimoto, Y. (2003). Cell death regulation by the Bcl-2 protein family in the mitochondria. *J. Cell. Physiol.* **195,** 158–167.

Uhler, T., Firm, D., Pakzaban, P., and Isacson, O. (1994). The effects of mega-dose methylprednisolone and U-78517F on glutamate receptor mediated toxicity in the rat neostriatum. *Neurosurgery* **34,** 122–127.

van den Buuse, M., van Acker, S. A., Fluttert, M. F., and de Kloet, E. R. (2002). Involvement of corticosterone in cardiovascular responses to an open field novelty stressor in freely moving rats. *Physiol. Behav.* **75,** 207–215.

Van Eekelen, J. A. M., Jiang, W., de Kloet, E. R., and Bohn, M. C. (1988). Distribution of the mineraiocorticoid and the glucocorticoid receptor mRNAs in the rat hippocampus. *J. Neurosci. Res.* **21,** 88–94.

Van Eekelen, J. A. M., Bohn, M. C., and de Kloet, E. R. (1991). Postnatal ontogeny of mineralocorticoid and glucocorticoid receptor gene expression in regions of the rat tel- and diencephalon. *Dev. Brain Res.* **61,** 33–43.

Van Haarst, A. D., Oitzl, M. S., and de Kloet, E. R. (1997). Facilitation of feedback inhibition through blockade of glucocorticoid receptors in the hippocampus. *Neurochem. Res.* **22,** 1323–1328.

Vazquez, D. M., Lopez, J. F., Morano, M. I., Kwak, S. P., Watson, S. J., and Akil, H. (1998). Alpha, beta and gamma mineralocorticoid receptor messenger ribonucleic acid splice variants: Differential expression and rapid regulation in the developing hippocampus. *Endocrinology* **139,** 3165–3177.

Viggiano, D. (2008). The hyperactive syndrome: Metanalysis of genetic alterations, pharmacological treatments and brain lesions which increase locomotor activity. *Behav. Brain Res.* **194,** 1–14.

Virgin, C. E. Jr., Ha, T. P., Packan, D. R., Tombaugh, G. C., Yang, S. H., Horner, H. C., and Sapolsky, R. M. (1991). Glucocorticoids inhibit glucose transport and glutamate uptake in hippocampal astrocytes: Implications for glucocorticoid neurotoxicity. *J. Neurochem.* **57,** 1422–1428.

Walton, M., Connor, B., Lawlor, P., Young, D., Sirimanne, E., Gluckman, P., Cole, G., and Dragunow, M. (1999). Neuronal death and survival in two models of hypoxic-ischaemic brain damage. *Brain Res. Rev.* **29,** 137–168.

Wei, Q., Lu, X.-Y., Liu, L., Schafer, G., Shieh, K.-R., Burke, S., Robinson, T. E., Watson, S. J., Seasholtz, A. F., and Akil, H. (2004). Glucocorticoid receptor overexpression in forebrain: A mouse model of increased emotional lability. *Proc. Natl. Acad. Sci. USA* **101,** 11851–11856.

Welberg, L. A. M., Seckl, J. R., and Holmes, M. C. (2001). Prenatal glucocorticoid programming of brain corticosteroid receptors and corticotrophin-releasing hormone: Possible implications for behaviour. *Neuroscience* **104,** 71–79.

Woolley, C. S., Gould, E., Sakai, R. R., Spencer, R. L., and McEwen, B. S. (1991). Effects of aldosterone or RU28362 treatment on adrenalectomy-induced cell death in the dentate gyrus of the adult rat. *Brain Res.* **554,** 312–315.

Yi, S.-J., Masters, J. N., and Baram, T. Z. (1994). Glucocorticoid receptor mRNA ontogeny in the fetal and postnatal rat forebrain. *Mol. Cell. Neurosci.* **5,** 385–393.

Zennaro, M.-C.h., Le Menuet, D., and Lombes, M. (1996). Characterization of the human mineralocorticoid receptor gene V-regulatory region: Evidence for differential hormonal regulation of two alternative promoters via nonclassical mechanisms. *Mol. Endocrinol.* **10,** 1549–1580.

CHAPTER TWENTY-ONE

# GLUCOCORTICOIDS AND LITHIUM IN ADULT HIPPOCAMPAL NEUROGENESIS

Shuken Boku, Shin Nakagawa, *and* Tsukasa Koyama

## Contents




## Abstract

Adult hippocampal neurogenesis is decreased in rodent models for stress-related disorders partly through an elevated level of glucocorticoids (GCs). On the other hand, lithium (Li), a mood stabilizer and an inhibitor of GSK-3$\beta$, increases adult hippocampal neurogenesis. However, it remains unclear whether GCs-induced decrease can be recovered by Li or not. Recently we established the culture system of adult rat dentate gyrus-derived neural precursor cell (ADP) and examined GCs and Li actions on ADP proliferation. GCs decreased ADP proliferation and Li recovered it. Both cyclin Dl expression and nuclear $\beta$-catenin are also reciprocally regulated by GCs and Li. In addition, GCs activated GSK-3$\beta$. Therefore, GSK-3$\beta$/$\beta$-catenin pathway may be important in the reciprocal actions of GCs and Li on ADP proliferation. In this manuscript, we review the past literature and our study and summarize what is currently known about the effects of GCs and Li on adult hippocampal neurogenesis. © 2010 Elsevier Inc.


Department of Psychiatry, Hokkaido University Graduate School of Medicine, Sapporo, Japan

*Vitamins and Hormones,* Volume 82
ISSN 0083-6729, DOI: 10.1016/S0083-6729(10)82021-7

## I. Introduction

Mood disorders, including major depression and bipolar disorder, are severe, chronic illness which affects 8–12% or 1–6% of general population, respectively (Judd and Akiskal, 2003; Lopez *et al.*, 2006). Although now we have a broad range of drugs such as mood stabilizers like lithium (Li) or antidepressants like fluoxetione, the effectiveness of such drugs on mood disorders is still unsatisfactory. Thus, we can regard them as the most prevalent and costly brain disorders and it is urgent issue to understand their pathophysiology and develop more effective treatments for them.

Mood disorders are often triggered by considerable psychosocial stress. It has been well established that elevated levels of glucocorticoids (GCs) constitute one of the causal events in stress-related disorders (de Kloet *et al.*, 2005). Receptors for GCs such as glucocorticoid and mineralcorticoid receptors (GR and MR) are mainly expressed in the hippocampus in adult brain (Sapolsky *et al.*, 2000). Neurogenesis occurs in adult human hippocampus (Eriksson *et al.*, 1998) and is highly regulated by a variety of environmental, endocrinical, and pharmacological stimuli (Sahay and Hen, 2007; Warner-Schmidt and Duman, 2006). Therefore, these facts make us focus on the involvement of adult hippocampal neurogenesis in the pathophysiology and/or drug therapy of mood disorders.

This manuscript begins with an introduction of adult hippocampal neurogenesis and then refers to important findings about the effects of GCs and Li on adult hippocampal neurogenesis. Finally, the reciprocal action mechanism of GCs and Li for adult hippocampal neurogenesis will be discussed on the basis of our study.

## II. Differential Steps of Adult Hippocampal Neurogenesis

Neurogenesis has been identified in adult brain of various species, including mouse, rat, guinea pig, primate, and human (Altman and Das, 1967; Cameron *et al.*, 1993; Eriksson *et al.*, 1998; Gould *et al.*, 1999; Kempermann *et al.*, 1997). It mainly occurs in two discrete brain regions such as the subventricular zone (SVZ) (Alvarez-Buylla and Garcia-Verdugo, 2002) and the subgranular zone (SGZ) of the dentate gyrus (DG) in the hippocampus (Kempermann *et al.*, 2006). In adult human, neurogenesis has been shown only in the DG (Eriksson *et al.*, 1998).

Adult hippocampal neurogenesis to neural linage can be divided into two phases: precursor cell phase and postmitotic maturation phase. On the basis of the expression pattern of marker proteins, four distinct stages (type-1

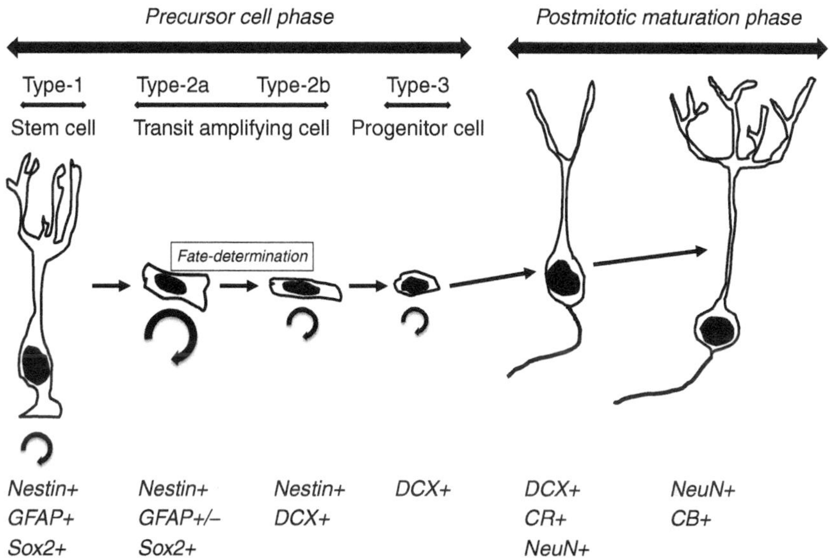


**Figure 21.1** *Differentiation phases in adult hippocampal neurogenesis.* See text for details. The determination of cell fate is assumed to be made in type-2a cells. In this figure we present only neural lineage. GFAP, glial fibrillary acidic protein; SOX2, sex determining region Y-box2; DCX, doublecortin; CR, calretinin; CB, calbindin.

cells, type-2a cells, type-2b cells and type-3 cells) have been identified in precursor cell phase (Fig. 21.1; Kempermann *et al.*, 2004; Steiner *et al.*, 2006). It remains unclear when the decision of cell fate is made. However, it must occur on the level of type-2a cells because the marker proteins of immature neuron, including Prox1, NeuroD, doublecortin, and PSA-NCAM, is expressed in type-2b cells but not in type-2a cells. Both type-1 cells and type-2 cells can respond to extrinsic stimuli such as voluntary wheel running or antidepressants (Encinas *et al.*, 2006; Huttmann *et al.*, 2003; Kronenberg *et al.*, 2003; Kunze *et al.*, 2006; Nakagawa *et al.*, 2002; Segi-Nishida *et al.*, 2008). However, type-2a cells may be the main target of extrinsic stimuli increasing the proliferation of neural precursor cells in adult DG because the burden of cell proliferation lies on type-2a cells in precursor cell phase (Kronenberg *et al.*, 2003).

In postmitotic maturation phase, the dendrite development is initiated (Zhao *et al.*, 2006) and the postmitotic markers such as NeuN are expressed (Brandt *et al.*, 2003). However, the number of NeuN-positive new immature neuron is decreased dramatically within a few days (Brandt *et al.*, 2003). This process has been shown to be apoptotic (Biebl *et al.*, 2000; Kuhn *et al.*, 2005) and regulated by NMDA-type glutamate receptor-mediated input activity (Tashiro *et al.*, 2006). Following to this elimination process, new immature neurons translocate from SGZ into the molecular cell layer and come to rest in the lower third of the granular cell layer. Then, the upper two-thirds of the DG emerge to be occupied predominantly by new immature neurons derived from neural precursor cells in SGZ (Ahn and Joyner, 2005; Laplagne *et al.*, 2006). The number of excitatory synapses grows in following two mouths and plateaus. Further structural alterations occur for months.

## III. Stress and Glucocorticoids Actions on Adult Hippocampal Neurogenesis

Stress is the most notorious negative regulator of adult hippocampal neurogenesis. Acute stress such as resident-intruder model, predator odor, restraint, or electrical foot shocks, dramatically decreases cell proliferation in adult dentate gyrus (Gould *et al.*, 1997; Marberg and Duman, 2003; Pham *et al.*, 2003; Tanapat *et al.*, 2001). In addition to acute stress, it has been shown that chronic mild stress, which recapitulates the behavioral characteristics of depression rather than acute stress (Willner, 1990), also decreases adult hippocampal neurogenesis (Alonso *et al.*, 2004; Jayatissa *et al.*, 2006; Silva *et al.*, 2008).

Stressful events activate the hypothalamic–pituitary–adrenal (HPA) axis in animals including human. First, the paraventricular nucleus of the hypothalamus secretes corticotrophin-releasing factor (CRF), which stimulates the release of adrenocorticotropic hormone (ACTH) from the anterior pituitary. Next, ACTH stimulates the release of GCs from the adrenal grand and the levels of GCs in blood and cerebrospinal fluid is elevated. Although the elevated levels of them are reversed by the negative-feedback loop in healthy subjects, it does not often work in depressive subjects and the hyperactivity of HPA axis and the elevated levels of GCs are continued (de Kloet *et al.*, 2005). Therefore, GCs must be one of the mediators of stress and may be involved in stress action on adult hippocampal neurogenesis. Actually, psychosocial stress paradigms elevate the levels of them in animal model (Fuchs and Flügge, 1998) and a sustained increase in plasma GCs suppresses proliferation of neural precursor cells in the dentate gyrus (Ambrogini *et al.*, 2002; Gould *et al.*, 1992).

Many studies of GCs action on adult hippocampal neurogenesis have focused attention on cell proliferation and have shown that GCs decrease cell proliferation in adult dentate gyrus (Cameron and McKay, 1999; Gould *et al.*, 1997). Therefore, we also focus attention on GCs action on the proliferation of neural precursor cells in adult DG in this manuscript.

Both GR and MR are mainly expressed on the nuclear membrane and acts as transcription factors when a ligand binds to them (Sapolsky *et al.*, 2000). It has been shown that GR is expressed in all stages including type-1 cells and type-2a cells and that MR is expressed only in mature cells (Garcia *et al.*, 2004).

In addition, dexamethasone (DEX), a specific agonist of GR, decreases cell proliferation in adult hippocampus (Kim *et al.*, 2004a) and mifepristone, a specific antagonist of GR, recovers corticosterone-induced decrease of cell proliferation in adult hippocampus (Mayer *et al.*, 2006). These studies suggest that GR is involved in cell proliferation rather than MR in adult hippocampal neurogenesis.

To investigate the direct effects of various factors and drugs, we recently established the *in vitro* culture system of adult rat dentate gyrus-derived neural precursor cell (ADP) (Boku *et al.*, 2009). The expression pattern of marker proteins in ADP and its limited proliferation potency indicates that ADP corresponds to type-2a cell. In addition, ADP expresses GR but not MR as in the case with type-2a cell *in vivo*. We examined the direct effects of DEX on ADP and found that DEX decreased ADP proliferation on dose-dependent manner (Boku *et al.*, 2009). Therefore, GCs may directly decrease the proliferation of neural precursor cells.

## IV. Lithium Action on Adult Hippocampal Neurogenesis

Li is a common mood stabilizer and used for the treatment of bipolar disorder. It is also often used for the augmentation therapy of refractory depression. However, the mechanism underlying the therapeutic effects of Li on mood disorders is poorly understood. It has been well established that Li increases adult hippocampal neurogenesis in rodents (Chen *et al.*, 2000; Kim *et al.*, 2004b; Son *et al.*, 2003). On the other hand, stress decreases adult hippocampal neurogenesis as described above. We examined the effect of Li on the proliferation of ADP (Boku *et al.*, 2009). The results showed that only Li had no effect on ADP proliferation. Interestingly, Li could recover DEX-induced decrease of ADP proliferation. Taken together, it is assumed that the therapeutic effects of Li on mood disorders are at least partly mediated by its effects on adult hippocampal neurogenesis.

Two molecules are well known as the targets for Li as their inhibitors: glycogen synthase kinase-3$\beta$ (GSK-3$\beta$) (Klein and Melton, 1996; Stambolic *et al.*, 1996) and inositol monophosphatase (Atack *et al.*, 1995). GSK-3$\beta$ is a key regulator of $\beta$-catenin pathway (also known as canonical Wnt pathway) (Aberle *et al.*, 1997; Orford *et al.*, 1997). Moreover, the activation of $\beta$-catenin pathway increases cell proliferation through promoting cyclin Dl expression in tumor-derived cell line (Shtutman *et al.*, 1999; Tetsu and McCormick, 1999). These studies suggest that GSK-3$\beta$ may be essential for Li action on adult hippocampal neurogenesis. Actually, recent studies have shown that GSK-3$\beta$ and $\beta$-catenin pathway is involved in adult hippocampal neurogenesis both *in vivo* (Eom and Jope, 2009; Lie *et al.*, 2005) and *in vitro* (Wexler *et al.*, 2008). In addition, we have also shown that DEX decreases nuclear *β*-catenin and the expression of cyclin Dl in ADP. Conversely, Li recovers them (Boku *et al.*, 2009). Taken together, Li may increase adult hippocampal neurogenesis through inhibiting GSK-3*β* and following activation of *β*-catenin pathway.

Our results described above also suggest that DEX may activate GSK-3*β*. However, it is not known whether GCs are involved in the regulation of GSK-3*β* and *β*-catenin pathway. To elucidate it, we examined the effects of DEX on the phosphorylation of $Ser^9$ and $Tyr^{216}$ of GSK-3*β* in ADP because the activity of GSK-3*β* is regulated by two phosphorylated residues: $Ser^9$ to render it inactive (Cross *et al.*, 1995) and $Tyr^{216}$ to render it active (Hughes *et al.*, 1993). DEX had no effect on the phosphorylation of $Ser^9$ but remarkably increased that of $Tyr^{216}$. On the other hand, Li had no effect on both of them. In addition, DEX had no effect on the expression of GSK-3*β* (Boku *et al.*, 2009). Therefore, DEX is considered to inhibit *β*-catenin pathway through activating GSK-3*β* and Li is considered to recover DEX-induced inactivation of *β*-catenin pathway through inhibiting activated GSK-3*β* (Fig. 21.2).

## V. Concluding Remarks

GCs are key mediators of stress and Li is commonly used for the treatment of stress-related disorders. It has been well established that adult hippocampal neurogenesis is involved in the therapeutic action of drugs for stress-related disorders. Moreover, many evidences for the involvement of Cs and Li in the regulation of adult hippocampal neurogenesis have been accumulated. Therefore, we can consider that GCs and Li reciprocally regulate adult hippocampal neurogenesis. Our recent study is the first study showed that GCs and Li reciprocally regulated the proliferation of adult G-derived neural precursor cell. In addition, we found that GSK-3*β* and *β*-catenin pathway was involved in it. Our study was performed *in vitro* culture system. Therefore, further *in vivo* experiments are necessary for confirming our hypothesis about the mechanism of the reciprocal effects of glucocorticoids and Li on adult hippocampal neurogenesis.

In contrast to us, Wexler *et al.* have shown that Li increases the proliferation of neural precursor cells derived from adult entire hippocampus without DEX (Wexler *et al.*, 2008). The discrepancy between our result and Wexler's one might be due to the difference of the source and character of cells as well as culture condition. We have no answer regarding which culture condition and reactivity to Li is closer to those of *in vivo* neural precursor cells in adult DG. To answer this question is necessary for a further understanding of the direct effects of Li on neural precursor cell in

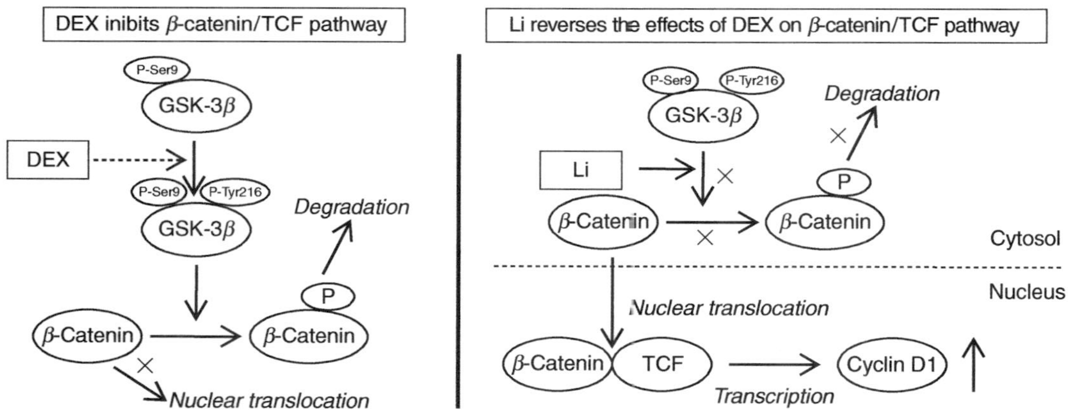


**Figure 21.2** Schematic drawing of the proposed mechanism underlying the reciprocal effects of glucocorticoids and Li on ADP proliferation. DEX activates GSK-3$\beta$ through increasing Tyr$^{216}$ phosphorylation on it and activated GSK-3$\beta$ phosphorylates $\beta$-catenin. Then, $\beta$-catenin is degraded and the expression of cyclin Dl is decreased. On the other hand, Li inhibits DEX-activated GSK-3$\beta$ and decreases the phosphorylation of $\beta$-catenin. Then, the nuclear translocation of $\beta$-catenin is decreased and the transcription of cyclin Dl gene is increased.

adult DG. Nonetheless, these studies suggest that Li may directly affect the proliferation of neural precursor cell in adult DG.

Our study suggests that the expression of cyclin Dl is decreased by DEX through inactivating $\beta$-catenin pathway. However, there is a possibility that GR directly represses the transcription of cyclin Dl because GR is a transcription factor that can promote or repress the transcription of various genes through direct binding to their promoters (Schoneveld *et al.*, 2004). Our results do not exclude this possibility, and this direct mechanism could regulate ADP proliferation in cooperation with $\beta$-catenin pathway.

In this manuscript we have focused attention on the proliferation of neural precursor cell and not referred to neural differentiation and cell survival (antiapoptosis) of neural precursor cell, both are also essential components of neurogenesis. Several studies have shown that GCs decrease the rate of neural differentiation (Wong and Herbert, 2006) and cell survival (Heine *et al.*, 2004). In addition, Li increases the rate of neural differentiation (Kim *et al.*, 2004a,b) or cell survival (Chen *et al.*, 2000) in adult hippocampus. We also have shown that Li promotes both neural differentiation and cell survival in ADP (our unpublished data). Therefore, both GCs and Li can affect not only proliferation but also neural differentiation and cell survival of neural precursor cell in adult hippocampus. To elucidate, the functional significance of each component in adult hippocampus is expected to discover the new target for drugs to mood disorders.

## ACKNOWLEDGMENTS

We thank Masuda, T., Nishikawa, H., Kato, A., and Inoue, T. for their careful reading and helpful comments on this manuscript.

## REFERENCES

Aberle, H., Bauer, A., Stappert, J., Kispert, A., and Kemler, R. (1997). Betacatenin is a target for the ubiquitin-proteasome pathway. *EMBO J.* **16,** 3797–3804.

Ahn, S., and Joyner, A. L. (2005). In vivo analysis of quiescent adult neural stem cells responding to Sonic Hedgehog. *Nature* **437,** 894–897.

Alonso, R., Griebel, G., Pavone, G., Stemmelin, J., Le Fur, G., and Soubrie, P. (2004). Blockade of CRF(l) or V(lb) receptors reverses stress-induced suppression of neurogenesis in a mouse model of depression. *Mol. Psychiatry* **9,** 278–286.

Altman, J., and Das, G. D. (1967). Postnatal neurogenesis in the guinea-pig. *Nature* **214,** 1098–1101.

Alvarez-Buylla, A., and Garcia-Verdugo, J. M. (2002). Neurogenesis in the adult subventricular zone. *J. Neurosci.* **22,** 629–634.

Ambrogini, R., Orsini, L., Mancini, C., Ferri, P., Bardanti, I., and Cuppini, R. (2002). Persistently high corticosterone levels but not normal circadian fluctuations of the hormone affect cell proliferation in the adult rat dentate gyrus. *Neuroendocrinology* **76,** 366–372.

Atack, J. R., Broughton, H. B., and Pollack, S. J. (1995). Inositol monophosphatase—A putative target for $Li^+$ in the treatment of bipolar disorder. *Trends Neurosci.* **18,** 343–349.
Biebl, M., Cooper, C. M., Winkler, J., and Kuhn, H. G. (2000). Analysis of neurogenesis and programmed cell death reveals a self-renewing capacity in the adult rat brain. *Neurosci. Lett.* **291,** 17–20.
Boku, S., Nakagawa, S., Masuda, T., Nishikawa, H., Kato, A., Kitaichi, Y., Inoue, T., and Koyama, T. (2009). Glucocorticoids and lithium reciprocally regulate the proliferation of adult dentate gyrus-derived neural precursor cells through GSK-3$\beta$ and $\beta$-catenin/TCF pathway. *Neuropsychopharmacology* **34,** 805–815.
Brandt, M. D., Jessberger, S., Steiner, B., Kronenberg, G., Reuter, K., Bick-Sander, A., von der Behrens, W., and Kempermann, G. (2003). Transient calretinin expression defines early postmitotic step of neuronal differentiation in adult hippocampal neurogenesis of mice. *Mol. Cell. Neurosci.* **24,** 603–613.
Cameron, H. A., and McKay, R. D. G. (1999). Restoring production of hippocampal neurons in old age. *Nat. Neurosci.* **2,** 894–897.
Cameron, H. A., Wooley, C. S., McEwen, B. S., and Gould, E. (1993). Differentiation of new born neurons and glia in the dentate gyrus of the adult rat. *Neuroscience* **56,** 337–344.
Chen, G., Rajkowska, G., Du, F., Seraji-Bozorgzad, N., and Manji, H. K. (2000). Enhancement of hippocampal neurogenesis by lithium. *J. Neurochem.* **75,** 1729–1734.
Cross, D. A., Alessi, D. R., Cohen, P., Andjelkovich, M., and Hemmings, B. A. (1995). Inhibition of glycogen synthase kinase-3 by insulin mediated by protein kinase B. *Nature* **378,** 785–789.
de Kloet, E. R., Joëls, M., and Holsboer, F. (2005). Stress and the brain: From adaptation to disease. *Nat. Rev. Neurosci.* **6,** 463–475.
Encinas, J. M., Vaahtokari, A., and Enikolopov, G. (2006). Fluoxetine targets early progenitor cells in the adult brain. *Proc. Natl. Acad. Sci. USA* **103,** 8233–8238.
Eom, T. Y., and Jope, R. S. (2009). Blocked inhibitory serine-phosphorylation of glycogen synthase kinase-3alpha/beta impairs in vivo neural precursor cell proliferation. *Biol. Psychiatry* **66,** 494–502.
Eriksson, P. S., Perfilieva, E., Bjork-Eriksson, T., Alborn, A. M., Nordborg, C., Peterson, D. A., and Gage, F. H. (1998). Neurogenesis in the adult hippocampus. *Nat. Med.* **4,** 1313–1317.
Fuchs, E., and Flügge, G. (1998). Stress, glucocorticoids and structural plasticity of the hippocampus. *Neurosci. Biobehav. Rev.* **23,** 295–300.
Garcia, A., Steiner, B., Kronenberg, G., Bick-Sander, A., and Kempermann, G. (2004). Age-dependent expression of glucocorticoid and mineralcorticoid receptors on neural precursor cell populations in the adult murine hippocampus. *Aging Cell* **3,** 363–371.
Gould, E., Cameron, H. A., Daniels, D. C., Woolley, C. S., and McEwen, B. S. (1992). Adrenal hormones suppress cell division in the adult rat dentate gyrus. *J. Neurosci.* **12,** 3642–3650.
Gould, E., McEwen, B. S., Tanapat, P., Galea, L. A. M., and Fuchs, E. (1997). Neurogenesis in the dentate gyrus of the adult tree shrew is regulated by psychosocial stress and NMDA receptor activation. *J. Neurosci.* **17,** 2492–2498.
Gould, E., Reeves, A. J., Fallah, M., Tanapat, P., Gross, C. G., and Fuchs, E. (1999). Hippocampal neurogenesis in adult Old World primates. *Proc. Natl. Acad. Sci. USA* **96,** 5263–5267.
Heine, V. M., Maslam, S., Zareno, J., Joëls, M., and Lucassen, P. J. (2004). Suppressed proliferation and apoptotic changes in the rat dentate gyrus after acute and chronic stress are reversible. *Eur. J. Neurosci.* **19,** 131–144.
Hughes, K., Nikolakaki, E., Plyte, S. E., Totty, N. F., and Woodgett, J. R. (1993). Modulation of the glycogen synthase kinase-3 family by tyrosine phosphorylation. *EMBO J.* **12,** 803–808.

Huttmann, K., Sadgrove, M., Wallraff, A., Hinterkeuser, S., Steinhauser, C., and Gray, W. P. (2003). Seizures preferentially stimulate proliferation of radial glia-like astrocytes in the adult dentate gyrus: Functinal and immunocytochemical analysis. *Eur. J. Neurosci.* **18,** 2769–2778.

Jayatissa, M. N., Bisgaard, C., Tingström, A., Papp, M., and Wiborg, O. (2006). Hippocampal cytogenesis correlates to escitalopram-mediated recovery in chronic mild stress rat model of depression. *Neuropsychopharmacology* **31,** 2395–2404.

Judd, L. L., and Akiskal, H. S. (2003). The prevalence and disability of bipolar spectrum disorders in the US population: Re-analysis of the ECA database taking into account subthreshold casaes. *J. Affect. Disord.* **73,** 123–131.

Kempermann, G., Kuhn, H. G., and Gage, F. H. (1997). More hippocampal neurons in the adult mice living in an enriched environment. *Nature* **386,** 493–495.

Kempermann, G., Jessberger, S., Steiner, B., and Kronenberg, G. (2004). Milestones of neuronal development in the adult hippocampus. *Trends Neurosci.* **27,** 447–452.

Kempermann, G., Chesler, E. J., Lu, L., Williams, R. W., and Gage, F. H. (2006). Natural variation and genetic covariance in the adult hippocampal neurogenesis. *Proc. Natl. Acad. Sci. USA* **103,** 780–785.

Kim, J. B., Ju, J. Y., Kim, J. H., Kim, T. Y., Yang, B. H., Lee, Y. S., and Son, H. (2004a). Dexamethasone inhibits proliferation of adult hippocampal neurogenesis in vivo and in vitro. *Brain Res.* **1027,** 1–10.

Kim, J. S., Chang, M. Y., Yu, I. T., Kim, J. H., Lee, S. H., Lee, Y. S., and Son, H. (2004b). Lithium selectively increases neuronal differentiation of hippocampal neural progenitor cells both in vitro and in vivo. *J. Neurochem.* **89,** 324–336.

Klein, P. S., and Melton, D. A. (1996). A molecular mechanism for the effect of lithium on development. *Proc. Natl. Acad. Sci. USA* **93,** 8455–8459.

Kronenberg, G., Reuter, K., Steiner, B., Brandt, M. D., Jessberger, S., Yamaguchi, M., and Kempermann, G. (2003). Subpopulations of proliferating cells of the adult hippocampus respond differently to physiologic neurogenic stimuli. *J. Comp. Neurol.* **467,** 455–463.

Kuhn, H. G., Biebl, M., Wilhelm, D., Li, M., Friedlander, R. M., and Winkler, J. (2005). Increased generation of granule cell in adult Bcl-2-overexpressing mice: A role for cell death during continued hippocampal neurogenesis. *Eur. J. Neurosci.* **22,** 1907–1915.

Kunze, A., Grass, S., Witte, O. W., Yamaguchi, M., Kempermann, G., and Redecker, C. (2006). Proliferative response of distinct hippocampal progenitor cell populations after cortical infarcts in the adult brain. *Neurobiol. Dis.* **21,** 324–332.

Laplagne, D. A., Esposito, M. S., Piatti, V. C., Morgenstern, N. A., Zhao, C., van Praag, H., Gage, F. H., and Schinder, A. F. (2006). Functional convergence of neurons generated in the developing and adult hippocampus. *Plos Biol.* **4,** e409.

Lie, D. C., Colamarino, S. A., Song, H. J., Desire, L., Mira, H., Consiglio, A., Lein, E. S., Jessberger, S., Lansford, H., Dearie, A. R., and Gage, F. H. (2005). Wnt signaling regulates adult hippocampal neurogenesis. *Nature* **437,** 1370–1375.

Lopez, A. D., Mathers, C. D., Ezzati, M., Jamison, D. T., and Murray, C. J. (2006). Global and regional burden of disease and risk factors, 2001: Systemic analysis of population health data. *Lancet* **367,** 1747–1757.

Marberg, J. E., and Duman, R. S. (2003). Cell proliferation in the adult hippocampus is decreased by inescapable stress: Reversal by fluoxetine treatment. *Neuropsychopharmacology* **28,** 1562–1571.

Mayer, J. L., Klumpers, L., Maslam, S., de Kloet, E. R., Joels, M., and Lucassen, P. J. (2006). Brief treatment with the glucocorticoid receptor antagonist mifepristone normalize the corticosterone-induced reduction of adult hippocampal neurogenesis. *J. Neuroendocrinol.* **18,** 629–631.

Nakagawa, S., Kim, J. E., Lee, R., Malberg, J. E., Chen, J., Steffen, C., Zhang, Y. J., Nestler, E. J., and Duman, R. S. (2002). Regulation of neurogenesis in adult mouse hippocampus by cAMP and the cAMP response element-binding protein. *J. Neurosci.* **22,** 3673–3682.

Orford, K., Crockett, C., Jensen, J. P., Weissman, A. M., and Byers, S. W. (1997). Serine phosphorylation-regulated ubiquitination and degradation of beta-catenin. *J. Biol. Chem.* **272,** 24375–24378.

Pham, K., Nacher, J., Hof, P. R., and McEwen, B. S. (2003). Repeated restraint stress suppresses neurogenesis and induces biphasic PSA-NCAM expression in the adult rat dentate gyrus. *Eur. J. Neurosci.* **17,** 879–886.

Sahay, A., and Hen, R. (2007). Adult hippocampal neurogenesis in depression. *Nat. Neurosci.* **10,** 1110–1115.

Sapolsky, R. M., Romero, L. M., and Munck, A. U. (2000). How do glucocorticoids influence stress response? Integrating permissive, suppressive, stimulatory, and preparative actions. *Endocr. Rev.* **21,** 55–89.

Schoneveld, O. J. L. M., Gaemers, L., and Lamers, W. H. (2004). Mechanisms of glucocorticoid signaling. *Biochim. Biophys. Acta* **1680,** 114–128.

Segi-Nishida, E., Warner-Schmidt, J. L., and Duman, R. S. (2008). Electroconvulsive seizure and VEGF increase the proliferation of neural stem-like cells in rat hippocampus. *Proc. Natl. Acad. Sci. USA* **105,** 11352–11357.

Shtutman, M., Zhurinsky, J., Simcha, I., Albanese, C., D'amico, M., Pestell, R., and Ben-Ze'ev, A. (1999). The cyclin Dl gene is a target of the $\beta$-catenin/LEF-l pathway. *Proc. Natl. Acad. Sci. USA* **96,** 5522–5527.

Silva, R., Mesquita, A. R., Bessa, J., Sousa, J. C., Sotiropoulos, I., Leão, P., Almeida, O. F., and Sousa, N. (2008). Lithium blocks stress-induced changes in depressive-like behavior and hippocampal cell fate: The role of glycogen-synthase-kinase-3beta. *Neuroscience* **152,** 656–669.

Son, H., Yu, I. T., Hwang, S. J., Kim, J. S., Lee, S. H., Lee, Y. S., and Kaang, B. K. (2003). Lithium enhances long-term potentiation independently of hippocampal neurogenesis in the rat dentate gyrus. *J. Neurochem.* **85,** 872–881.

Stambolic, V., Ruel, L., and Woodgett, J. R. (1996). Lithium inhibits glycogen synthase kinase-3 activity and mimics wingless signaling in intact cells. *Curr. Biol.* **6,** 1644–1668.

Steiner, B., Klempin, F., Wang, L., Kott, M., Kettenmann, H., and Kempermann, G. (2006). Type-2 cells as link between glial and neuronal linage in adult hippocampal neurogenesis. *Glia* **54,** 805–814.

Tanapat, P., Hastings, N. B., Rydel, T. A., Galea, L. A., and Gould, E. (2001). Exposure to fox odor inhibits cell proliferation in the hippocampus of adult rats via an adrenal hormone-dependent mechanism. *J. Comp. Neurol.* **437,** 496–504.

Tashiro, A., Sandler, V. M., Toni, N., Zhao, C., and Gage, E. H. (2006). NMDA-receptor-mediated, cell-specific integration of new neurons in adult dentate gyrus. *Nature* **442,** 929–933.

Tetsu, O., and McCormick, F. (1999). $\beta$-Catenin regulates expression of cyclin D1 in colon carcinoma cells. *Nature* **398,** 422–426.

Warner-Schmidt, J. L., and Duman, R. S. (2006). Hippocampal neurogenesis: Opposing effects of stress and antidepressant treatment. *Hippocampus* **16,** 239–249.

Wexler, E. M., Geschwind, D. H., and Palmer, T. D. (2008). Lithium regulates adult hippocampal progenitor development through canonical Wnt pathway activation. *Mol. Psychiatry* **13,** 285–292.

Willner, P. (1990). Animal models of depression: An overview. *Pharmacol. Ther.* **45,** 425–455.

Wong, E. Y. H., and Herbert, J. (2006). Raised circulating corticosterone inhibits neuronal differentiation of progenitor cells in the adult hippocampus. *Neuroscience* **137,** 83–92.

Zhao, C., Teng, E. M., Summers, R. G. Jr., Ming, G. L., and Gage, F. H. (2006). Distinct morphological stages of dentate granule neuron maturation in the adult mouse hippocampus. *J. Neurosci.* **26,** 3–11.

# Index

## A

## B



## C

## H

## I



## K



## L

FSC
www.fsc.org
MIX
Papier | Fördert gute Waldnutzung
FSC® C083411

Kolibri 360 GmbH
Ferdinand-Jühlke-Straße 7
99095 Erfurt, Deutschland
produktsicherheit@kolibri360.de